THE ALKALOIDS

Chemistry and Pharmacology

Volume XXI

THE ALKALOIDS
Chemistry and Pharmacology

Edited by
Arnold Brossi
National Institutes of Health
Bethesda, Maryland

Founding Editor
R. H. F. Manske

VOLUME XXI

1983

ACADEMIC PRESS
A Subsidiary of Harcourt Brace Jovanovich, Publishers
New York ● London
Paris ● San Diego ● San Francisco ● São Paulo ● Sydney ● Tokyo ● Toronto

ACADEMIC PRESS, INC.
111 Fifth Avenue, New York, New York 10003

United Kingdom Edition published by
ACADEMIC PRESS, INC. (LONDON) LTD.
24/28 Oval Road, London NW1 7DX

LIBRARY OF CONGRESS CATALOG CARD NUMBER: 50-5522

ISBN 0-12-469521-3

PRINTED IN THE UNITED STATES OF AMERICA

83 84 85 86   9 8 7 6 5 4 3 2 1

CONTENTS

Chapter 1. Acridone Alkaloids: Experimental Antitumor Activity of Acronycine

KOERT GERZON AND GORDON H. SVOBODA

Chapter 2. The Quinazolinocarboline Alkaloids

JAN BERGMAN

CONTENTS

Chapter 3. Isoquinolinequinones from Actinomycetes and Sponges

TADASHI ARAI AND AKINORI KUBO

Chapter 4. Camptothecin

JUN-CHAO CAI AND C. RICHARD HUTCHINSON

Chapter 5. Amphibian Alkaloids

BERNHARD WITKOP AND EDDA GÖSSINGER

Chapter 6. Simple Isoquinoline Alkaloids

JAN LUNDSTRÖM

Chapter 7. Mammalian Alkaloids

MICHAEL A. COLLINS

LIST OF CONTRIBUTORS

Numbers in parentheses indicate the pages on which the authors' contributions begin.

TADASHI ARAI (55), Department of Antibiotics, Research Institute for Chemo-biodynamics, Chiba University, Chiba, Japan

JAN BERGMAN (29), Department of Organic Chemistry, Royal Institute of Technology, S-100 44 Stockholm, Sweden

JUN-CHAO CAI* (101), School of Pharmacy, University of Wisconsin, Madison, Wisconsin 53706

MICHAEL A. COLLINS (329), Department of Biochemistry and Biophysics, Loyola University of Chicago, Stritch School of Medicine, Maywood, Illinois 60153

EDDA GÖSSINGER (139), Institut für Organische Chemie, Universität Wien, Vienna, Austria

KOERT GERZON (1), Department of Pharmacology, Indiana University School of Medicine, Indianapolis, Indiana 46223

C. RICHARD HUTCHINSON (101), School of Pharmacy, University of Wisconsin, Madison, Wisconsin 53706

AKINORI KUBO (55), Department of Organic Chemistry, Meiji College of Pharmacy, Tokyo, Japan

JAN LUNDSTRÖM (255), Astra Pharmaceuticals, Södertälje, Sweden and Department of Pharmacognosy, Faculty of Pharmacy, Uppsala University, Uppsala, Sweden

GORDON H. SVOBODA (1), Indianapolis, Indiana 46226

BERNHARD WITKOP (139), National Institutes of Health, Bethesda, Maryland 20205

*Present address: Department of Medicinal Chemistry, Shanghai Institute of Materia Medica, Academia Sinica, Shanghai, 200031, People's Republic of China.

PREFACE

Twenty volumes of *The Alkaloids* have appeared, sixteen of which were edited by its founder, the late R. H. F. Manske, and four by R. G. A. Rodrigo. Devoted to reviews of nitrogen-containing plant constituents, *The Alkaloids* has been received with great interest by scientists investigating the botany, chemistry, pharmacology, characterization, biosynthesis, taxonomy, and medical uses of plant alkaloids. This editor does not intend to alter the policy or appearance of a periodical of high scientific quality that is indispensable to experts in the field.

The chapters devoted to simple isoquinoline alkaloids and quinazolinocarboline alkaloids, organized in the traditional manner, represent updated reviews on two important groups of plant alkaloids. The chapters on camptothecines and acridone alkaloids focus on two groups of plant alkaloids that have attracted attention in recent years as antitumor agents. To broaden its scope, however, future contributions will focus not only on plant alkaloids, but on chemically related, nitrogen-containing substances originating from such other sources as mammals, amphibians, fish, insects, microorganisms, and oceanic plants. This is illustrated here with reviews on amphibious alkaloids, mammalian alkaloids, and isoquinolinequinones from actinomycetes and sponges. In addition to covering all aspects of the recent chemistry of this group of substances, the chapter on amphibious alkaloids presents in detail the biochemical and pharmacological properties of these interesting toxins. It is expected that the broadened scope of *The Alkaloids* will not adversely affect its objectives, but will rather enhance its quality through contributions of interest to a wider scientific audience. The positive response of the scientific community at large to the planned changes is most gratifying.

The editor is very pleased to present as contributors here and in future volumes a group of internationally recognized experts in the field.

Arnold Brossi

CONTENTS OF PREVIOUS VOLUMES

Contents of Volume IV (1954)

edited by R. H. F. Manske and H. L. Holmes

Contents of Volume V (1955)

edited by R. H. F. Manske

Contents of Volume VI (1960)

edited by R. H. F. Manske

Contents of Volume VII (1960)

edited by R. H. F. Manske and H. L. Holmes

Contents of Volume VIII (1965)

edited by R. H. F. Manske and H. L. Holmes

Contents of Volume IX (1967)

edited by R. H. F. Manske and H. L. Holmes

Contents of Volume X (1967)

edited by R. H. F. Manske and H. L. Holmes

Contents of Volume XI (1968)

edited by R. H. F. Manske and H. L. Holmes

Contents of Volume XII (1970)

edited by R. H. F. Manske and H. L. Holmes

Contents of Volume XIII (1971)

edited by R. H. F. Manske and H. L. Holmes

Contents of Volume XIV (1973)

edited by R. H. F. Manske and H. L. Holmes

Contents of Volume XV (1975)

edited by R. H. F. Manske and H. L. Holmes

Contents of Volume XVI (1977)

edited by R. H. F. Manske and H. L. Holmes

Contents of Volume XVII (1979)

edited by R. H. F. Manske and H. L. Holmes

Contents of Volume XVIII (1981)

edited by R. H. F. Manske and R. G. A. Rodrigo

Contents of Volume XIX (1981)

edited by R. H. F. Manske and R. G. A. Rodrigo

Contents of Volume XX (1981)

edited by R. H. F. Manske and R. G. A. Rodrigo

—— CHAPTER 1 ——

ACRIDONE ALKALOIDS: EXPERIMENTAL ANTITUMOR ACTIVITY OF ACRONYCINE

KOERT GERZON

Department of Pharmacology, Indiana University School of Medicine, Indianapolis, Indiana

AND

GORDON H. SVOBODA

Indianapolis, Indiana

I. Historical Introduction

In the review of acridine alkaloids by J. R. Price, which appeared just 30 years ago in Volume II of this treatise (*1*), the author recognized a relatively small group of acridone alkaloids present in the bark and leaves of certain Rutaceae species found in northern Australia's tropical rain forests. These acridone alkaloids reported in the period from 1948 to 1952 by Price (*1, 2*), Lahey (*2*), and their associates at the Universities of Sydney and Melbourne, include alkylated derivatives of (a) 1,3-dihydroxy-*N*-methylacridone [e.g., acronycine (*1, 2*) and 1,3-dimethoxy-*N*-methylacridone (1)], of (b) 1,2,3-trihydroxy-*N*-methylacridone (evoxanthine) (*1, 3*), and (c) 1,2,3,4-tetrahydroxy-*N*-methylacridone (melicopine, melicopidine, and melicopicine) (*1, 2*), all of which have an *O*-1-methyl substituent in common while differing in the respective pattern of alkylation at O-2, O-3, and O-4. Two further

THE ALKALOIDS, VOL. XXI

ISBN 0-12-469521-3

TABLE I
Acridone Alkaloids of Six Rutaceae Species

Alkaloid	Molecular formula	CH$_3$O group(s) location	Methylenedioxy location	Substituent at N	Rutaceae species[a]	Reference
Acronycine	C$_{20}$H$_{19}$O$_3$N$_1$	1 (C-1)[b]	—[c]	—CH$_3$	Acr. b.	1, 2, 5, 6
1,3-Dimethoxy-N-methyl-acridone	C$_{16}$H$_{15}$O$_3$N$_1$	2 (C-1, C-3)	—[c]	—CH$_3$	Acr. b.	1
Evoxanthine	C$_{16}$H$_{13}$O$_4$N$_1$	1 (C-1)	C-2, C-3	—CH$_3$	Ev. x., Ev. a.	1, 3, 8
Melicopine	C$_{17}$H$_{15}$O$_5$N$_1$	2 (C-1, C-2)	C-3, C-4	—CH$_3$	Acr. b., Mel. f. / Act. ac.	1[d] / 3
Melicopidine	C$_{17}$H$_{15}$O$_5$N$_1$	2 (C-1, C-4)	C-2, C-3	—CH$_3$	Acr. b., Mel. f. / Ev. x., Ev. a.	6 / 9
Melicopicine	C$_{18}$H$_{19}$O$_5$N$_1$	4 (C 1-4)	—[c]	—CH$_3$	Acr. b., Mel. f.	10
Evoxanthidine	C$_{15}$H$_{11}$O$_4$N$_1$	1 (C-1)	C-2, C-3	—H	Ev. x.	1, 3, 8, 11
Xanthevodine	C$_{16}$H$_{13}$O$_5$N$_1$	2 (C-1, C-4)	C-2, C-3	—H	Ev. x.	1, 3, 11
Noracronycine[e]	C$_{19}$H$_{17}$O$_3$N$_1$	—	—[c]	—CH$_3$	Gly. p.	4
De-N-methylacronycine[e]	C$_{19}$H$_{17}$O$_3$N$_1$	1 (C-1)	—[c]	—H	Gly. p.	4
De-N-methylnor-acronycine[e]	C$_{18}$H$_{15}$O$_3$N$_1$	—	—	—H	Gly. p.	4
Normelicopine[e]	C$_{16}$H$_{13}$O$_5$N$_1$[f]	1 (C-2)	C-3, C-4	—CH$_3$	Acr. b.	5, 6
Normelicopidine[e]	C$_{16}$H$_{13}$O$_5$N$_1$	1 (C-4)	C-2, C-3	—CH$_3$	Acr. b.	5, 6
Normelicopicine[e]	C$_{17}$H$_{17}$O$_5$N$_1$[f]	3 (C 2-4)	—	—CH$_3$	Acr. b.	5, 6

[a] Abbreviations for botanical species studied: *Acronychia baueri* Schott = Acr. b.; *Evodia xanthoxyloides* F. Muell. = Ev. x.; *Evodia alata* F. Muell. = Ev. a.; *Acronychia acidula* F. Muell. = Acr. ac.; *Melicope fareana* Engl. = Mel. f.; and *Glycosmis pentaphylla* (Retz.) Correa. = Gly. p.

[b] C-1 in the acridone numbering system (see first column) corresponds to C-6 in the acronycine system (see Fig. 1).

[c] In acronycine a *gem*-dimethylchromene ring system is located at the C-3–C-4 bond site (Figure 1).

[d] References 1, 3, 6, 9, 10 refer to the group of melicope alkaloids.

[e] The designation "nor" for de-*O*-1-methyl analogs of parent alkaloids is the one used in Reference 1.

[f] No analytical data confirming the molecular formulas of these natural products are given, but conversion of the parent natural *O*-1-methyl product to the nor species under acidic conditions has been reported (1,4).

members of this group reported by the Australian workers are evoxanthidine and xanthevodine (*1, 3*), shown to be the de-*N*-methyl congeners of evoxanthine and melicopidine, respectively.

In 1966 three further simpler acridone alkaloids, close relatives of acronycine, were obtained from a Rutaceae species in India (*4*). In the same year, the Lilly Research Laboratories reported the isolation of the de-*O*-1-methyl congeners of melicopine, melicopidine, and melicopicine from the bark of the Australian *Acronychia baueri* Schott (*5, 6*). The possibility of inadvertent, facile conversion of the natural *O*-1-methyl alkaloid to the de-*O*-methyl analog under the influence of acid conditions used in isolation procedures had already been recognized (*7*).

The above acridone agents are listed in Table I together with pertinent structural features, species of botanical origin, and references. The structures of representative members of the group and the numbering used are presented in Scheme 1 and Figure 1.

As the 1952 review of acridine alkaloids (*1*) adequately describes the main knowledge of these alkaloids at that time, the present discussion of acridine alkaloids concerns itself with the chemical and structural information obtained since 1952 and, in particular, with the unique biological activity of acronycine in experimental cancer chemotherapy.

It is noted parenthetically that the alkaloid acronycidine (*1, 2*), which

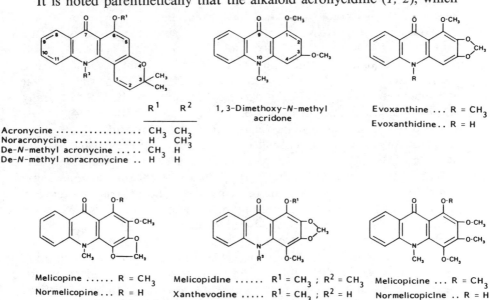

	R^1	R^2
Acronycine	CH_3	CH_3
Noracronycine	H	CH_3
De-*N*-methyl acronycine	CH_3	H
De-*N*-methyl noracronycine ..	H	H

1,3-Dimethoxy-*N*-methyl acridone

Evoxanthine ... R = CH_3
Evoxanthidine .. R = H

Melicopine ...... R = CH_3
Normelicopine ... R = H

Melicopidine ...... R^1 = CH_3 ; R^2 = CH_3
Xanthevodine ..... R^1 = CH_3 ; R^2 = H
Normelicopidine.... R^1 = H ; R^2 = CH_3

Melicopicine ... R = CH_3
Normelicopicine .. R = H

SCHEME 1. *Principal acridone alkaloids of Rutaceae.*

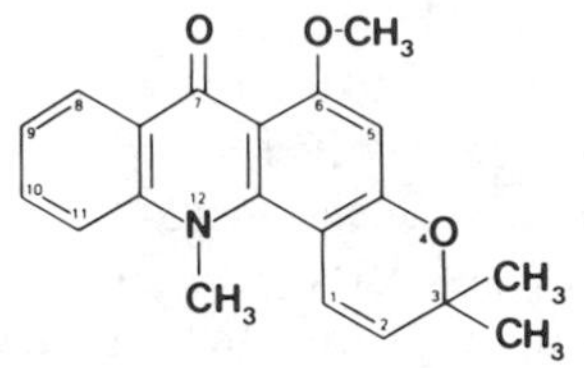

FIG. 1. Molecular structure of acronycine.

co-occurs with acronycine, possesses a furoquinoline-type structure (*12*) and thus falls outside the scope of the present review. The same applies *mutatis mutendi* to other close-named co-occurring Rutaceae alkaloids. A chapter entitled "Quinoline Alkaloids Related to Anthranilic Acid" has appeared in the Volume XVII of this treatise (*13*).

II. Physical and Chemical Properties

The physical and chemical properties of the acridone alkaloids that have been adequately described in 1952 (*1*) are mentioned briefly in the present discussion because of their relevance to the natural occurrence, isolation, and biological activities of the alkaloids. As pointed out by J. R. Price, the acridone alkaloids are very weak bases and, though soluble in strong, aqueous mineral acid (e.g., 10% HCl), they are liberated and deposited as the free bases at lower acid concentrations. The possibility of oxonium salt formation as an aid in solubilizing the alkaloids has been postulated. Also, frequent use of acidic conditions in isolating these alkaloids suggests that adequate caution be exercised in making the distinction between naturally occurring alkloids and artifacts produced during isolation (*1, 4*). In the original isolation of acronycine (*2, 14*) such caution fortunately was exercised, thus preserving the possibility of detecting its biological activity at a later date (*5*).

The conversion of *O*-1-methylacridone alkaloids to the corresponding "nor" (de-*O*-methyl) products has been reported to occur at varying temperatures under acid conditions. Heating melicopine in 5% aqueous ethanolic solution under reflux for 1 hr yields normelicopine (free base), red needles melting at 235.5–236.5°C. Noracronycine, which had been obtained as a bright yellow solid, mp 200.5–201°C, by heating acronycine hydrochloride at 130°C, also was isolated without exposure to acid from the yellow root bark of *Glycosmis pentaphylla* (T Retz.) Correa (*4*) growing in India. Therefore, noracronycine may be considered to be a naturally occurring alkaloid. It is readily methylated with dimethyl sulfate and potassium carbonate in acetone to regenerate the parent alkaloid acronycine. The lack of solubility of the alkaloids under physiological conditions may have been responsible for the initial lack of information on biological activity of acridone alkaloids as a class (*15*).

Aided by the emergence of modern techniques for testing compounds with low intrinsic solubility and prompted by a report on mouse behavior studies indicating central nervous system activity of a defatted ethanolic extract obtained from the bark of *A. baueri* (*5*), the attention of one of the authors (G. H. S.) was directed to the alkaloids of this plant. A phytochemical investigation of *A. baueri* was therefore initiated in the middle 1960s (*5*). Discovery of striking activity against several murine tumor systems and identification of acronycine as the active agent in the course of this investigation (*5*) led to a detailed study of the chemistry, synthesis, and biological and toxicological properties of this alkaloid, eventually resulting in a recommendation for its clinical evaluation.

III. Structure of Acronycine: Principal Syntheses

The generally accepted "angular" structure of acronycine, designated 3,12-dihydro-6-methoxy-3,3,12-trimethyl-7*H*-pyrano[2,3-*c*]acridin-7-one (earlier 1-methoxy-2′,2′,10-trimethyl-pyrano[5′,6′,3,4]acridone) (*4*), is shown in Fig. 1. This angular formulation rather than the alternate "linear" formulation was proposed as the preferred structure on the basis of degradative studies by Lahey and co-workers (*14, 16, 17*). The angular formulation (Fig. 1) was further supported by Indian (*4*) and Australian workers (*18*), and finally confirmed by X-ray crystallographic analysis (*19*) and by chemical syntheses (*20–23*).

Oxidative degradation (*17, 18*) resulting in cleavage of the pyran ring yielded 1,3-dimethoxy-*N*-methylacridone-4(?)-carboxylic acid (*17*) and corresponding methyl ester (*18*), which was found to differ from the authentic, synthetic 2-carboxylic acid methyl ester analog (*18*). This finding effectively eliminated the linear formulation, but although it supported the angular structure, it did not provide direct chemical proof of the latter. Furthermore, indirect support for the angular formulation resulted from the work of Govindachari and associates (*4*) that involved the preparation of 6-tosylnoracronycine from natural noracronycine. The tosyl ester on hydrogenolysis with Raney nickel gave a 6-deoxy product that had been concurrently tetrahydrogenated in the unsubstituted aromatic ring. A detailed inspection of the NMR spectrum of this deoxy product revealed that it could only have been derived from the angular formula. In the course of these studies (*4*), the presence of a peak at m/e 292 (M − 15) in the mass spectrum of noracronycine was noted that had an intensity three times that of the parent ion m/e 307 (M⁺). This mass spectral behavior, shown previously to be characteristic for 2,2-dimethylchromenes (24), was attributed to the formation of benzopyrilium ions.

Finally, single-crystal X-ray diffraction analysis of 5-bromo-1,2-dihydroacronycine (14) has been reported to confirm fully the angular formulation (19).

Direct chemical evidence for this formulation (Fig. 1) was subsequently obtained by the synthesis of acronycine at the Lilly Research Laboratories, which was accomplished along three interrelated routes (20, 21). A further synthesis of acronycine in excellent yield at the University of Sydney (22, 23) facilitates the preparation of acronycine and acronycine analogs for biological examination (vide infra).

One of the three syntheses reported by the Lilly group (21) follows a pathway (Scheme 2) resembling one initially postulated for the phytosynthesis of acridone alkaloids by Hughes and Ritchie (25) in which anthranilic acid and trihydroxybenzene, and — in the case of acronycine — mevalonic acid are the presumed precursors. Thus, 1,3-dihydroxy-9-acridone (1), itself prepared in yields of 25% by the reaction of 1,3,5-trihydroxybenzene and anthranilic acid in butanol solution in the presence of zinc chloride, was allowed to react with 1-chloro-3-methyl-2-butene in trifluoroacetic acid with zinc chloride as the catalyst. The single product, 6-hydroxy-3,3-dimethyl-2,3-dihydro-7($12H$)-1H-pyrano[2,3-c]acridine (2, R = H), was obtained in 20% yield and without prior purification converted to acronycine in two steps. Methylation with methyl iodide and potassium carbonate in refluxing acetone gave 2,3-dihydronoracronycine (2, R = CH₃), identical with an authentic sample prepared from noracronycine (16). Dehydrogenation, utilizing 2,3-dichloro-5,6-dicyanobenzoquinone in refluxing dioxane, produced noracronycine (3, R = H) which was converted to acronycine (3, R = CH₃) by a published methylation procedure (14).

Proceeding from the same starting material, 1,3-dihydroxy-9-acridone (1), Hlubucek et al. (22, 23) allowed the reaction of 1 with 3-chloro-3-methylbut-1-yne in dimethylformamide (DMF) to proceed at elevated temperature in the presence of potassium carbonate and sodium iodide (Scheme 2). The product, bisnoracronycine (4) was obtained in 85% yield*. Methylation of 4 with dimethyl sulfate–sodium hydride in DMF gave acronycine (3, R = CH₃). No trace of alternate linear material (5) was detected in 4. In the course of this elegant synthetic work (23), the physical proximity of the N—CH₃ group and the H-1 atom of the chromene ring was conclusively demonstrated by the observation of nuclear Overhauser effects in the NMR spectra of acronycine and noracronycine, thereby again confirming the angular formulation.

---

* Bisnoracronycine is referred to in Table I as de-N-methylnoracronycine.

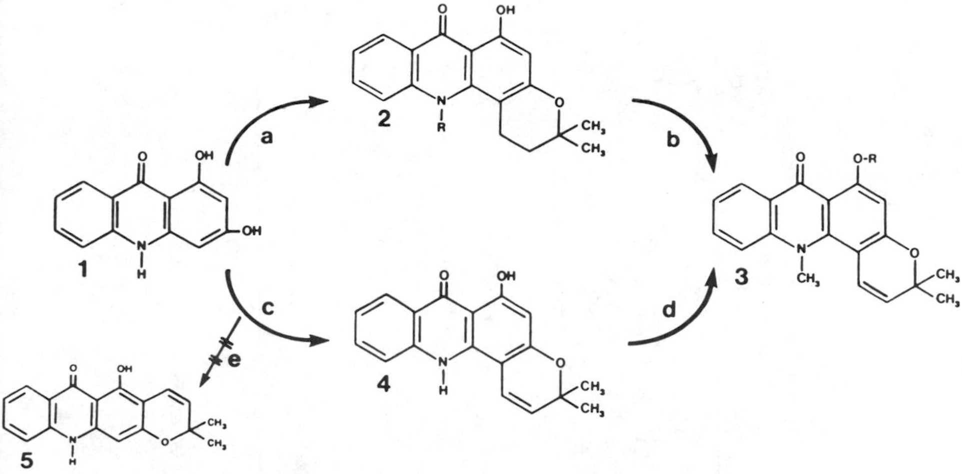

a) 1 + 1-chloro-3-methyl-2-butene in CF_3COOH + $ZnCl_2$
b) 2,3-Dichloro-5,6-dicyano-benzoquinone;
c) 1 + 3-chloro-3-methyl-1-butyne;
d) $(CH_3)_2SO_4$ and NaH;
e) Not formed.

SCHEME 2. *Synthesis of acronycine.* [*From Beck* et al. *(21) and Hlubucek* et al. *(23)*].

IV. Experimental Antitumor Activity of Acronycine

The scrub ash or scrub yellowwood *A. baueri* Schott. (Family Rutaceae) is one of approximately 20 species of the genus *Acronychia* native to Australia and tropical Asia. This tree is found in Queensland and New South Wales where it grows to a height of 50–60 ft. No references to its medicinal usefulness were found prior to 1966. The wood, however, is said to be excellent for mallet and chisel handles (*27*). Continuing the tradition at the Lilly Research Laboratories of searching for new antitumor substances among alkaloids obtained from selected higher order plants, a tradition which had contributed materially to the codiscovery of the *Catharanthus* ("Vinca") agents vinblastine and vincristine from the leaves of the periwinkle (*Catharanthus rosea*) (*28, 29*), this search was extended to the acridone alkaloids reported to be present in the bark of *A. baueri* (*5, 6*).

Initial evaluation of a defatted ethanolic extract of the bark showed it to be inactive in the primary tumor screen. This screen then included the P-1534 leukemia, the murine leukemia uniquely responsible for the detection of activity of the Vinca alkaloids (*28*), the Mecca lymphosarcoma, and the adenocarcinoma 755. These and further murine tumor models, as well as their manipulation in the assessment for antitumor activity, have been

TABLE II

EXPERIMENTAL TUMOR SPECTRUM OF ACRONYCINE[a,b]

Tumor	Dose (ip, mg/kg/day)	Average weight change (g, T/C)	Average life span (days, T/C)	Percent activity[c]
B 82 leukemia	37.5 × 1 × 7	−1.4/+0.4	23.7/14.6	61
C1498 leukemia	28 × 1 × 10	−1.4/+0.8	31.5/17.6	79(7)
P1534 leukemia	30 × 1 × 10	−0.2/−0.7	16.5/18.2	0[d]
L5178Y leukemia	28 × 1 × 10	+2.2/+3.2	24.2/15.0	62
AKR leukemia	28 × 1 × 10	+0.1/+0.8	38.3/21.5	78(5)
Ehrlich ascites	30 × 1 × 10	+5.6/+7.8	21.8/18.4	0(1)
			Average tumor size (mm, T/C)	
Sarcoma 180	30 × 1 × 10	+3.2/+6.0	7.1/11.9	40(9)
Mecca lymphosarcoma[e]	30 × 1 × 10	−0.4/+2.5	6.2/16.9	63(7)
Ridgeway osteogenic sarcoma	48 × 1 × 9	−0.6/+3.4	0/9.6	100(10)
X5563 myeloma	30 × 1 × 8	+0.1/+0.3	0/9.1	100(8)
Adrenocarcinoma 755	30 × 1 × 10	−0.5/+1.9	11.9/19.7	40(10)
Shionogi carcinoma 115	36 × 1 × 9	+1.4/+1.4	0/15.3	100(7)
S91 melanoma	36 × 1 × 9	−1.4/+0.1	0/14.1	100(4)

[a] From Svoboda et al. (5).

[b] Ten mice per group in each assay unless otherwise noted. T/C = treated/control.

[c] The number in parentheses indicates survivors on solid tumors, indefinite survivors at 45 days on leukemia or ascitic tests.

[d] P1534 Leukemia is highly sensitive to vinblastine and vincristine (28).

[e] Seven animals studied.

adequately described in reports from the Lilly Research Laboratories (5, 28, 30, 31).

The extraction scheme used (5, 6) yielded a large amount of fat-soluble extractive from which the triterpene lupeol was isolated. Extraction of the defatted drug with ether and subsequent concentration to small volume resulted in the deposition of crystals consisting of a mixture of the acridone alkaloids, melicopine, and acronycine. Evaluation of the mixed crystals revealed significant activity in the C-1498 leukemia, the plasma cell tumor X-5563, and the adenocarcinoma 755. Fractional crystallization yielded the pure alkaloids and subsequent testing indicated that the antitumor activity resided in acronycine, melicopine being inactive.

When tested against a large number of murine tumors, acronycine proved to possess the broadest experimental antitumor spectrum of any alkaloid isolated from higher order plants up to 1965, a statement perhaps still true at the time of this writing when restricted to products derived from higher plants. In addition to possessing significant activity against 12 of 17 experimental tumor models (see Table II), acronycine also was found to be active against the experimental LPC-1 plasma cell tumor used at the Chemotherapy Evaluation Branch of the National Cancer Institute (32).

The experimental antitumor activity of acronycine is held to be extraordinary not only because of its broad spectrum but also because of the following four activity qualities.

1. Several of the responsive experimental murine tumors are unique in some of their characteristics: the C-1498 myelogenous leukemia responds to few, if any, of the known clinically active agents to the extent it responds to acronycine; the X-5563 plasma cell tumor has several properties that relate to those of multiple myeloma in human patients; and the Shionogi carcinoma 115 is a tumor model characterized by its dependency on androgen.

2. Acronycine is active not only when given by intraperitoneal injection but also when administered orally at dose levels not differing appreciably from those effective parenterally (see Table III).

3. When drug administration is withheld for several days after tumor implantation, acronycine has the ability to prolong life in animals in which tumors (e.g., C-1498 and X-5563) are well established (5).

4. Structure–activity relationships in the acronycine group are extraordinarily rigid; of the naturally occurring related acridone alkaloids and acronycine analogs prepared from acronycine (6) or by synthesis (33, 34), only acronycine—with perhaps one possible exception (34)—possessed activity against the tumor models used (see Scheme 3).

The lack of aqueous solubility of acronycine evidently has not prevented the alkaloid from expressing its potential experimental antitumor activity

TABLE III
ACTIVITY OF ACRONYCINE VERSUS C-1498 LEUKEMIA VIA VARIOUS ROUTES OF ADMINISTRATION[a]

Route	Dosage (mg/kg × 1 × 10)	Average weight change (T/C)[b]	Average life span (days, T/C)	Percent prolongation	Indefinite survivors
Intraperitoneal	20	−1.1/+0.8	20.0/17.6	65	6
	30	−1.2/+0.8	31.0/17.6	76	3
	75	−2.8/−0.2	23.0/13.3	72	—
	90	−2.7/+0.1	29.2/17.6	81	2
Oral	45	−1.4/+0.4	31.4/17.9	76	5
	60	−1.8/+0.4	31.3/17.9	75	7

[a] From Svoboda et al. (5).
[b] T/C = treated/control.

(5, 6). A suitable suspension was prepared by grinding the alkaloids with small amounts of Emulphor (General Aniline and Film Corp., Melrose Park, Ill.), a nonionic dispersant to form a uniform suspension which was then diluted with additional dispersant to the desired concentration for ip injection.

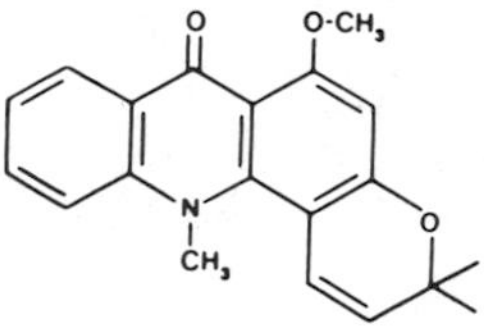

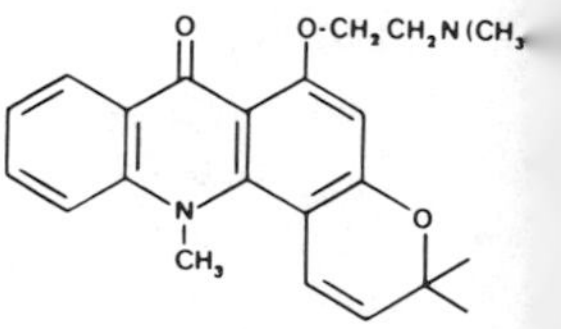

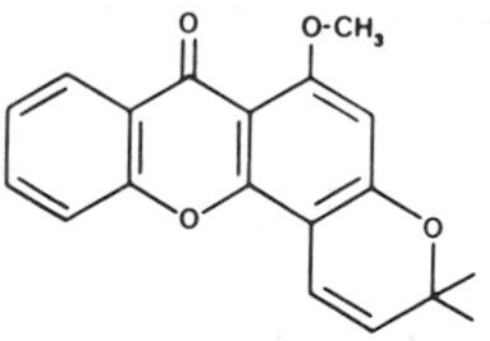

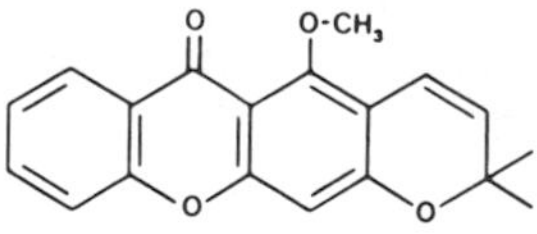

a "Angular" Acronycine structure; b "Linear" structure;
5, 6, and 7 synthetic analogs (see text); 8 Psorospermin[35, 36]

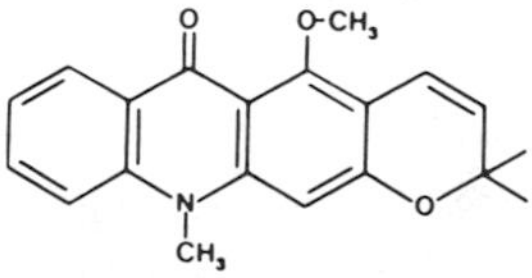

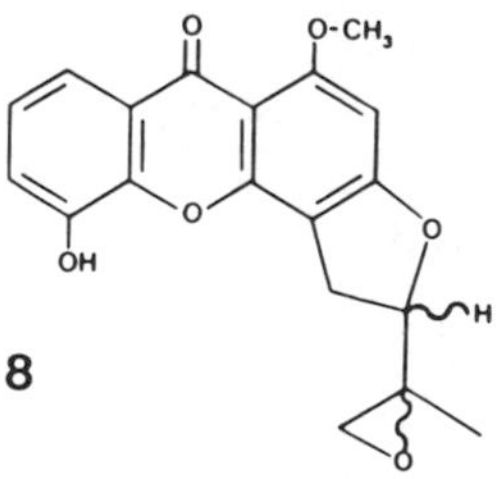

SCHEME 3. Acronycine and synthetic analogs. [From Schneider et al. (34)].

V. Structure–Activity Relationships (SAR)

The essence of structure–activity relationships (SAR) among acronycine derivatives, for one reason or another, is a case of structure–inactivity relationships. Thus, the one-step derivatives prepared in 1949 including 1,2-dihydroacronycine, noracronycine, 2-nitroacronycine, and the alkaline hydrolysis product acronycine phenolate, all lack activity against the C-1498 leukemia and X-5563 murine tumor models (6) which are most responsive to acronycine itself. Neither normelicopine (see Scheme 1) nor any of the other acridone alkaloids found in Rutaceae species (Scheme 1) had activity in these two tumor systems. Moreover, of a group of synthetic substituted acridines or acridones none was active (6). It must be noted that of these several inactive compounds none possessed the unperturbed dimethylchromene ring system of acronycine.

A group of synthetic, somewhat remote analogs of acronycine reported by Liska (33) include 2,2,5,7,7,10-hexamethylbenzo[1,2-b;4,5-b']dipyran, a nonsymmetrical benzobischroman that regrettably was not converted to the bischromene analog. The biological assay of this synthetic group of acronycine relatives performed at the Cancer Chemotherapy National Serice Center reportedly (33) failed to demonstrate activity comparable to that of acronycine. Since the L1210 leukemia, which is not responsive to acronycine itself (5, 6), was unrealistically used as the tumor model, it is possible that potential activity remained undetected.

The synthesis and biological activity of a further group of synthetic acronycine analogs have been prepared at the Hoffmann–La Roche Research Laboratories (34) from the appropriately substituted acridones by the Hlubucek synthesis (23). The authors concluded that "none of the analogs and derivatives prepared showed enhanced activity against experimental tumor models in mice and rats." However, an inspection of the data (34) leads to a more hopeful conclusion. For example, O-dimethylaminoethyl-noracronycine (5, Scheme 3), a product in which the dimethylaminoethyl side chain replaces the O-6-methyl group of acronycine, caused a 52% decrease of tumor size in mice bearing the solid form of sarcoma 180. A 60% prolongation of lives of mice inoculated with Ehrlich ascites was attained following oral treatment with 5 at a (low) dose of 10 mg/kg per day for 8 days. Acronycine itself is effective against C-1498 leukemia (Table II) at dose levels of 30–40 mg/kg given by the intraperitoneal route for 10 days and at slightly higher doses (45–60 mg/kg) given orally for the same period. Against sarcoma 180, acronycine (30 mg/kg for 10 days) has been reported (5, 6) active, 9 out of 10 mice surviving throughout the experiment.

Although a strict comparison between 5 and acronycine in the same tumor model(s) cannot be deduced from the above information, it appears that the base-substituted acronycine analog **5** cannot be considered to be inactive. The possible activity of **5** against the C-1498 leukemia and the X-5563 plasma cell murine models thus remains a matter for further inquiry.

Compounds of special interest are the linear acronycine structure **b** and the pair of xanthone congeners **6** and **7** corresponding, respectively, to the angular and linear structures discussed for acronycine. Since neither **6**, **7**, nor **b** are active against the sarcoma 180 or Ehrlich ascites model, it must be tentatively concluded that in acronycine the *gem*-dimethylchromene moiety and its specific site of attachment to the acridone system of the molecule, are prerequisite structural features for activity. To confirm this conclusion, assay of the angular xanthone congener **7** against C-1498 leukemia and X-5563 plasma cell tumor, again, is a matter of considerable interest.

Raising yet another paradox, in a recent review of natural products as medicinal agents a striking correspondence in molecular structure between acronycine and psorospermin (**8**), an antitumor agent isolated by Kupchan and associates from *Psorospermum febrifugum* (!) Tanzania (*35*), was pointed out and enlarged upon by the author (*36*). A certain phytogenetic kinship between the respective isoprenoid C_5-moieties in these two products is quite apparent. What is particularly surprising is the demonstrated activity of psorospermin (**8**) against P388 murine leukemia at a dose level of 8 mg/kg and against the CD (mammary) and C_6 (colon) tumor models (*35, 36*), especially when contrasted with the inactivity of the angular xanthone (**6**) (*34*). This paradox is deepened in noting the closer structural correspondence of acronycine with the angular xanthone (**6**) than with psorospermin (**8**).

Further nonactive analogs mentioned in this report (*34*) include 9-chloroacronycine and the products in which the 6-methoxy function of acronycine has been replaced by ethoxy-, allyloxy-, and ethyloxycarbonyl-methoxy-($-O-CH_2COOC_2H_5$).

To summarize SAR reflections, the activity reported for the single, effective acronycine analog (**5**) again tends to indicate the positive chemotherapeutic benefit that occasionally results from introducing a basic functionality into a molecule of potential or established antitumor activity. Such benefits have been reported to accompany, for example, the introduction of a diethylaminopropylamino substituent at C-1 in ellipticine (*37*) and to be associated with the two hydroxyethylaminoethylamino substituents in the anthracenedione (DHAQ, NSC 279836)* (*38*).

Although information gained from the present SAR reflections upon acronycine congeners often can be most helpful, even more informative is the reinterpretation of such relationships at the time when precious feedback information from clinical trial experience with the selected candidate be-

---

* 1,4-Dihydroxy-5,8-bis [[2-[(2-hydroxyethyl)amino]-ethyl]-amino]-9,10-anthracene

comes available. Indeed, it is the valuable feedback from clinical experience that enables the experimentalist to return for a thoughtful and—one hopes —fruitful session at the designing board.

In a novel approach to acronycine analogs for antitumor assay, Brannon and associates at the Lilly Research Laboratories (*39*) and Rosazza *et al.* at the University of Iowa (*40*) subjected acronycine to modification by metabolizing microbiological cultures.

From a group of about 500 cultures, the first team selected cultures of *Aspergillus alleaceus* and *Streptomyces spectabilicus* for this purpose, while the Iowa group chose the culture of *Cunninghamella echinulata* (NRLL 3655) to obtain, hopefully, microbial transformation products with antitumor activity superior to that of acronycine. While addition of acronycine to a metabolizing culture of *A. alleaceus* yielded 9-hydroxyacronycine, 3-hydroxy-methylacronycine was formed by cultures of *S. spectabilicus* (*39*). Neither was active against the X-5563 plasma cell tumor nor the C-1498 murine leukemia at dose levels where acronycine expresses its activity. Metabolites obtained by the Iowa group using *C. echinulata* included 9-hydroxyacronycine as the major product, a metabolite also obtained in mammalian metabolism studies (*vide infra*). Additional metabolites were 11-hydroxyacronycine, 9,11-dihydroxyacronycine, and 3-hydroxymethyl-11-hydroxyacronycine (for structures see Scheme 4 in Section VI).

Rosazza has discussed the basic biochemical aspects of metabolism, comparing microbial and mammalian cell systems, stressing similarities in the mechanism and the nature of the products of these diverse systems of metabolism (*40*).

VI. Mammalian Metabolism

As an integral part of the preclinical studies of acronycine, the metabolism of the alkaloid was investigated in four mammalian species (*41*). In these experiments involving mice, rats, guinea pigs, and dogs, several metabolites were isolated from urine, feces (or bile), and serum samples and their structures determined. The metabolic profiles observed in these four species showed extensive overlap. Differences were noted as well, especially in the case of the guinea pig, in which animal O-demethylation plays a prominent role (see Table IV). When clinical urine samples from patients receiving orally administered acronycine (100 mg. capsule daily) became available for study (*vide infra*), a close correspondence of the metabolic profile in humans and mice was noted (Table IV).

These metabolism studies utilized acronycine ^{14}C-labeled in either the $O-CH_3$ or $N-CH_3$ function. Metabolites present in urine, bile, or blood

TABLE IV

ACRONYCINE METABOLITES IN LABORATORY ANIMALS AND IN HUMANS[a]

Compound	Structure number	Animal species				
		Rat	Guinea pig	Dog	Mouse	Human
9-Hydroxyacronycine[b]	9	x	x	x	x	x
11-Hydroxyacronycine	10	x	x	x	x	x
3-Hydroxymethylacronycine[c]	11	x			x	
11-Hydroxy-3-hydroxymethyl-acronycine	12	x		x	x	x
9,11-Dihydroxyacronycine	13	x			x	x
11-Hydroxy-O-demethyl-acronycine	14		x			

[a] From Betts et al. (41).
[b] Metabolite 9 also found as a major microbial transformation product of Acronycine by A. allaceus (39) and C. echinulate (40).
[c] Metabolite 11 formed by S. spectabilis (39). Both metabolites 9 and 11 were inactive against C-1498 leukemia and X-5563 plasma cell tumor (39).

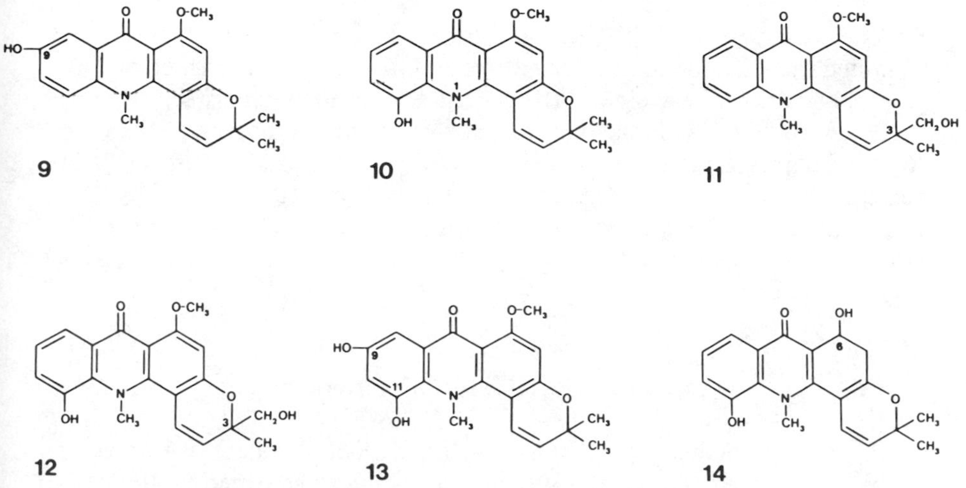

SCHEME 4. *Mammalian metabolites of acronycine.* [*From Sullivan* et al. (41)].

samples were acidified to pH 5.5 and under the appropriate conditions were exposed to the action of β-glucuronidase and sulfatase enzyme preparations. The extraction, purification, and characterization employed standard methods, e.g., suitable derivatization (CH_2N_2), GLC-MS and NMR techniques (*41*).

In the rat, the dog, and the human, O-demethylation was found to be an insignificant route of metabolism, but it was a major route of metabolism in the guinea pig (50%) and occurred to a lesser extent (23%) in the mouse. In the absence of significant O-demethylation in the rat, [*O-methyl-*[14]C] acronycine was used to advantage in this rodent and injected by the ip route. While only 12.3% of ^{14}C-labeled products was detected in the urine after 24 hr, nearly 90% was recovered from the feces in 48 hr. Only 12% of the ^{14}C administered was recovered from the rat bile following oral administration. To conclude from the latter experiment that this lower level in the bile is an indication that acronycine is poorly absorbed from the gastrointestinal tract, would be to discredit the reported efficacy of oral acronycine administration (Table III) in the C-1498 leukemia and in the X5563 plasma cell tumor (*5*).

Among the metabolites, 9-(**9**) and 11-hydroxyacronycine (**10**) were shown to occur in the four mammalian species and in human urine samples (Table IV, Scheme 4). 3-Hydroxymethylacronycine (**11**) was identified in bile samples from the rat and the mouse. Three further metabolites were identified as 11-hydroxy-3-hydroxymethyl-(**12**), 9,11-dihydroxy-(**13**), and, in the guinea pig only, 11-hydroxy-*O*-demethylacronycine (**14**). The structures of the metabolites are given in Scheme 4 and the distribution of metabolites

9-14 is listed in Table IV. In view of the inactivity of the metabolites obtained from microbial transformation and from mammalian metabolism [e.g., 11-hydroxyacronycine (10)] against the tumor systems sensitive to acronycine (39-41), the activity of psorospermin (8), having a hydroxyl function corresponding to that in 10, is worthy of note. A close inspection of the qualitative and the quantitative aspects of the metabolism study (41) indicates that small amounts of phosphorylated metabolites may have eluded detection.

VII. Preclinical Studies of Acronycine

The recommendation of acronycine for clinical evaluation was based mainly on these considerations: (a) exceptionally broad experimental anti-tumor spectrum that included activity against C-1498 leukemia and the X-5563 (5) and LPC-1 (32) plasma cell tumors; (b) activity on delayed treatment; (c) activity by the oral route at favorable dose levels; (d) novel type of molecular structure differing from that of other effective agents; and finally, (e) indication of a novel type of action mechanism (see following section).

Acute and chronic studies performed in rats, dogs, and monkeys at the Toxicology Division, Lilly Research Laboratories (42) and at the National Cancer Institute (NCI), Bethesda (32) completed the required preclinical investigation of acronycine.

In these toxicological studies the drug was administered orally unless indicated otherwise. In acute studies, the LD_{50} in mice ranged from 350 to 680 mg/kg and in rats from 560 to 675 mg/kg as a single dose. At toxic dose levels the signs of acute toxicity noted were hypoactivity, anorexia, weakness, and death. Most of the deaths that occurred were delayed. Cats tolerated a single dose of 100 mg/kg. Of special interest was the observation that acronycine is 2-3 times more toxic by the intraperitoneal route of administration than by the oral route.

In chronic studies, groups of rats fed diets containing 0.1 or 0.25% of drug for 1 month showed poor weight gain, anemia, leukopenia, and death. Groups of mice were fed a diet containing 0.0 (control), 0.006, 0.0125, 0.025, 0.05, and 0.1% acronycine for 3 months. At the lowest two levels no toxic effects were observed. Toxic effects observed at the higher dose levels included decreased weight gain, reduction in erythrocyte and leukocyte count, increased liver weight, thymic atrophy, and a few deaths.

Groups of dogs were given oral doses of 0.25, 0.5, 1.0, and 2.5 mg/kg of the drug for 3 months. At the two lower dose levels no toxic signs were noted, except for transient anorexia. Toxic signs observed at the two highest dose levels included anorexia, emesis, hypoactivity, and lowered erythrocyte count. At these higher levels, tests revealed an increase in glutamate pyruvate transaminase activity, in alkaline phosphatase, and blood urea nitrogen (BUN) levels. Deaths were seen at the highest dose level only. Finally, groups of Rhesus monkeys tolerated daily oral doses of 0.5 and 1–2 mg/kg for 15 days. Elevation of alkaline phosphatase levels was noted as the primary evidence of toxicity.

In the course of a Phase I clinical investigation initiated at the Lilly Laboratories for Clinical Research, dose-ranging studies in adult patients established a recommended total oral daily dose of 200 mg to further evaluate acronycine (43). The main side effects frequently but not consistently seen at this and higher dose levels (up to 500 mg total daily dose) consisted of nausea and vomiting.

In subsequent Phase I and early Phase II trials conducted under the auspices of the Lilly Clinical Research Laboratories and NCI (NSC 403169) (33), suggestive evidence for acronycine efficacy was seen in the treatment of a number of patients with multiple myeloma (43). Nausea and vomiting were encountered as dose-limiting side effects, seen more frequently at total doses above 200 mg per day.

Additional, limited Phase II trials of acronycine have been conducted. In these clinical trials, patients with multiple myeloma were treated for varied periods of time with acronycine at oral dose levels aimed at producing minimal side effects. At the time of this writing conclusive confirmation of acronycine effectiveness against multiple myeloma has not been reported in the medical literature (43).

With increased water solubility of acronycine as the stated objective, the preparation of O-acetylacronycinium perchlorate as a soluble prodrug form of acronycine was accomplished (44, 45). The reasonably short half-life of about 5 min, determined for this salt under aqueous conditions approximating the in vivo situation, indicates that this prodrug might have suitable qualifications for iv injection. A similar highly water-soluble prodrug O-methylacronycinium fluorosulfonate was reported by Smithwick et al. (46).

Antitumor assays of these prodrugs against the C-1498 or X-5563 murine model systems have not been reported. For this reason, no clear evidence is available to demonstrate possible superiority over the broad efficacy shown by acronycine (Table II).

Suffness and Douros in a brief appraisal of acronycine (47) point to an "incongruity" in the assessment of "broadest" antitumor activity spectrum by Svoboda et al. (5) and that of "inactivity" in tumor systems (P388 and L-1210 leukemias, B-16 melanoma) at the NCI. These systems are held to have significant predictive merit for clinical efficacy (47).

VIII. Mechanism of Action

One reason for recommending acronycine for clinical evaluation was the notion that the mechanism of its antitumor activity differed from that of other clinically active agents. Perusal of the studies quoted below tends to reinforce this view. A classification of antitumor agents generally accepted by cancer chemotherapists recognizes the following types of action: (a) agents acting as antimetabolites, (b) agents such as alkylators, intercalators, and others binding to, or interacting with, DNA and/or DNA function, (c) agents influencing hormonal control of cellular function, (d) agents affecting mitosis and binding to tubulin, and (e) other (48). Acronycine does not appear to fit in any of the categories (a–d).

The critical question of whether the striking experimental antitumor action of acronycine can translate into clinical efficacy transcends the issue of this individual agent. The question evolves into the broader issue of asking whether this or other agents with similar (acronycine-type) mechanism of action hold promise of clinical efficacy.

A matter of growing interest is the relationship of the mechanism of antitumor action to its potential for carcinogenicity (*vide infra*).

Effects of acronycine on mammalian tumor cells in culture have been reported (6, 49–53) from the Cancer Research Centre, University of British Columbia (49–51, 53), the Michigan Cancer Foundation, Detroit (52), and the Lilly Research Laboratories (6). The latter report also lists some *in vivo* experiments including general endocrine and hormone antagonistic action.

Two important, though negative, observations serve to outline acronycine's mode of action. The drug appears neither to interact with DNA nor to affect DNA function, certainly not at dose levels at which it does affect RNA synthesis (49, 50). Also, acronycine has been reported not to arrest mammalian cells (L5178Y leukemia cells, Chinese hamster lung cells, others) in mitosis (5, 49). Gradual cessation of mitotic activity has been noted (51).

In the presence of acronycine the population growth of L5178Y leukemia cells in culture was inhibited at drug concentrations that rapidly and appreciably inhibited incorporation of labeled uridine, cytidine, and other nucleosides into RNA (50). Acronycine added to cultured IRC 741 rat leukemia cells causes dose-dependent binucleation, possibly through interference with cytokinesis (53). Furthermore, tests showed that the reduced incorporation of nucleosides into RNA is not due to drug interference with nucleoside phopshorylation or with nucleotide incorporation into RNA, but is caused by inhibition of nucleoside transport across plasma membranes and membranes of subcellular organelles, such as Golgi complexes and mitochondria (51, 52). Uptake of formate into L5178Y cells was also reduced by acronycine (50).

In terms of time course of action, cytological effects of acronycine, e.g., interference with nucleoside transport and binding of membrane probes, occurred within minutes after exposure to the drug (52), whereas reduction of cytokinesis and reduction of cell proliferation become effective only in a matter of hours. It was concluded that the delayed actions noted here are secondary effects resulting from acronycine incorporation into membranes (52), or, in the words of Low and Auersperg "acronycine interferes *primarily* with structure, function and/or turnover of cell membrane components, thereby changing the fluidity of the plasma membrane" (53).

Although the demonstrated inhibition of nucleoside transport serves as a clue of acronycine's mode of action, insight into the molecular mechanism of such action still remains elusive. Direct evidence for a molecular basis of acronycine action appears to be lacking. A search for indirect evidence involves two approaches. The first explores the comparison of acronycine's action with that of two other cytoactive agents cytochalasin B (51, 53, 54) and cyclic adenosine 3′,5′-monophosphate (cAMP) (51, 55), which share some characteristics with acronycine. The second approach favors another look at the SAR of acronycine and related xanthones **6, 7,** and psorospermin (**8**).

Initial observations in mammalian cell cultures by Tan and Auersperg (51) suggested that acronycine causes a reduction in cell density and a change from epithelial to fusiform cell shape, cytological effects noted before in response to cAMP (55). On further comparison, however, it appeared that differences outweighed similarities because cAMP was found to have little effect on cell density of the three mammalian cell lines studied and did not reduce the rate of mitotic activity (51). In a further comparison of the respective cytological effects caused by acronycine and by cytochalasin B, the differences again were found to outweigh similarities (51, 53).

Low and Auersperg subsequently drew attention to the lipophilic nature common to both acronycine and cytochalasin B. The authors considered the lipophilicity of both agents as the basis for similarity of some of the cytological effects induced by the alkaloids (53). Both acronycine and cytochalasin B do interfere with nucleoside transport, with melanosome translocation, and with cell cleavage, the last action resulting in the emergence of binucleated cells. However, cytochalasin B effects appear in less than a minute, whereas the effects induced by acronycine, known to enter the cell lines studied within seconds after exposure, are delayed by hours. Similarly, acronycine appears to inhibit later stages of the cell's furrowing progression and daughter cell formation. Cytochalasin B, on the other hand, interfered with initiation and early progression of the cleavage furrow (53).

In view of these and other differences the authors conclude that "in spite of a resemblance in chemical properties and several similar end-effects, the primary mode of action of the two agents differ" (53). It would appear that the lipophilic nature of acronycine and cytochalasin B, cited by these authors, is common to several antitumor agents, vinblastine (log $P = 2.9$) (56), maytansine (57), and several others (48) with widely differing action mechanisms. Therefore, lipophilicity per se may have only limited relevance in defining the mode of drug action.

Returning for a second look at the SAR of a small group of close relatives, the activity of acronycine against responsive tumor models (Table II) when contrasted with inactivity reported (34) for the linear, isomeric product (Scheme 3, structure b) with quite similar chemical functionality, suggests a strict requirement for the angular molecular structure. Such dependence on a specific structure would seem more compatible with a specific structure-based, drug-receptor type of mechanism (antimetabolite role?), than with a less precise type of action based on a physical characteristic, such as lipophilicity or inadequate solubility. The latter characteristic becomes important when insoluble material is administered parenterally and the possible local precipitation of microparticles of the drug may go unnoticed.

The quest for a specific molecular mechanism for acronycine remains elusive and little insight is gained from the xanthone triad **6**, **7**, and **8**. Xanthone **6**, the structural analog of acronycine, reported to be inactive deserves testing against acronycine-sensitive tumor models for verification of its status. The activity of psorospermin (**8**) against the P388 leukemia is of considerable interest, as its isoprenoid moiety is congruent and isomeric with that of the acronycine metabolites **11** (3-hydroxymethyl-) and **12** (11-hydroxy-3-hydroxymethylacronycine) (Scheme 4, Table IV).

A technique of gaining information about the molecular basis by attempting to reverse acronycine action with specific metabolites has not been explored. Such reversal studies have yielded striking information on metabolite analogy and enzyme involvement, which has on occasion resulted in valuable knowledge of molecular mechanism of action. The case of mycophenolic acid (58) in its relation to guanylic acid and that of pyrazofurin (59) in its relation to orotidylic acid illustrates the merit of reversal studies.

Interest in this phenomenon of biochemical equivalence of seemingly different molecules (60) [e.g., lincomycin and erythromycin (61)] has been much enhanced by discovery of a structural kinship existing between certain natural drug products [e.g., morphine and endogenous peptide factors (62)]. Whether or not equivalence of acronycine with an endogenous metabolite can be shown to exist depends on future research in this area (36).

Having accepted membrane effects and related nucleoside transport inhibition as the possible mode of action, we now turn to the issue concerning potential for carcinogenicity of antitumor agents (63–65), including acronycine. This issue asks for close attention because of the medical and the scientific (mechanistic) implications involved. These two implications were briefly assessed in a recent review of *Catharanthus* (Vinca) alkaloids: "While a number of clinically effective anti-cancer agents express their activity by interacting—covalently or non-covalently—with DNA or with DNA function and in so doing carry a potential risk of carcinogenesis for the patient, the no less efficacious alkaloids, Vinblastine and Vincristine, interacting (non-covalently) with tubulin and arresting mitosis, are thought not to carry this risk" (*66*). In an experimental verification of this concept, the lack of carcinogenicity of vinblastine has been demonstrated in an *in vivo* model (*65*). Publications reporting lack of mutagenicity (*67*) and the absence of carcinogenic hazard (*64*) of Vinca therapy have appeared. The therapeutic merit of noncarcinogenicity and the tubulin-binding action mechanism of the Vinca alkaloids are thought to be closely linked qualities (*66*).

As cited above (*50*), acronycine does not interact with DNA and primary effects of the drug on DNA synthesis or DNA function were not observed. It appears plausible, therefore, to expect that acronycine activity be of a noncarcinogenic nature.

In a recent review Weisburger and Williams, noting the differences in carcinogenic mechanisms of various agents, proposed that carcinogens be classified in two broad types: namely, genotoxic agents causing DNA damage in a direct manner, and epigenetic agents, operating by an oncogenic mechanism of a nongenetic nature (*68*). Whereas the genotoxic carcinogen or its activated metabolite (proximal carcinogen) interacts directly with DNA damaging the cell's genotype, the epigenetic-type oncogen acting by a number of mechanisms still largely unknown does not appear to convert a normal cell to a neoplastic one, but instead may support completion of an earlier initiation event in the history of the cell.

The authors further suggest that oncogens be divided into eight subclasses, three of the genotoxic type (direct-acting, procarcinogen, and inorganic carcinogen agents) and five of the epigenetic type (hormones, immunosuppressors, promotors, cocarcinogens, and solid-state carcinogens) (*68*). No information is available identifying acronycine with any of the last four subclasses, whereas the role of an hormonal agent is thought to be unlikely for acronycine. In an experiment designed to estimate estrogenic activity in the immature rat, acronycine was rated as having neglible (4%) activity relative to estradiol (**b**).

On the basis of the above evidence, lack of carcinogenic potential for acronycine's activity indeed appeared plausible.

In a report of NCI, Carcinogenesis Testing Program, acronycine was termed to be carcinogenic in rats upon chronic administration (*69*). The drug was given for 52(?) weeks by the intraperitoneal route of administration in a vehicle of 0.05% polysorbate 80 in phosphate-buffered saline. Sarcomas and other tumors of the peritoneum were noted in both males and females.

Tumors of the mammary gland were seen in females and osteosarcomas in males.

At the time of this writing, the NCI report is available in abstract form only. A quantitative appraisal of the incidence and potency of carcinogenicity reported for acronycine, therefore, is not feasible at this time. The choice of the ip route of injection of the drug at elevated (toxic) dose levels is deemed unfortunate, especially in view of the oral use of the drug in the clinical trials done under the sponsorship of Eli Lilly Company and NCI (*32*).

To summarize this discussion of mode of action, the need of additional research (*36*) in the area of acronycine and congeners is apparent. A definition of the molecular basis of its action represents one objective of such research. Another objective is the relation of the drug's molecular mechanism of action as antitumor agent to its reported potential for carcinogenicity (*69*). An accurate assessment of incidence and potency (*70*) of animal carcinogenicity — preferably determined by the oral route — would help to serve as quantitative predictor of human risk (*70*).

IX. Summary and Conclusions

The acridone class of alkaloids comprising less than two dozen members has been surveyed with special emphasis on experimental cancer chemotherapy studies of acronycine. Synthesis of acronycine, its experimental antitumor spectrum, and toxicological studies were discussed as an introduction to the clinical evaluation of acronycine in patients with multiple myeloma. The mode of action of acronycine reportedly involves its incorporation into cellular and subcellular membranes resulting in inhibition of nucleoside transport. A report issued by the NCI, Carcinogenesis Testing Program, described the formation of tumors in male and female rats given the alkaloid by ip injection (*69*). Acronycine was reported to be carcinogenic. Since the drug does not interact with DNA, the possibility that the carcinogenicity is of an epigenetic rather than a genotoxic nature has been mentioned (*68*).

The possibility of further research in the acronycine area has been encouraged by noting the activity against the P388 leukemia and other tumor models by the related xanthone psorospermin (**8**). The desirability of reversal studies using specific metabolites (nucleosides, etc.) has been mentioned.

On the basis of its mechanism of nucleoside transport inhibition, it is conceivable that combination studies with agents such as pyrazofurin (*60*), *N*-phosphonoacetyl-L-aspartate (PALA) (*71*), and other inhibitors of nucleoside synthetic pathways may prove fruitful. The activity of these latter agents is said to be hampered sometimes by a competition of the "salvage" pathway (*72*). Possibly, favorable, synergistically acting combinations could be exploited to suggest favorable clinical regimens for trial.

A final point concerns the policy of selecting a drug as clinical candidate, enunciated in a brief discussion of acronycine by colleagues at the NCI (*47*). In the discussion it was stated that acronycine presently would not be recommended for clinical evaluation at the Institute's Drug Development Branch because it lacks activity against the four primary tumor models rightly held to be of high predictive merit for clinical efficacy.

While fully agreeing with the selection of these four tumor models, P388 leukemia, leukemia L 1210, B-16 melanoma, and Lewis lung carcinoma and with the significant studies of Venditti and associates (*73*) on which the selection of these four models is based, it is felt that the too rigid interpretation of this selection as exemplified in the NCI Statement (*47*) is not in the best interest of drug development. That an antitumor agent may be selected because of other activities, even when not active against all or any of the four models, is exemplified convincingly by the institute's earlier decision to support and participate in the clinical evaluation of acronycine (*32*). A second example is seen in the selection of PALA (*71*) for clinical evaluation which was based on its appreciable activity against the Lewis lung carcinoma and on its novel mode of action (*74*).

For these reasons the selection of acronycine for clinical evaluation which was based on its broad antitumor spectrum (*6*) as well as on its novel mechanism of action (*49–53*) and which received NCI support because of its activity against the PC-1 plasma cell tumor (*32*), is thought to be a justifiable one.

Appendix

Three publications not reported in the present discussion will be mentioned briefly. A 1982 patent (*75*) cites the evaluation of acronycine in combatting multiple myeloma in patients. Reduction of clinical signs (serum protein levels) were reported for a male patient treated with oral acronycine for a period of 50 days. Reduction of pain of the spine was observed also in this patient. Subjective improvement (e.g., full reduction of bone pain) was seen in another patient treated for 90 days. At the dose of 200 mg/m² of body surface used in this trial, toxic side effects were minimal.

Effects of acronycine on the cell cycle of L cells in culture have been reported (*76*). Twenty-four-hour treatment with 10 μg/ml resulted in an accumulation of 46.6% of the cells in the $(G_2 + M)$ phase, compared to 12.1% in control experiments. These results do not differ materially from the cytological effects reported before (*49–53*).

The synthesis of 7-thionoacronycine has been accomplished from acronycine with tetraphosphorus decasulfide. Test results of this sulfur analog, synthesized as a pro-drug, were not reported (77).

Addendum

Two recent publications dealing with acridone alkaloids by workers in Taiwan have appeared in Heterocycles (78, 79). Isolation, characterization, and structure determination of five acridone alkaloids (Scheme 5) from the root and stem bark of *Glycosmis citrifolia* (Willd.) Lindl. was reported in detail. These alkaloids include glycofoline (**15**), glyfoline (**16**), glycocitrine I (**17**), glycocitrine II (**18a**), and O_3-methylglycocitrine II (**18b**), of which glyfoline (**16**) is the highest oxygenated acridone obtained from natural sources.

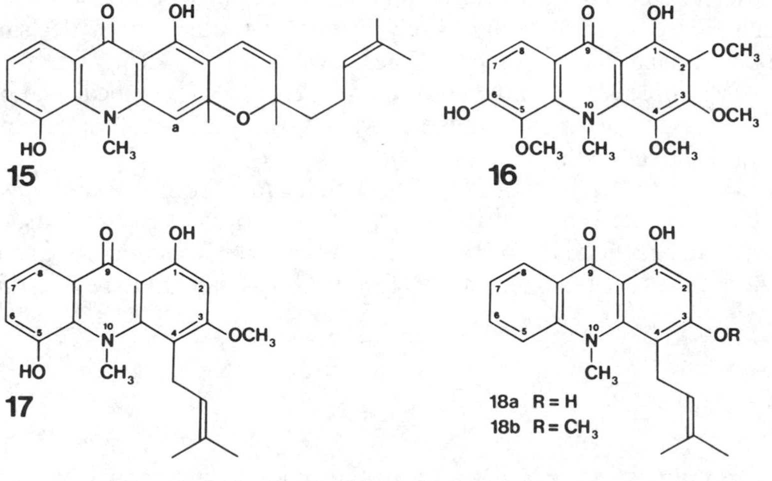

SCHEME 5. *Acridone alkaloids of* Glycosmis citrifolia. [*From Wu and Furukawa* (78) *and Wu* et al. (79)].

Glycosmis citrifolia (Willd.) Lindl. (*G. Cochinchinensis* Pierre), a genus of Rutaceae, is a wild shrub claimed to serve as folk medicine in the treatment of scabies, boils, and ulcers (78).

The structure determination of glycofoline (**15**) was accomplished with the aid of chemical and physical methods, including UV and IR spectral analysis, inspection of the ^{1}H- and ^{13}C-NMR. spectra, and mass spectral examination. To distinguish between the linear (**15**) and the angular orientation (as in acronycine, Fig. 1) of the pyran ring, the authors examined the nuclear Overhauser enhancement (nOe) for the N-methyl signal at $\delta 4.00$.

Irradiation of this signal produced a 20% enhancement of the signal at $\delta 6.23$ assigned to H-a in **15**.

From a biogenetic point of view, it is not immediately clear why in the case of glycofoline (**15**), a first example of a naturally occurring monoterpenoid acridone alkaloid, alkylation with the geranyl unit should take place at C-2 (acridone numbering: Scheme 1 and formula **16**), when in acronycine (Fig. 1) prenylation evidently occurs at C-4 (acridone numbering). The authors have not addressed this issue and no chemical (degradative) evidence in support of the linear structure is presently available. No information concerning biological activity of glycofoline (**15**) or the other alkaloids (**16, 17, and 18**) was reported (*78, 79*). The synthesis of C-5 desoxyglycofoline has been reported earlier (*80*).

Structure assignments for glyfoline (**16**), glycocitrine I (**17**), glycocitrine II (**18a**) and its methyl ether (**18b**) were established by the physical chemical methods mentioned above. The location of the free phenolic group at C-6 in glyfoline (**16**), a hexa-oxygenated *N*-methyl-9-acridone, was demonstrated by means of an nOe experiment with its methoxymethyl ether at the C-6 hydroxy function. Irradiation at $\delta 5.36$, the frequency corresponding to the methylene protons of the methoxymethyl ether function, resulted in a 12% enhancement of the signal at $\delta 7.14$ assigned to H-7 in **16**. Methylation of glycocitrine II (**18a**) with diazomethane gave the monomethyl ether identical with this alkaloid (**18b**) isolated from the plant.

From the standpoint of SAR interest, the presence or absence of biological activity in glycofoline (**15**) is of interest, especially in relation to the acronycine metabolite (**14**, Table IV) which shares with glycofoline a similar substitution at C-11 (hydroxy group) and at C-6 (de-*O*-methyl function) (acronycine numbering, Fig. 1).

For the sake of clarity, it is noted that the numbering system (not shown here) used for glycofoline in reference 78, a repetition of the acridone numbering system, differs from the one used in this chapter for acronycine (Fig. 1).

ACKNOWLEDGMENTS

Koert Gerzon wishes to express his appreciation to spokesmen of the Lilly Research Laboratories for providing information on the toxicological and clinical studies with acronycine.

The assistance generously given by Brenda Heady in the preparation of the manuscript is acknowledged with gratitude. We wish to thank Deborah Eads Staley for drawing the art work.

The help given by Dr. John M. Cassady in pointing to the relevance of psorospermin activity to acronycine research is gratefully acknowledged.

REFERENCES

1. J. R. Price, *in* "The Alkaloids" (R. H. F. Manske and H. L. Holmes, eds.), Vol. 2, p. 353. Academic Press, New York, 1952.
2. G. K. Hughes, F. N. Lahey, J. R. Price, and L. J. Webb, *Nature* (*London*) **162**, 223 (1948).
3. G. K. Hughes and K. G. Neill, *Aust. J. Sci. Res. Ser. A* **2**, 429 (1949).
4. T. R. Govindachari, B. R. Pai, and P. S. Subramaniam, Tetrahedron **22**, 3245 (1966).
5. G. H. Svoboda, G. A. Poore, P. J. Simpson, and G. A. Boder, *J. Pharm. Sci.* **55**, 758 (1966).
6. G. H. Svoboda, *Lloydia* **29** (3), 217 (1966).
7. W. D. Crow and J. R. Price, *Aust. J. Sci. Res., Ser. A* **2**, 255 (1949).
8. K. G. Neill, M. Sc. Thesis, University of Sydney, 1949.
9. J. R. Price, *Aust. J. Sci. Res., Ser. A* **2**, 249 (1949).
10. F. N. Lahey and W. C. Thomas, *Aust. J. Sci. Res.,* Ser. A **2**, 423 (1949).
11. G. K. Hughes, K. G. Neill, and E. Ritchie, *Aust. J. Sci. Res.,* Ser. A **5**, 401 (1952).
12. F. N. Lahey, J. A. Lamberton, and J. R. Price, *Aust. J. Sci. Res.,* Ser. A **3**, 155 (1950).
13. M. F. Grundon, *in* "The Alkaloids" (R. H. F. Manske, and R. G. A. Rodrigo, eds.), Vol. 17, p. 105. Academic Press, New York, 1979.
14. R. D. Brown, L. D. Drummond, F. N. Lahey, and W. C. Thomas, *Aust. J. Sci. Res., Ser. A* **2**, 622 (1949).
15. N. R. Farnsworth, L. K. Henry, G. H. Svoboda, R. N. Blomster, M. J. Yates, and K. L. Euler, *Lloydia* **29**, 101 (1966).
16. R. D. Brown and F. N. Lahey, *Aust. J. Sci. Res., Ser. A* **5**, 593 (1952), see footnote on page 607.
17. L. J. Drummond and F. N. Lahey, *Aust. J. Sci. Res., Ser. A* **2**, 630 (1949).
18. P. L. Mac Donald and A. V. Robertson, *Aust. J. Chem.* **19**, 275 (1966).
19. J. Z. Gougatas and B. A. Kaski; unpublished experiments (see reference 21 of this chapter).
20. J. R. Beck, R. N. Booher, A. C. Brown, R. Kwok, and A. Pohland, *J. Am. Chem. Soc.* **89**, 3934 (1967).
21. J. R. Beck, R. Kwok, R. N. Booher, A. C. Brown, L. E. Patterson, P. Pranc, B. Rocker, and A. Pohland, *J. Amer. Chem. Soc.* **90**, 4706 (1968).
22. J. Hlubucek, E. Ritchie, and W. C. Taylor, *Chem. Ind.* 1809 (1969).
23. J. Hlubucek, E. Ritchie, and W. C. Taylor, *Aust. J. Chem.* **23**, 1881 (1970).
24. C. S. Barnes, J. L. Occolowitz *et al., Tetrahedron Lett.* **281** (1963).
25. G. K. Hughes and E. Ritchie, J. *Aust. Sci. Res., Ser. A* **4**, 423 (1951).
27. J. W. Audas, *in* "Native Trees of Australia," p. 136. Whitcombe and Tombs, Melbourne Australia.
28. I. S. Johnson, H. F. Wright, G. H. Svoboda, and J. Vlantis, *Cancer Res.* **20**, 1016 (1960).
29. J. H. Cutts, C. T. Beer, and R. L. Noble, *Cancer Res.* **20**, 1023 (1960).
30. I. S. Johnson, J. G. Armstrong, M. Gorman, and J. P. Burnett, *Cancer Res.* **23**, 1390 (1963).
31. G. H. Svoboda, *Lloydia* **24**, 173 (1961).
32. S. K. Carter, *Cancer Chemother. Rep. Part 3* **2** (1), 81 (1971).
33. K. J. Liska, *J. Med. Chem.* **15**, 1177 (1972).
34. J. Schneider, L. Evans, E. Grunberg, and R. I. Fryer, *J. Med. Chem.* **15**, 266 (1972).
35. S. M. Kupchan, *J. Nat. Prod.* **43**, 296 (1980).
36. J. M. Cassady, C.-J. Chang, and J. L. McLaughlin, *in* "Natural Products as Medicinal Agents" (J. L. Beal and E. Reinhardt, eds.), p. 93. Hippokrates Verlag, Stuttgart, 1981.
37. C. Ducrocq, F. Wendling, M. Tourbez-Perrin, C. Rivalle, P. Tambourin, F. Pochon, E. Bisagni, and J. C. Chermann, *J. Med. Chem.* **23**, 1212 (1980).
38. R. K. Johnson, R. K.-Y. Zee-Cheng, W. W. Lee, E. M. Acton, D. W. Henry, and C. C. Cheng, *Cancer Treat. Rep.* **63**, 425 (1979).

39. D. R. Brannon, D. R. Horton, and G. H. Svoboda, *J. Med. Chem.* **17**, 653 (1974).
40. R. E. Betts, D. E. Walters, and J. P. Rosazza, *J. Med. Chem.* **17**, 599 (1974).
41. H. R. Sullivan, R. E. Billings, J. L. Occolowitz, J. E. Boaz, F. J. Marshall, and R. E. McMahon, *J. Med. Chem.* **15**, 904 (1970).
42. Toxicology Division, Lilly Research Laboratories, Greenfield, Indiana, unpublished experiments.
43. Lilly Laboratories for Clinical Research, Indianapolis, Indiana, personal communication to Koert Gerzon, January, 1982.
44. A. B. Hansen, B. Kreilgard, C-H. Huang, and A. J. Repta, *J. Pharm. Sci.,* **67**, 237 (1980).
45. T. Higuchi, A. J. Repta, and D. W. A. Bourne, U.S. Patent 3,943,137 (1976).
46. E. L. Smithwick, U.S. Patent, 3,843,658 (1974).
47. M. S. Suffness and J. D. Douros, *in* "Anti-Cancer Agents based on Natural Product Models" (J. M. Cassady and J. D. Douros, eds.), p. 472. Academic Press, New York, 1980.
48. W. B. Pratt and R. W. Rundon, "The Anti-Cancer Drugs," p. 44. Oxford University Press, New York and Oxford, 1979.
49. P. W. Gout, B. P. Dunn, and C. T. Beer, *J. Cell Physiol.* **78**, 127 (1971).
50. B. P. Dunn, P. W. Gout, and C. T. Beer, *Cancer Res.* **33**, 2310 (1973).
51. P. Tan and N. Auersperg, *Cancer Res.* **33**, 2320 (1973).
52. D. Kessel, *Biochem. Pharmacol.* **26**, 1077 (1977).
53. R. S. Low and N. Auersperg, *Exp. Cell Res.* **131**, 15 (1981).
54. S. E. Malavista, *Nature (London)* **234**, 354 (1971).
55. A. W. Hsie and T. T. Puck, *Proc. Natl. Acad. Sci. U.S.A.* **68**, 358, 361 (1971).
56. R. A. Conrad, G. J. Cullinan, K. Gerzon, and G. A. Poore, *J. Med. Chem.* **22**, 391 (1979).
57. J. M. Kupchan, Y. Komoda, W. A. Court, G. J. Thomas, R. M. Smith, A. Karim, C. J. Gilmore, R. C. Haltiwanger, R. F. Bryan, *J. Am. Chem. Soc.* **95**, 1354 (1972).
58. J. C. Cline, J. D. Nelson, K. Gerzon, R. H. Williams, and D. C. DeLong, *Appl. Microbiol.* **18**, 14 (1969).
59. F. J. Streightoff, J. D. Nelson, J. C. Cline, K. Gerzon, M. Hoehn, R. H. Williams, M. Gorman, and D. C. DeLong, *Ninth Conf. Antimicrobial Agents and Chemother.* **8**, *Washington, D.C.,* October, 1969, Abstract No. 18.
60. K. Gerzon, D. C. Delong, and J. C. Cline, *Pure Appl. Chem.* **28**, 489 (1971).
61. B. Weisblum, *in* "Drug Action and Drug Resistance in Bacteria" (S. Mitsuhashi, ed.), p. 233. University Park Press, Baltimore, Tokyo, 1971.
62. A. P. Feinberg, I. Creese, and S. H. Snyder, *Proc. Natl. Acad. Sci. U.S.A.* **73** (11), 4215 (1976).
63. D. Schmähl., *Cancer* **40**, (4), 1927 (1977).
64. E. K. Weisburger, *Cancer* **40** (4), Supplement.
65. D. Schmähl and H. Osswald, *Arzneim. Forsch.* **20**, 1461 (1970).
66. K. Gerzon, *in* "Anti-cancer Agents based on Natural Products Models" (J. M. Cassady and J. D. Douros, eds.), p. 274. Academic Press, New York, 1980.
67. Y. Seino, M. Nagao, T. Yahagi, A. Hoshi, T. Kawachi, and T. Sugimura, *Cancer Res.* **38**, 2148 (1978).
68. J. H. Weisburger and G. M. Williams, *Science* **214**, 401 (1981).
69. Carcinog. Test Program, N.C.I., Bethesda, Md., Bio-assay of Acronycine for Possible Carcinogenicity, CAS No. 7008-42-6. Reported in *Chem. Abstracts,* **90**, 097501[h] (1979); NIOSH Registry, Suspected Carcinogens, 2nd Edition, Cincinnati, Ohio 45226, Dec. 1976, Code No. UQ03300.
70. E. Crouch and R. Wilson, *J. Toxicol. Environ. Health,* **5**, 1095 (1979).
71. R. K. Johnson, *Biochem. Pharmacol.* **26**, 81 (1977).
72. A. W. Murray, *Ann. Rev. Biochem.* **40**, 773 (1971).

73. J. Venditti, *in* "Pharmacological Basis of Cancer Chemotherapy" p. 245. Williams and Wilkins, Baltimore, Md., 1975.
74. K. D. Collins and G. R. Stark, *J. Biol. Chem.* **246**, 6599 (1971).
75. G. H. Svoboda, U.S. Patent, 4,309,431, Jan. 5, 1982; continuation of U.S. Patent, 3,985,899.
76. S. B. Reddy, W. A. Linden, F. Zywietz, H. Baisch, and U. Struck, *Arzneim. Forsch.* **27**, 1549 (1977).
77. J. R. Dimmocks, A. J. Repta, J. J. Kaminski, *J. Pharm. Sci.* **68**, 36 (1979).
78. T.-S. Wu and H. Furukawa, *Heterocycles* **19** (5), 825 (1982).
79. T.-S. Wu, H. Furukawa, and C.-S. Kuoh, *Heterocycles* **19** (6), 1047 (1982).
80. W. M. Bandaranayake, M. J. Begley, B. O. Brown, D. G. Clarke, L. Crombie, and D. Whiting, *J. Chem. Soc., Perkin Trans. 1,* 998 (1974).

—— CHAPTER 2 ——

THE QUINAZOLINOCARBOLINE ALKALOIDS

JAN BERGMAN

Department of Organic Chemistry, Royal Institute of Technology,
Stockholm, Sweden

I. Introduction and Nomenclature

The first known representatives (*1, 2*) of the quinazolinocarboline alkaloids were rutaecarpine (**1**) and evodiamine (**2**) (both from *Evodia rutaecarpa* Hook. f. et Thoms.). More recently, more than 20 other representatives have been isolated.

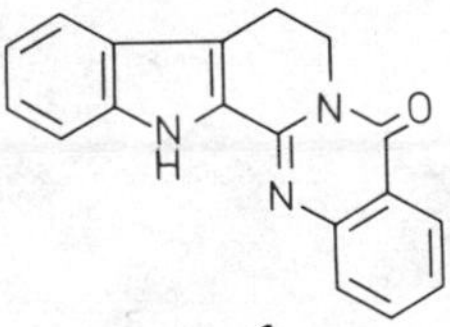

1 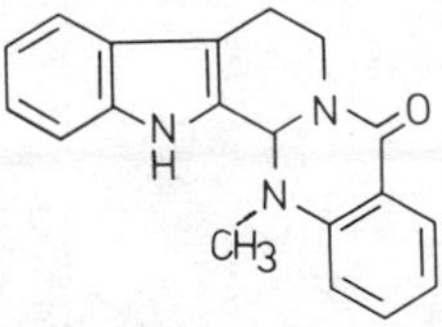   **2**

This chapter is an extension of the review by Manske in Volume VIII, p. 55. The long-established name quinazolinocarboline (*3*) will be retained in this review although since 1960 Chemical Abstracts has favored the name

THE ALKALOIDS, VOL. XXI
Copyright © 1983 by Academic Press, Inc.
All rights of reproduction in any form reserved.
ISBN 0-12-469521-3

TABLE I
QUINAZOLINOCARBOLINE ALKALOIDS OCCURRING IN PLANTS

Plant	Alkaloid	Composition	Melting point (°C)	Formula	Reference
Evodia rutaecarpa Hook. f. et Thom.	Rutaecarpine	$C_{18}H_{13}N_3O$	258	**1**	*6, 14*
	Evodiamine	$C_{19}H_{17}N_3O$	278	**2**	*6*
	13*b*,14-Dihydrorutaecarpine	$C_{18}H_{15}N_3O$	214	**8**	*6*
	14-Formyl-13*b*,14-dihydrorutaecarpine	$C_{19}H_{15}N_3O_2$	280	**30**	*6*
	13,13*b*-Dehydroevodiamine	$C_{19}H_{15}N_3O$	189	**37**	*41*
	7-Carboxyevodiamine	$C_{20}H_{17}N_3O_3$		**9**	*7*
Hortia arborea Engl.	Rhetsinine	$C_{19}H_{17}N_3O_2$	196 dec	**27**	*14, 40*
	Hortiacine	$C_{19}H_{15}N_3O_2$	252	**14**	*23*
	Hortiamine	$C_{20}H_{17}N_3O_2$	208	**13**	*23*
	Rutaecarpine				*23*
Hortia braziliana Vel.	Hortiamine				*12*
	Base	$C_{20}H_{17}N_3O_2$	269		*49*
	Nine other bases				*49*
Hortia badinii	Hortiacine				*69*
	Rutaecarpine				*69*
Hortia longifolia	Hortiacine				*70*

Euxylophora paraensis Huber	Euxylophoricine A	$C_{20}H_{17}N_3O_3$	295	**15**	*35*
	Euxylophoricine B	$C_{20}H_{17}N_3O_3$	310	**16**	*35*
	Euxylophoricine C	$C_{19}H_{13}N_3O_3$	310	**17**	*36*
	Euxylophoricine D	$C_{21}H_{19}N_3O_4$	293	**24**	*37*
	Euxylophoricine E	$C_{21}H_{17}N_3O_4$	290	**22**	*37*
	Euxylophoricine F	$C_{19}H_{15}N_3O_3$	226	**23**	*51*
	Euxylophorine A	$C_{21}H_{19}N_3O_3$	227	**18**	*35*
	Euxylophorine B	$C_{21}H_{17}N_3O_3$	268	**19**	*36*
	Euxylophorine C	$C_{22}H_{21}N_3O_4$	207	**20**	*37*
	Euxylophorine D	$C_{22}H_{19}N_3O_4$	256	**21**	*37*
	1-Hydroxyrutaecarpine	$C_{18}H_{13}N_3O_2$	318	**25**	*50*
	Paraensine	$C_{24}H_{21}N_3O_3$	281	**26**	*9*
Zanthoxylum rhetsa Dc. (*Z. budrunga* Wall)	Rhetsine = (±)-Evodiamine	$C_{19}H_{17}N_3O$	277	**2**	*44, 46, 47*
Zanthoxylum oxyphyllum Edg.	Evodiamine				*12*
Zanthoxylum pluviatile Hartl.	Rhetsinine				*12*
Araliopsis tabouensis Aubrev. et Pellegr	Rutaecarpine				*68*
	Evodiamine				*68*
	Rhetsinine				*68*

(continued)

a Structures:

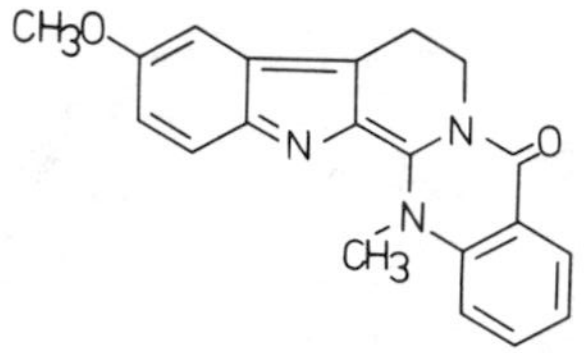

13 Hortiamine

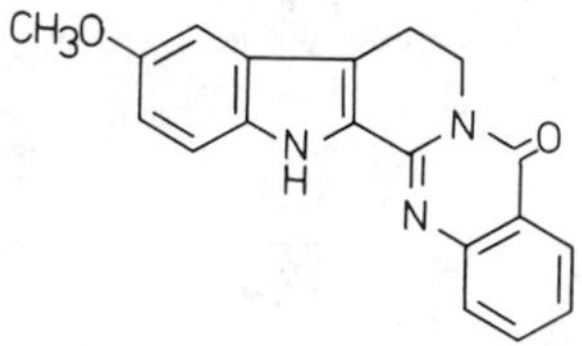

14 Hortiacine

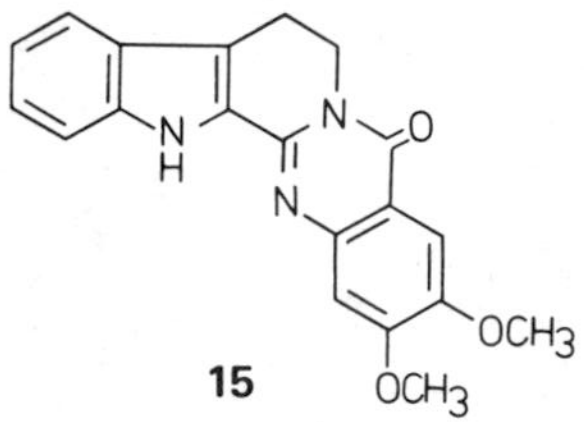

15 Euxylophoricine A

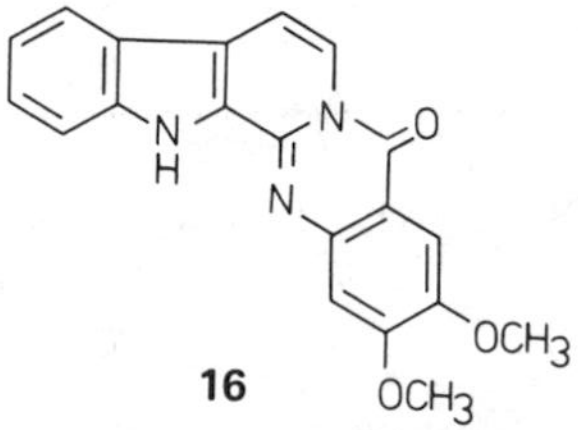

16 Euxylophoricine B

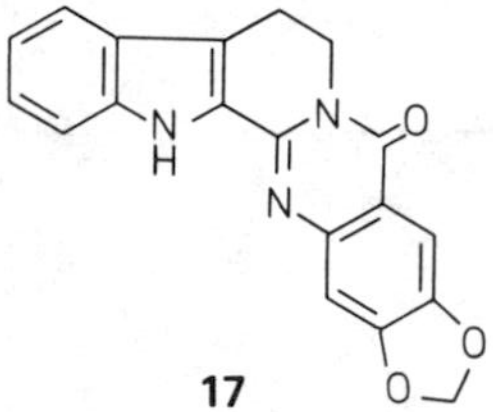

17 Euxylophoricine C

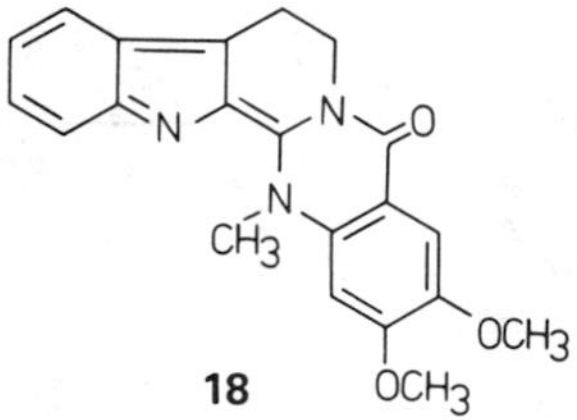

18 Euxylophorine A

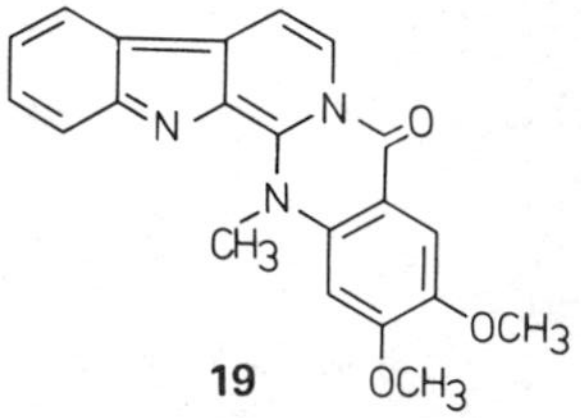

19 Euxylophorine B

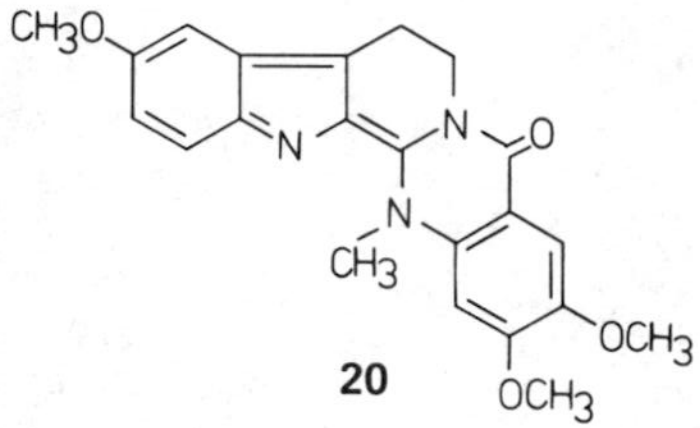

20 Euxylophorine C

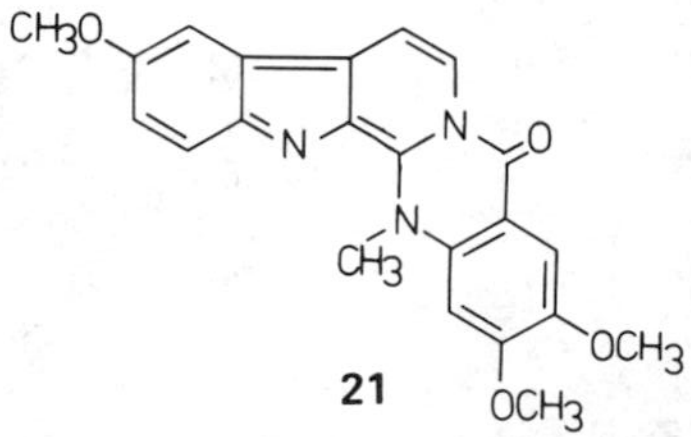

21 Euxylophorine D

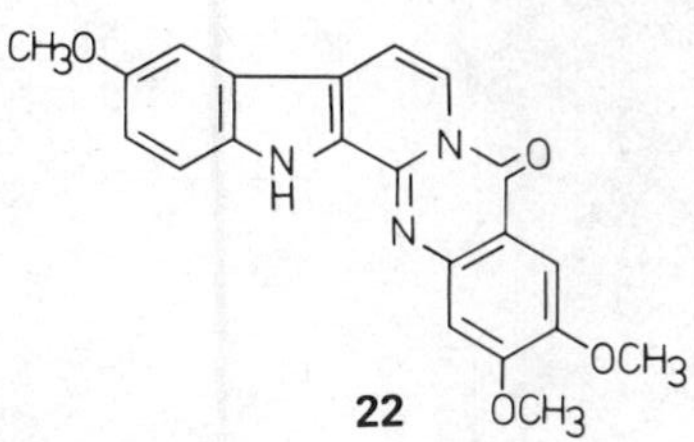

24

Euxylophoricine D

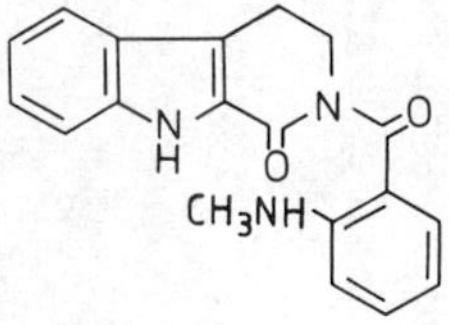

27

Rhetsinine

23

Euxylophoricine F

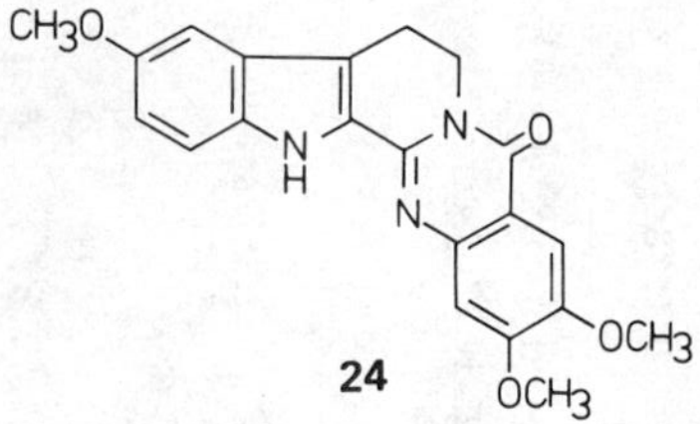

26

Paraensin

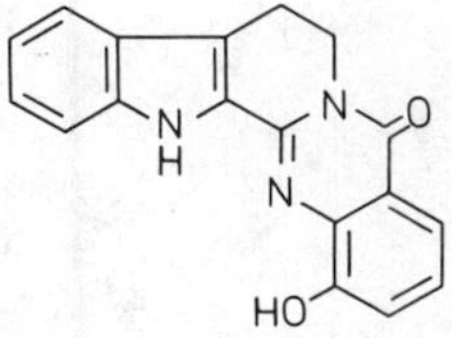

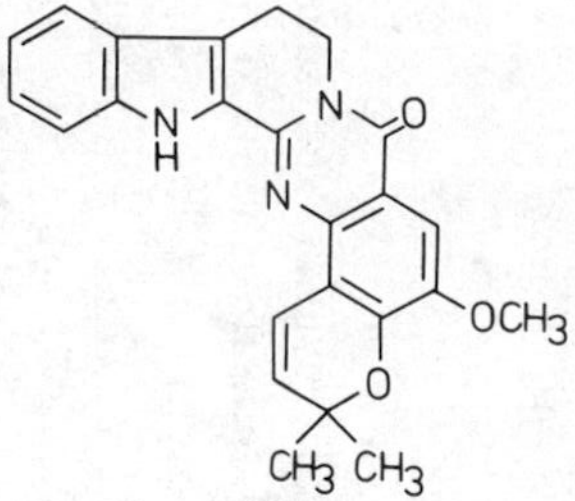

22

Euxylophoricine E

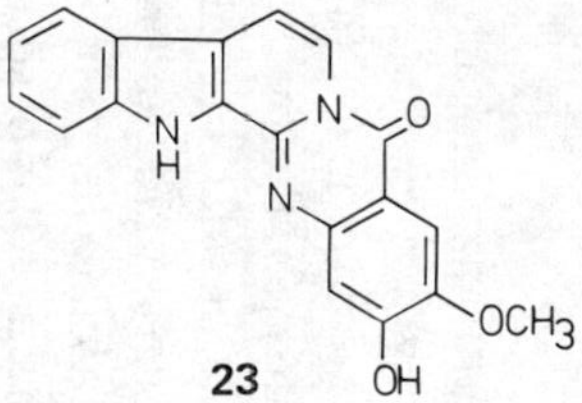

25

33

indolo [2',3': 3,4]pyrido[2,1-*b*]quinazoline (**3**) for the parent system. Hence, the systematic name for rutaecarpine is 8,13-dihydroindolo[2',3': 3,4]pyrido[2,1-*b*]quinazolin-5(7*H*)-one. The preferred trivial name for this alkaloid is rutaecarpine rather than the more "modern" rutecarpine, used in *Chemical Abstracts*.

The given numbering system is currently used by most authors although other systems (*4, 5*) are still prevalent. The rings are designated as indicated in formula **4**.

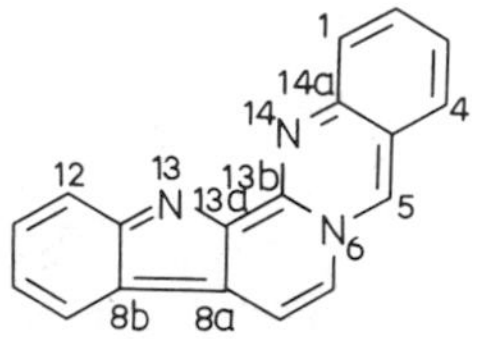

3

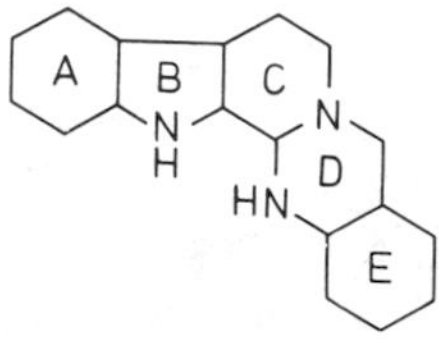

4

Since Manske's review (1965) several new quinazolinocarboline alkaloids have been isolated, and the most notable new representatives include the biogenetically significant 13*b*,14-dihydrorutaecarpine (**8**) (*6*), 7-carboxyevodiamine (**9**) (*7, 8*) as well as paraensine (**26**) (*9*), the first alkaloid of this class that incorporates an isoprenoid unit. Considerable advances in the synthesis of the quinazolinocarbolines have been made and no less than six new routes to rutaecarpine have been recorded. The so-called retro mass spectrometric synthesis developed by Kametani (*10*) is of special interest, at least from a theoretical point of view.

This principle is illustrated in Scheme 1: Evodiamine (**2**) shows two

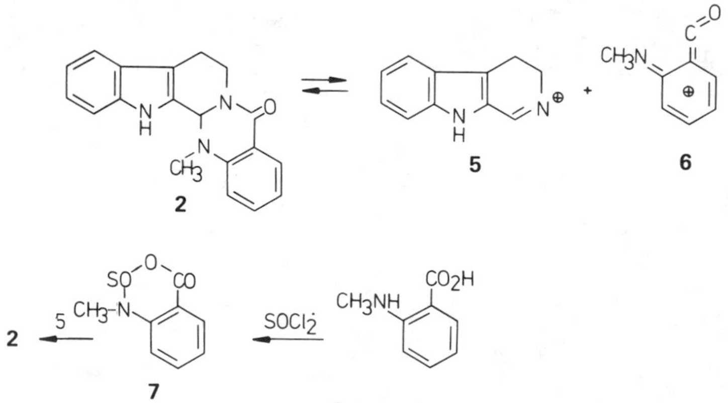

Scheme 1

characteristic ions in its mass spectrum, (5) and (6), formed by a retro-Diels–Alder cleavage of ring D. This behavior indicates that 2 might be synthesized from 3,4-dihydro-β-carboline (5) and the iminoketene (6). Thus, treatment of 5 with the unstable sulfinamide anhydride (7) in dry benzene at room temperature regioselectively afforded 2 in a reasonable yield.

For practical purposes, however, a new cyclization principle (*11*) involving elimination of CF_3^- in the final step is recommended (See Section III, A).

Too few systematic studies of the spectroscopic properties of the quinazolinocarboline alkaloids have been made. This unfortunate situation partly explains why too many relevant compounds have at first been assigned incorrect structures.

II. Occurrence

The Rutaceae are one of a few families of higher plants known to be able to synthesize the quinazoline nucleus; others include the Zygophyllaceae, Acanthaceae, and Saxifragaceae (*12*). The simplest quinazoline found in the Rutaceae is glycosminine (2-benzyl-4(3*H*)-quinazolinone) found in *Glycosmis arborea* DC. and *Ruta graveolens.*

The more complex quinazolinocarbolines occur in the genera *Evodia, Hortia, Zanthoxylum,* and *Euxylophora,* all members of the Rutaceae, and are restricted to only a few species. Most of the other species in these genera that have been examined yield furoquinoline and acridine alkaloids (*12*). Table I is a record of the occurrences of the quinazolinocarboline alkaloids, with their empirical formulas and melting points. They are all optically inactive except (+)-evodiamine $[\alpha]_D^{15} + 352°$ (acetone), (+)-7-carboxyevodiamine (9) $[\alpha]_D^{20} + 441°$ ($c = 1$, $CHCl_3$), (−)-13b,14-dihydrorutaecarpine (8) $[\alpha]_D^{28} - 564°$ ($c = 0.355$, DMF) and (+)-14-formyl of 13b,14-dihydrorutaecarpine $[\alpha]_D^{28} + 260°$ ($c = 0.535$, DMF). The absolute configurations (*7, 8*) of the first two chiral alkaloids mentioned are known.

Racemic evodiamine has been isolated from several plants and is often referred to as rhetsine (*13*). The most intensely studied plant is *Evodia rutaecarpa* Hook f. et Thoms. from which besides rutaecarpine, evodiamine,

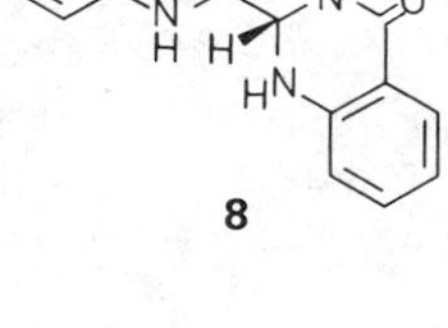

8 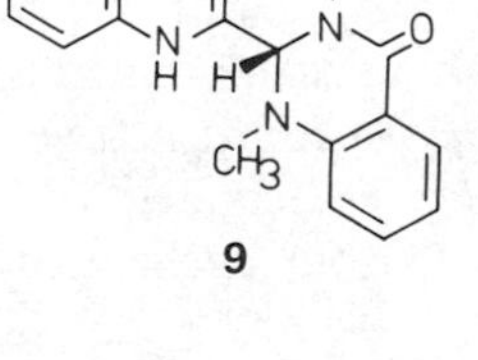   9

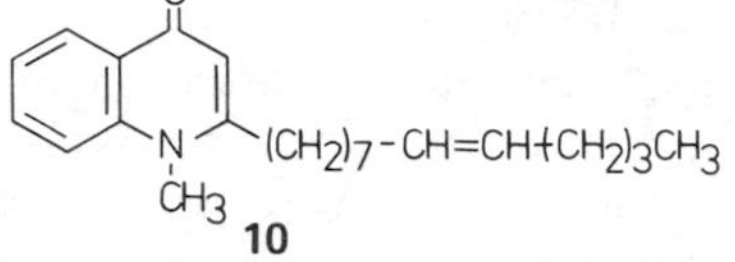

Evocarpine

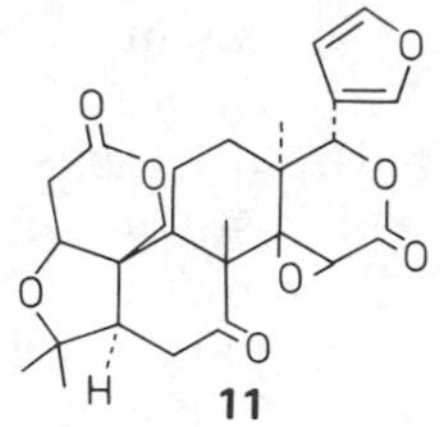

10

Limonin

11

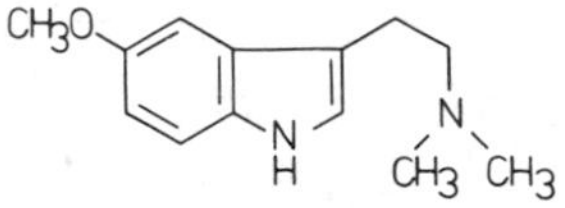

12

and related alkaloids, an impressive variety of natural products, such as evocarpine **10** (*14, 15*) (for further related compounds, see Volume XVII, p. 179), limonin (**11**) (*16–18*), also called evodin before its identification with limonin (*17*), and *N,N*-dimethyl-5-methoxytryptamine (**12**) (*18*) have been isolated. In an early paper (*19*) a compound (wuchuyine) with the composition $C_{10}H_{13}O_2N$ (mp 237°, $[\alpha]_D$ – 18°) was described. All attempts (*14, 20*) to reisolate wuchuyine from *Evodia rutaecarpa* have failed and nothing is known about its structure.

III. Structure and Synthesis

A. Rutaecarpine

The first suggested structure of rutaecarpine was based on the erroneous identification of one of the products of its fission with alkali (*2,21*). The fission fragment ultimately proved to be tryptamine and this finding led to the correct structure (**1**), since the second fragment had already been recognized as anthranilic acid. An early synthesis was achieved by condensing 1-keto-1,2,3,4-tetrahydro-β-carboline with methyl anthranilate in the presence of phosphorus trichloride (*22*). Later, Pachter (*23*) found that PCl₃ is inefficient and POCl₃ should be the preferred reagent (Scheme 2A). A minor variation is given in Scheme 2B, but 2-nitrobenzyl chloride has been used recently instead of 2-nitrobenzoyl chloride in this synthesis (*24*).

Several other syntheses have been recorded (*11, 25–34*). A recent synthesis by Kametani (*25a*) is given in Scheme 3, and another developed by Bergman (*11*) is depicted in Scheme 4. Another approach (Scheme 5) involving a Fischer cyclization of 6-phenylhydrazono-6,7,8,9-tetrahydro-

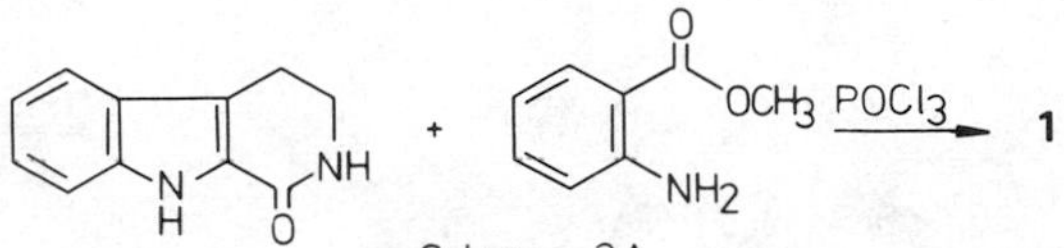

Scheme 2A

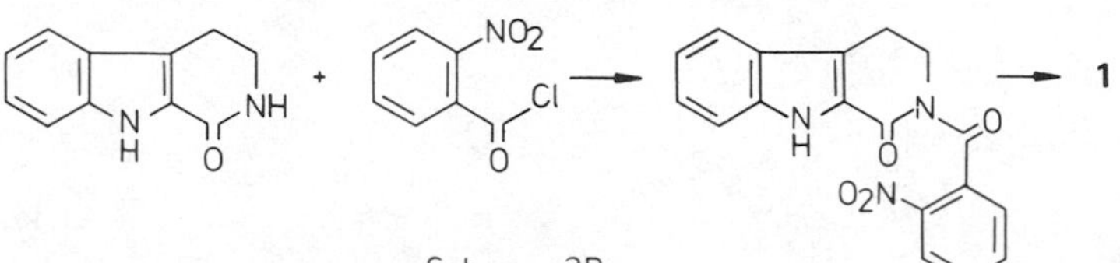

Scheme 2B

11H-pyrido[2,1-b]quinazolin-11-one is offered by Hermecz et al. (34). The so-called retro mass spectral synthesis (cf. Scheme 1) has also been used for the synthesis of rutaecarpine (25b).

Some of the problems involved in Kametani's synthesis (25a) are the low yield of 1 coupled with the necessary chromatography of the multicomponent mixture, the long reaction time, and a relatively low yield in the final cyclization step. The cyclization step can be considered to involve the transformations: 28 → 29 → 30 → 8 → 1.

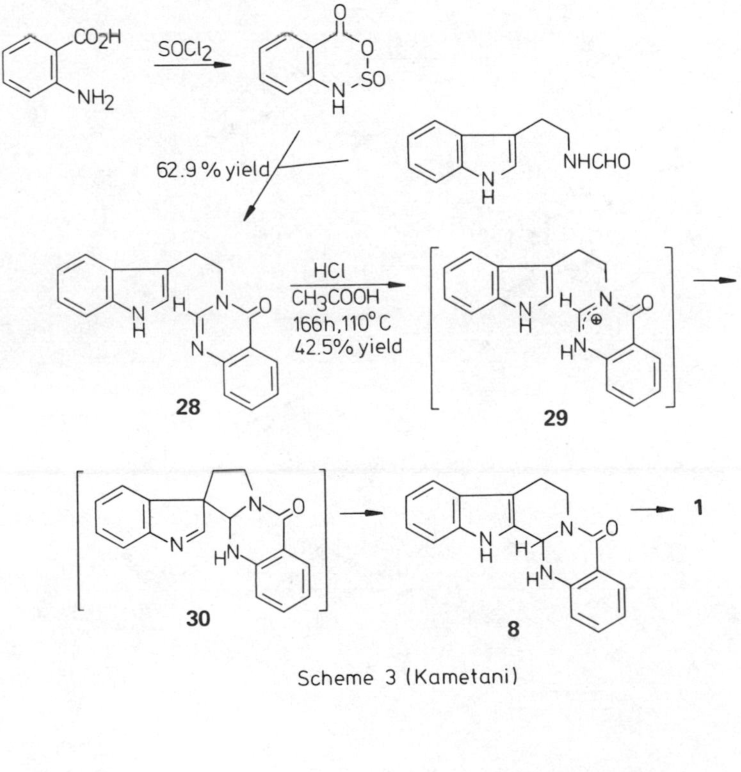

Scheme 3 (Kametani)

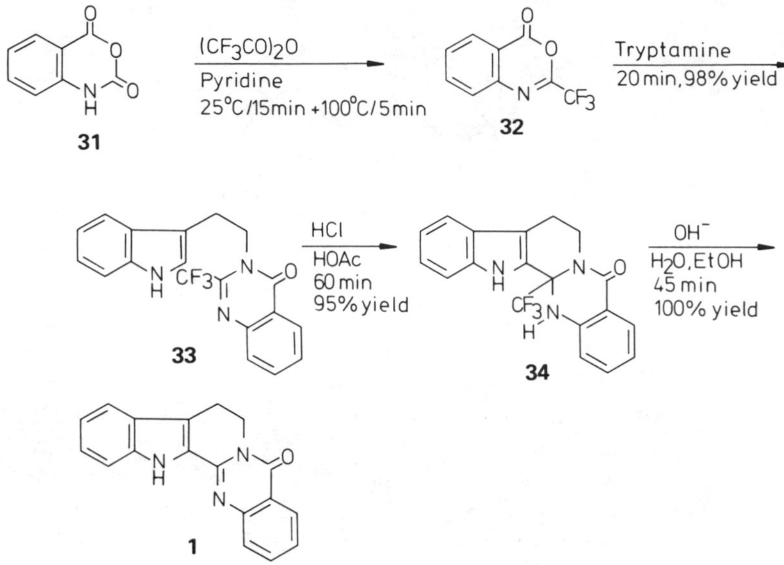

31 → (CF₃CO)₂O / Pyridine / 25°C/15min +100°C/5min → **32** → Tryptamine / 20 min, 98% yield →

33 → HCl / HOAc / 60 min / 95% yield → **34** → OH⁻ / H₂O, EtOH / 45 min / 100% yield →

1

Scheme 4 (Bergman)

2 Br₂

Ph–N₂⁺Cl⁻

Ph–NHNH₂

Z

E

PPA, 0.5 h, 180°C

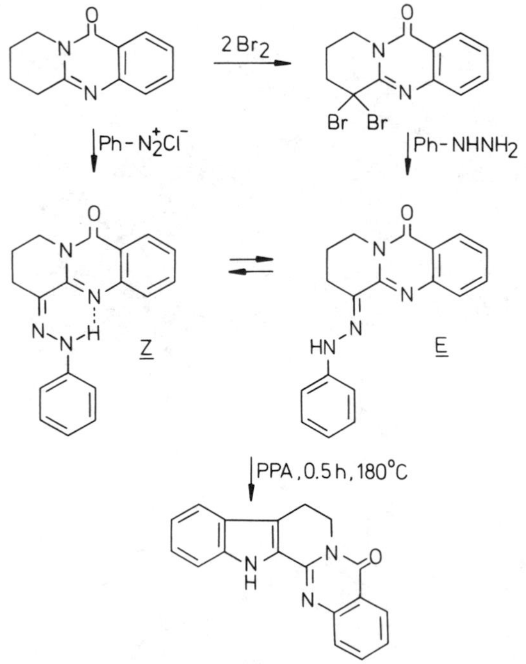

Scheme 5

The reactions indicated in Scheme 4, were based on the following considerations. The aza-stabilized carbonium ion formed on protonation of **33** should be a much better electrophile than **29** due to the presence of the strongly electron-withdrawing CF₃ group. In the final step, generation of an anion under mild conditions, followed by a displacement of CF_3^- should be expected to yield (and indeed does) a much cleaner reaction mixture than the final dehydrogenation step required in Scheme 3. The fact that hortiacine (i.e., 10-methoxyrutaecarpine) could be similarly prepared indicates the generality of this synthesis of quinazolinocarboline alkaloids (*11*).

All the syntheses discussed so far rely on tryptamine or tryptamine derivatives as starting materials and the common principle is to connect the C and D rings. The new approach by Hermecz (*34*) connects the B and C rings in the crucial Fischer indolization step. The fact that phenylhydrazines rather than tryptamines are required as starting materials should make this route attractive for the synthesis of rutaecarpine derivatives substituted in the A ring (see Scheme 5).

B. 13*b*,14-DIHYDRORUTAECARPINE

By extensive column chromatography of an extract from *Evodia rutaecarpa,* Kamikado *et al.* (*6*) have recently isolated (−)13*b*,14-dihydrorutaecarpine (**8**) $[\alpha]_D^{28}$ −564° (*c* = 0.355, DMF) and (+)-14-formyl-13*b*,14-dihydrorutaecarpine $[\alpha]_D^{28}$ +260° (*c* = 0.535, DMF). The structures were established by the synthetic routes shown in Scheme 6. Dehydrogenation of **8** with MnO₂ yielded rutaecarpine (**1**).

Another synthetic route to **8** exploring a similar type of chemistry as outlined in Schemes 3 and 4 has recently been developed (Scheme 7) by

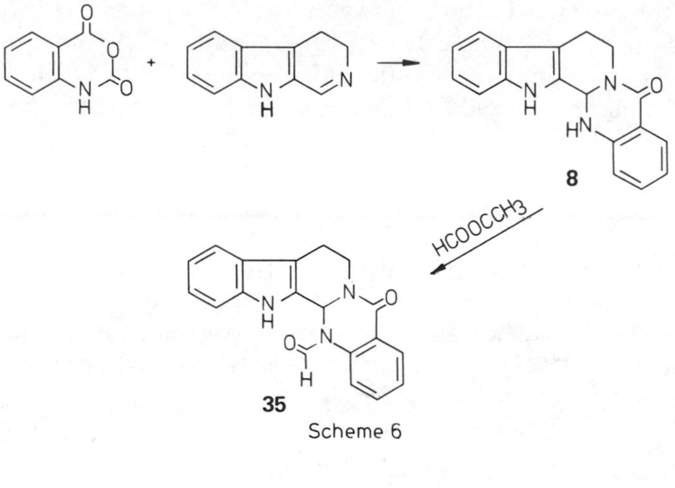

Scheme 6

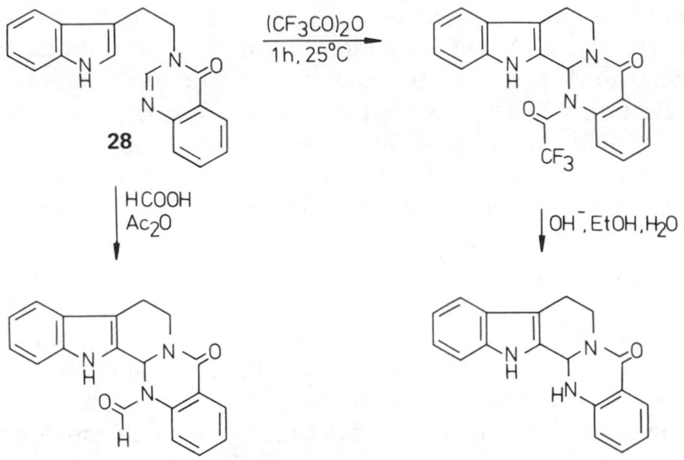

Scheme 7

Bergman (*26*). In this route the cyclization induced by HCOOH/CH₂O, and the conversion of **28** to **2** is of interest. In the original paper (*28*) this cyclization was incorrectly described as a reductive alkylation (**8 → 2**). Compound **28** is readily prepared from anthranoyl tryptamide and triethyl orthoformate (*11*).

In 1976 Waterman (*5*) isolated from the root bark of *Zanthoxylum flavum* an optically active compound, ($[\alpha]_D^{25} - 21$ ($c = 0.1$ CHCl₃) mp 168°, with the composition $C_{18}H_{15}N_3O$ and suggested it to be **8**. However, the reported carbonyl absorption (1680 cm⁻¹) in the IR spectrum is clearly incompatible with a dihydroquinazolinone structure. (Authentic 13*b*,14-dihydrorutae-carpine absorbs at 1633 cm⁻¹.) However, chemical evidence (LiAlH₄ reduction and methylation studies) does suggest that the compound isolated by Waterman is closely related to the quinazolinocarboline alkaloids. In this connection it should be added that the earlier studies (e.g., **1** + LiAlH₄) were erratic (*27, 28*) and should be reinvestigated (*26*).

C. 7,8-DEHYDRORUTAECARPINE

Although 7,8-dehydrorutaecarpine (**36**) has been reported (*24*) as an impurity in samples of rutaecarpine, it has not yet been identified as a natural product. On the other hand, three derivatives euxylophoricine E (**22**), euxylophoricine F (**23**), and euxylophoricine B (**16**), have been isolated from *Euxylophora paraensis* Hub. (*35–39*).

The 7,8-dehydrogenation has been effected by Se (300°, 10hr), DDQ (refluxing benzene) and MnO_2 (*23*). The melting point (154–155°) reported for **36** by Danieli *et al.* (*40*) seems to be in error. Repetition (*26*) of the experiment (**1** + DDQ) yielded, in fact, a product, mp. 280–281°, in complete agreement with the product obtained from **1** and MnO_2(*24*).

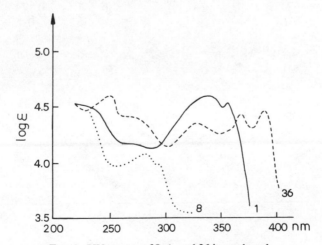

36

Finally, it should be noted that the relatively recently isolated alkaloids euxylophorine B (**19**) and euxylophorine D (**21**) are 2,3-dimethoxy and 2,3,10-trimethoxy derivatives, respectively, of 7,8,13,13*b*-tetradehydroevodiamine.

Rutaecarpine (**1**), 7,8-dehydrorutaecarpine (**36**), 13*b*,14-dihydrorutaecarpine (**8**) can be readily distinguished by UV spectroscopy (Fig. 1).

D. Evodiamine, 13, 13*b*-Dehydroevodiamine, and 7-Carboxyevodiamine

(+)-Evodiamine, a common congener of rutaecarpine, is readily transformed into a ring-opened stable hydrate (isoevodiamine, mp 155°) which on dehydration with acetic anhydride, generates optically inactive (±)-evodiamine (rhetsine) (*1,2,42–45*). Degradation with alkali generates dihydronor-

Fig. 1. UV spectra of **8**, **1**, and **36** in methanol.

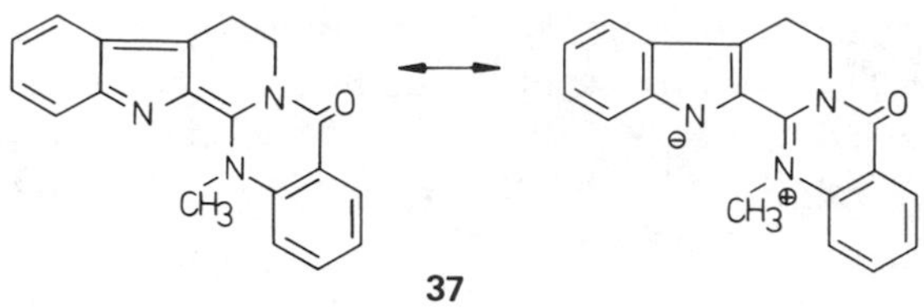

37

harman and *N*-methylanthranilic acid. An early synthesis involving the condensation of *N*-methylisatoic anhydride with tryptamine followed by heating with ethyl *o*-formate confirmed the structure of evodiamine (**2**) (*43*).

A more recent synthesis of evodiamine has been developed by Danieli and Palmisano (*33*). Thus, condensation of tryptamine with *N*-methylanthranilic acid (CBr$_4$/PPh$_3$, refluxing toluene, 5 hr) followed by intramolecular amidomethylation (36% CH$_2$O, HCl, refluxing MeOH, 3 hr) gave rise to 1,2-dihydro-1-methyl-3-(3-indolyl)ethyl-4(3*H*)-quinazolinone (**38**) in 84% overall yield. Dehydrogenation of **38** with Hg(OAc)$_2$ afforded (±)-evodiamine (**2**) in 92% isolated yield. The formation of (**2**) takes place via intramolecular acid-catalyzed interaction of the electrophilic carbon of the intermediate 3,4-dihydro-4-oxoquinazolinium salt (**39**) with the nucleophilic indole moiety (Scheme 8).

Oxidation of evodiamine (**2**) with MnO$_2$ yielded rutaecarpine (**1**) in 85% yield, whereas dehydrogenation with DDQ or Tl(OAc)$_3$ afforded the zwitterionic dehydroevodiamine (**37**). It was also reported that the conversion (**2** → **1**) previously effected (*42, 46b*) by thermolysis gave erratic and low yields. Actually heating evodiamine (sealed tube 210°) or refluxing in DMF

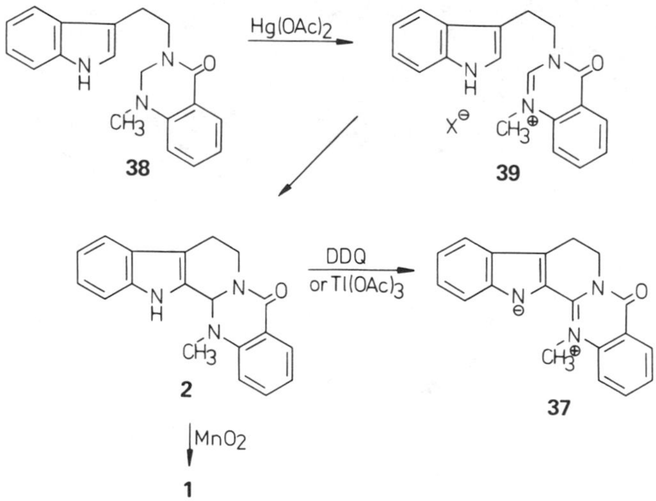

Scheme 8

gave first dehydroevodiamine (< 5%) which then rearranged to a 1 : 1 mixture of rutaecarpine and 13-methylrutaecarpine. Dehydrogenation of **37** with DDQ yielded, as expected (cf. Section III, C), a 7,8-dehydrogenated derivative (*33*).

The transformation **38** → **39** → **2** can also be reversed; thus, reflux of **2** with conc. HCl in ethanol for 1 hr yielded **39** (as the hydrochloride), which could be reduced with NaBH₄ to **38**. It might be added that prolonged heating of evodiamine with concentrated HCl yields a blue violet color (*48*), the nature of which is still not known, although a *2,2′,2″*-tris-indolyl methane derivative should be a good guess.

Dehydroevodiamine (**37**) was isolated from the leaves of *E. rutaecarpa* by Nakasato (*41*) and the structure established with the reactions shown in Scheme 9. The product from oxidation of evodiamine with KMnO₄ was first studied by Asahina (*42*), who assigned structure **40** rather than the correct ring-opened isomeric structure **27**. These facts seem to have been considered simultaneously by Nakasato (*41*) and by Pachter (*13*). Interestingly, compound **27**, rhetsinine, has been isolated (*44, 46*) from e.g., *Zanthoxylum Rhetsa,* although some doubts seem to exist regarding its status as either a true natural product or an artifact. In this connection the reported (*47*) facile photochemical transformation (**2** → **27**) is of interest. Equilibria similar to

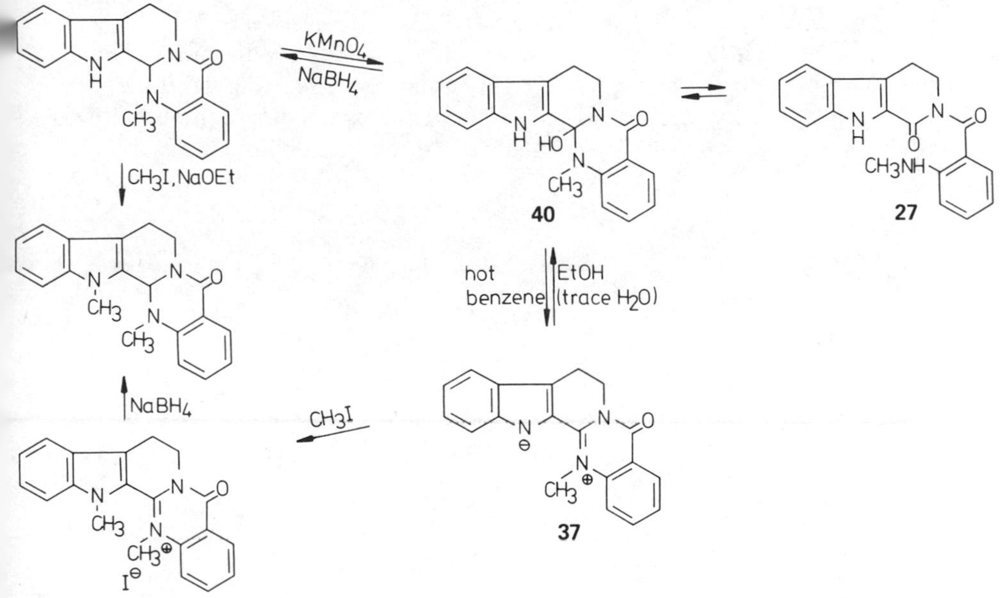

Scheme 9

those given in Scheme 9 have been advocated (*38*) in the euxylophorine series. Even in this case the covalent hydrate (the pseudobase) could not be isolated, although a pseudobase of **39** was described previously by Asahina (*42*).

The interesting compound (**9**), retaining the carboxylic group of the biogenetic precursor tryptophan, has been isolated recently by Palmisano and Danieli (*7, 8*). Treatment of the acidic chromatographic residue from an extract of *E. rutaecarpa* with diazomethane, followed by repeated preparative TLC yielded a small amount of the corresponding ester (8 mg from 3.5 kg of dried fruits). The location of the carbomethoxy group at C-7 and the absolute configuration of this center were established by synthesizing **43a** from (3*S*)-3-carbomethoxy-3,4-dihydro-3*H*-pyrido[3,4-*b*]indole hydrochloride and *N*-methylanthranilic acid in THF containing 1 Eq of pyridine and in the presence of PPh$_3$/CBr$_4$. This condensation yields only the "natural" isomer and not the expected mixture of 13*b*-epimers. This fact was explained in terms of preferred steric and stereoelectronic requirements in the transition state(s). Furthermore, no epimerization of **43a** occurred in MeOH, MeONa (room temperature, 5 hr), nor in HOAc (reflux, 6 hr).

Another approach yielded a small amount of the unnatural isomer (**43b**) together with the natural isomer and is outlined in Scheme 10. The unnatural isomer readily epimerized (e.g., by heating or by acid treatment), presumably via a 13*a*,13*b* ring opening. The absolute configuration of (+)-evodiamine could be determined by a correlation with **43a** (*8*).

As discussed earlier (Scheme 1), the mass spectrum of evodiamine exhibits two strong fragment ions at *m/e* 169 and *m/e* 134, respectively (*66*). 13*b*,14-Dihydrorutaecarpine (**8**) shows an analogous fragmentation (*6*), whereas the spectrum of **28**, as expected, is dominated by the peaks at *m/e*

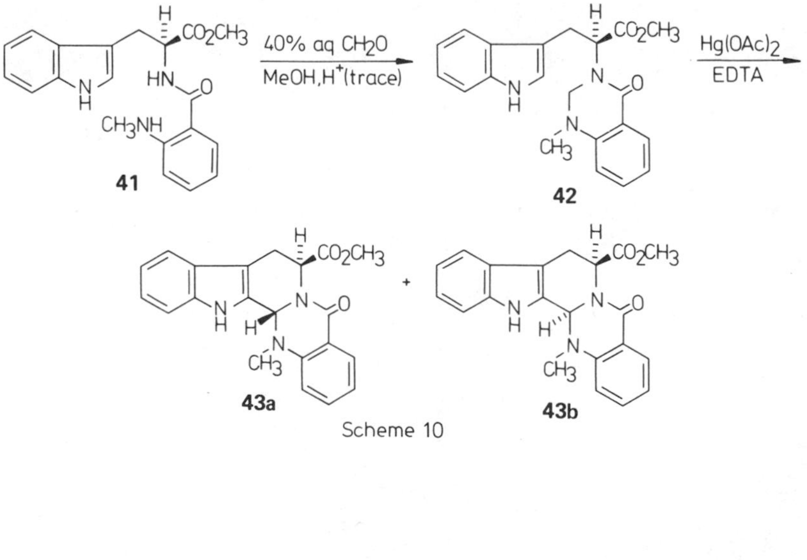

Scheme 10

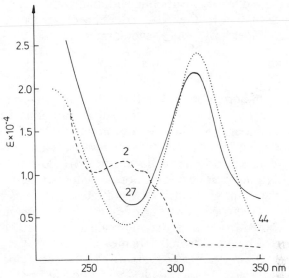

FIG. 2. UV spectra of **2**, **27**, and **44** in CH₃CN.

143 and m/e 130 (cleavage of the tryptamine moiety). In the original paper
(67) this cleavage was described as "surprizing" because the structure of the
studied compound had been incorrectly assigned as **8**, rather than **28**.

E. RHETSININE AND RHETSINE

These two bases have been isolated (46, 47) from *Evodia rutaecarpa,*
Xanthoxylum rhetsa, and a few other species. The originally suggested
structures (47) necessitated revisions. Rhetsinine (**27**) was recognized in 1961
(13) as being identical with the long-known (42) product (earlier called
hydroxyevodiamine), obtained by mild permanganate oxidation of evodiamine (cf. Section III, D). The IR spectrum, however, showed no bands due to
hydroxyl, but well-defined bands ascribable to carbonyl and to NH.

Furthermore, the UV spectrum of rhetsine (Fig. 2) agreed nicely with
that of the model compound 2-benzoyl-1-oxo-1,2,3,4-tetrahydropyrido-
[3,4-b]indole (**44**). Finally, a synthesis of rhetsinine (**27**) was easily achieved

44

by condensing 3,4-dihydro-β-carboline with methyl N-methyl anthranilate in the presence of POCl₃. Catalytic reduction of rhetsinine (27) yielded rhetsine, which is identical with (±)-evodiamine (*13*).

F. HORTIAMINE AND HORTIACINE

The structure of hortiamine (13) follows from the fact that alkali fission generates 6-methoxy-1-oxo-1,2,3,4-tetrahydro-β-carboline and N-methyl-anthranilic acid and a POCl₃-induced recombination of these compounds regenerates the red alkaloid (*23*) which readily forms a yellow hydrate and even an ethanolate. The structure of hortiacine (14) follows from the fact that it is formed from hortiamine hydrochloride by loss of methyl chloride on heating. It is the 10-methoxy derivative of one of its congeners, namely, rutaecarpine (1). Hydrolysis of hortiamine gives rise to 6-methoxyrhetsinine (45). A new synthetic route to hortiacine analogous to that for rutaecarpine has been reported by Bergman (cf. Scheme 4).

Brief heating of 6-methoxyrhetsinine with acetic anhydride yielded (*49*), besides two normal acetylation products, a cyclized and rearranged product called isohortiamine to which structure 46 was assigned. No mechanistic details of this interesting transformation are known. Even more interesting is the fact that there is evidence available suggesting that the major alkaloid in *Hortia braziliana* has the same ring system as isohortiamine. It was further-more suggested that the methoxy group is located in the E ring rather than in the A ring. No further studies of these interesting problems seem to have appeared.

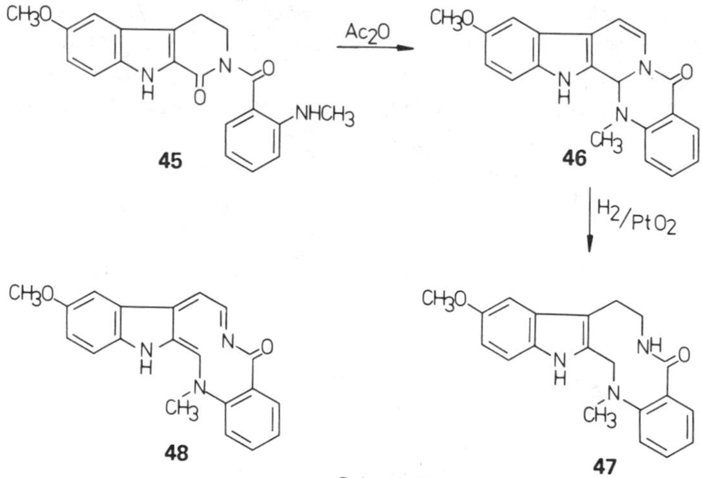

Scheme 11

Hydrogenation of isohortiamine in the presence of Adams catalyst yielded a colorless, neutral compound containing four additional hydrogen atoms, and suggested structure (**47**). In this transformation the valence tautomer **48** of **46** might be an intermediate (*49*) (Scheme 11).

G. Simple Quinazolinocarboline Alkaloids from *Euxylophora paraensis* Hub.

Several quinazolinocarboline alkaloids have been isolated from the yellow bark of *Euxylophora paraensis* Hub. (*35 – 37, 50, 51*) Typical representatives are the colorless euxylophoricine A (**15**), the orangered euxylophorine A (**18**), and its yellow 7,8-dehydroderivative euxylophorine B (**19**). All the other alkaloids isolated are given in Table I.

The chemistry of these alkaloids, their synthesis, and chemical transformations have close parallels (including the equilibria in Scheme 9) with rutaecarpine and evodiamine and will not be discussed here.

In connection with the structure elucidation of **25** (*51*), the spectral analysis indicated the presence of an OH group in the E ring. Methylation (MeI–K_2CO_3 in anhydrous acetone) of the alkaloid resulted not only in methylation of the OH group but also, somewhat surprisingly, in an N_{13}-methylation. This dimethylated compound could be identified with a product obtained by condensation of 3,4-dihydro-β-carboline with 3-methoxyanthranilic acid in the presence of $POCl_3$ (cf. Scheme 2a) followed by N_{13}-methylation. All the possible isomers were also synthesized. Finally, it was found that condensation of 3-hydroxyanthranilic acid with 3,4-dihydro-β-carboline yielded 1-hydroxyrutaecarpine (*51*).

When a sample of the red alkaloid euxylophorine C (**20**) was dissolved in boiling benzene and allowed to stand without exclusion of moisture, a pale yellow solid separated, that in close analogy with the conversion (**2** → **40** → **27**) discussed in Section III, D, could be identified as the trimethoxyrhetsinine (**49**). The IR spectrum in Nujol showed two NH bands at 3360 and 3260 cm^{-1}, as well as two carbonyl bands at 1660 and 1650 cm^{-1}.

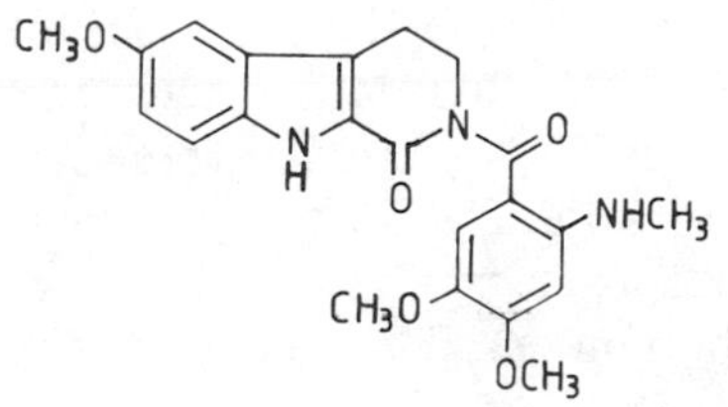

49

The UV spectrum (in CH_3CN) showed two maxima at 318 and 395 nm.

Both euxylophoricine A and euxylophoricine C have recently (71) been synthesized according to the principle outlined in Scheme 1.

H. Paraensine

In addition to the alkaloids discussed in the previous section, a more complex alkaloid with the composition $C_{24}H_{21}O_3N_3$ and named paraensine (26), has been isolated from the bark of *Euxylophora paraensis* Hub. (9). The structure was verified by the synthesis outlined in Scheme 12. The dihydroparaensine (50) formed was found to be identical with the dihydroderivative of the natural product (26).

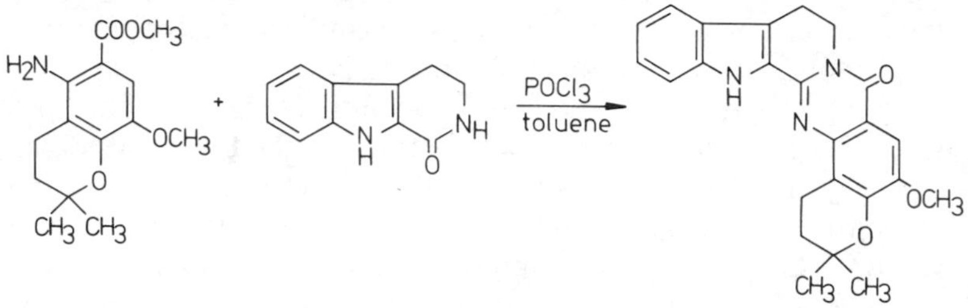

Scheme 12

IV. Biosynthesis

Around 1966–1967, Yamazaki *et al.* (52, 53) reported the results of some feeding experiments with [3-^{14}C]tryptophan, sodium [^{14}C]formate (or methionine-^{14}CH$_3$) and [^{3}H]anthranilic acid with *E. rutaecarpa*. Degradation studies showed that all labels were incorporated into the two alkaloids studied rutaecarpine and evodiamine. The C_1 unit from methionine or formate was introduced to the 13b position of both alkaloids as well as into the N-methyl group of evodiamine. The incorporation ratio of the C_1 unit into evodiamine was lower than in rutaecarpine, suggesting that evodiamine is not formed by N-methylation of rutaecarpine (or 13b,14-dihydrorutaecarpin, which was not known at this time). The fact that 7-carboxyevodiamine (9) but not 7-carboxyrutaecarpine has been isolated from *E. rutaecarpa* is also of interest in this context. The conclusions made by Yamazaki are summarized in Scheme 13.

It should be noted that the involvement of 3,4-dihydro-β-carboline has not

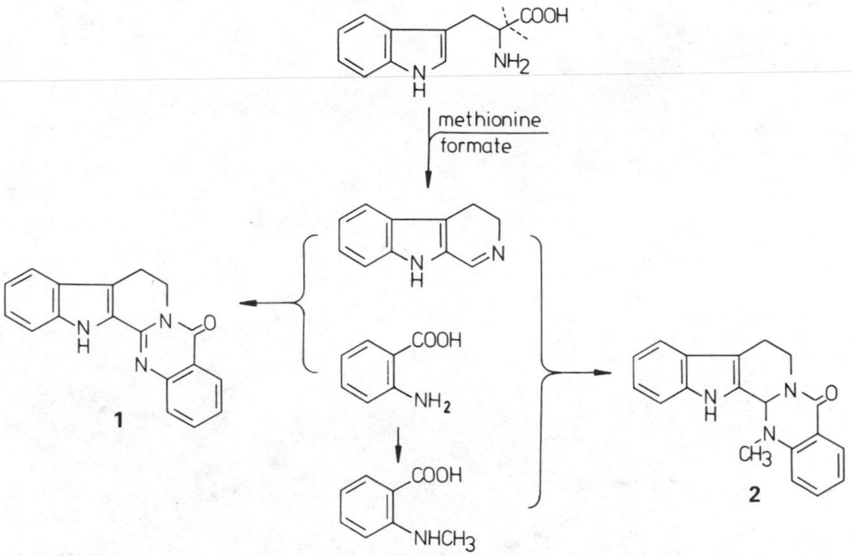

Scheme 13

been proved and recently Takagi *et al.* (*18*) have discussed the possible role of *N*-methylanthranoyltryptamide in the biosynthesis of evodiamine.

No studies of the biogenetic relations of 13*b*,14-dihydrorutaecarpine, evodiamine, and 13-formyldihydrorutaecarpine have been reported but should be of considerable interest. It might be added that 13*b*,14-dihydrorutaecarpine has been suggested (1971, i.e., before its isolation) as the key intermediate in the biosynthesis of the quinazolinocarboline alkaloids (*54, 55*).

V. Pharmacological Properties of Quinazolinocarboline Alkaloids and Synthetic Analogs

In spite of the fact that the dried fruits of *Evodia rutaecarpa* under the name Wu-Chu-Yu has a long tradition in Chinese medicine (*56*), only limited information is available about the pharmacology of pure quinazolinocarboline alkaloids. Traditionally, Wu-Chu-Yu has been administered for the treatment of headache, abdominal pain, dysentery, cholera, worm infestations, and postpartum disturbances (*19, 48, 57*). In modern literature rutaecarpine has been reported (*58*) to increase the arterial pressure. Furthermore, hortiamine has been reported (*49*) to be hypotensive. In a recent paper

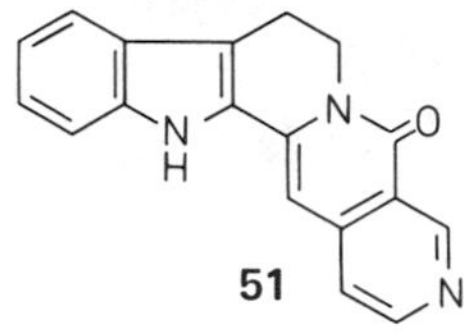

51
nauclefine

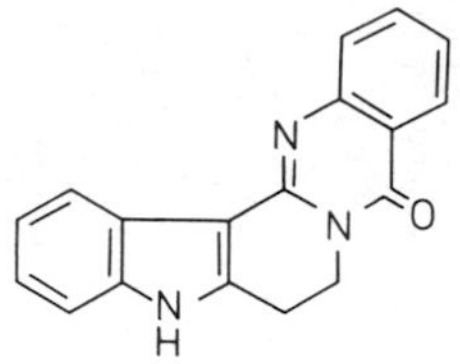

52

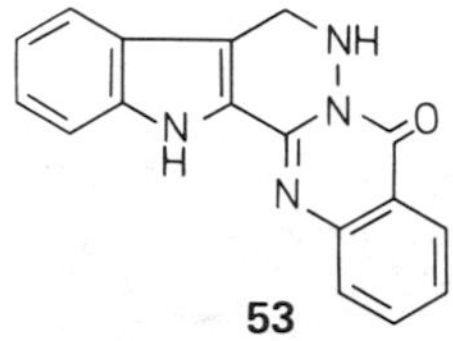

53

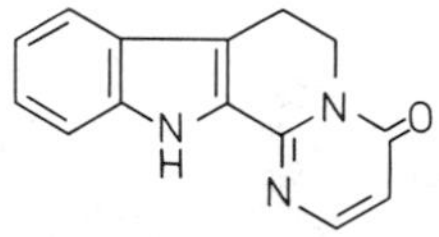

54

rutaecarpine, 13*b*,14-dihydrorutaecarpine, and 13-formyldihydrorutaecar-pine were reported (*6*) to be active against the silkworm larvae (*Bombyx mori* L.) No published data are available concerning cancer screening of quinazolinocarboline alkaloids. In this connection it might be added that

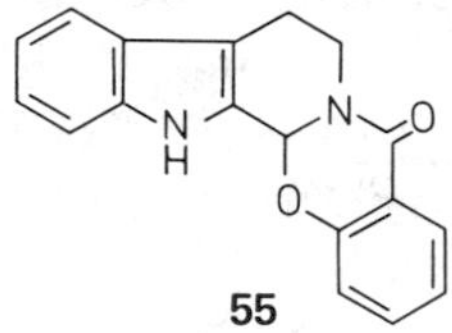

55

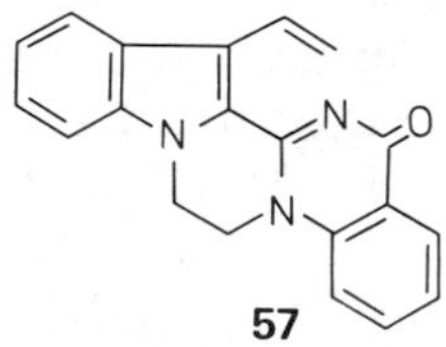

57

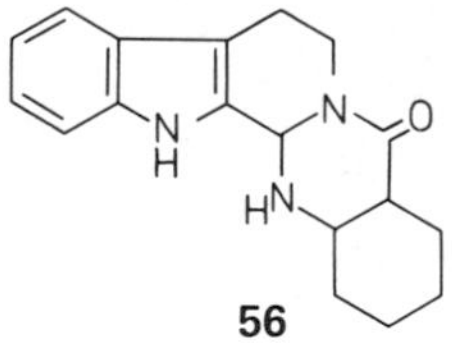

56

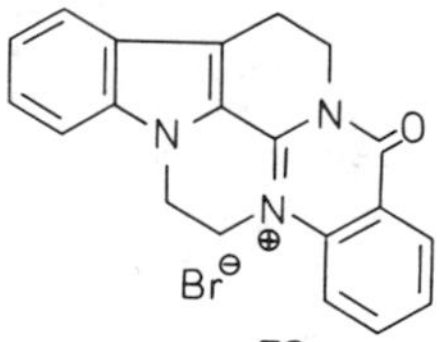

58

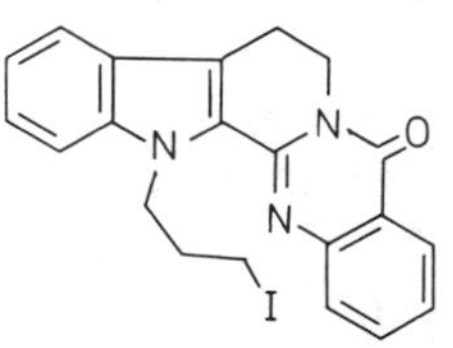

59

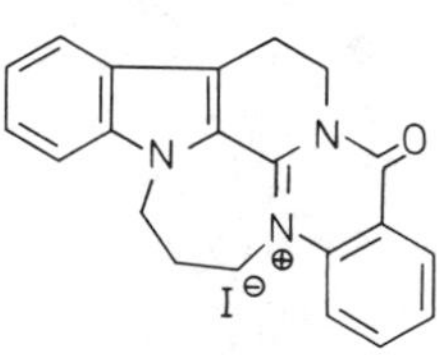

60

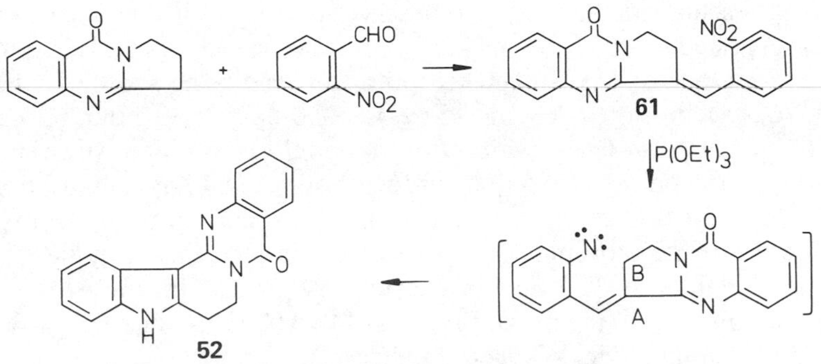

Scheme 14

nauclefine (**51**), an isomer of rutaecarpine, has been reported (*59, 60*) to possess antileukaemic properties.

In spite of the scant information on the pharmacological properties of the quinazolinocarboline alkaloids many analogs have been prepared and compounds **52–60** are representative (*59–65*).

Deoxygenation of the nitro compound (**61**) with P(OEt)$_3$ yielded pseudo-rutaecarpine (**52**) in 6% yield (Scheme 14). The ring expansion is induced by an intramolecular attack on the double bond of the nitrene formed (*61*). As expected no ring expansion which should have afforded rutaecarpine was

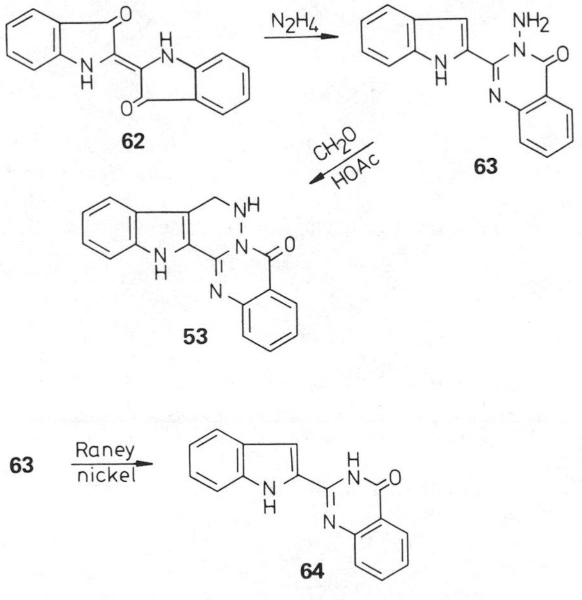

Scheme 15

observed. Pseudorutaecarpine also deserves interest as it was once (errone-ously) suggested (*2*) to represent the structure of rutaecarpine.

The 7-aza analog (**53**) of rutaecarpine can readily be synthesized from indigo (**62**) according to Scheme 15 (*62*). It should be noted that **63** can be readily converted to **64**, a compound that could be an interesting starting material for the synthesis of quinazolinocarboline alkaloids (e.g., by directed alkylations). In this connection the formation of **57** and **58** from **1** by treatment with 1,2-dibromoethane in acetone in the presence of K_2CO_3 is of considerable interest. The yield of the primary product (**58**) was low due to ready ring-opening to the major product (**57**). Alkylation of **1** with 1,3-diio-dopropane similarly yielded **59** and **60** (*66*).

Recently, Kong (*72, 73*) has reported about the uterotonic activity of 13,13b-dehydroevodiamine (**37**) and rutaecarpine (**1**). Both compounds were tested as hydrochlorides. The more soluble salt, that of (**37**), was studied in more detail and it was found that a threshold dose can potentiate the uterine contracting action of acetylcholine, serotonin, oxytocin, and PGF_{2a}. The uterotonic action and the potentiation effect of **37** can be blocked by indomethacin and mepacrine, which do suggest that the action of **37** may be mediated through PG synthesis.

In connection with these results it is of interest to note that the unripe fruit of *Evodia rutaecarpa* has a long tradition (*56*) for the treatment of "female reproductive disorders."

ACKNOWLEDGMENTS

I would like to thank Drs. B. Danieli, G. Palmisano, J. Kökösi, and I. Hermecz for helpful discussions and provision of manuscripts prior to publication. Thanks are also due to Dr. L. Gou-Qiang for help with the Chinese literature, Dr. Karen Jernstedt for linguistic improve-ments, and to Mr. H. Challis and Mrs. Ingvor Larsson for invaluable technical assistance.

REFERENCES

1. Y. Asahina and K. Kashiwaki, *J. Pharm. Soc. Japan,* No. 405, 1293 (1915); *Chem. Abstr.* **10,** 607 (1916).
2. Y. Asahina, and S. Mayeda, *J. Pharm. Soc. Japan,* No. 416, 871 (1916); *Chem. Abstr.* **11,** 332 (1917).
3. L. Marion, *in* "The Alkaloids" (R. H. F. Manske and H. L. Holmes, eds.), Vol. 2, p. 402. Academic Press, New York, 1954.
4. G. Toth, K. Horvath-Dora, and O. Clauder, *Liebigs Ann. Chem.,* 529 (1977) and earlier papers in this series.
5. P. G. Waterman, *Phytochemistry* **15,** 578 (1976).
6. T. Kamikado, S. Murakoshi, and S. Tamura, *Agric. Biol. Chem.* **42,** 1515 (1978).

7. B. Danieli, G. Lesma, and G. Palmisani, *Experientia* **35,** 156 (1979).
8. B. Danieli, G. Lesma, and G. Palmisani, *Chem. Comm.* 1092 (1982).
9. B. Danieli, P. Manitto, F. Ronchetti, G. Russo, and G. Ferrari, *Experientia* **28,** 249 (1972).
10a. T. Kametani, and K. Fukumoto, *Heterocycles* **7,** 615 (1977).
10b. T. Kametani, and K. Fukumoto, *Shitsuryo Bunseki* **29,** 39 (1981).
11. J. Bergman, and S. Bergman, *Heterocycles* **16,** 347 (1981).
12. P. G. Waterman, *Biochem. Syst. Ecol.* **3,** 149 (1975).
13. I. J. Pachter, and G. Suld, *J. Org. Chem.* **25,** 1680 (1960).
14. R. Tschesche, and W. Werner, *Tetrahedron* **23,** 1873 (1967).
15. T. Kamikado, C.-F. Chang, S. Murakoshi, A. Sakurai, and S. Tamura, *Agric. Biol. Chem.* **40,** 605 (1976).
16. D. H. R. Barton, S. K. Oradgab, S. Sternhall, and J. F. Templeton, *J. Chem. Soc.* 255 (1962).
16b. Y. Hirose, *J. Pharm. Bull.* **11,** 535 (1963).
17. M. S. Schechter, and H. L. Haller, *J. Am. Chem. Soc.* **62,** 1307 (1940).
18. S. Takagi, T. Akiyama, T. Kinoshita, U. Sankawa, and S. Shibata, *Shoyakugaku Zasshi* **33,** 30 (1979).
19. A. L. Chen, and K. K. Chen, *J. Am. Pharm. Assoc.* **22,** 716 (1933).
20. J. H. Chu, *Science Record China* **4,** 279 (1951).
21. Y. Asahina, and S. Hayeda, *J. Pharm. Soc. Japan* No. 405, 1293 (1915); *Chem. Abstr.* **11,** 332 (1917).
22. Y. Asahina, R. H. F. Manske, and R. Robinson, *J. Chem. Soc.* 1708 (1927).
23. I. J. Pachter, R. F. Raffauf, G. E. Ullyot, and O. Ribeiro, *J. Am. Chem. Soc.* **82,** 5187 (1960).
24. H. Möhrle, C. Kamper, and R. Schmid, *Arch. Pharm.* **313,** 990 (1980).
25a. T. Kametani, C. Van Loc, T. Higa, M. Koizumi, M. Ihara, and K. Fukumoto, *J. Am. Chem. Soc.* **99,** 2306 (1977).
25b. T. Kametani, C. Higa, C. Van Loc, M. Ihara, M. Koizumi, and K. Fukumoto, *J. Am. Chem. Soc.* **98,** 6186 (1976).
26. J. Bergman, and S. Bergman, unpublished results.
27. O. Clauder, and K. Horvath-Dora, *Acta Chim. Acad. Sci. Hung.* **72,** 221 (1972).
28. K. Horvath-Dora, O. Clauder, *Acta Chim. Acad. Sci. Hung.* **84,** 93 (1975).
29. C. Schöpf, and H. Steuer, *Ann.* **558,** 124 (1947).
30. S. Petersen, and E. Tietze, *Liebigs Ann. Chem.* **623,** 166 (1959).
31. Y. Asahina, T. Irie, and T. Ohta, *J. Pharm. Soc. Japan* No. 543, 541 (1927); *Chem. Abstr.* **21,** 3054 (1927).
32. T. Ohta, *J. Pharm. Soc. Japan* **60,** 311 (1940); *Chem. Abstr.* **34,** 7291 (1940).
33. B. Danieli, and G. Palmisano, *Heterocycles* **9,** 803 (1978).
34. J. Kökösi, I. Hermecz, G. Szasz, and Z. Meszaros, *Tetrahedron Lett.* 4861 (1981).
35. L. Canonica, B. Danieli, P. Manitto, G. Russo, and G. Ferrari, *Tetrahedron Lett.* 4865 (1968).
36. B. Danieli, P. Manitto, F. Ronchetti, G. Russo, and G. Ferrari, *Phytochemistry* **11,** 1833 (1972).
37. B. Danieli, G. Palmisano, G. Russo, and G. Ferrari, *Phytochemistry* **12,** 2521 (1973).
38. B. Danieli, G. Lesma, and G. Palmisano, *Heterocycles* **12,** 353 (1979).
39. B. Danieli, G. Lesma, and G. Palmisano, *Heterocycles* **12,** 1433 (1979).
40. B. Danieli, and G. Palmisano, *Gazz. Chim. Ital.* **105,** 45 (1975).
41. T. Nakasato, S. Asada, and K. Marui, *Yakugaku Zasshi* **82,** 619 (1961).
42. Y. Asahina, and T. Ohta, *J. Pharm. Soc. Japan* No. 530, 293 (1926); *Chem. Abstr.* **21,** 2134 (1927).
43. Y. Asahina, and T. Ohta, *Ber.* **61,** 310 (1928).

44. K. W. Gopinath, T. R. Govindachari, and U. R. Rao, *Tetrahedron* **8**, 293 (1960).
45. B. Danieli, and G. Palmisano, *Gazz. Chim. Ital.* **105**, 99 (1975).
46a. A. Chatterjee, S. Bose, and C. Ghosh, *Tetrahedron* **7**, 257 (1959).
46b. A. Chatterjee, and J. Mitra, *Sci. Cult.* (Calcutta), **25**, 493 (1960).
47. M. Yamazaki, and T. Kawana, *Yakugaku Zasshi* **87**, 608 (1967).
48. Y. Asahina, *Acta Phytochim.* **1**, 67 (1923).
49. I. J. Pachter, R. J. Mohrbacher, and D. E. Zacharias, *J. Am. Chem. Soc.* **83**, 635 (1961).
50. B. Danieli, G. Palmisano, G. Rainoldi, and G. Russi, *Phytochemistry* **13**, 1603 (1974).
51. B. Danieli, C. Farachi, and G. Palmisano, *Phytochemistry* **15**, 1095 (1976).
52. M. Yamazaki, and A. Ikuta, *Tetrahedron Lett.* 3221 (1966).
53. M. Yamazaki, A. Ikuta, T. Mori, and T. Kawana, *Tetrahedron Lett.* 3317 (1967).
54. M. Luckner, *Herba Hung.* **10**, 27 (1971).
55. S. Johne, and D. Gröger, *Herba Hung.* **10**, 65 (1971).
56. S. C. Li, *Pentsau Kang Mu*, Chapter 32 (1596).
57. Ming-Tao Li and Ho-I Huang, *Yo Hsueh Hsueh Pao* **13**, 266 (1966).
58. M. Raymond-Hamet, *Comptes rendus* **220**, 792 (1945).
59. Atta-ur-Rahman and M. Ghazala, *Z. Naturforsch.* **37B**, 762 (1982).
60. G. R. Lenz, *Synthesis,* 517 (1978).
61. T. Kametani, T. Yamanaka, and K. Nyu, *J. Heterocyc. Chem.* **9**, 1281 (1972).
62. J. Bergman, and N. Eklund, *Chem. Scr.* **19**, 193 (1982).
63. Z. Pal, K. Horvath-Dora, G. Toth, J. Tamas, and O. Clauder, *Acta Chim Acad. Sci. Hung.* **99**, 43 (1979).
64. T. Kametani, T. Higa, D. Van Loc, M. Ihara, and K. Fukumoto, *Chem. Pharm. Bull.* **25**, 2735 (1977).
65. T. Kametani, T. Ohsawa, M. Ihara, and K. Fukumoto, *Chem. Pharm. Bull.* **26**, 1922 (1978).
66. B. Danieli, and G. Palmisano, *J. Heterocyc. Chem.* **14**, 839 (1977).
67. J. Tamas, Gy. Bujtas, K. Horvath-Dora, and O. Clauder, *Acta Chim. Acad. Sci. Hung.* **89**, 85 (1976).
68. F. Fish, I. A. Meshal, and P. G. Waterman, *Planta Medica* **29**, 310 (1976).
69. D. de Corrêa, O. R. Gottlieb, and A. P. de Padua, *Phytochemistry* **14**, 2059 (1975).
70. D. de Corrêa, O. R. Gottlieb, A. P. de Padua, and A. I. da Rocha, *Rev. Latinoamer. Quim.* **7**, 43 (1976).
71. T. Kametani, C. Van Loc, T. Higa, M. Ihara, and K. Fukumoto, *J. Chem. Soc. Trans. 1,* 2347 (1977).
72. C. L. King, Y. C. Kong, N. S. Wong, H. W. Yeung, H. H. S. Fong, and U. Sankawa, *J. Nat. Prod.* **51**, 245 (1980).
73. Y. C. Kong, *Adv. Pharmacol. Ther. Proc. Int. Congr., 8th,* 239 (1982).

—CHAPTER 3—

ISOQUINOLINEQUINONES FROM ACTINOMYCETES AND SPONGES

TADASHI ARAI

Department of Antibiotics
Research Institute for Chemobiodynamics
Chiba University
Chiba, Japan

AND

AKINORI KUBO

Department of Organic Chemistry
Meiji College of Pharmacy
Tokyo, Japan

THE ALKALOIDS, VOL. XXI

ISBN 0-12-469521-3

I. Introduction and Occurrence

During the past few years several naturally occurring isoquinolinequinones have been isolated from Actinomycetes and from marine sponges, whereas they have not been found in plant sources.

Arai and co-workers have isolated satellite antibiotics, named mimosamycin (1), saframycins A – E (2), S (3), and mimocin (4) from *Streptomyces lavendulae.*

In 1977, the structure of mimosamycin was determined as 2,6-dimethyl-7-methoxy-3,5,8-trioxo-2,3,5,8-tetrahydroisoquinoline (21) by an X-ray crystallographic study and synthesis (5). 3,5,8-Trioxo-2,3,5,8-tetrahydroisoquinoline proved to be the first naturally occurring ring system.

Among the saframycins, the structure of saframycin C (2) was first elucidated by an X-ray crystallographic analysis of its hydrobromide (6). Subsequently, the structures of saframycins B (1), A (3), and S (8) have been deduced on the basis of the ^{13}C-NMR spectral data and chemical evidence. The saframycins were found to have two 1,2,3,4-tetrahydroisoquinolinequinone moieties in common. It is noteworthy that such heterocyclic quinone antibiotics include mitomycin C (7), streptonigrin (8), and naphthyridinomycin (18).

Naphthyridinomycin (18) was isolated from *Streptomyces lusitanus* and its structure was established by an X-ray crystallographic study (9). More recently, structurally closely related SF-1739 HP (19) and naphthocyanidine (20) were obtained from *Streptomyces griseoplanus* (10).

Even more interesting is the recent discovery of two very similar isoquinolinequinone antibiotics, mimocin (32) and renierone (41) from *Streptomyces lavendulae* (4) and from a marine sponge *Reniera* sp. (11), respectively. Further studies of *Reniera* sp. led to the isolation of mimosamycin (21), renieramycins A – D (14 – 17), *N*-formyl-1,2-dihydrorenierone (56), *O*-demethylrenierone (57), and 1,6-dimethyl-7-methoxy-5,8-dihydroisoquinoline-5,8-dione (58) (12). The structures of the renieramycins were determined by analysis of the ^{1}H-NMR data. It is interesting to note that renieramycins and renierone are found to contain an angelate ester side chain in place of the pyruvamide side chain of saframycins and mimocin (32). These isoquinolinequinone antibiotics are shown in Table I.

In this chapter, these antibiotics are divided into four types represented by saframycin C (2), renieramycin A (14), naphthyridinomycin (18), and mimosamycin (21), and will be described including their spectral data, chemical properties, and syntheses. In particular their biological activities are described in Section VI.

Although biosynthesis of these antibiotics is quite interesting, no detailed investigations with labeled precursors have been reported.

TABLE I

ISOQUINOLINEQUINONE ANTIBIOTICS

Name	Empirical formula	Melting point (°C)	$[\alpha]_D{}^a$	Reference
Saframycin A	$C_{29}H_{30}N_4O_8$	122–126°	+18.2 (m)	2
Saframycin B	$C_{28}H_{31}N_3O_8$	108–109°	−54.4 (m)	2
Saframycin C	$C_{29}H_{33}N_3O_9$	143–146°	−20.8 (m)	2
Saframycin D	$C_{28}H_{31}N_3O_9$	150–154°	+141.0 (m)	2
Saframycin E	$C_{28}H_{33}N_3O_9$	146–148°	−37.3 (m)	2
Saframycin S	$C_{28}H_{31}N_3O_9$	107–115° (dec.)	+32.5 (m)	3
Renieramycin A	$C_{30}H_{34}N_2O_9$		−36.3 (m)	12
Renieramycin B	$C_{32}H_{38}N_2O_9$		−32.2 (m)	12
Renieramycin C	$C_{30}H_{32}N_2O_{10}$		−89.2 (m)	12
Renieramycin D	$C_{32}H_{36}N_2O_{10}$		−100.7 (m)	12
Naphthyridinomycin	$C_{21}H_{27}N_3O_6$	108–110° (dec.)	+69.4 (c)	9
SF-1739 HP	$C_{20}H_{25}N_3O_6$		+64.0 (m)	10
Naphthocyanidine	$C_{21}H_{24}N_4O_5$		+76.0 (m)	10
Mimosamycin	$C_{12}H_{11}NO_4$	227–231°		1
Mimocin	$C_{15}H_{14}N_2O_5$	189–191° (dec.)		4
Renierone	$C_{17}H_{17}NO_5$	91.5–92.5°		11
N-Formyl-1,2-dihydrorenierone	$C_{18}H_{19}NO_6$		−227.0 (m)	12
O-Demethylrenierone	$C_{16}H_{15}NO_5$	135–136°		12
1,6-Dimethyl-7-methoxy-5,8-dihydroisoquinoline-5,8-dione	$C_{12}H_{11}NO_3$	188–190° (dec.)		12

^{a} Rotation measured in c = chloroform; m = methanol.

II. Saframycin-Type Antibiotics

The safran-colored saframycins are satellite antibiotics that are coproduced in trace quantities with streptothricin by *S. lavendulae* No. 314. To date, saframycins A, B, C, D, E (*2*), and S (*3*) have been isolated and characterized as summarized in Table I.

Of these six saframycin-type antibiotics, the structure of saframycin C (**2**) was the first to be elucidated by an X-ray crystallographic analysis of the corresponding hydrobromide (*6*). The structures of saframycins B (**1**), A (**3**), and S (**8**) have been determined by ^{13}C-NMR studies and by chemical means. The structures of saframycins D and E have not been established yet.

A. Saframycins B and C

Saframycins B and C were the main antibiotics isolated from the basic fractions of the methylene chloride extracts of the culture broth and purified by silica gel, LH-20 Sephadex column chromatography, and preparative TLC.

Saframycin B (1) [$C_{28}H_{31}N_3O_8$; mp 108–109° (ethyl ether); $[\alpha]_D$ –54.4° (c = 1.0, MeOH)] (2) showed UV maxima at 269 (logε 4.35) and 368 (logε 3.13) nm, and IR bands at 3430, 1720, 1690, 1660, and 1620 cm⁻¹. The ¹H-NMR spectrum (CDCl₃, 100 MHz) showed signals at δ 1.90 (s), 1.98 (s), 2.23 (s), 2.28 (s), 4.00 (2 × s) due to six methyl groups. Its ¹³C-NMR spectrum (CDCl₃) revealed the nature of the methyl groups [δ 8.6 (2 × C—CH₃), 24.2 (COCH₃), 41.2 (N—CH₃), 60.9 (2 × OCH₃)] and showed six pairs of singlets at δ 127–187 ascribed to the quaternary aromatic and carbonyl carbons (Table II) (6). Comparisons of these ¹³C-NMR data with those of mimosamycin (21) (vide infra) and 3,6-dimethoxythymoquinone (13) revealed the presence of two 3-methoxy-2-methyl-p-benzoquinone moieties in saframycin B (1).

Saframycin C (2) [$C_{29}H_{33}N_3O_9$; mp 143–146° (ethyl ether); $[\alpha]_D$ –20.8° (c = 1.0, MeOH)] (2) showed UV maxima at 266.5 (logε 4.32), 368 (logε 3.19) nm (Fig. 1) and IR bands at 3400, 1720, 1685, 1655, and 1615 cm⁻¹. The ¹H-NMR spectrum (CDCl₃, 100 MHz) included seven methyl singlets at δ 1.86, 2.00, 2.38, 2.44, 3.46, 3.96 (2 ×s). Its ¹³C-NMR spectrum (CDCl₃) revealed the nature of all the methyl groups [δ 8.7, 9.0 (2 ×C—CH₃), 24.3 (COCH₃), 42.3 (N—CH₃), 59.3, 60.9, and 61.0 (3 ×OCH₃)] (Table II) (6). Other features included the presence of the characteristic six pairs of singlets at δ 127–187, which were almost identical to those of saframycin B (1).

These spectral data indicated that 2 had the same carbon skeleton as 1 and an additional methoxyl group (δ 59.3) which should be located at the methylene carbon (δ 22.7) adjacent to the one of p-quinone moieties in 1.

The complete structure and stereochemistry of saframycin C (2) was established by an X-ray crystallographic study of its hydrobromide. The crystals were found to have the monoclinic space group $P2_1$, with a =11.819 (3), b = 19.644 (5), c =7.650 (3) Å, β =114.7° (6).

From these results, the structure of saframycin B was determined as 14-demethoxysaframycin C. Saframycins B and C therefore have the structures of 1 and 2 respectively, except for their absolute configuration.

B. Saframycin A

Saframycin A (3) [$C_{29}H_{30}N_4O_8$; mp 122–126° (ethyl ether); $[\alpha]_D$ + 18.2° (c =1.0, MeOH)] (2) was obtained as a yellow amorphous powder from the neutral fractions of the methylene chloride extracts of the culture filtrate and

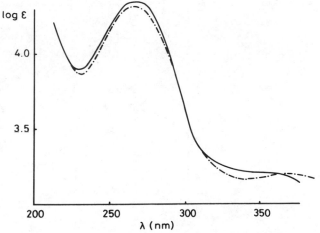

1 R = H, R' = H
2 R = H, R' = OCH₃
3 R = CN, R' = H

purified by silica gel and LH-20 Sephadex column chromatography, and finally by preparative TLC (2).

The production of saframycins was as low as 0.01 μg/ml of the culture filtrate in the initial stage of studies. However, the significantly increased production of saframycin A was accomplished by addition of sodium cyanide to the culture broth and by maintaining the pH below 5.5 after peak production of the antibiotic (3) (*vide infra*).

Saframycin A (3) gave UV maxima at 267 (logε 4.34) and 370 (logε 3.15, inf.) nm, which were very close to those of saframycins B (1) and C (2) (Fig. 1).

The IR spectrum showed absorption bands at 3400, 1716, 1685, 1660, and 1615 cm⁻¹, and the ¹H-NMR spectrum (CDCl₃, 100 MHz) showed signals at δ 1.90 (s), 1.98 (s), 2.24 (s), 2.30 (s), 4.04 (2 ×s) for six methyl groups. The ¹³C-NMR spectrum data (CDCl₃) revealed the nature of all the methyl groups [δ 8.8 (2 × C−CH₃), 24.3 (COCH₃), 41.7 (N−CH₃), 61.0, and 61.1 (2 × OCH₃)] and displayed the characteristic six pairs of signals at δ 128 – 187 due

FIG. 1. UV spectra of saframycin A (3) (——) and saframycin C (2) (— · — · —).

TABLE II

^{13}C-Chemical Shifts (δ) of Mimosamycin (21), Saframycins B (1), C (2), and A (3) in CDCl$_3$[a]

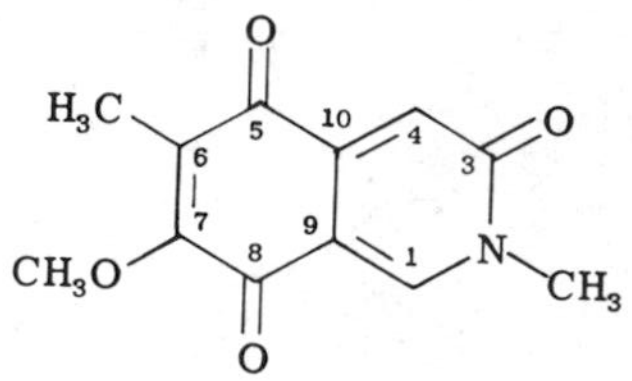

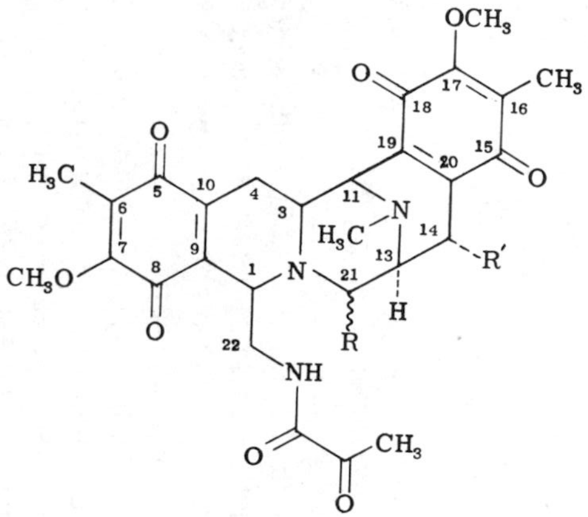

1 R = H, R' = H
2 R = H, R' = OCH$_3$
3 R = CN, R' = H

Carbon No.	21
5	183.5
6	133.6
7	159.5
8	177.3
9	111.3
10	138.9
6—CH₃	9.6
7—OCH₃	61.3
N—CH₃	38.5

Carbon No.	1	2	3
5 or 15	185.7 or 187.0	185.5 or 186.6	185.2 or 186.5
6 or 16	127.7 or 129.2	127.9 or 130.7	128.3 or 129.2
7 or 17	155.5 or 156.1	155.4 or 156.1	155.6 or 155.9
8 or 18	181.3 or 182.8	181.3 or 183.2	180.8 or 183.4
9 or 19	136.3 or 136.6	136.6 or 136.6	135.6 or 135.6
10 or 20	141.6 or 142.8	141.5 or 141.6	141.2 or 141.6
6 or 16 —CH₃	8.6 or 8.6	8.7 or 9.0	8.7 or 8.7
7 or 17 —OCH₃	60.9 or 60.9	60.9 or 61.0	61.0 or 61.1
N—CH₃	41.2	42.3	41.6
1 or 11	52.2 or 54.8	55.2 or 55.7	54.0 or 54.3
3 or 13	56.9 or 57.4	57.6 or 58.0	54.6 or 56.3
4	25.6 (t)	25.5 (t)	25.1 (t)
14	22.7 (t)	71.9 (d)	21.6 (t)
14—OCH₃		59.3	
21	58.7 (t)	55.7 (t)	58.3 (d)
22	40.4 (t)	40.7 (t)	40.7 (t)
NHCO	160.1	160.2	160.2
COCH₃	24.2, 196.5	24.3, 196.5	24.3, 196.7
CN			116.7 (s)

[a] Natural abundance ¹H noise-decoupled (PND) and off-resonance ¹³C FT NMR-spectra were recorded on a Jeol FX-60 and FX-100 FT NMR spectrometer. TMS served as an internal reference (δ 0).

to the quaternary aromatic and carbonyl carbons, and an unusual signal at δ 116.7 (s) (Table II) (14). From these spectral data, saframycin A (3) must have the same carbon skeleton as saframycins B (1) and C (2) and comparison of the empirical formula of saframycin A (3) with that of saframycin B ($C_{28}H_{31}N_3O_8$) (1) suggested that one cyano group is substituted for one hydrogen atom of 1. This assignment was substantiated by the ^{13}C-NMR spectrum (δ 116.7) and was confirmed by examination of ^{13}C-NMR data of the nitril signals of model compounds (14).

However, in its IR spectrum, no absorption band was observed in the C≡N region. This result could be explained by assuming that in the IR spectrum, the introduction of the oxygenated groups into the molecule results in a quenching of the C≡N absorption intensity (15). Furthermore, acid hydrolysis of saframycin A (3) was carried out in 0.1 N H_2SO_4 at 100° for 20 min and 3 released about 1 mole equivalent HCN as determined by the Epstein method (16). This presents additional evidence of the cyano group in the molecule. The exact location of the cyano group was determined by ^{13}C-NMR studies (14). The ^{13}C-NMR data of 3 excluded the possibility of the cyano group being localized at the C-4, C-14, C-22 methylene carbons, or the C-1, C-3, C-11, C-13 methine carbons adjacent to tertiary nitrogens. The above results and the absence of a signal corresponding to the C-21 methylene carbon [δ 58.7 (t) in 1] and the presence of a methine carbon at δ 58.3 (d) enabled the localization of the cyano group at the C-21 position in 1. Moreover, the assumption that saframycin A (3) contains the cyano group at C-21 was supported by the observation that the C-14 methylene carbon moved to δ 21.6 (t) in saframycin A (3) from δ 22.7 (t) in saframycin B (1) because of its γ position (Table II) (17, 18).

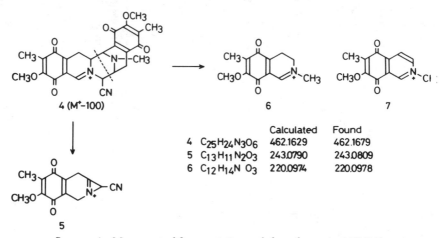

		Calculated	Found
4	$C_{25}H_{24}N_3O_6$	462.1629	462.1679
5	$C_{13}H_{11}N_2O_3$	243.0790	243.0809
6	$C_{12}H_{14}N\ O_3$	220.0974	220.0978

SCHEME 1. *Mass spectral fragmentation path for saframycin A (3) (14).*

The high resolution mass spectrum confirmed the above conclusion (*14*). The fragment ions m/z 462 ($M^+ - 100$) (**4**), 220 (**6**) and 218 (**7**) were derived from the basic skeleton of saframycin-type antibiotics. Of particular significance was the presence of the fragment ion m/z 243 (**5**) corresponding to the ion containing a cyano group at C-21 in the molecule. (Scheme 1, structures **4–7**).

Finally, treatment of saframycin A (**3**) with mineral acid afforded saframycin S (21-hydroxysaframycin B) (**8**) (*3*), thus settling the position of the cyano substituent (*vide infra*). Consequently, saframycin A has the structure represented by **3**, 21-cyanosaframycin B. However, the assignment of the orientation of the cyano group remains undetermined.

C. ¹H-NMR Analysis and Conformations of Saframycins A and C

Lown *et al.* carried out detailed analysis of the high-field (400 MHz) ¹H-NMR spectra of saframycins A (**3**) and C (**2**) in $CDCl_3$, C_6D_6, and in aqueous DMSO solutions, and all the proton chemical shifts and coupling constants in each excluding the methyl groups have been assigned unambiguously by selective double-irradiation experiments (Table III) (*19*).

Furthermore, from a combination of the magnitude of the coupling constants and NOE studies it was possible to assign the molecular geometry and average conformations in solution as shown in Table III.

In the NMR spectrum of saframycin A (**3**), the coupling constant ($J_{4,5} = 3.0$ Hz) between H-4 and H-5 over five bonds was consistent with such long-range coupling along a W path, and this established unambiguously stereochemical assignment of H-4 and H-5 to be trans.

Double irradiation of the signal at δ 3.90 due to H-8 simplified the ddd signal at δ 3.02 (H-7) to a dd as well as the ddd signal at δ 2.57 (H-11) to dd through a W-type long-range coupling effect. Irradiation of the dd signal at δ 2.31 (H-9) collapsed the H-10 signal at δ 1.68 as well as the ddd signal at δ 2.57 (H-11) to a dd. Irradiation of the H-10 signal at δ 1.68 caused the H-9 signal at δ 2.31 to change from a dd to a d and the H-11 signal at δ 2.57 from a ddd to a dd.

Finally, irradiation of the H-11 signal at δ 2.57 led to the collapse of the d signal of H-12 at δ 2.90 to an s and double irradiation of H-12d in turn permitted assignment of the coupling constant $J_{11,12} = 2.5$ Hz (Table III).

Although the orientation of the cyano group has not been conclusively established on the basis of these coupling constants, the α-orientation was deduced by chemical evidence which explains the easy elimination of the cyano group on mild acid treatment (*3, 14*). Therefore, this requires that the lone pair on N-2 be trans to the cyano group.

TABLE III

HIGH FIELD ^{1}H-NMR SPECTRA OF SAFRAMYCINS A (3) AND C (2)

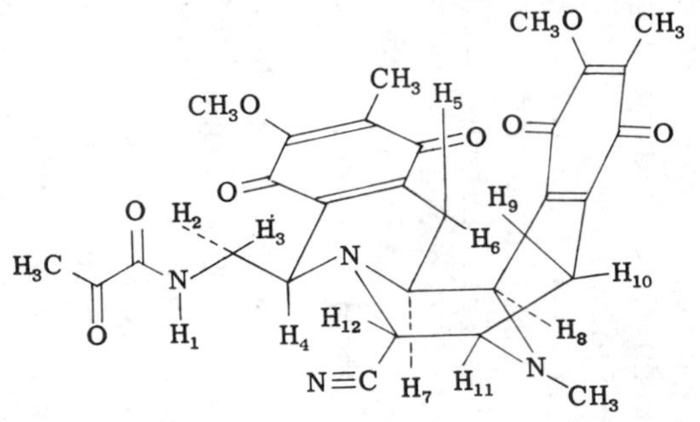

CH$_3$O CH$_3$

OCH$_3$

H$_9$ H$_8$

H$_6$ N—CH$_3$

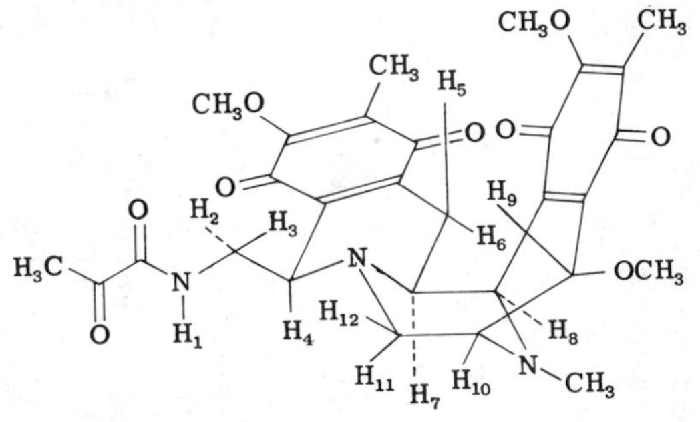

CH$_3$ H$_5$ H$_{10}$

H$_3$C N H$_{12}$ H$_{11}$ H$_7$

CH$_3$O H$_3$ H$_4$

H$_2$ N H$_1$

2

CH$_3$O CH$_3$

H$_{10}$

H$_9$ H$_8$

H$_6$ N—CH$_3$

CH$_3$ H$_5$ H$_{11}$

H$_3$C N≡C H$_7$

CH$_3$O H$_3$ H$_{12}$

H$_2$ N H$_4$

H$_1$

3

Proton chemical shifts and coupling constants for saframycin A in C$_6$D$_6$ and CDCl$_3$

Proton	Chemical shifts[a] δ(ppm)		Coupling constants[b] (Hz)
	C$_6$D$_6$	CDCl$_3$	
1	6.33	6.70	$J(1-2) = 9.5$
			$J(1-3) = 4.0$
2	3.76	3.84	$J(2-1) = 9.5$
			$J(2-3) = 14.0$
			$J(2-4) = 1.5$

Proton chemical shifts and coupling constants for saframycin C in C$_6$D$_6$ and CDCl$_3$

Proton	Chemical shifts[a] δ(ppm)		Coupling constants[b] (Hz)
	C$_6$D$_6$	CDCl$_3$	
1	6.64	6.74	$J(1-2) = 9.5$
			$J(1-3) = 3.0$
2	3.89	3.77	$J(2-1) = 9.5$
			$J(2-3) = 14.0$
			$J(2-4) = 1.5$

Position	δ	δ	J (Hz)
3	2.50	3.26	$J(3-1) = 4.0$, $J(3-2) = 14.0$, $J(3-4) = 4.0$
4	3.87	3.98	$J(4-2) = 1.5$, $J(4-3) = 4.0$, $J(4-5) = 3.0$, $J(4-6) = 0.0$
5	1.32	1.28	$J(5-4) = 3.0$, $J(5-6) = 17.0$, $J(5-7) = 11.0$
6	2.89	2.87	$J(6-5) = 17.0$, $J(6-7) = 3.0$, $J(6-4) = 0.0$
7	3.02	3.14	$J(7-5) = 11.0$, $J(7-6) = 3.0$, $J(7-8) = 3.0$
8	3.90	4.06	$J(8-7) = 3.0$, $J(8-11) = 2.5$
9	2.31	2.83	$J(9-10) = 20.5$, $J(9-11) = 7.5$
10	1.68	2.30	$J(10-9) = 20.5$, $J(10-11) < 0.5$
11	2.57	3.44	$J(11-9) = 7.5$, $J(11-10) < 0.5$, $J(11-12) = 2.5$
12	2.90	3.99	$J(11-12) = 2.5$

Position	δ	δ	J (Hz)
3	2.64	3.20	$J(3-1) = 3.0$, $J(3-2) = 14.0$, $J(3-4) = 4.0$
4	3.38	3.63	$J(4-2) = 1.5$, $J(4-3) = 4.0$, $J(4-5) = 3.0$
5	1.36	1.18	$J(5-4) = 3.0$, $J(5-6) = 17.5$, $J(5-7) = 11.0$
6	2.91	2.75	$J(6-5) = 17.5$, $J(6-7) = 3.0$
7	2.40	2.67	$J(7-5) = 11.0$, $J(7-6) = 3.0$, $J(7-8) = 3.0$
8	4.10	4.10	$J(8-7) = 3.0$, $J(8-10) = 1.0$
9	3.79	3.85	$J(9-10) < 0.5$
10	2.94	3.27	$J(10-9) < 0.5$, $J(10-8) = 1.0$, $J(10-11) = 2.0$, $J(10-12) = 3.0$
11	2.36	2.83	$J(11-10) = 2.0$, $J(11-12) = 11.0$
12	2.38	3.03	$J(12-10) = 3.0$, $J(12-11) = 11.0$

[a] The chemical shifts were obtained at 400 MHz and are given relative to internal TMS as reference.

[b] The coupling constants were obtained by double irradiation experiments.

A similar analysis of the high-field NMR spectrum of saframycin C (2) in C_6D_6 and $CDCl_3$ gave the accurate chemical shifts and coupling constants using extensive double irradiation experiments (Table III). Noteworthy is the unusual upfield H-5 ddd signal at δ 1.32 for saframycin A (3) and δ 1.36 for saframycin C (2). This result could be explained by assuming the two shielding effects, i.e., the conformation of both antibiotics predicts that H-5 be placed in the strong shielding zone of the quinone ring (adjacent to H-8) and in close proximity to the lone pair of the bridging central nitrogen atom (N-2). Also noteworthy is the preferential protonation of N-12 over N-2.

Examination of the spectra of both antibiotics in DMSO-d_6–D_2O–CF_3COOD revealed that the signals of H-8, H-11, and N-Me in saframycin A (3) and H-8, H-10, and N-Me in saframycin C (2) were shifted relatively more downfield (0.91, 0.95, and 0.66 ppm, respectively for 3 and 0.82, 0.80, and 0.60 ppm, respectively for 2) compared to H-7 and H-12 in 3 and H-7, H-11, and H-12 in 2, and the signal of H-4 in both antibiotics was hardly affected, indicating preferential protonation of N-12 over N-2 in both antibiotics.

D. SAFRAMYCIN S

Saframycin S (8) [$C_{28}H_{31}N_3O_9$; mp 107–115° (dec.); $[\alpha]_D$ +32.5° (c = 0.5, MeOH)] (3) was obtained as a dark yellow powder from the basic fractions of the chloroform extracts of the culture broth and has been proposed as a precursor to saframycin A (3) in the biosynthetic process involving the intermediacy of the iminium salt 9 that subsequently undergoes nucleophilic addition or displacement by cyanide anion (3).

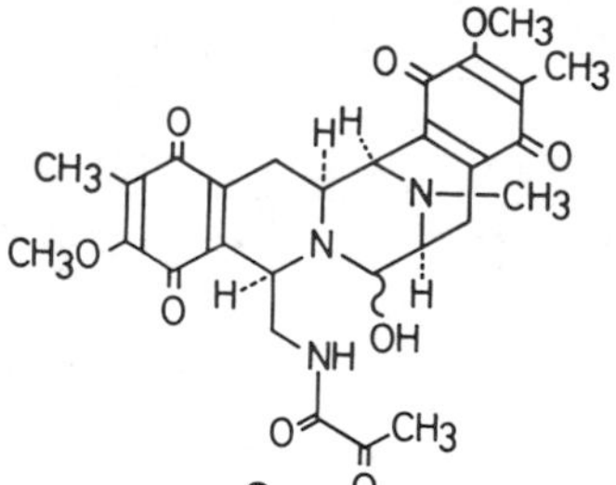

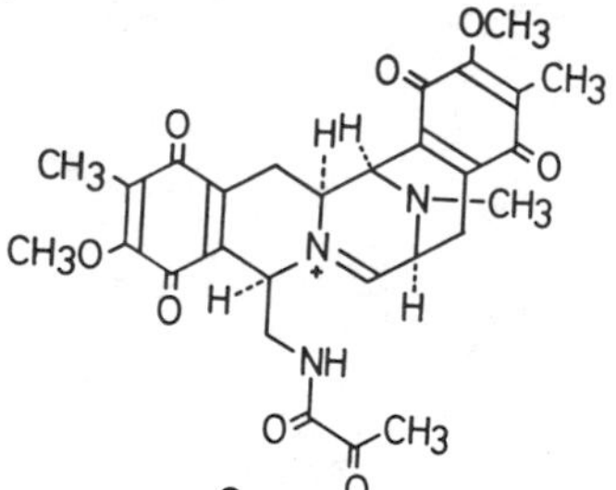

Saframycins A (3) and S (8) have been interconverted chemically. Thus, treatment of saframycin A (3) with 0.1 N H_2SO_4 at 120° for 40 min afforded saframycin S (8) in 40.5% yield and the conversion of 8 to 3 was accomplished in 60.1% yield in the presence of NaCN in phosphate buffer (pH 7.0) solution (Scheme 2, structure 8) (3).

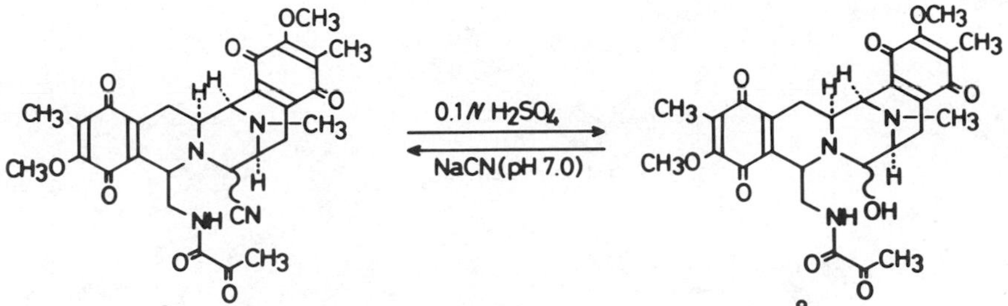

SCHEME 2. *Interconversion of saframycins A (3) and S (8)* (2).

The UV spectrum showed absorption maximum at 268 (logε 4.21) nm, which was very similar to that of saframycin A (3). The IR spectrum showed absorption bands at 3400, 1720, 1680, and 1650 cm⁻¹. ¹H- and ¹³C-NMR spectra revealed the presence of two C-methyl groups (δ 1.94, 2.06; 8.8), one COCH₃ group (δ 2.30; 24.3), one N-methyl group (δ 2.52; 42.5), and two C-methoxyl groups (δ 4.06; 61.1).

These spectral data indicated that saframycin S (8) had the same carbon skeleton as saframycins A (3), B (1), and C (2). Furthermore, comparing the ¹³C-NMR spectra of saframycins A (3) and S (8), the signals at δ 116.7 (s) and 58.3 (d), which were assigned to the N—CH—CN group (α-aminonitrile) (*18*) in the former disappeared, while a new signal appeared at δ 82.1 (d) in the latter indicating the presence of a N—CH—OH group (α-carbinolamine) (*20*) in saframycin S (8). This indicated the C-21 cyano group in saframycin A (3) was substituted with a hydroxyl group yielding saframycin S (8). The structure of saframycin S therefore has been established as 8, 21-hydroxy-saframycin B.

E. SYNTHESIS

The total synthesis of saframycins has not yet appeared in the literature. Recently, Kurihara and Mishima reported a synthetic approach to the C, D, E ring system of the parent pentacyclic framework of saframycin skeleton starting with readily available amino acid (Scheme 3, structures 10 – 13) (*21*).

The key steps in this work were the preparation of 3,4-dihydro-4-ethoxy-carbonyl-2-pyrazinone (12) and the transformation of 12 to tricyclic 1,5-imino-3-benzazocine (13). Phenylalanylsarcocinal dimethyl acetal (11) that was prepared by condensation of N-ethoxycarbonyl phenylalanine (10) with sarcocinal dimethyl acetal in excellent yield, was converted to 3,4-dihydro-4-ethoxycarbonyl-1-methyl-2-pyrazinone (12) in the presence of a catalytic amount of hydrochloric acid in acetonitrile. Heating of 12 in trifluoroacetic acid was found to cause the acid-induced intramolecular cyclization, yield-

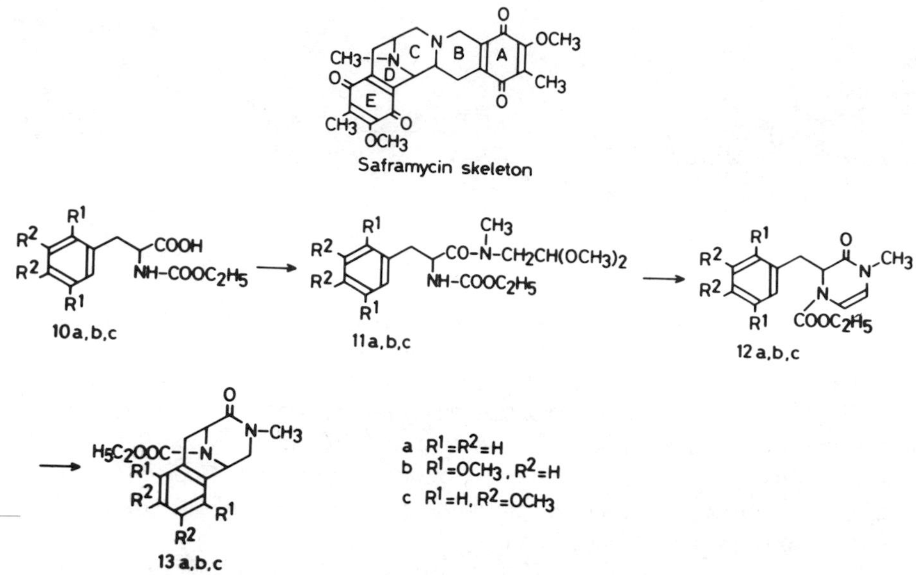

Saframycin skeleton

a $R^1 = R^2 = H$
b $R^1 = OCH_3$, $R^2 = H$
c $R^1 = H$, $R^2 = OCH_3$

SCHEME 3. *Synthesis of the saframycin skeleton* (12).

ing regioselectively 1,5-ethoxycarbonylimino-3-methyl-1,2,5,6-tetrahydro-3-benzazocin-4(3H)-one (**13**) in quantitative yield.

It is noteworthy that compound **13b** should be a suitable intermediate for preparing the tricyclic quinone, and therefore this approach might serve well for the construction of the saframycin skeleton.

F. FERMENTATION AND BIOSYNTHESIS

Actinomycetes, streptomyces, and streptoverticillium in particular, have been well known to be excellent biological sources of antibiotics. After an extensive search for new antibiotics among streptomyces and streptoverticillium, however, it has become considerably more difficult to discover antibiotics of newer groups among these microorganisms. In an attempt to find novel classes of antibiotics, the search for antibiotics from rare actinomycetes, precursor feeding, mutasynthesis, and chemical and biological modification of known antibiotics and others have been explored lately.

Saframycins are novel satellite antibiotics that are simultaneously produced with streptothricin by *Streptomyces lavendulae*. Streptothricin is a historic aminoglycoside antibiotic first isolated by Waksman *et al.* in 1942 (*22*). The search for such satellite antibiotics among known species of streptomyces coupled with *in vitro* screening procedures seems to hold a fair chance of isolating unprecedented groups of antibiotics apt to be missed by conventional screening procedures. However, difficulty is inherent in the development of such satellite antibiotics for chemotherapeutic agents because of their extremely low production in the culture of the producing microorganism. Increased production of the antibiotic was attained to some extent by improving culture conditions, i.e., extraordinarily dissolved oxygen tension in the culture and modification of medium ingredients.

However, a satisfactory way of circumventing the low production level was found quite unexpectedly with the observation that cyanide was directly incorporated in the cyano residue of saframycin A (**3**). *Streptomyces lavendulae,* a saframycin-producing strain, was grown in the culture medium until the production of the antibiotic reached maximum. The culture filtrate was prepared and incubated with Na^{14}CN for 1 hr, followed by extaction with ethyl acetate. Figure 2 showes a UV scanning profile and distribution of radioactivity on the thin layer chromatogram of the extracts. The incubation of the culture filtrate with Na^{14}CN resulted in increments in the absorbance as well as association of radioactivity at the position of R_f value 0.55 representing saframycin A (**3**). Moreover, the molar amount of ^{14}CN asso-

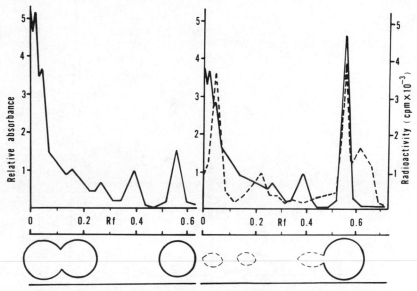

Fig. 2. Chromatographic profile and bioautogram of extract of culture filtrate after incubation with or without Na^{14}CN. The equal portion of culture filtrate (pH 5.8) was incubated with (*right*) or without (*left*) 0.33 mM Na^{14}CN for 1 hr at room temperature and the extracts with ethyl acetate were analyzed by thin layer chromatography. ——, UV absorption; ---, ^{14}C radioactivity. Each thin layer plate was applied to bioautography on *Bacillus subtilis,* and the zone of inhibition was shown.

ciated with saframycin A (3) was equivalent to the molar increment of 3 induced by the addition of NaCN. These results indicated the presence of an unidentified substance, later determined to be decyanosaframycin A or saframycin S (21-hydroxysaframycin B) (8), in the culture filtrate which stoichiometrically reacted with cyanide to give saframycin A (3). An additional striking feature revealed by bioautography shown in Fig. 2 was that both spots located at R_f values 0 and about 0.18 disappeared from the culture filtrate treated with NaCN. This result suggested that decyanosaframycin A exists in plural forms. The possible structure that can be written for decyano-saframycin A is the iminium cation (9) or the α-carbinolamine (8) (*vide supra*). Decyanosaframycin A was a highly active compound but difficult to isolate because of its unstability. The proof of direct incorporation of cyanide in the cyano residue of saframycin A (3) was obtained by the use of ^{13}C NMR. [13]CN-labeled saframycin A (3) when subjected to ^{13}C NMR clearly demon-

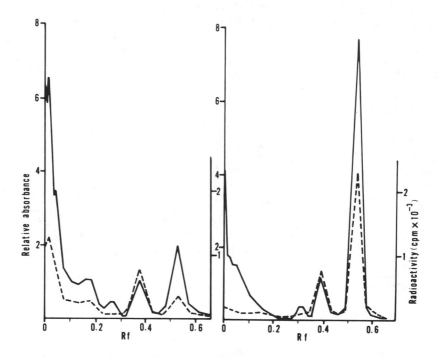

FIG. 3. Biosynthetic labeling of saframycin A (3) with [14C]tyrosine. At 5 hr prior to the termination of fermentation, [14C]tyrosine (50 μCi/flask) was added to the culture. Then, the culture filtrate was incubated with (*right*) or without (*left*) 1 mM NaCN for 1 hr and analyzed as described in the legend of Fig. 2.

strated enrichment of the signal at δ 118, which was assigned as cyano carbon in the previous study (*14*).

Since dimeric quinone skeletons common to saframycins were indicative of the dimeric form of tyrosine cyclizing to yield 2,5-dioxopiperazine as a biosynthetic precursor, [^{14}C]tyrosine was added to the fermentation broth at the time corresponding to the initiation of antibiotic synthesis. The uptake of radioactivity by saframycin A (**3**) was observed (Fig. 3, left). Subsequent treatment of the culture filtrate with NaCN revealed that the disappearance of radioactivity at an R_f value below 0.2 was accompanied with the increase in radioactivity associated with saframycin A (**3**) (Fig. 3, right). There is ample evidence that decyanosaframycin A shares the heterocyclic quinone skeletons. On the basis of the above experimental results, significant increase of saframycin A (**3**) potency could be attained by addition of NaCN to the culture broth. On the other hand, a drastic degradation of saframycin A (**3**) in the culture was successfully prevented by maintaining the pH below 5.5 after peak production of the antibiotic. The combined application of these improvements resulted in approximately a 1000-fold increase in saframycin A (**3**) production as compared with the parental level (*3*).

These findings seem to pave the way for obtaining novel antibiotics, that otherwise are quite unstable and difficult to isolate, by the secondary treatments of the microbial cultures. To quote a concrete example, antibiotics SF-1739 HP (**19**) and naphthocyanidine (**20**) were prepared by treatment of unstable antibiotic SF-1739 with mineral acid followed by similar addition of cyanide anion. (*10*)

III. Renieramycin-Type Antibiotics

Renieramycins A–D (**14–17**) have been isolated as minor metabolites from marine sponge *Reniera* sp. together with major metabolite, renierone (**41**) (*vide infra*) (*12*). The structures of the renieramycins were deduced from the spectral data, particularly the ^{1}H-NMR data, and found to have a "dimeric" quinone structure similar to saframycin antibiotics. The name *renieramycin*, presumably christened after saframycin is *nomen dubidum* from the viewpoint of antibiotic nomenclature, because the suffix *mycin* has been preferentially used to designate antibiotics derived from actinomycetes.

A. RENIERAMYCINS A AND B

The major dimer, renieramycin A (**14**) [$C_{30}H_{34}N_2O_9$; $[\alpha]_D$ – 36.3° (c = 0.16, MeOH] (*12*) showed UV maxima at 268 (ϵ 15800) and 365 (ϵ 1370) nm, and IR bands at 3160 (hydrogen-bonded hydroxyl group), 1720 (α, β-unsatu-

TABLE IV

^{1}H-NMR Chemical Shifts (δ) for Renieramycins A–D (**14–17**)^{a,b}

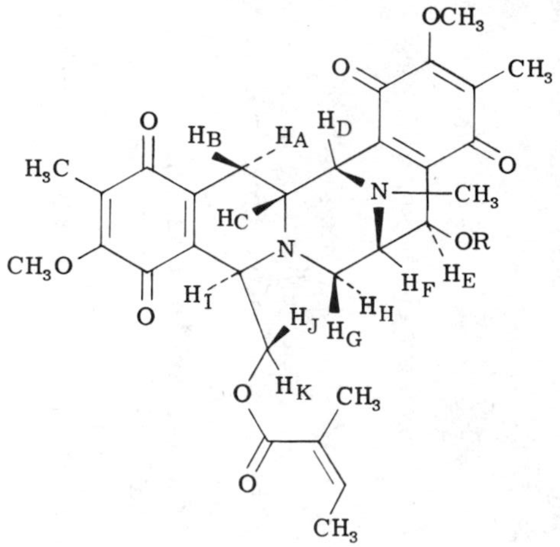

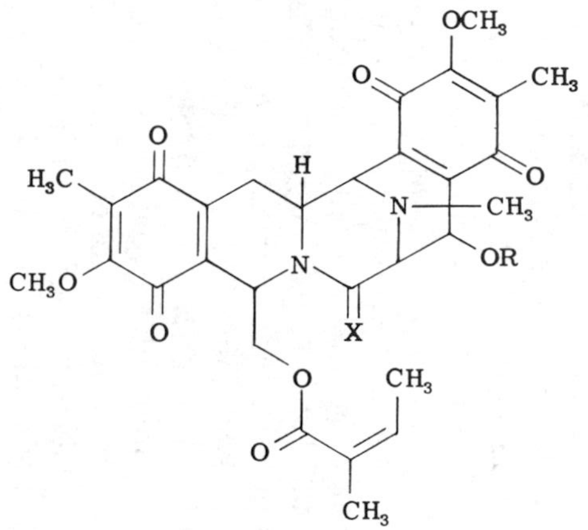

14 R = H, X = H$_2$
15 R = Et, X = H$_2$
16 R = H, X = O
17 R = Et, X = O

14–17

	14	15	16	17
H$_A$	1.26 (17, 11, 3)	1.23 (18, 11, 3)	1.41 (17, 12, 2)	1.41 (17, 12, 1)
H$_B$	2.75 (17, 2.5, 1)	2.74 (18, 3)	3.02 (17, 2.5)	3.02 (17, 2.5)

H_C	2.64 (11, 2.5, 2.5)	2.62 (11, 3, 2.5)	3.86 (12, 3.5, 2.5)	~3.87
H_D	4.04 (2.5, 1)	4.04 (3, 1)	4.19 (3.5, 1)	4.22 (2)
H_E	4.44 (2)	4.05 (s)	4.78 (br s)	4.36 (br s)
H_F	3.18 ($w_{1/2}$ =5)	3.19 (br s)	3.73 (s)	3.67 (br s)
H_G	2.71 (11, 3.5)	2.76 (11, 3.5)		
H_H	3.18 (11)	3.10 (11, 2)		
H_I	3.60 ($w_{1/2}$ =9)	3.61 (br s)	5.48 (br s)	5.49 (br s)
H_J	4.19 (11.5, 2)	4.27 (11.5, 2.5)	4.72 (11.5, 2.5)	4.63 (12, 2.5)
H_K	4.49 (11.5, 3)	4.32 (11.5, 3)	4.36 (11.5, 2.5)	4.42 (12, 2.5)
ring A OCH_3	4.00*c	4.00*c	4.01*c	3.97*c
CH_3	1.91**c	1.90**c	1.93**c	1.92**c
ring E OCH_3	4.01*c	3.95*c	4.05*c	4.04*c
CH_3	1.92**c	1.93**c	1.95**c	1.95**c
$N-CH_3$	2.43	2.48	2.61	2.62
angelate CH_3	1.55 (br s)	1.56 (br s)	1.50 (br s)	1.48 (br s)
CH_3	1.78 (d, 7)	1.79 (d, 7)	1.69 (d, 7)	1.68 (d, 7)
H	5.92 (br q, 7)	5.91 (br q, 7)	5.90 (br q, 7)	5.89 (br q, 7)
OEt		1.19 (t, 7)		1.23 (t, 7)
		3.76 (q, 7)		3.87 (q, 7)

[a] From Frincke and Faulkner (12).

[b] Coupling constants (Hz) in parentheses.

[c] Asterisk and double asterisk numbers may be exchanged.

rated ester), 1660 (quinone) cm^{-1}. The ^{1}H-NMR spectrum (360 MHz) showed signals at δ 1.91 (s), 1.92 (s), 4.00 (s), 4.01 (s) due to the methyl and methoxyl groups on the two quinone rings, at δ 2.43 (s) due to the N-methyl group, and revealed the presence of the angelate ester [δ 1.55 (br s, 3H), 1.78 (d, 3H, $J = 7$ Hz) and 5.92 (br q, 1H, $J = 7$ Hz)]. The remaining proton chemical shifts and coupling constants were assigned by spin-decoupling experiments (Table IV) (12). Interestingly, these data were almost identical with those of the saframycins (Table III) (vide supra) (19).

The relatively large homoallylic coupling ($J_{H_A,H_I} = 3$ Hz) between the signals at δ 1.26 (H$_A$) and 3.60 (H$_I$) indicated that both protons were orthogonal to the plane of the quinone ring, suggesting a pseudo boat conformation for the nitrogen-containing ring B. Additionally, the long-range coupling ($J_{H_D,H_F} = 1$ Hz) between signals at δ 4.04 (H$_D$) and 3.18 (H$_F$) was expected for the bridgehead protons in a bicyclo[3.3.1] ring system. The absence of coupling between the signals at δ 3.18 and 4.44 due to H$_F$ and H$_E$, respectively, required a dihedral angle of 80–90° for these protons. Furthermore, examination of the detailed nuclear Overhauser enhancement difference spectra (NOEDS) led to the definition of the geometry of renieramycin A (14) as shown in Table IV (structures 14–17). The large upfield shift of the H$_A$ signal at δ 1.26 relative to the H$_B$ signal at δ 2.75 served to confirm the renieramycin A geometry indicating H$_A$ proton was placed in the ring current of the quinone ring E. It is noteworthy that the ring system and the relative stereochemistry of the BC and CD ring junctions of renieramycin A (14) was identical with that of the saframycins (vide supra).

Renieramycin B (15) [C$_{32}$H$_{38}$N$_2$O$_9$; [α]$_D$ $-32.2°$ ($c = 0.15$, MeOH)] (12) exhibited the UV maxima at 268 (ϵ 17400) and 365 (ϵ 1460) nm, and IR bands at 1715, 1660, 1645, and 1620 cm^{-1}.

Comparison of the ^{1}H-NMR data (Table IV) with those of renieramycin A (14) indicated that renieramycin B (15) had the same carbon skeleton as 14. Furthermore, the ^{1}H-NMR spectrum showed signals at δ 1.19 (t, 3H, $J = 7$ Hz) and 3.76 (q, 2H, $J = 7$ Hz) due to an ethoxyl group and lacked a signal due to a hydroxyl group. The structure of renieramycin B (15) was deduced as the ethyl ether of renieramycin A (14). Renieramycin B (15) was probably an artifact resulting from solvent exchange during storage of the sponge in ethanol.

B. RENIERAMYCINS C AND D

Renieramycin C (16) [C$_{30}$H$_{32}$N$_2$O$_{10}$; [α]$_D$ $-89.2°$ ($c = 0.065$, MeOH)] (12) showed UV absorptions at 266 (ϵ 14900) and 360 (ϵ 2160) nm. Its IR spectrum showed absorption bands at 3200, 1720, 1680, 1660, 1650, and 1620 cm^{-1}. The presence of the amide band at 1650 cm^{-1} suggested that renieramycin C (16) was a tertiary amide related to renieramycin A (14). The site of the tertiary amide was then determined by analysis of the ^{1}H-NMR spectrum (Table IV).

The downfield shift of the H_I signal from δ 3.60 in 14 to 5.48 (br s, 1H) in 16 was consistent with the change from an adjacent tertiary amine to a tertiary amide and was attributed to a change of geometry that caused H_I to be situated further away from the quinone ring E. Additionally, the signal at δ 1.41 (ddd, 1H, J = 17, 12, 2 Hz) was assigned to H_A that was placed in the ring current of the quinone ring E, implying that the stereochemistry of the BC and CD ring junctions in 16 was the same as in renieramycin A (14). Furthermore, the ^{1}H-NMR signals at δ 4.78 and 3.73 due to H_E and H_F, respectively, appeared as a broad singlet indicating that the configuration of the hydroxyl group in 16 was the same as in renieramycin A (14).

Renieramycin D (17) [$C_{32}H_{36}N_2O_{10}$; $[\alpha]_D$ −100.7° (c =0.092, MeOH)] (12) showed UV [λ_{max} 264 (ϵ 16100) and 370 (ϵ 1450) nm], IR [1720, 1680, 1665, 1645, and 1620 cm^{-1}] and ^{1}H-NMR data (Table IV) essentially identical with those of renieramycin C (16).

Examination of ^{1}H-NMR spectral data led to the conclusion that reniera-mycin D (17) was the ethyl ether of 16.

IV. Naphtyridinomycin-Type Antibiotics

A. NAPHTHYRIDINOMYCIN

1. Structure

A new broad-spectrum antibiotic naphthyridinomycin was obtained in crystalline form from the culture filtrate of *Streptomyces lusitanus* AY B-1026 (9, 23), isolated from a soil sample collected in Easter Island (Rapa Nui).

The ruby-red-colored naphthyridinomycin (18) [$C_{21}H_{27}N_3O_6$; mp 108–110° (dec.) (ethyl ether); $[\alpha]_D^{25}$ +69.4° (c = 1, CHCl$_3$)] (9) was isolated by the general procedure for a lipophilic basic substance, and from 160 liters of fermentation broth 1.6 g of pure naphthyridinomycin crystals was routinely obtained (23).

The UV spectrum showed a characteristic absorption maximum at 270($E_{1\,cm}^{1\%}$ 248.5) nm and the IR spectrum (CHCl$_3$) showed absorption bands at 3000, 2940, 2880, 2845, 1715, 1690, 1650, 1604, and 1495 cm^{-1}. The ^{1}H-NMR spectrum indicated the presence of one *C*-methyl group, one *N*-methyl group, and one *O*-methyl group.

The structure of naphthyridinomycin (18) was completed by X-ray crystallography (24). The crystal belongs to the orthorhombic system, a =11.038 (1), b =19.560 (2), c =9.255 (1) Å and the space group is $P2_12_12_1$.

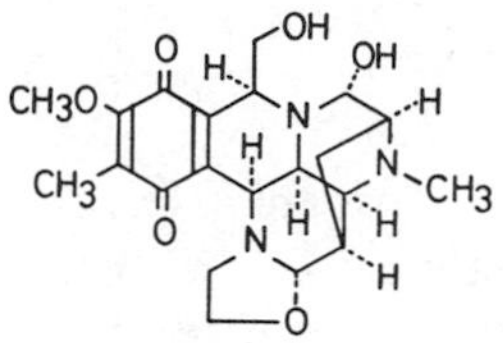

18

A noteworthy feature of the naphthyridinomycin molecule is the electron-rich N (6), N (11), and O (6B2) cavity which can be created by a 180° rotation of the H (6B2) proton around the C (6B1)—O (6B2) bond (Fig. 4).

2. Synthesis

The total synthesis of naphthyridinomycin (**18**) had never been completed until now. However, in 1981 the studies directed toward the total synthesis of the antibiotic were presented at the 28th IUPAC meeting by D. A. Evans' group of The California Institute of Technology (*25*).

B. SF-1739 HP AND NAPHTHOCYANIDINE

A quite unstable antibiotic, SF-1739 exhibited high activity to bacteria and tumor cells and was isolated from the culture filtrate of *Streptomyces griseoplanus* SF-1739 (*26*).

Treatment of the antibiotic SF-1739 with mineral acid followed by addition of cyanide anion afforded two semisynthetic antitumor antibiotics, SF-1739 HP (**19**) and naphthocyanidine (**20**) (*10*).

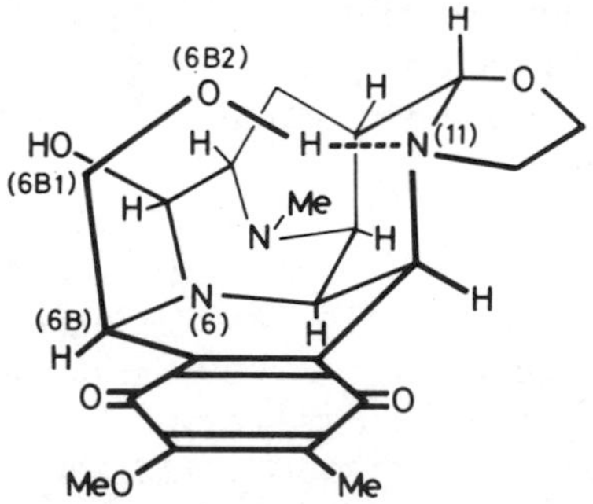

FIG. 4. Structure of naphthyridinomycin (**18**) showing the electron-rich N (6), N (11) and O (6B2) cavity. From Kluepfel *et al.* (*23*).

1. SF-1739 HP

SF-1739 HP (**19**) [$C_{20}H_{25}N_3O_6$; $[\alpha]_D^{25}$ +64.0° ($c=0.5$, MeOH)] was obtained as a purple powder and was soluble in water and methanol but insoluble in chloroform, ethyl acetate, and ethyl ether (*10*). The UV spectrum of **19** in methanol showed the absorption maxima at 275 ($E_{1cm}^{1\%}$ 165) and 330 (shoulder) nm. The IR spectrum (KBr) showed the absorption bands at 3400, 2950, 2900, 1650, 1530, 1390, 1240, 1180, and 1000 cm^{-1}. The ^{1}H-NMR spectrum of **19** in D_2O indicated the presence of one *C*-methyl

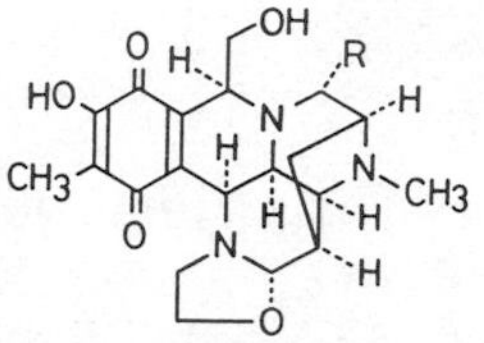

19 R=OH
20 R=CN

group and one *N*-methyl group. The ^{13}C-NMR spectrum in D_2O showed resonances at δ 8.5 (C—CH$_3$), 28.7, 34.2, 40.5 (N—CH$_3$), 48.0, 50.0, 51.3, 53.4, 60.5, 60.6, 62.4, 66.5, 79.3 (N—CH—OH), 93.8, 114.1, 137.9, 146.6, 169.2, and 185.3, 186.8 (quinone CO).

The structure of SF-1739 HP was deduced as **19** by comparison of the ^{13}C-NMR spectra with those of naphthocyanidine (**20**) which was produced by treatment of SF-1739 HP (**19**) with potassium cyanide in methanol (*vide infra*).

2. Naphthocyanidine

The reddish-yellow-colored naphthocyanidine (**20**) [$C_{21}H_{24}N_4O_5$; $[\alpha]_D^{25}$ +76.0° ($c=0.5$, MeOH)] was prepared in 58.8% yield by treating SF-1739 HP (**19**) with potassium cyanide in methanol at room temperature for 30 min (*10*). The UV spectrum in methanol showed the maxima at 273 ($E_{1cm}^{1\%}$ 300) and 395 ($E_{1cm}^{1\%}$ 32) nm. The IR spectrum (KBr) showed absorption bands at 3400, 2970, 2900, 2240, 1670, 1550, 1400, 1240, 1180, and 1050 cm^{-1}. The ^{1}H-NMR spectrum of **20** in D_2O indicated the presence of one *C*-methyl group and one *N*-methyl group. The ^{13}C-NMR spectrum in CDCl$_3$ revealed all 21 carbons as follows: δ 8.1 (C—CH$_3$), 29.2, 35.3, 41.2 (N—CH$_3$), 48.1, 50.1, 53.5, 54.4, 56.6, 59.9, 60.6, 61.8, 62.7, 93.4, 113.8, 117.6 (CN), 136.6, 146.0, 164.3, and 183.9, 186.4 (quinone CO).

These ^{13}C-NMR data of **20** showed a close similarity to that of SF-1739 HP (**19**) except for a lack of signal at δ 79.3 (N—CH—OH) in **19** and for the appearance of a new signal at 117.6 (CN) in **20**. This indicated that naphthocyanidine (**20**) was produced by replacement of the hydroxyl group in **19** by a cyanide ion.

Note that the same chemical transformation has been observed in the conversion of saframycin A (**3**) to saframycin S (**8**) (*vide supra*) (*3*). Furthermore, comparison of these ^{13}C-NMR data with those of saframycin antibiotics (Table II) indicated the presence of a 3-hydroxy-2-methyl-*p*-benzoquinone moiety in **19** and **20**. Finally, the structure of naphthocyanidine (**20**) was established by an X-ray crystallographic analysis (*10*). The crystals of **20** were orthorhombic, space group $P2_12_12_1$, $a = 13.800$, $b = 19.243$, $c = 9.272$ Å, $\alpha = \beta = \gamma = 90°$, and $Z = 4$.

V. Mimosamycin-Type Antibiotics

A. MIMOSAMYCIN

1. Structure

The mimosa-yellow-colored mimosamycin (**21**) [$C_{12}H_{11}NO_4$; mp 227–231° (ethyl acetate – ether)] (*1*) is a neutral substance obtained from the organic solvent extract of the culture filtrate of *Streptomyces lavendulae* No. 314.

The UV spectrum [λ_{max} 317 (logϵ 4.14), 396 (logϵ 3.56) nm] of mimosamycin (**21**) (Fig. 5) was novel, bearing no similarity to the spectra of

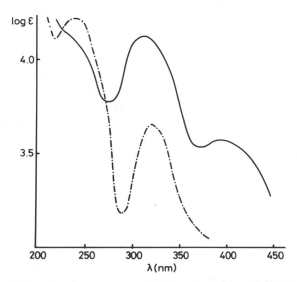

FIG. 5. The UV spectra of mimosamycin (**21**) (———) and mimocin (**32**) (—·——·—).

saframycin antibiotics. The IR spectrum (KBr) of mimosamycin (**21**) showed carbonyl absorptions at 1685, 1655, and 1635 cm^{-1}. The ^{1}H-NMR spectrum (CDCl$_3$) revealed all 11 protons as follows: a C-methyl singlet at δ 2.10, an N-methyl singlet at δ 3.69, an O-methyl singlet at δ 4.20, and two aromatic one-proton singlets at δ 7.12 and 8.28. The structure of mimosamycin was then completely clarified as 2,6-dimethyl-7-methoxy-3,5,8-trioxo-2,3,5,8-tetrahydroisoquinoline (**21**) by an X-ray crystallographic study (*27*). The crystals have strong twinning forms and were hard to separate as single crystals. The space group is $P2_1/c$, with $Z=4$, $a = 10.799$ (4), $b = 14.302$ (4), $c = 6.898$ (2) Å; $\beta = 100.25$ (2)°.

It became clear that mimosamycin (**21**) belonged to an unusual heterocyclic quinone (*28*) and a thus-far unprecedented structural class of antibiotics, and this was finally confirmed by total synthesis (*5*) (*vide infra*).

The ^{1}H- and ^{13}C-NMR, and mass spectroscopic data of mimosamycin (**21**) were studied in detail (*29*). The ^{1}H-NMR spectrum indicated the presence of three methyl groups, showing singlets at δ 2.10, 3.69, and 4.20 due to the C-methyl, N-methyl, and O-methyl groups, respectively, as mentioned earlier. Irradiation of the signal at δ 3.69 caused a 17.6% increase in the integrated intensity of the signal at δ 8.28. The observed nuclear Overhauser effect (NOE) indicated that the signals at δ 3.69 and 8.28 are due to the N-methyl and to the proton at C-1, respectively.

The O-methyl group resonates at lower field (δ 4.20) than the usual range (δ 3.4–4.0) and this is explained in terms of the compression effect of the methyl group. The C-4 proton resonated at δ 7.12 and coupled with the C-1 proton ($J_{H_1,H_4} = 0.5$ Hz).

The signals of the ^{13}C-NMR spectrum were assigned as shown in Table V.

TABLE V
^{13}C-NMR Chemical Shifts (δ) for Mimosamycin (**21**) and 5,8-Dihydro-7-Methoxy-6-methyl-3-morphilino-5,8-dioxoisoquinoline (**22**) in CDCl$_3$

Compound	C_1	C_3	C_4	C_5	C_6	C_7	C_8	C_9	C_{10}	CH$_3$	OCH$_3$
21	142.1	162.8	116.7	183.5	133.6	159.5	177.3	111.3	138.9	9.6	61.3
22	149.3	161.2	100.8	185.3	130.5	158.8	179.1	115.8	139.1	9.2	61.2

The assignment of the carbon chemical shifts was based on the comparison of the ^{13}C-NMR spectrum of a model compound (**22**) (*29*). The signal at δ 133.6 was assigned to the C-6 carbon because it was coupled with the signal of the C-methyl protons at δ 2.10. The signal at δ 162.8 was assigned to the lactam carbonyl carbon at C-3, and δ 183.5 and δ 177.3 were ascribed to the quinone carbonyl carbon at C-5 and C-8, respectively, on the basis that the C-8 carbon resonance would shift upfield under the influence of the methoxyl group attached to the vicinal carbon (*30*).

The mass spectrum showed strong peaks at the following positions: m/z 233 (M$^+$), 218 (M$^+$ − CH$_3$), 205 (M$^+$ − CO), 190 (M$^+$ − CH$_3$ − CO), 177 (M$^+$ − 2 CO), 162 (M$^+$ − CH$_3$ − 2 CO), 149 (M$^+$ − 3 CO), and 134 (M$^+$ − CH$_3$ − 3 CO) (*29*).

The proposed major fragmentation pathway of mimosamycin (**21**) is given in Scheme 4. A sequence of fragmentation is the loss of a methyl radical to produce the peak at m/z 218 (*a*) and the subsequent loss of three carbonyls [m/z 190 (*b*), 162 (*c* or *c'*), 134 (*d*)]. The alternative sequence of fragmentation is the successive loss of three carbonyls (m/z 205, 177, 149) from the molecular ion without the loss of the methyl radical. This fragmentation would initially give ion *e*.

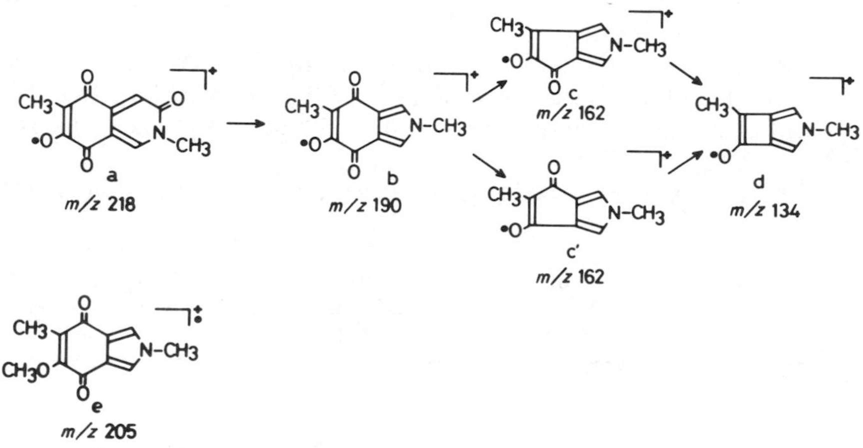

SCHEME 4. *Mass spectral fragmentation path for mimosamycin* (21).

2. Synthesis

The synthesis of mimosamycin (**21**) has been achieved by the Sankyo Laboratories group (Scheme 5) (*5*). The key step in the total synthesis was the introduction of the oxygen and/or potential oxygen functional groups at C-3, -5, -7, and -8 positions of the isoquinoline nucleus.

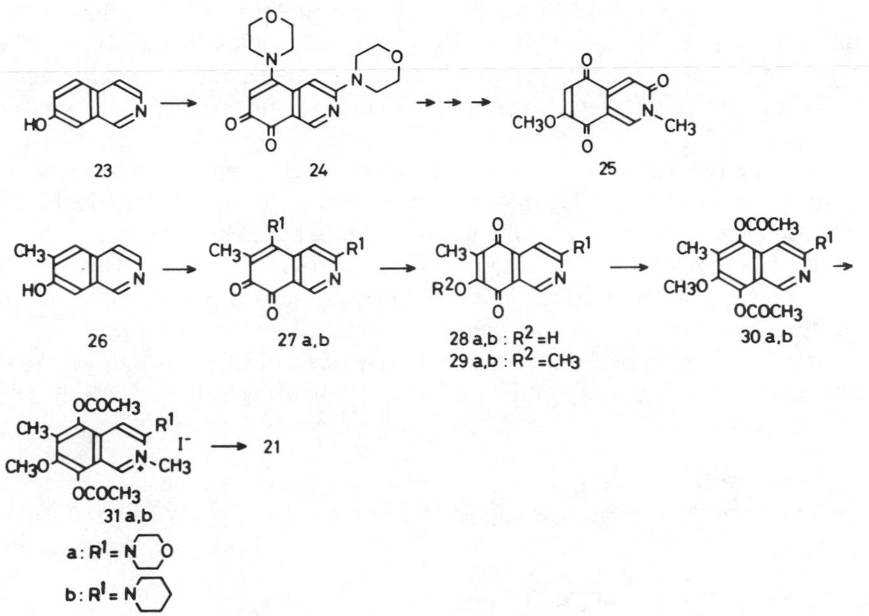

SCHEME 5. *Synthesis of mimosamycin* (**21**) (5).

Some years ago, Tsizin described 7-isoquinolinol (**23**) affording 7,8-dihy-dro-3,5-dimorpholino-7,8-dioxoisoquinoline (**24**) by cupric-catalyzed oxidation with oxygen in the presence of morpholine (*31*). Tsizin's method was applied initially to the synthesis of a model compound 7-methoxy-2-methyl-3,5,8-trioxo-3,5,8-trihydroisoquinoline (**25**) (*32*), and finally to the total synthesis of mimosamycin (**21**). Thus, copper-catalyzed autoxidation of 7-hydroxy-6-methylisoquinoline (**26**), which was readily prepared in 80% overall yield in five steps from 3-methoxy-4-methylbenzaldehyde, in the presence of morpholine and piperidine afforded 7,8-dihydro-3,5-dimorpholino-7,8-dioxo-6-methylisoquinoline (**27a**) and 7,8-dihydro-7,8-dioxo-3,5-dipiperidino-6-methylisoquinoline (**27b**) in 59% and 73% yields, respectively. By aerial oxidation, the oxygen necessary to complete all functional groups for the synthesis were provided in one step at the desired C-3, -5, -7, and -8 positions. Conversion of **27a** and **27b** to the corresponding *p*-quinones (**28a** and **28b**) was effected by warming a methanolic solution of **27a** and **27b** in the presence of a catalytic amount of concentrated sulfuric acid or sodium hydroxide. Reductive acetylation of **29a** and **29b,** which were prepared by treatment of **28a** and **28b** with diazomethane, with zinc–acetic acid and acetic anhydride afforded the diacetate of hydroquinones **30a** and **30b** in good yields.

Alkylation of **30a** and **30b** with methyl iodide in DMF at 100° afforded the quaternary salts **31a** and **31b** in 89% and 80% yields, respectively. The formation of **31** as the sole product indicated N-methylation occurs preferentially at the isoquinoline ring nitrogen and not at the exocyclic morpholine nitrogen (*33*).

Treatment of methiodide (**31a**) with silver oxide in methanol gave rise to natural mimosamycin (**21**) in one step in 49% yield, and alternatively, on basic treatment **31b** provided the natural product in 29% yield. Under these conditions, a reaction of three steps was assumed to proceed via (i) hydrolysis of the acetoxyl groups, (ii) oxidation of the resultant hydroquinone, and (iii) substitution of the amino (morpholine or piperidine) group with a hydroxyl group. The identity of the synthetic material and natural mimosamycin (**21**) was established by mixed mp and the comparison of their IR, UV, NMR, and MS spectra.

B. Mimocin

1. Structure

Mimocin (**32**) [$C_{15}H_{14}N_2O_5$; mp 189–191° (dec.) (ethyl ether)] (*4*) is a yellow crystalline material obtained in a very small amount from the same source as mimosamycin (**21**) (*vide supra*).

The UV spectrum [λ_{max} 242 (logε 4.23), 320 (logε 3.65) nm] of mimocin (**32**) was close to that of mimosamycin (**21**) (Fig. 5) (*vide supra*) and the IR spectrum (KBr) showed an N—H absorption at 3380 cm⁻¹ and carbonyl absorptions at 1720, 1685, and 1665 cm⁻¹.

The FT-¹H-NMR spectrum (CDCl₃) showed the presence of three methyl groups [δ 2.09 (s, aro CH₃), 2.52 (s, COCH₃) and 4.17 (s, aro OCH₃)]; a benzylic methylene group [δ 5.10 (d, *J* = 5 Hz), collapsed to a singlet by D₂O treatment]; C-4 and C-3 isoquinoline protons [δ 7.92 (d, *J* = 5 Hz) and 8.94 (d, *J* = 5 Hz)]; and N—H proton [δ 8.57 (br, s), D₂O exchangeable] (Fig. 6).

High-resolution mass spectrum of mimocin (**32**) showed characteristic peaks at the following positions: *m/z* 302 (M⁺), 259 (M⁺ − COCH₃), 216

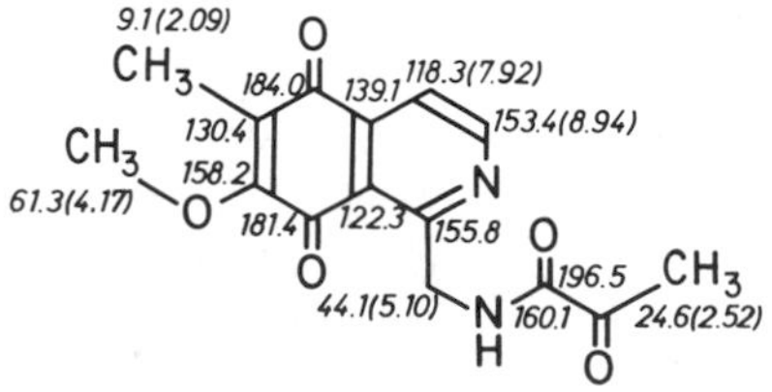

FIG. 6. ¹³C-NMR assignments for mimocin (**32**) (¹H-NMR assignment in parenthesis). From Kubo and Nakahara (*41*).

($M^+ - COCH_3 - CONH$). These spectral data are consistent with the presence of a 5,8-dihydro-5,8-dioxo-7-methoxy-6-methylisoquinoline and of a $-CH_2NHCOCOCH_3$ grouping that is common to all saframycin group antibiotics (*vide supra*). Therefore, the tentative structure **32** has been given to mimocin and conclusive proof of the structure was provided by a total synthesis.

2. Synthesis

The synthesis of mimocin (**32**) has been reported starting from 7-methoxy-6-methyl-5,8-isoquinolinedione (**33**), and was prepared in 58% overall yield from 7-methoxy-6-methylisoquinoline (**5**) in three steps (Scheme 6) (*4*).

Reductive acetylation of **33** gave the hydroquinone diacetate (**34**) in 82% yield, that was then converted in 63% yield to 1-cyano-5,8-hydroquinone diacetate (**35**) in the usual manner (*34*). Catalytic hydrogenation of **35** over 10% Pd–C in methanol containing hydrogen chloride afforded the sensitive 1-aminomethyl-5,8-hydroquinone diacetate (**36**) isolated as its dihydrochloride. Treatment of the dihydrochloride of **36** with pyruvic acid in α,α-dichloromethyl methyl ether (*35*) without solvent (40–50°, 30 min), followed by a basic workup afforded the desired mimocin (**32**) directly although in low yield. The sequence **35** → **36** → **32** was obviously unsatisfactory and a more efficient method for the synthesis of mimocin (**32**) was developed by Kubo and co-workers (*36*).

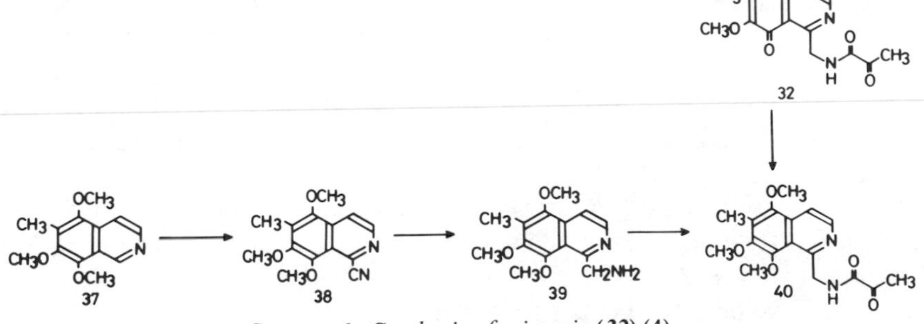

SCHEME 6. *Synthesis of mimocin (32) (4).*

Recently cerium (II) ammonium nitrate (CAN) has been used for oxidative demethylation of dimethyl ethers of 1,4-hydroquinones to give p-qui-
nones (37) and this method was applied to the synthesis of mimocin (32)
(Scheme 6). 5,7,8-Trimethoxy-6-methylisoquinoline (37), which was obtained from 2,6-dimethoxytoluene in 43% overall yield in ten steps, was
converted to the 1-cyano derivative (38) in 95% yield as mentioned above.
Catalytic hydrogenation of 38 over 10% Pd–C in the presence of hydrochloric acid gave rise to 1-aminomethyl-6-methyl-5,7,8-trimethoxyisoquinoline
(39), and the free base of 39 treated successively with pyruvoyl chloride was
freshly prepared from pyruvic acid and α,α-dichloromethyl methyl ether at
50–60° for 10 min, to give 6-methyl-1-pyruvoylaminomethyl-5,7,8-tri-
methylisoquinoline (40) in 28% yield. Oxidative demethylation of 40 with
CAN in the presence of pyridine-2,6-dicarboxylic acid N-oxide afforded an
83.1% yield of mimocin (32) and its methoxyorthoquinone isomer. The
isolated yields of 32 and its orthoquinone isomer were 31.0% and 52.1%
yields, respectively. The synthetic specimen was identical with natural
mimocin (32) by mixed mp and by the comparison of their UV, IR, ¹H
NMR, and mass spectra. The ¹³C-NMR assignments for the synthetic
mimocin (32) (36) are recorded in Fig. 6.

C. RENIERONE

1. Structure

Renierone (41) [$C_{17}H_{17}NO_5$; mp 91.5–92.5°] (11) was isolated from the
ethanolic extracts of an intense blue sponge, *Reniera* species. Renierone (41)
was found to be the major antibacterial metabolite (0.03% dry weight).

The IR spectrum showed an unsaturated ester (1705 cm⁻¹) and a quinone
moiety (1650 cm⁻¹), while the UV spectrum [λ_{max} 214 (ϵ 20000), 312 (ϵ 4000)
nm] indicated the quinone must be further conjugated. The acid moiety of
the ester was determined as angelic acid from the ¹H-NMR spectrum [δ 1.99
(d, 3H, J = 1.5 Hz), 2.04 (dd, 3H, J = 7, 1.5 Hz), and 6.10 (q, 1H, J = 7 Hz)]
and comparison of ¹³C-NMR signals with those of methyl angelate (38). In
addition, the ¹H-NMR spectrum indicated the presence of methyl groups [δ
2.09 (s, aro CH₃) and 4.15 (s, aro OCH₃)]; a benzylic methylene bearing the
ester [δ 5.78 (s)]; β and α protons on an isoquinoline ring [δ 7.87 (d, J = 5 Hz)
and 8.91 (d, J = 5 Hz)]. The signals of the ¹³C-NMR spectrum were assigned
as shown in Fig. 7.

The structural elucidation was completed by single crystal X-ray diffraction analysis (11). Renierone (41) crystallized in the monoclinic crystal and
the space group is $P2_1/c$, with a = 7.574 (2), b = 7.940 (1), c = 25.624 (4) Å; β =
92.06 (5)°. Renierone (41) is the first member of a unique class of sponge
metabolites (39).

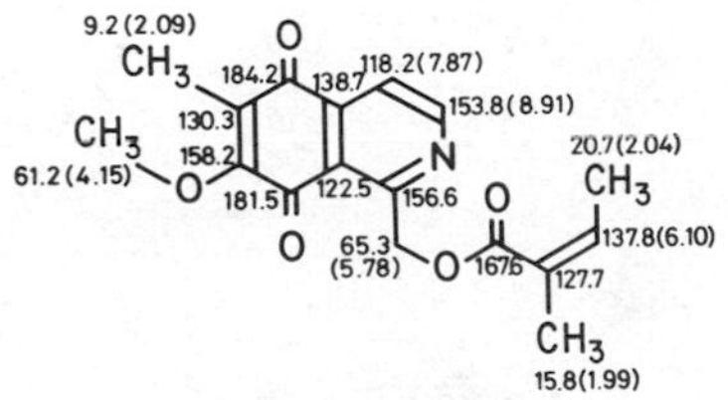

FIG. 7. ¹³C-NMR assignments for renierone (**41**) (¹H-NMR assignments in parentheses).

2. Synthesis

Independent synthesis of renierone (**41**) have been reported by Danishefsky *et al.* (*40*) and by Kubo and Nakahara (*41*).

A key step in the Danishefsky synthesis (Scheme 7) (*40*) was the new synthesis of tetrahydroisoquinoline derivatives using the intramolecular Ben-Ishai reaction (*42*). The nitrile (**42**) obtained in 70–75% overall yield from 2,3,6-trimethoxytoluene in three steps was reduced with borane–THF followed by acylation of the resultant amine with benzylchloroformate giving the carbamate (**43**) in 50% yield.

Reaction of **43** with glyoxylic acid gave the adduct that was then treated directly with dichloroacetic acid (room temperature, 16 hr) to yield the acid. Esterification of the crude acid with diazomethane was accompanied by hydrogenolysis to afford the tetrahydroisoquinoline derivative (**44**) from **43** in 80% overall yield.

Dehydrogenation of **44** with choloranil at 150° gave the isoquinoline (**45**) in 60–65% yield followed by reduction with diisobutyl aluminum hydride affording the carbinol (**46**) in 50–70% yield. Treatment of **46** with angelic acid under the Ziegler acylating condition (DCC–DMAP/ether) (*43*) afforded the angelate ester (**47**) in 55% yield along with the corresponding tiglate ester in 17% yield.

Oxidative demethylation of **47** with argentic oxide (AgO) in 6*N* HNO₃–dioxane (*44*) afforded a 90% yield of renierone (**41**) and its methoxyorthoquinone isomer (**48**). The isolated yields of **41** and **48** were 52% and 38%, respectively (Scheme 7). Treatment of **48** with aqueous sulfuric acid in dioxane–acetone (*45*) followed by reaction with Ag₂O–CH₃I/CHCl₃ (*46*) afforded renierone (**41**) (combined yield starting with **47** was 83%). The synthetic specimen was found to be identical (IR, ¹H NMR, MS, TLC) with natural renierone.

An alternative synthesis of renierone (**41**) has been achieved (*41*) starting from 7-methoxy-6-methylisoquinoline, which was used previously in conjunction with the synthesis of mimosamycin (**21**) and mimocin (**32**) (*vide supra*) (Scheme 8) (*41*). 7-Methoxy-6-methyl-8-nitroisoquinoline (**50**) (*47*)

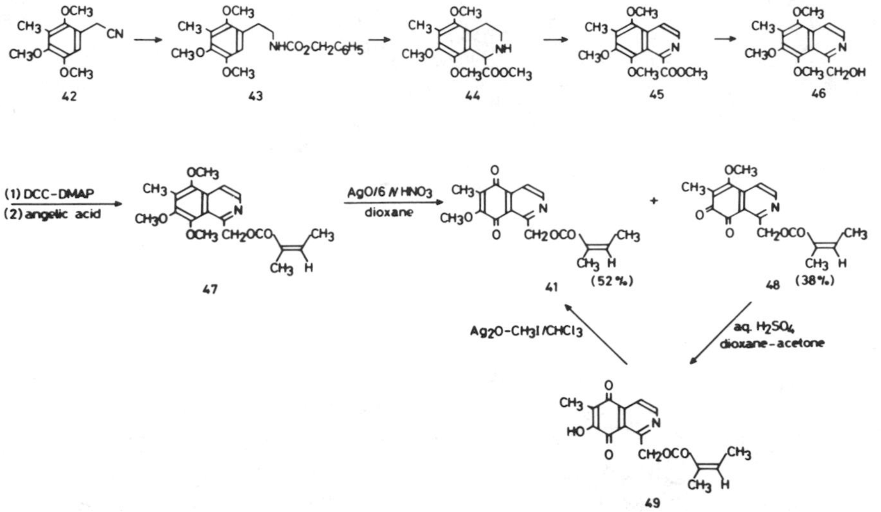

SCHEME 7. *Synthesis of renierone (41)*.

obtained by nitration of 7-methoxy-6-methylisoquinoline was converted to the Reissert compound (51) (*48*) in 40% yield. Treatment of **51** with phenyl lithium in dioxane–ether at −20°, followed by the reaction with gaseous formaldehyde (*49*) gave a 61% yield of the 1-carbinol benzoate (**52**) which upon hydrolysis by 2% NaOH–EtOH afforded the 1-carbinol (**53**) in 90.3% yield. Fremy's salt oxidation (*50*) of the aminocarbinol (**54**), obtained in 92.1% yield by catalytic reduction of **53**, provided 5,8-dioxoisoquinolyl-

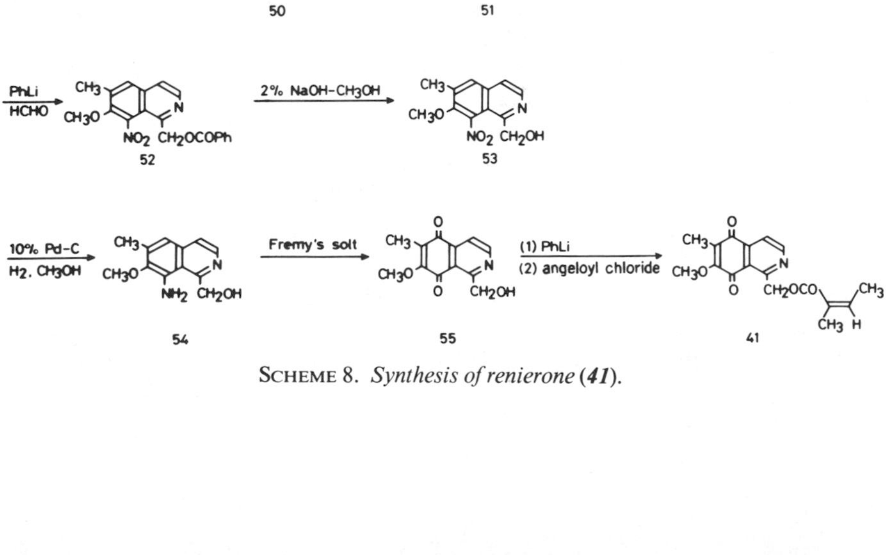

SCHEME 8. *Synthesis of renierone (41)*.

carbinol (55) in 63.7% yield. Treatment of 55 with phenyl lithium in dioxane–ether at −20° followed by the reaction with angeloyl chloride gave renierone (41) in 37%, which was identified with the natural renierone by comparison of their IR, ^{1}H- and ^{13}C-NMR spectra.

D. N-Formyl-1,2-dihydrorenierone, O-Demethylrenierone, and 1,6-Dimethyl-7-methoxy-5,8-dihydroisoquinoline-5,8-dione

Further studies of the metabolites of the sponge *Reniera* sp. have resulted in the isolation of three novel isoquinoline quinone antibiotics (*12*).

1. N-Formyl-1,2-dihydrorenierone

N-Formyl-1,2-dihydrorenierone (56) [$C_{18}H_{19}NO_6$; [α]$_D$ −227.0° (c = 0.023, MeOH)] (*12*) was obtained as a noncrystalline red solid. The UV spectrum showed maxima at 216 (ε 31000), 265 (ε 16500), 340 (ε 6000) and 515 (ε 3500) nm. The IR spectrum showed absorption bands at 1715 (α,β-unsaturated ester) and 1650 (formamide and quinone) cm^{-1}. The mass spectrum of 56 was almost identical to that of renierone (41) with the exception of the presence of a small molecular ion peak at m/z 345. Examination of both the ^{1}H- and ^{13}C-NMR spectra revealed that 56 was a 2:1 mixture of two stereoisomers.

The signals of the ^{1}H-NMR spectrum were assigned to the major stereo-isomer (56a) and the minor rotamer (56b) (Table VI). The ^{1}H-NMR spectrum of the major isomer 56a showed the signals at δ 4.21 (dd, 1H, J = 12, 3 Hz), 4.37 (dd, 1H, J = 12, 4 Hz), and 5.99 (dd, 1H, J = 4, 3 Hz) which were assigned to C-9 methylene protons bearing the angelate ester and to the C-1 methine proton of N-substituted 1,2-dihydroisoquinoline, respectively. The structure of 56 was confirmed by chemical transformation into renierone (41).

Thus, a dichloromethane solution of 56 was treated with 0.01 N aqueous methanolic sodium hydroxide solution in a two-phase reaction to afford natural renierone (41).

2. O-Demethylrenierone and 1,6-Dimethyl-7-methoxy-5,8-dihydroisoquinoline-5,8-dione

O-Demethylrenierone (57) [$C_{16}H_{15}NO_5$; mp 135–136°] was obtained as beige crystals (*12*). The IR spectrum showed absorption bands at 3450 (hydroxyl group), 1715 (α,β-unsaturated ester) and 1660 (quinone) cm^{-1}. The absence of a methoxyl signal in the ^{1}H-NMR spectrum indicated that the 7-methoxyl group in renierone (41) was replaced by a hydroxyl group in 57.

TABLE VI

^{1}H-NMR Spectra of Renierone (**41**) and *N*-Formyl-1,2-dihydrorenierone (**56a** and **56b**)^a

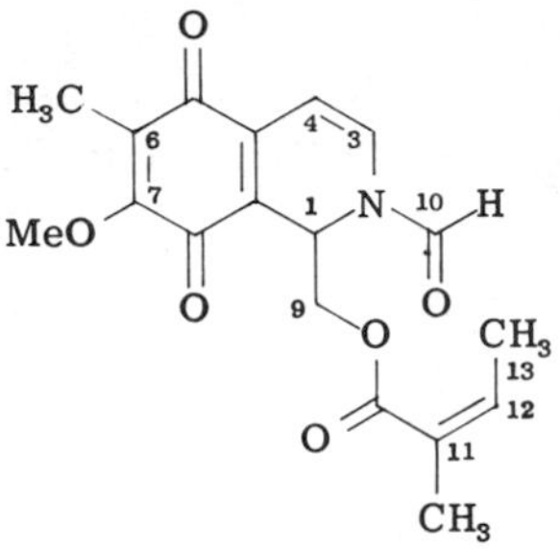

56a

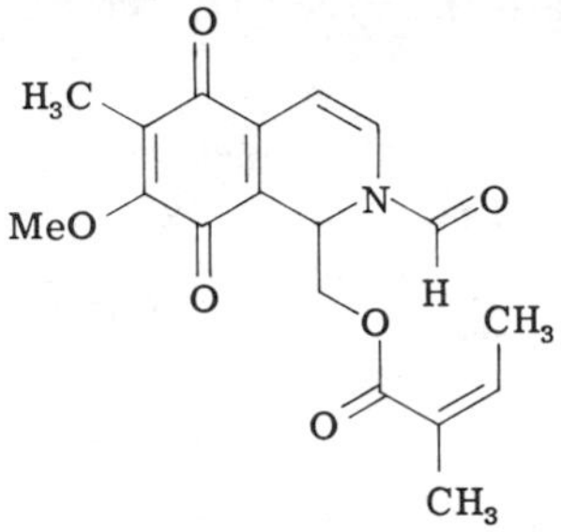

56b

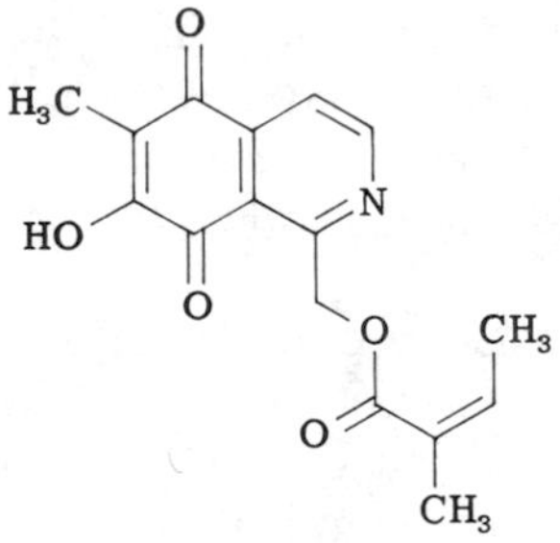

57

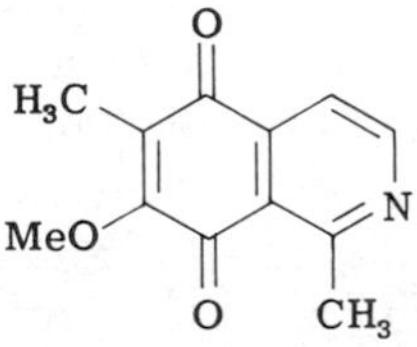

58

H at C no.	41	56a	56b
1		5.99 (dd, 4, 3)	5.37 (dd, 9, 4)
3	8.91 (d, 5)	6.92 (d, 8)	7.45 (d, 8)
4	7.87 (d, 5)	6.03 (d, 8)	6.25 (d, 8)
6-Me	2.09 (s)	1.95 (s)	1.98 (s)
7-OMe	4.15 (s)	4.05 (s)	4.07 (s)
9	5.78 (s)	4.37 (dd, 12, 4)	4.24 (dd, 12, 9)
9		4.21 (dd, 12, 3)	3.91 (dd, 12, 4)
10		8.43 (s)	8.22 (s)
11	1.99 (bs)	1.77 (br s)	1.87 (br s)
12	6.10 (q, 7)	6.06 (q, 7)	6.15 (q, 7)
13	2.04 (d, 7)	1.91 (d, 7)	2.00 (d, 7)

^{a} Multiplicities and coupling constants in parentheses.

The structure of **57** ws established by converting it to the natural renierone **(41)** by treatment of **57** with excess ethereal diazomethane.

1,6-Dimethyl-7-methoxy-5,8-dihydroisoquinoline-5,8-dione **(58)** [C_{12}-$H_{11}NO_3$; mp 188–190° (dec.)] was also isolated together with *O*-demethylrenierone **(57)** (*12*). The IR spectrum exhibited absorption band at 1675 (quinone) cm^{-1}. Comparison of the ¹H-NMR spectrum of **58** with that of renierone **(41)** suggested that the angelate ester side chain and the methylene group at C-1 had been replaced by a methyl group [δ 2.94 (s, 3H)]. Furthermore, comparison of ¹³C-NMR spectral data with that of renierone **(41)** confirmed the structural assignment.

VI. Biological Activity

A. BIOLOGICAL ACTIVITY OF SAFRAMYCINS WITH SPECIAL REFERENCE TO ACTION MECHANISM

Among saframycin antibiotics, saframycin A **(3)** proved to be most biologically active. Saframycin A **(3)** was found to inhibit the growth of a number of gram-positive bacteria, the most sensitive being *Corynebacterium diphtheriae*. The minimal inhibitory concentrations for *Staphylococcus aureus* and *C. diphtheriae* were 0.1 μg/ml and less than 0.003 μg/ml, respectively. The antibiotic was also active on some gram-negative bacteria such as *Escherichia coli, Shigella dysenteriae, Klebsiella pneumoniae*, and *Brucella abortus*, although much less potent than on gram-positive bacteria. Saframycins B **(1)**, C **(2)**, D, and E were 100–1000 times less active than saframycin A **(3)**; they were inactive on fungi. Saframycin S **(8)**, or decyanosaframycin A, was more biologically active, which means to possess higher antibacterial, antitumor activities, and also higher toxicity (*51*).

The acute toxicity of saframycin A **(3)** was determined with mice. The LD$_{50}$s of saframycin A **(3)** for *ddY* mice by single injection were 4.9 mg/kg intraperitoneally, and 3.3 mg/kg intravenously. The toxicity was less with C3H/He mice, and the LD$_{50}$s were 10.5 mg/kg intraperitoneally and 9.7 mg/kg intraveneously. No sign of toxicity of saframycin C **(2)** was observed in *ddY* mice up to 15 mg/kg by intraperitoneal administration.

Saframycin A **(3)** showed marked inhibition of Ehrlich ascites carcinoma at the dose of 1 mg/kg/day intraperitoneally from days 1 to 4 or from days 1 to 6, and the percentages of 60-day survivors were 60% and 70%, respectively. A significant prolongation of mean survival time was also observed with mice bearing P388 leukemia and receiving saframycin A **(3)** intraperitoneal treatment at doses of 0.75–1.0 mg/kg for 10 consecutive days, and the maximum increase of life span obtained was 119% (*52*). The results were similar with mouse lymphoma L5178Y (54). The antibiotic proved to be less effective on leukemia L1210 and showed a marginal effect on B16 melanoma. Ninety percent inhibition of Sarcoma-180 solid tumor which was comparable to the effect of mitomycin C was obtained with 1.25 mg/kg/day intraperitoneal administration of saframycin A (3) for 7 days (53). Human breast tumor xenograft MX-1 in nude mice also regressed and treated over control (T/C) value of 46% was obtained by intraperitoneal administration of 1.5 mg/kg/day of saframycin A (3) for 5 consecutive days (54).

One of the characteristic features of saframycin A (3) is its low toxicity to immunologically competent organs and cells. When mice bearing Ehrlich ascites tumor were treated and cured with saframycin A (3), the mice rejected the rechallenge of the same tumor. One of the reasons for this establishment of tumor immunity by saframycin A (3) treatment might be ascribed to its low toxicity to the bone marrow. When the distribution of [^{14}C]tyrosine-labeled saframycin A (3) was measured in tissues of tumor-bearing mice after intraveneous or intraperitoneal administration, high initial concentrations were observed in the small intestine, kidney, and liver. Concentrations in the liver and kidney fell off rapidly followed by a steep rise in the small intestine. Moderate initial concentrations were obtained in the lung and spleen, while radioactivity was not detected in the brain and such immune system organs as the thymus, lymph nodes, and bone marrow (53).

The action mechanism of saframycin A (3) is quite intriguing with reference to its structure and is described in more detail below (55).

With L1210 leukemic cells, RNA synthesis was most sensitive to the drug and was significantly inhibited at a concentration of 0.2 μg/ml. At concentrations higher than 2 μg/ml synthesis of DNA was also affected, accompanied with the complete inhibition of RNA synthesis; protein synthesis was least sensistive. With *Bacillus subtilis,* the relation between DNA and RNA synthesis as to sensitivity against saframycin A (3) was reversed, i.e., the synthesis of DNA was more sensitive than that of RNA. The reason for this discrepancy has not been elucidated.

In order to examine whether saframycin A (3) inhibits RNA synthesis in a specific manner as exemplified by actinomycin D, RNA extracted from nucleolar (pre-rRNA) or nucleoplasmic (hnRNA) fraction of L1210 leukemic cells, which were treated with saframycin A (3) followed by pulse labeling with [^{3}H]uridine, was analyzed by polyacrylamide–agarose gel electrophoresis. As shown in Fig. 8, pulse-labeled RNA from nucleolar fraction of untreated cells appeared as a sharp predominant peak at 45 S with a shoulder at 32 S, while that from nucleoplasm showed a broad distribution of radioactivity ranging from 18 to 90 S with a peak at 55 S. Two significant features of the action of the antibiotic were that it did not show any selectivity in the inhibition of nucleolar and nucleoplasmic RNA syntheses and newly

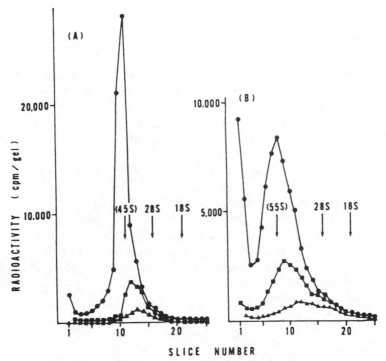

FIG. 8. Electrophoretic profile of nucleolar and nucleoplasmic RNA from saframycin A treated or untreated L 1210 cells; 3×10^7 cells in a total volume of 20 ml were incubated in the presence or absence of saframycin A for 20 min and labeled with [³H]uridine for 10 min. RNA extracted from nucleolar (A) or nucleoplasmic (B) fraction was separated by electrophoresis on a 2% polyacrylamide – 0.5% agarose gel. Radioactivity in each slice of gel was then determined as described in text. The positions of 18 S and 28 S rRNA are shown by arrows. Also presented are values calculated as described in text. Saframycin A: 0 μg/ml (●), 0.25 μg/ml (■), or 0.50 μg/ml (△).

synthesized RNA products obtained from the antibiotic-treated cells exhibited shorter chain length as doses were increased. On the other hand, the antibiotic failed to affect RNA synthesis in the *in vitro* transcription system employing calf thymus DNA, *Escherichia coli* RNA polymerase, and all other prerequisites for transcription. This problem was solved by the fact that reducing agents such as dithiothreitol and sodium borohydride were required for the action of saframycin A (**3**) *in vitro*.

In an attempt to elucidate whether the target molecule for saframycin A (**3**) was DNA template or RNA polymerase, DNA or RNA polymerase was preincubated with the antibiotic in the presence of dithiothreitol and, thereafter, the polymerization reaction was initiated. The presence of DNA template in the preincubation mixture caused a greater inhibition than the presence of RNA polymerase. The primary site of saframycin A (**3**) action was thus determined as the retardation of DNA template activity.

Figure 9 summarizes the effect of saframycin A (**3**) on various DNA including artificial DNAs. Saframycin A (**3**) inhibited the template activity of calf thymus DNA by 10–90% in a concentration range of 0.017–50 μg/ml, whereas the inhibition was greatly diminished toward denatured DNA. This finding suggests that saframycin A (**3**) has a higher affinity toward DNA in native double-helical configulation. DNA isolated from *B. subtilis* also showed a behavior similar to native calf thymus DNA. As to base specificity, saframycin A (**3**) showed significant inhibition with poly (dG) · poly (dC), while the template activity of poly (dA) · poly (dT) or poly (dI) · poly (dC) was not affected at all. These results strongly suggested that the base to which saframycin A (**3**) binds was guanine. However, saframycin A (**3**) did not bind to a single-stranded polymer such as poly (dG) or poly (G) (*56*). Binding of the antibiotic to DNA studied further by using two kinds of labeled antibiotics (*56*). As stated in the section on biosynthesis (Section II, F), the isoquinolinequinone skeleton was assumed to be synthesized from tyrosine

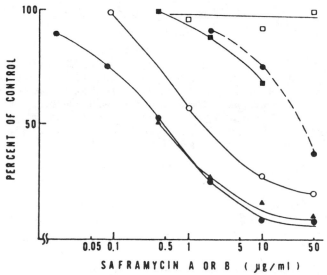

FIG. 9. Dose dependency of saframycin A or B for the inhibition of various DNA templates. Native (●—●) or denatured (■) calf thymus DNA, poly (dG) · poly (dC) (○), poly (dA) · poly (dT) (□), and *B. subtilis* DNA (▲) were preincubated with varying doses of saframycin A in the presence of 2 mm DTT and 50 mm Tris-HCl (pH 7.9). The result obtained from the experiment using saframycin B and native calf thymus DNA was also included (●---●). Preincubation was carried out for 30 min except for 1 hr in the case of *B. subtilis* DNA. The values obtained from the experiment without the drug served as control (100%).

and the cyano residue was directly incorporated in the molecule. These facts enabled us to prepare two kinds of labeled saframycin A (3), [14C]tyrosine or 14CN-labeled saframycin A (3), which greatly helped to elucidate the reaction of the antibiotic with DNA because the cyano residue uniquely found in saframycin A (3) seemed to play an important role in the biological function.

[14C]Tyrosine-labeled saframycin A (3) was incubated with calf thymus DNA in the presence of dithiothreitol, followed by precipitation of the DNA with cold ethanol. Figure 10 shows that radioactivity retained in the DNA fraction progressively increased as a function of the incubation time, offering good evidence that the drug has a reactivity toward DNA. The sum of radioactivity recovered from precipitate and from supernatant accounted for the total radioactivity input. Therefore, for all practical purposes, the radioactivity retained in DNA was determined by counting the amount of radioactivity remaining in the supernatant. No reaction was evoked by omitting dithiothreitol from the reaction mixture. Thus, the reduction of the drug was an obligatory step for the interaction.

On the other hand, when 14CN-labeled saframycin A (3) replaced [14C]tyrosine-labeled saframycin A (3) under conditions allowing the drug to bind to DNA, none of the radioactivity was associated with DNA, indicating that the cyano residue was lost in the final drug–DNA complex. Nevertheless, the residual radioactivity detected in the supernatant solution did not

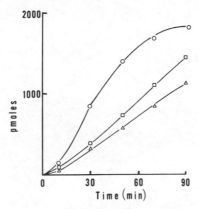

FIG. 10. Absence of cyano residue in saframycin A–DNA complex. [14C]Tyrosine- or 14CN-labeled saframycin A at a concentration of 9 μM was incubated with 400 μg/ml of calf thymus DNA, 2 mM dithiothreitol, 100 mM KCl, and 50 mM sodium acetate, pH 5.6 at 37°C. At the times indicated, DNA was precipitated by the addition of 2 volumes of cold ethanol and the radioactivity found in the supernatant and in precipitate was determined. The radioactivity of 14CN-labeled saframycin A was not recovered from the DNA fraction and, therefore, was not shown. O——O, [14C]Tyrosine-labeled saframycin A bound to DNA; □——□, radioactivity of 14CN lost from the complete reaction mixture; △——△, radioactivity of 14CN lost from the reaction mixture from which DNA was omitted.

account for the total radioactivity input. This conflicting result was a reflection of the phenomenon that the cyano residue was converted to a volatile form, presumably to HCN. The liberation of ^{14}CN from the incubation mixture occurred even when ^{14}CN-labeled saframycin A (3) was reduced by dithiothreitol in the absence of DNA. Therefore, it is reasonable to assume that the reduction triggers the release of the cyano residue from saframycin A (3). The resultant iminium cation (9) or α-carbinolamine (8) is assumed to be the actual species involved in the interaction with DNA. This was confirmed by the fact that dithiothreitol-inducible binding of saframycin A (3) to DNA was blocked by excess cyanide into the reaction mixture regardless of the time of addition. The role of the quinone skeleton that was also reduced by dithiothreitol was assessed by studying the reactivity of [^{14}C]tyrosine-labeled decyanosaframycin A toward DNA in the presence or absence of the reducing agent. The reactivity was made equivalent to that of the reduced form of saframycin A (3) only in the presence of dithiothreitol (Fig. 11), and the hydroquinone form proved to play also an important role in enhancing the drug–DNA reactivity. The fact that saframycin A (3) is unreactive with poly (dI) · poly (dC) while it binds to phage T4 DNA of which O–6 guanines

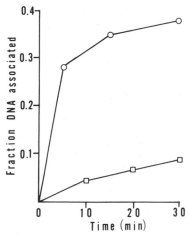

FIG. 11. Reactivity of decyanosaframycin A toward DNA in the presence or absence of dithiothreitol. Fermentation, biosynthetic labeling of saframycins with [^{14}C]tyrosine, and thin layer chromatography were the same as in the legend of Fig. 3. The fraction having an R_F value below 0.2 in Fig. 3 was used for the source of decyanosaframycin A. Although this fraction covers saframycin B, C, and abundant other UV-absorbing materials, the disappearance of radioactivity from this fraction after the addition of NaCN indicates the majority of the radioactivity present in this fraction was derived from decyanosaframycin A. An aliquot of crude preparation of [^{14}C]tyrosine-labeled decyanosaframycin A (2400 cpm) was incubated with 400 µg/ml of calf thymus DNA, 100 mM KCl, and 50 mM sodium acetate, pH 5.6, in the absence (□——□) or presence (○——○) of 2 mM dithiothreitol.

are glycosylated (*57*), provided compelling evidence that the antibiotic fits into the minor groove of DNA (*58*).

The cyano residue of the antibiotic is released on reduction and the resultant α-carbinolamine (**8**) might function as the reactive group on saframycin A (**3**). Guanine is apparently the reactive base on DNA and a possible position for alkylation was assumed to be the N-2 position. Saframycin A (**3**) would be held also to DNA by a secondary strong hydrogen-binding between the phenolic group at C-8 or C-15 of the antibiotic. The structure proposed for the saframycin A – DNA adduct is shown in Fig. 12.

B. BIOLOGICAL ACTIVITY OF RENIERAMYCIN-TYPE ANTIBIOTICS

Antimicrobial activity of renieramycins (**14 – 17**) were only determined by the paper disk method (*12*).

The antibiotics were moderately active against gram-positive bacteria, such as *B. subtilis, S. aureus* and slightly active against gram-negative bacteria.

C. BIOLOGICAL ACTIVITY OF NAPHTHYRIDINOMYCIN-TYPE ANTIBIOTICS

Naphthyridinomycin (**18**) was active against a variety of microorganisms including gram-positive and -negative bacteria. However, gram-positive bacteria were much more sensitive. The antibiotic did not exhibit any significant activity against pathogenic yeast and the dermatophytes. The intraperitoneal injection of an aqueous solution of 3.125 mg/kg of naphthyridinomycin (**18**) killed mice in 24 – 48 hr (*23*). The primary effect of naphthyridinomycin (**18**) on *E. coli* was assumed to be on the synthesis of

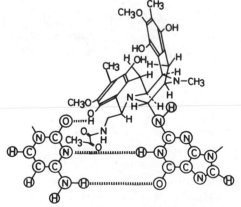

FIG. 12. Proposed structure of saframycin – DNA adduct.

DNA which was strongly inhibited within a few minutes of exposure to low concentrations of the antibiotic (59).

Naphthocyanidine (20) is reported to inhibit the growth of *Staphylococcus aureus, Bacillus subtilis, Escherichia coli,* and *Pseudomonas aeruginosa* at concentrations of 0.2, 25.0, 6.25, and 6.25, respectively (10). Intravenous LD_{50} of naphtyocyanidine (20) as determined with mice was 24.3 mg/kg. When administered intraperitoneally, 8 mg/kg/day of naphthocyanidine (20) for 3 consecutive days prolonged the life span of mice bearing P388 leukemia up to 183.8%. SF-1739 HP (19) was found to be more toxic than naphthocyanidine (20) and its LD_{50} was determined to be 2.0 mg/kg intravenously (10). On the other hand, SF-1739 HP (19) proved to have almost the same antimicrobial activity as that of naphthocyanidine (20). The antibiotic was more active on P388 leukemia by the same treatment regimen as that of naphthocyanidine (20) and an increase of life span of over 368.6% was obtained (10).

D. BIOLOGICAL ACTIVITY OF MIMOSAMYCIN-TYPE ANTIBIOTICS

Mimosamycin (21) showed weak antibacterial activity against some gram-positive bacteria. Mycobacteria were found most sensitive to this antibiotic (1a). *Mycobacterium tuberculosis* H37Rv and a *Mycobacterium tuberculosis* streptomycin–resistant strain were inhibited at concentrations of 0.78 μg/ml and 1.56 μg/ml in Sauton's synthetic medium, respectively, while *Mycobact. intercellulare* was more resistant being inhibited from 6.25 to 12.5 μg/ml. Synthetic isoquinolinequinone inhibited the growth of *Bacillus subtiltis, Staphylococcus aureus,* and *Trichophyton mentagrophytes* at concentrations from 10 to 25 μg/ml, while mimocin (32) and renierone (41) are less biologically active.

N-Formyl-1,2-dihydrorenierone (56), *O*-demethylrenierone (57), and 1,6-dimethyl-7-methoxy-5,8-dihydroisoquinone-5,8-dione (50) were also moderately active against gram-positive bacteria (12).

Addendum

1. Quite recently, the first stereocontrolled total synthesis of (±)-saframycin B (1) has been achieved by Fukuyama and Sachleben (60). The benzobicyclo[3.3.1]ring system 60 was constructed from a diastereomeric mixture of amino acid derivatives 59 via double cyclization in three steps[(1) O_3, 50% MeOH—CH_2Cl_2, −78°, Me_2S workup; (2) DBU, CH_2Cl_2, 0°; (3) HCOOH, 60°] and obtained in 74% yield. Compound 60 was converted into the desired pentacyclic compound

61 in a six-step sequence[(1) H_2, Raney Ni-W2, EtOH, 100°, 1000 psi; (2) H_2, 37% HCHO−H_2O, Raney Ni-W2, EtOH, room temperature, 1000 psi; (3) AlH_3, THF, room temperature; (4) $CbzNHCH_2CHO$, CH_3CN, 70°; (5) H_2, 10% Pd-C, AcOH, room temperature, 1 atm; (6) $MeCOCOCl$, $PhNMe_2$, CH_2Cl_2, room temperature] in 41% overall yield. Oxidation of the phenol **61** with ceric ammonium nitrate afforded (±)-saframycin B (**1**) in 37% yield.

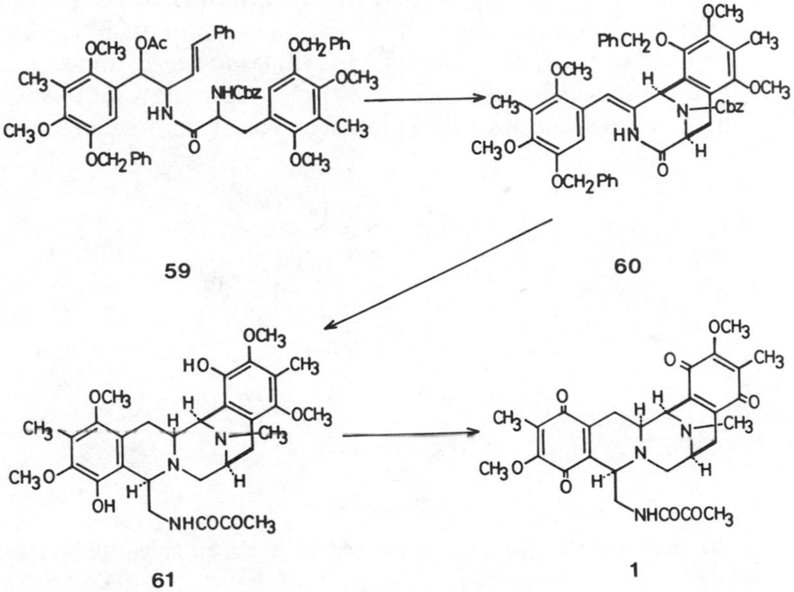

59

60

61

1

2. Recently the structures of novel saframycin-type antibiotics, namely saframycin D (**62**) (Table I) (*vide supra*), saframycin F (**63**) (*61*), safracin A (**64**) and safracin B (**65**) (*62*) have been elucidated by the spectroscopic data. Saframycin F (**63**) [$C_{29}H_{30}N_4O_9$] was isolated from

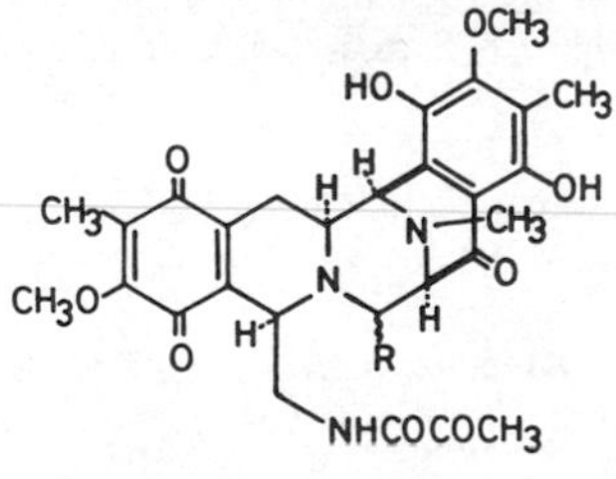

62   R= H

63   R= CN

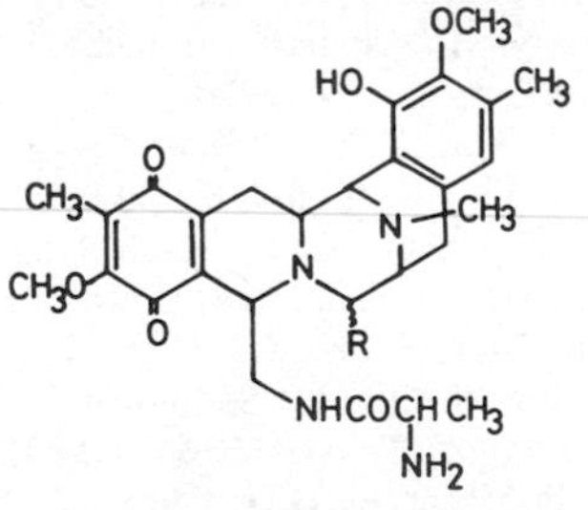

64   R= H

65   R= OH

the fermentation broth of *Streptomyces lavendulae* No. 314 and, safracin A (**64**) [$C_{28}H_{36}N_4O_6$] and safracin B (**65**) [$C_{28}H_{36}N_4O_7$] were isolated as hydrochloride from the culture filtrate of *Pseudomonas fluorescens* A_2-2. Safracin B (**65**) proved to be more biologically active and to have higher antitumor activity than safracin A (**64**).

3. A novel naphthyridinomycin-type antibiotic, cyanocycline A (**66**) [$C_{22}H_{26}N_4O_5$] (*63*) was isolated from the fermentation broth of *Streptomyces flavogriseus* No. 49 and its structure was elucidated as cyanonaphthyridinomycin by an X-ray crystallographic analysis, of material derived from naphthyridinomycin (**18**) (*vide supra*) and sodium cyanide in phosphate buffer pH 7.9 (*64*). Cyanocycline A (**66**) was found to have broad spectrum antimicrobial and antitumor activity (*63, 64*).

66

REFERENCES

1. T. Arai, K. Yazawa, Y. Mikami, A. Kubo, and K. Takahashi, *J. Antibiot.* (Tokyo) **29**, 398 (1976).
1a. Y. Mikami, K. Yokoyama, A. Omi, and T. Arai, *J. Antibiot.* (Tokyo) **29**, 408 (1976).
2. T. Arai, K. Takahashi, and A. Kubo, *J. Antibiot.* (Tokyo) **30**, 1015 (1977).
3. T. Arai, K. Takahashi, K. Ishiguro, and K. Yazawa, *J. Antibiot.* (Tokyo) **33**, 951 (1980).
4. A. Kubo, S. Nakahara, R. Iwata, K. Takahashi, and T. Arai, *Tetrahedron Lett.* 3207 (1980).
5. H. Fukumi, H. Kurihara, H. Hata, C. Tamura, H. Mishima, A. Kubo, and T. Arai, *Tetrahedron Lett.* 3825 (1977).
5a. H. Fukumi, H. Kurihara, and H. Mishima, *Chem. Pharm. Bull.* **26**, 2175 (1978).
6. T. Arai, K. Takahashi, A. Kubo, S. Nakahara, S. Sato, K. Aiba, and C. Tamura, *Tetrahedron Lett.* 2355 (1979).
7. R. W. Franck, *Fort. Chem. Org. Nat.* **38**, 1 (1979).
8. K. V. Rao, K. Biemann, and R. B. Woodward, *J. Am. Chem. Soc.* **85**, 2532 (1963).
8a. S. M. Weinreb, F. Z. Basha, S. Hibino, N. A. Khatri, D. Kim, W. E. Pye, and T.-T. Wu, *J. Am. Chem. Soc.* **104**, 536 (1982).
9. J. Sygusch, F. Brisse, S. Hanessian, and D. Kluepfel, *Tetrahedron Lett.* 4021 (1974).
10. J. Itoh, S. Omoto, S. Inouye, Y. Kodama, T. Hisamatsu, T. Niida, and Y. Ogawa, *J. Antibiot.* (Tokyo) **35**, 642 (1982).
11. D. E. McIntyre, D. J. Faulkner, D. Van Engen, and J. Clardy, *Tetrahedron Lett.* 4163 (1979).
12. J. M. Frincke and D. J. Faulkner, *J. Am. Chem. Soc.* **104**, 265 (1982).

13. L. F. Johnson and W. C. Jankowski, "Carbon-13 NMR Spectra," Wiley (Interscience), New York, 1972, No. 436.
14. T. Arai, K. Takahashi, S. Nakahara, and A. Kubo, *Experientia* **36**, 1025 (1980).
15. L. J. Bellamy, "The Infra-red Spectra of Complex Molecules," p. 264. Methuen, London, 1957.
16. J. Epstein, *Anal. Chem.* **19**, 272 (1947).
17. G. C. Levy and G. L. Nelson, "Carbon-13 Nuclear Magnetic Resonance for Organic Chemists," p. 48. Wiley (Interscience), New York, 1972.
18. B. Danieli, G. Palmisano, B. Gabetta, and E. M. Martinelli, *J. Chem. Soc., Perkin Trans. 1,* 601 (1980).
19. J. W. Lown, A. V. Joshua, and H. -H. Chen, *Can. J. Chem.* **59**, 2945 (1981).
20. A. Chatterjee, M. Chakrabarty, A. K. Ghosh, E. W. Hagaman, and E. Wenkert, *Tetrahedron Lett.* 3879 (1978).
20a. B. Danieli, G. Palmisano, and G. S. Ricca, *Tetrahedron Lett.* 4007 (1981).
21. H. Kurihara and H. Mishima, *Heterocycles* **17**, 191 (1982).
22. S. A. Waksman, H. B. Woodruff, *Proc. Soc. Exptl. Biol. Med.* **49**, 207 (1942).
23. D. Kluepfel, H. A. Baker, G. Piattoni, S. N. Sehgal, A. Sidorowicz, K. Singh, and C. Vézina, *J. Antibiot.* (Tokyo) **28**, 497 (1975).
24. J. Sygusch, F. Brisse, and S. Hanessian, *Acta Cryst.* B32, 1139 (1976).
25. Dr. D. A. Evans presented at the 28th International Union of Pure and Applied Chemistry meeting at Vancouver, Canada, August 17–21, 1981, program number OR 91.
26. H. Watanabe, T. Shomura, Y. Ogawa, Y. Kondo, K. Ohba, J. Yoshida, C. Moriyama, T. Tsuruoka, M. Kojima, S. Inouye, and T. Niida, *Sci. Reports Meiji Seika Kaisha* **16**, 20 (1976).
27. T. Hata, H. Fukumi, S. Sato, K. Aiba, and C. Tamura, *Acta Cryst. Sect.* B. **34**, 2899 (1978).
28. I. Baxter and B. A. Davis, *Q. Rev. Chem. Soc.* **25**, 239 (1971).
29. H. Fukumi, F. Maruyama, K. Yoshida, M. Arai, A. Kubo, and T. Arai, *J. Antibiot.* (Tokyo) **31**, 847 (1978).
30. G. Höfle, *Tetrahedron* **32**, 1431 (1976).
31. Yu. S. Tsizin, *Khim. Geterotsikl. Soedin.* (USSR) 1253 (1974).
31a. Yu. S. Tsizin and B. V. Lopatin, *Khim. Geterotsikl. Soedin.* (USSR) 500 (1977).
32. H. Fukumi, H. Kurihara, and H. Mishima, *J. Heterocycl. Chem.* **15**, 569 (1978).
33. R. Frampton, C. D. Johnson, and A. R. Katritzky, *Ann. Chem.* **749**, 12 (1971).
33a. D. L. Garmaise and G. Y. Paris, *Chem. Ind.* (*London*), 1645 (1967).
34. E. Ochiai and Z. Sai, *Yakugaku Zasshi* **65B**, 418 (1945).
35. H. C. J. Ottenheijm and J. H. M. de Man, *Synthesis,* 163 (1975).
35a. L. Heslinga and J. F. Arens, *Recl. Trav. Chim. Pays-Bas* **76**, 982 (1957).
36. A. Kubo, Y. Kitahara, S. Nakahara, and R. Numata, *Chem. Pharm. Bull.* **31**, 341 (1983).
37. P. Jacob, P. S. Callery, A. T. Shulgin, and N. Catagnoli, Jr., *J. Org. Chem.* **41**, 3627 (1976).
37a. L. Syper, K. Kloc, J. Młochowski, and Z. Szulc, *Synthesis* 521 (1979).
37b. L. Syper, K. Kloc, and J. Młochowski, *Tetrahedron* **36**, 123 (1980).
38. M. Brouwer and J. B. Stothers, *Can. J. Chem.* **50**, 601 (1972).
39. L. Minale, G. Cimino, S. de Stefano, and G. Sodano, *Fort. Chem. Org. Nat.* **33**, 1 (1976).
39a. D. J. Faulkner, *Tetrahedron* **33**, 1421 (1977).
40. S. Danishefsky, E. Berman, R. Cvetovich, and J. Minamikawa, *Tetrahedron Lett.* 4819 (1980).
41. A. Kubo and S. Nakahara, *Chem. Pharm. Bull.* **29**, 595 (1981).
42. D. Ben-Ishai, J. Altman, and N. Peled, *Tetrahedron* **33**, 2715 (1977).
43. F. E. Ziegler and G. D. Berger, *Synth. Commun.* **9**, 539 (1979).
44. C. D. Snyder and H. Rapoport, *J. Am. Chem. Soc.* **94**, 227 (1972).

45. C. A. Weber-Schilling and H. -W. Wanzlick, *Chem. Ber.* **104**, 1518 (1971).
46. W. B. Manning, T. P. Kelly, and G. M. Muschik, *Tetrahedron Lett.* 2629 (1980).
47. A. Kubo, N. Saito, S. Nakahara, and R. Iwata, *Angew. Chem. int. Ed. Engl.* **21**, 857 (1982).
48. B. C. Uff, J. R. Kershaw, and J. L. Neumeyer, *in* "Organic Syntheses," (G. H. Büchi, ed.) Vol. 56, p. 19. Wiley, New York, 1977.
49. H. W. Gibson, F. D. Popp, and A. Catala, *J. Heterocycl. Chem.* **1**, 251 (1964).
50. H. -J. Teuber and M. Hasselbach, *Chem. Ber.* **92**, 674 (1959).
51. Y. Mikami, K. Yokoyama, H. Tabeta, K. Nakagaki, and T. Arai, *J. Pharm. Dyn.* (*Tokyo*) **4**, 282 (1981).
52. T. Arai, K. Takahashi, K. Ishiguro, and Y. Mikami, *Gann* (*Tokyo*) **71**, 790 (1980).
53. K. Ishiguro and T. Arai, *Chiba Med. J.* (*Chiba*) **56**, 337 (1980).
54. T. Arai, unpublished data.
55. K. Ishiguro, S. Sakiyama, K. Takahashi, and T. Arai, *Biochemistry* **17**, 2545 (1978).
56. K. Ishiguro, K. Takahashi, K. Yazawa, S. Sakiyama, and T. Arai, *J. Biol. Chem.* **256**, 2162 (1981).
57. R. L. Erickson and W. Szybalski, *Virology* **22**, 111 (1964).
58. J. W. Lown, A. V. Joshua, and J. S. Lee, *Biochemistry* **21**, 419 (1982).
59. K. Singh, S. Sun, and D. Kluepfel, *in* "Development of Industrial Microbiology," (L. A. Underkofler, ed.) Vol. 17, p. 209. American Institute of Biological Sciences, Washington, D.C., 1976.
60. T. Fukuyama and R. A. Sachleben, *J. Am. Chem. Soc.* **104**, 4957 (1982).
61. A. Kubo, Y. Kitahara, N. Saito, S. Nakahara, K. Takahashi, and T. Arai, *Chem. Pharm. Bull.* **31**, 0000 (1983).
62. H. Matsumoto, T. Ogawa, T. Ikeda and T. Munakata, Abstracts of Papers, 102nd Annual Meeting of Pharmaceutical Society of Japan, Osaka, April, 1982, p. 568.
63. T. Hayashi, T. Noto, Y. Nawata, H. Okazaki, M. Sawada and K. Ando, *J. Antibiot.* (Tokyo) **35**, 771 (1982).
64. M. J. Zmijewski, Jr. and M. Goebel, *J. Antibiot.* (Tokyo) **35**, 524 (1982).

———CHAPTER 4———

CAMPTOTHECIN

JUN-CHAO CAI* and C. RICHARD HUTCHINSON

*School of Pharmacy, University of Wisconsin,
Madison, Wisconsin*

I. Introduction

Camptotheca acuminata Decne (Nyssaceae) is a tree distributed widely and abundantly in the southern part of China and was originally introduced to the United States in 1911 (*1*). During a plant antitumor screening program carried out under the auspices of the National Cancer Institute of the National Institutes of Health, it was found that the crude alcoholic extract of *C. acuminata* showed antitumor activity in animal tests.

A. STRUCTURE AND CHARACTERIZATION

Wall *et al.* isolated the antitumor alkaloid camptothecin (**1**) from the stem wood of this tree in combination with a bioassay method in 1966 (*2*). The structure of **1** was deduced from its spectral properties (UV, IR, ¹H NMR,

* Permanent Address: Shanghai Institute of Materia Medica, Chinese Academy of Sciences, Shanghai 200031, People's Republic of China

THE ALKALOIDS, VOL. XXI

MS), certain chemical properties (formation of mono-*O*-acetate, reaction with thionyl chloride and pyridine to give chlorocamptothecin, rapid saponification to a sodium salt that gave **1** on acidification, and reduction with sodium borohydride at room temperature to a lactol), and X-ray crystallographic analysis of its iodoacetate derivative (*2, 3, 4*). The latter technique established that **1** is a pyrrolo[3,4,-*b*]quinoline alkaloid, that is, 4(*S*)-4-ethyl-4-hydroxy-1*H*-pyrano-[3',4':6,7]indolizino[1,2,-*b*]quinoline-3,14(4H, 12H)-dione. As could be expected, rings ABCD and their substitutents (C-17, C-20, and the pyridone oxygen) are co-planar. Atoms C-21 and the lactone ring oxygen deviate from the plane 0.69 and 0.73 Å, respectively; thus ring E has a boat conformation (*4, 6*).

Shamma suggested using the Le Men–Taylor numbering system for **1**, based on the probable biogenetic relationship between camptothecin and the indole alkaloids, especially ajmalicine **2** (*5*).

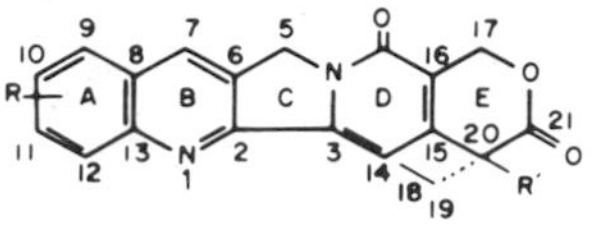

I   R=H, R'=OH
3   R=10-OH, R'=OH
4   R=10-OCH₃, R'=OH
5   R= 9-OCH₃, R'=OH
6   R=11-OH, R'=OH
7   R=11-OCH₃, R'=OH
8   R=H, R'=H

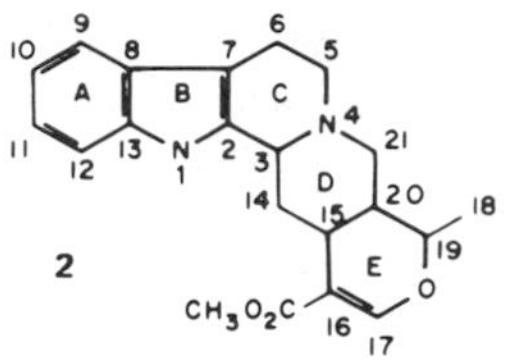

Camptothecin has two notable chemical properties: (1) Its lack of significant alkalinity causes it to behave as a neutral molecule (i.e., it does not form stable salts with mineral acids and gives negative tests with Dragendorff and Mayer reagents). Thus it is not an alkaloid in the usual sense of the definition (2). The presence of the C-20 tertiary alcohol imparts an unusual electrophilicity to the lactone carbonyl group, perhaps via a strong intramolecular H bond. This structural feature explains the behavior of **1** toward aqueous alkali, amine nucleophiles, and sodium borohydride. It also justifies the difficulty in preparing stable C-21 ester or amide derivatives of **1**; rapid reversion of such derivatives to **1** occurs by intramolecular attack of the C-17 primary alcohol at the electrophilic C-21 carbonyl group.

B. NATURAL ANALOGS AND SOURCES

Camptothecin and its analogs have been also found in some other plants. Further fractionation of the extract of the stem wood of *C. acuminata* resulted in the isolation of two minor compounds, 10-hydroxycamptothecin

TABLE I

CONTENTS OF CAMPTOTHECIN IN *Camptotheca acuminata*

Twigs	Stem bark	Roots	Root bark	Fruits
0.004%	0.01%	0.01%	0.02%	0.03%

(3) and 10-methoxycamptothecin (4), in 1969 by Wani and Wall (7). 9-Methoxycamptothecin (4), 1, and another related alkaloid mappicine (9) were isolated by Govindachari *et al.* from *Nothapodytes foetida* (Wight) Sleumer (Icacinaceae) (formerly, *Mappia foetida* Miers) (8, 9). Tafur *et al.* found 1 and 4 in *Ophiorrhiza mungos* Linn. (Rubiaceae) (10). Gunsekera *et al.* isolated 1 and 5 from *Ervatamia heyneana* (Wall) T. Cooke (Apocynaceae) (11). Hsu *et al.* investigated the alkaloidal components of the fruits of *C. acuminata* and found two new minor alkaloids, 11-hydroxycamptothecin (6) and 11-methoxycamptothecin (7), in addition to 1, 3, and 4 (12, 13). 20-Deoxycamptothecin (8) was also found in *C. acuminata* (12, 21).

The exact position of the hydroxy or methoxy group in these alkaloids was established by comparing their NMR spectra with those of the synthetic methoxypyrrolo[3,4-*b*]quinoline analogs 10 (7, 8, 12), and was confirmed subsequently by total synthesis (14, 15).

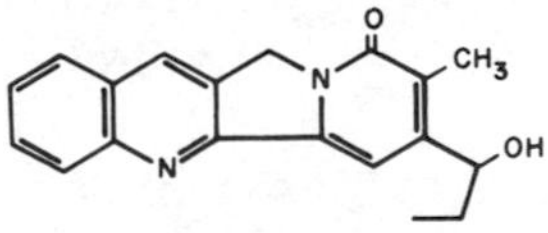

9 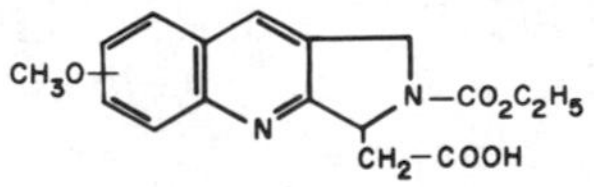   10

Hsu *et al.* examined the contents of camptothecin in different parts of *C. acuminata* and found that it was higher in the fruits than in other parts of the tree (Table I) (16).

Buta and Novak isolated camptothecin from dry, woody material of *C. acuminata* in 0.05% yield by gel permeation chromatography (17).

The contents of camptothecin in *N. foetida* was higher than that of other plants according to the reports of Govindachari (0.1%) (8a) and Agarwal (0.35%) (8b) (Table II).

TABLE II

CONTENTS OF CAMPTOTHECIN IN *Nothapodytes foetida*[a]

Stem	Bark	Roots	Leaves
0.06%	0.08%	0.1%	0.01%

[a] From Govindachari and Viswanathan (8a).

TABLE III
PHYSICAL CHARACTERISTICS OF CAMPTOTHECIN AND ITS ANALOGS

Compound	(°C)	Fluorescence under UV	$[\alpha]_D$	(c, solvent[a])	Reference
1	264–267 (dec)	Blue	+31.3	(N.A.,[b] A)	2
			+40	(0.475, A)	12
			+42	(1, A)	8a
			−139.5	(1.2, B)	8a
3	268–270	Red-yellow	−147	(0.674, B)	12
	266–267				7
4	254–255				7
	255–256 (dec)	Blue	N.A.[b]		10
	246–247 (dec)				12
5	258–260 (dec)	Yellow	−76.1	(0.5, B)	8a
6	327–330 (dec)	Dark red	−145.9	(0.19, B)	13
7	264–265 (dec)	Blue	N.A.[b]		12
8	272–274	Blue	+194	(0.58, C)	12
9	251–252	Bright green	N.A.[b]		9

[a] Solvent: A = $CHCl_3$:MeOH,4:1; B = pyridine; C = $CHCl_3$.
[b] N.A.: data not available.

All of the oxygenated analogs were found only as trace constituents: 9-methoxycamptothecin [0.002% (8a), 0.045% (8b)] in *N. foetida,* 10-methoxycamptothecin [0.005% (10)] in *O. mungos,* and 10-hydroxycamptothecin [0.002% (12)] in *C. acuminata.*

Chu *et al.* obtained 10-hydroxycamptothecin from camptothecin in 10% yield by biotransformation with fungi T-36 (18).

Sakato *et al.* obtained camptothecin from the leaves of *C. acuminata* by using a plant cell culture technique (19, 20). A medium containing 2,4-dichlorophenoxyacetic acid, kinetin, and gibberellic acid (GA_3) in the Murashige and Skoog basal medium was used. It was found that addition of tryptophan and phenylalanine markedly promoted the cell growth.

All the oxygenated camptothecin analogs showed antitumor activity in animal tests, and may be produced from 1 as a result of further metabolism in the plant.

Some physical characteristics of camptothecin and its analogs are given in Table III.

II. Total Synthesis

The announcement of camptothecin's structure in 1966 caused considerable excitement in the scientific community for two reasons: (1) The molecule represented a new heterocyclic ring system, and (2) it exhibited excellent biological activity in the *in vivo* rodent assay for antitumor activity.

There was an evident need for the development of practical synthetic routes to **1** and its analogs because of the scarcity of the natural source. Furthermore, the challenge to devise a general synthesis of the pyrrolo [3,4-*b*]quinoline ring system was not ignored by many research groups (*6,22,23*).

A. Total Synthesis of Camptothecin

1. Stork Synthesis

Stork and Schultz achieved the first total synthesis of *dl*-**1** in 1971 (*24*). A base-catalyzed Friedländer condensation of pyrrolidone **11** with 2-amino-benzaldehyde (**12**) gave the pyrrolo[3,4-*b*]quinoline acid **13** (rings ABC). Following hydrolysis and esterification, the resulting amino ester **14** was condensed with carbethoxyacetyl chloride to give the diester amide **15**, which was converted to the tetracyclic β-keto ester **16** by a Dieckmann cyclization. Hydrolysis and decarboxylation of **16** followed by reduction with sodium borohydride and elimination of water from the resulting β-hydroxylactam **17**, gave the desired dihydropyridone **18** (rings ABCD). The latter reacted efficiently at low temperature with the lithium anion of a protected α-hydroxybutyric acid ester to give the key pentacyclic compound **19** via intermediate **18a**. This transformation represented a new annulation method of an α,β-unsaturated carbonyl compound (*6*). Finally, hydrolysis of the ethyl ester of **19** followed by reduction with sodium borohydride gave the hemiacetal acid **20** which was converted to *dl*-**1** through acetate hydrolysis, hemiacetal reduction, and acidification to form the α-hydroxy lactone ring. The overall yield of *dl*-**1** was 1–2%.

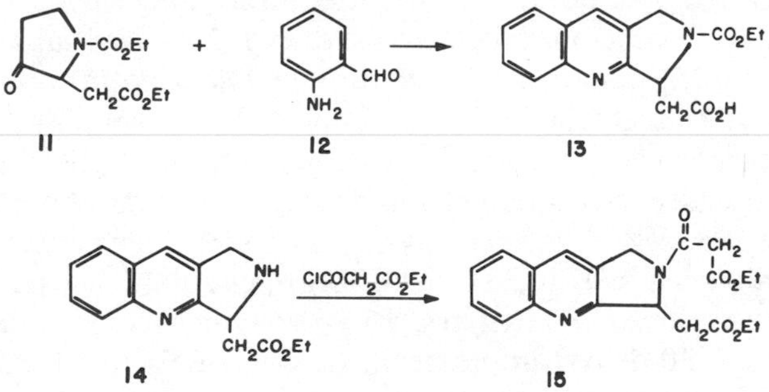

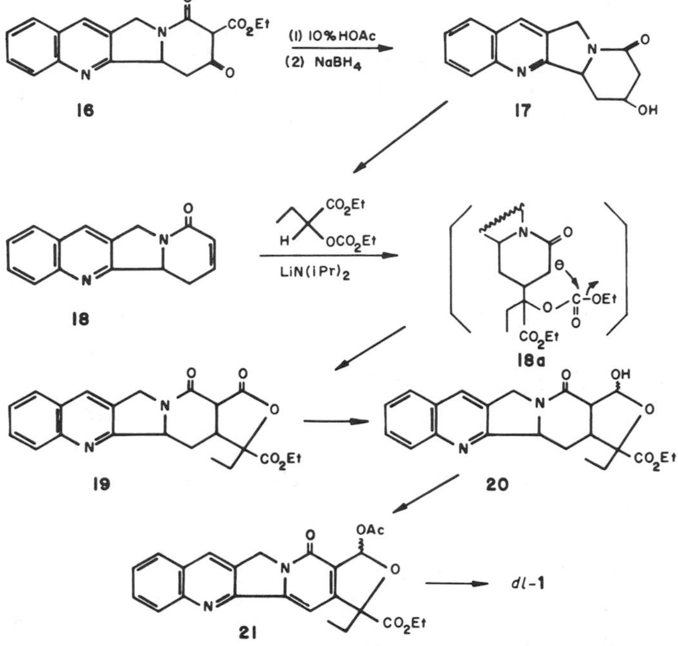

2. Danishefsky Synthesis

The report of the first synthesis of camptothecin preceded the description of the second total synthesis of *dl*-**1** by only 9 weeks. Danishefsky *et al.* (*25*) assembled the alkaloid by a Friedlander condensation approach in which the key pyridone **25** (rings ABCD) was prepared by a novel synthetic route to 4,6-disubstituted pyridones (*26*). The latter utilized the Michael addition of enamine **22** to dicarbethoxyallene to yield an 1,4,5,6-tetrasubstituted pyridone (**23**). Transformation of **23** to **25** through Dieckmann cyclization of an intermediate tetramethyl ester **24** was followed by hydrolysis – decarboxylation in aqueous acid, then a Friedländer condensation to give the tetracyclic diacid **25**. After selective esterification, the tetracyclic monomethyl ester **26** was obtained in 29% yield by pyrolysis over Cu(I)O. The latter compound was monoethylated to **27** in 20% yield. Treatment of **27** with paraformaldehyde in acidic solution resulted in C-16 hydroxymethylation and lactonization to give *dl*-20-deoxycamptothecin (*dl*-**8**) in 35% yield, along with a structurally isomeric minor product [isodeoxycamptothecin due to C-14 hydroxymethylation (*27*)]. Synthetic 20-deoxycamptothecin spontaneously oxidized to *dl*-1 on exposure of its solution to air. On a preparative scale, Danishefsky *et al.* (*25*) carried out this oxidation in 20% yield by treating a solution of the C-20 anion of *dl*-8 with hydrogen peroxide. The ease by which **8** gives **1** may have biogenetic significance.

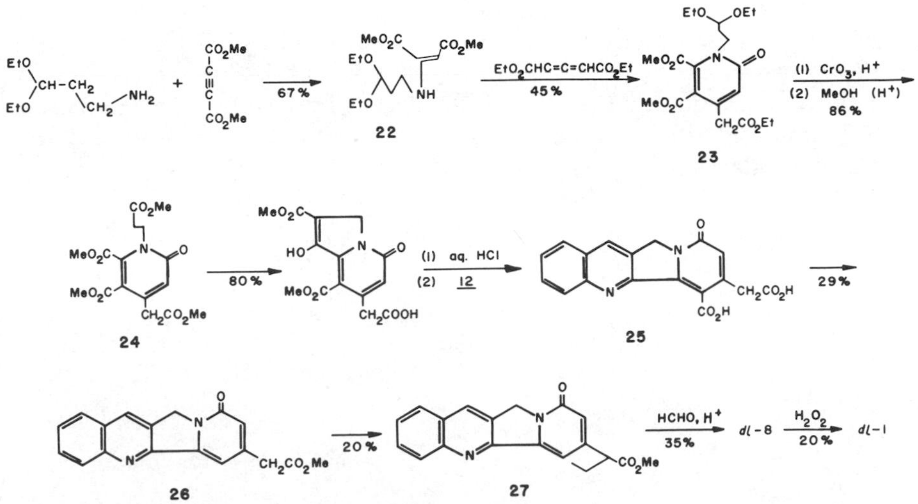

a. Quick Synthesis of 27. In 1977, Quick modified this synthesis by first using imine **28** as starting material to prepare pyridone **29**. This pyridone makes it possible to avoid the low-yield decarboxylation of **25** to **26**. In addition, the C-20 ethyl group was introduced at an early stage, i.e., ethylation of pyridone **29** to afford compound **30**, thus improving the yield of C-20 ethylation from 20 to 65%. According to this report, the overall yield from methyl pyruvate to tetracyclic ester **27** was 9% (*28*).

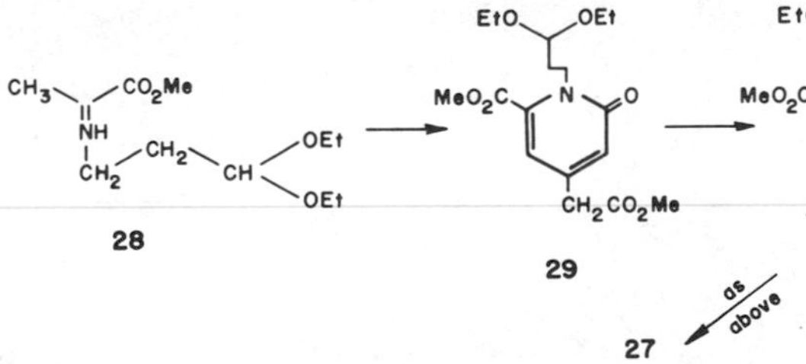

b. Büchi Synthesis of 27. In 1976, Büchi and Bradley devised a simple route for preparing tetracyclic ester **27** (*29*). Starting from ketoester **31**, readily available from methyl dimethoxyacetate and methyl butyrate, **33** was obtained in 82% yield by Wittig condensation. After debenzylation, **33** was converted to a compound similar to **36** but only in 56% yield. Consequently, **33** was rearranged to **34**, and the latter, after debenzylation, was condensed with tricyclic amine **35** to give **36** in 94% yield. Treatment of **36** with boron trifluoride and trifluoroacetic acid gave tetracyclic ester **27** in 41% yield (the overall yield was 30%).

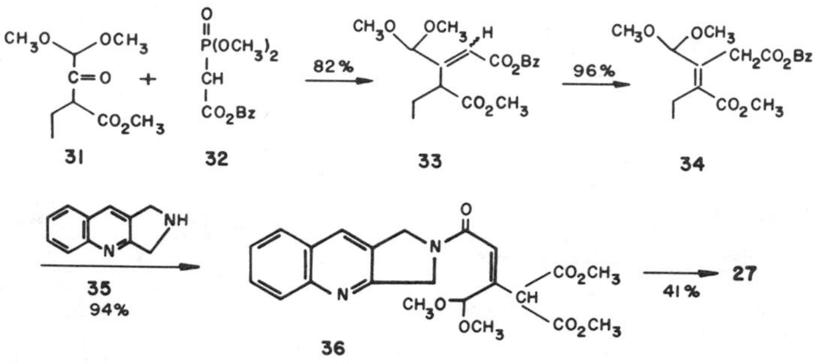

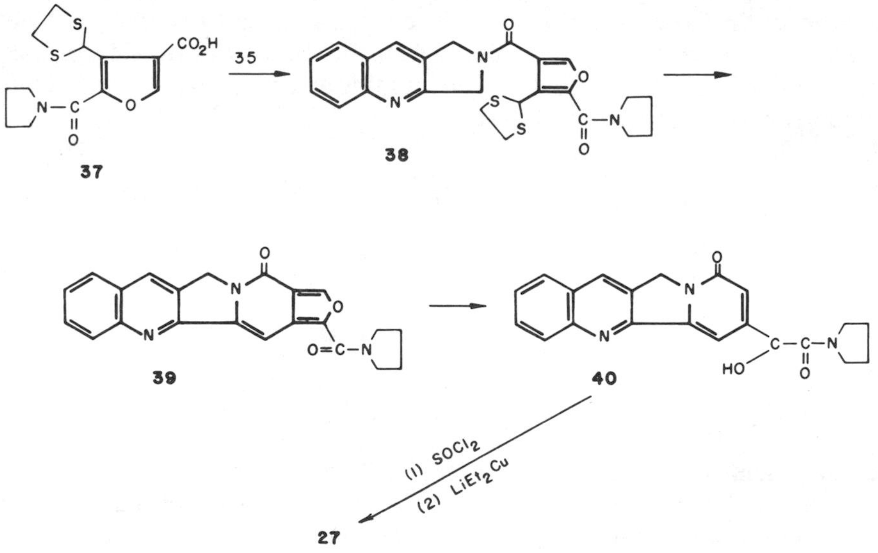

c. Kende Synthesis of 27. Kende *et al.* reported a novel synthesis of **27** in 1973 (*30*). Trisubstituted furan **37**, prepared from furfural dimethyl acetal via a six-step sequence of reactions in 19% yield, was condensed with **35** to give tricyclic amide **38**, which then was deprotected and cyclized to give pentacyclic amide **39**. The latter compound underwent a facile alkaline hydrolysis with loss of one carbon as CO_2 to give the hydroxyamide **40** in 74% yield, from which tetracyclic ester **27** was obtained through chlorination, ethylation, and methanolysis. The overall yield from furfural to **27** was about 3%.

3. Winterfeldt Biomimetic Synthesis

Winterfeldt *et al.* discovered that certain tetrahydro-β-carboline alkaloids could be autoxidized to pyrrolo[3,4-*b*]quinolines in strongly basic dimethyl-formamide solution (*31*). This observation led them to develop an imaginative total synthesis of camptothecin (*32*), whose strategy in part may parallel the alkaloid's biogenesis.

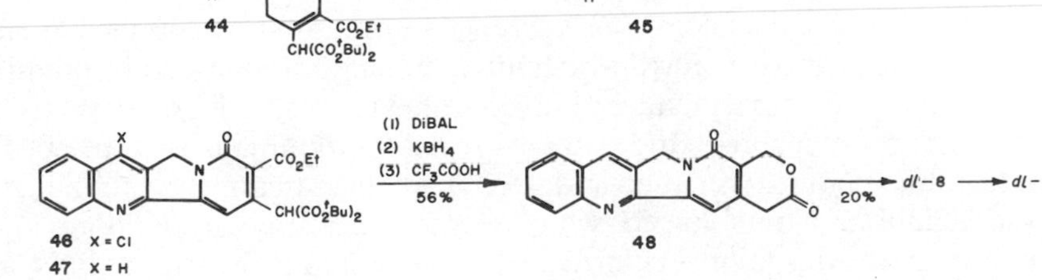

The Winterfeldt synthesis of *dl*-**1** employed the tetracyclic lactam **43** as the keystone for the formation of the pyrrolo[3,4-*b*]quinoline ring system. Lactam **43** was available from **41** via **42**. Thus, **41** was condensed with monoethyl malonate and cyclized in *t*-BuOK to obtain **42**, then the latter was reacted with diazomethane which resulted in the formation of the α,β-unsaturated lactam **43**. The 1,4-addition – elimination of sodium di-*tert*-butylmalonate to **43**, followed by autoxidation of the tetrahydro-β-carboline **44**, and treatment of **45** with thionyl chloride in DMF at room temperature, gave the 12-chloro-9-oxo-9,11-dihydroindolizino[1,2-*b*]quinoline **46** in good overall yield. Introduction of the 10a,6 double bond in **46** was postulated to involve a novel oxidative elimination of SOCl (*33*). After hydrogenolytic removal of the chlorine atom, the carbethoxy group in **47** enabled its chemoselective reduction to a primary alcohol by treatment with, first, diisobutylaluminum hydride at low temperature, then potassium borohydride, from which **48** resulted by treatment of the intermediate δ-hydroxy di-*tert*-butylmalonate with trifluoroacetic acid. Ethylation at C-20 of **48** proceeded poorly because of concurrent dialkylation at C-5 and C-20, but the resulting deoxycamptothecin was oxidized to *dl*-**1** quantitatively by oxygen in the presence of cupric chloride and aqueous dimethylamine.

Alternatively, **46** could also be converted to 7-chlorocamptothecin and then hydrogenolyzed (to remove the chlorine atom) to give *dl*-**1**.

Winterfeldt's group later reported the result of further developments in the synthetic chemistry of camptothecin that markedly improved the overall yield of their synthesis. The problem of C-5 ethylation was circumvented by ethylation of a monoisopropyl ester analog **49**, which was obtained by using isopropyl *tert*-butyl malonate instead of di-*tert*-butyl malonate in preparation of **44**. This modification made it possible to selectively hydrolyze the *tert*-butyl ester for subsequent decarboxylation. From compound **49**, 20-deoxycamptothecin could be obtained as before in 86% yield. The overall yield of *dl*-**1** from **41** was 13–14% (*34*).

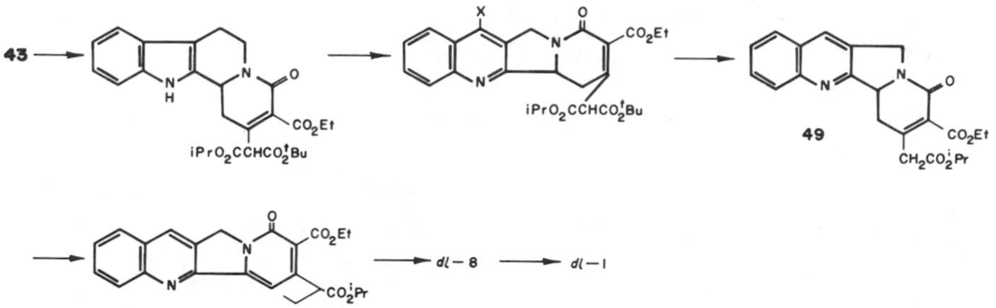

They also investigated the remarkable ease by which 20-deoxycamptothecin is autoxidized to *dl*-**1**, finding that this reaction has structural limitations (*35*). Only **8** and its ring DE analogs (as the N—CH$_3$ pyridone) could be oxidized at C-20 using their experimental conditions. Three other compounds (dimethyl 2-ethyl malonate, methyl 2-ethyl phenylacetate, and the benzene ring analog of rings DE of camptothecin) examined as models for this oxidation were completely unreactive. Clearly, the pyridone ring of **8** plays a vital role in the mechanism of C-20 autoxidation.

Kametani synthesis. Kametani *et al.* prepared methoxycarbonyl derivative **52**, the methyl analog of **47**, via an enamine annulation as the key step. Thus, 3,4-dihydro-1-methyl-β-carboline, prepared from acetyl tryptamine, was condensed with a mixture of unsaturated tetraesters followed by reduction with sodium borohydride to give the indolo[*a*]quinolizin-4-one **50**. The latter, after photooxygenation and subsequent base treatment, produced the indolizino[1,2-*b*]quinoline **51**, which was converted into the pyridone **52** via chlorination with thionyl chloride, dehalogenation by hydrogenolysis, and dehydrogenation with DDQ (*36*).

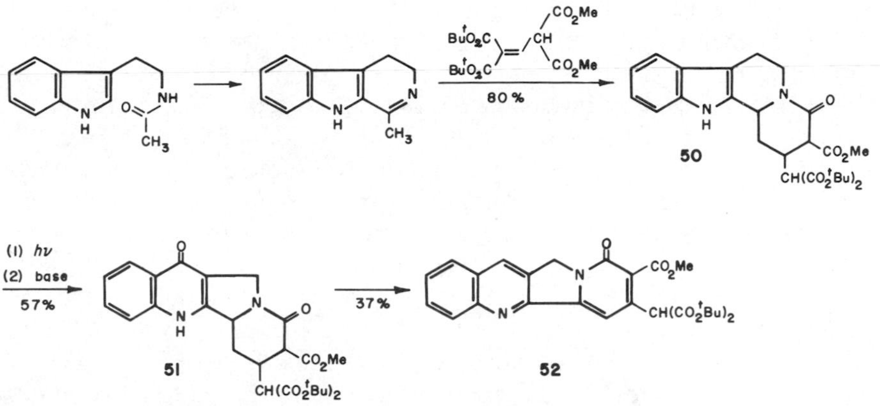

4. Rapoport Synthesis

The total synthesis of *dl*-1 by Rapoport *et al.* (*37*) is notable because of the impressive 15% overall yield from the starting material, pyridine-2,5-dicarboxylic acid (isocinchomeronic acid). In addition, these workers employed a novel rearrangement of a nipecotic acid to an α-methylene lactam in one key synthetic step (*38*).

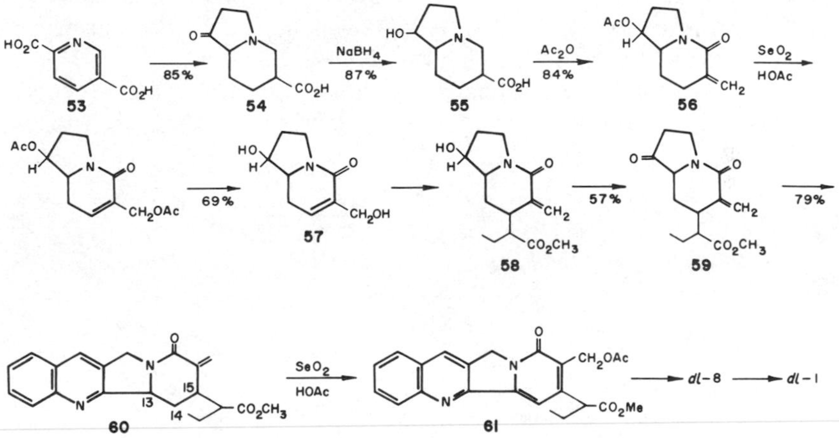

Nipecotic acid (54), prepared from 53 via esterification, reduction, alkylation, and Dieckmann cyclization in 85% yield, was reduced with sodium borohydride to the alcoholic β-amino acid 55. When refluxed with acetic anhydride, 55 underwent the α-methylene lactam rearrangement to afford the piperidone acetate 56 as a diastereoisomeric mixture. Rapoport *et al.* had planned to carry out this rearrangement at the tetracyclic stage after Friedlander condensation of 54 with 2-aminobenzaldehyde (12). Although the rearrangement could be done in 60% yield, the subsequent oxidation with selenium dioxide to the tetracyclic analog of **57** was not possible (complete aromatization occurred). Consequently, they oxidized **56** with selenium dioxide, hydrolyzed the product to **57**, then employed a Claisen ortho ester rearrangement to obtain a mixture of the free alcohol **58** and its butyrate ester. Hydrolysis of the crude reaction mixture followed by Pfitzner–Moffatt oxidation gave the α-methylene lactam **59**. Friedländer condensation of **59** with **12** gave the expected tetracyclic material **60**, whose oxidation with selenium dioxide simultaneously aromatized the D ring and introduced a C-17 acetoxy substituent. The acetoxy group probably was introduced by [3,3]sigmatropic rearrangement of a 3,14-dehydro-15-acetoxy derivative of **60**, because selenium dioxide oxidation (and also NBS) of a C-17 deacetoxy analog of **61** failed to occur. Compound **61** was converted to deoxycamptothecin by an acid-catalyzed hydrolysis–lactonization. Finally, the C-20 hydroxy group was introduced by using Winterfeldt's method.

5. Corey Synthesis of 20(S)-Camptothecin

The only synthesis of optically active 20(S)-camptothecin reported to date is that of Corey et al. (39). Although the overall yield of the Harvard group's synthesis was low, their strategy is typically novel. Their convergent synthetic approach brought together a chiral pseudoacid chloride (ring E) with a tricyclic amine (rings ABC, **35**) through which ring D was formed by cyclization of an intermediate α-aldehydo-*tert*-amide (**69**).

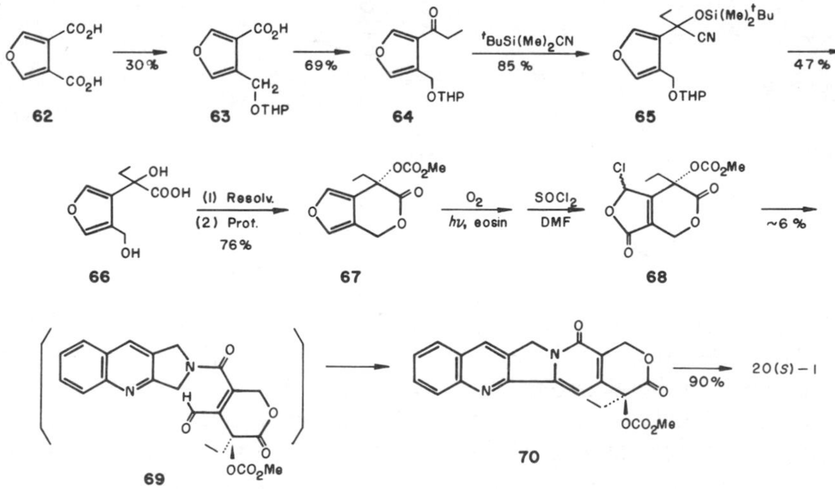

Resolution of the 3,4-disubstituted furan α-hydroxy acid **66** — prepared from 3,4-dicarboxyfuran by standard, but delicate, synthetic transformations in 8% yield—via its diastereomeric quinine salts and protection of the tertiary hydroxy in the lactonized form of (+)-**66** gave **67** in good yield. Photooxidation of **67** to the hemiacylal followed by its treatment with thionyl chloride in a catalytic amount of DMF gave a 2.5:1 mixture of pseudoacid chloride **68** and its undesired regioisomer. Condensation of this mixture with **35**, prepared from acridine in three steps (18%), in pyridine– acetonitrile (to **69**), followed by base-catalyzed cyclization gave 20(*S*)-20-methoxycarbonyl-camptothecin (**70**) in low yield. Deprotection of **70** by treatment with lithium mercaptide in HMPA gave 20(*S*)-**1** cleanly.

6. The Chinese and the Wall Synthesis

Chemists at the Shanghai Institute of Material Medica developed an efficient, general, and practical synthesis of *dl*-**1** and its analogs in 1976 (*40*), but the full details of their work did not appear in the available English literature until 1978 (*40b*). Since Wall *et al.* used a slightly modified version of the Chinese synthesis for the efficient preparation of *dl*-**1** and several of its analogs in 1980 (*41*), these two syntheses are discussed together. The new chemistry developed in these two laboratories allows the total synthesis of racemic camptothecin on a large scale, if this becomes advisable in the future.

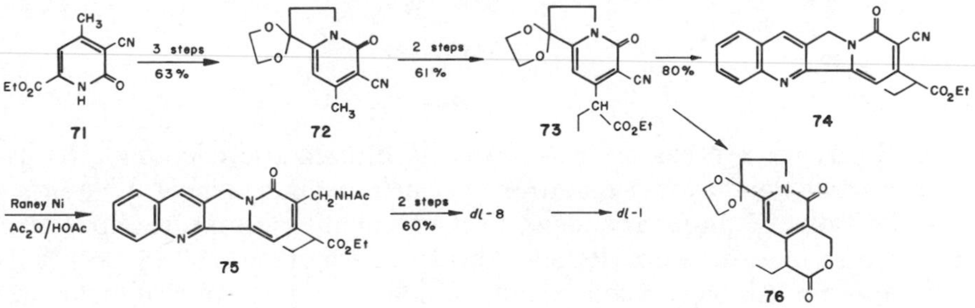

The Chinese/Wall campototothecin synthesis employed pyridone **71** (ring D) as starting material, available from the condensation of cyanoacetamide and the *O*-ethyl ether of ethyl acetopyruvate, to make indolizine **72** by a three-step sequence involving simultaneous Michael addition to methyl acrylate and Dieckmann cyclization, hydrolysis–decarboxylation in aqueous acid, and ketalization. The C-7 methyl of **72** was sufficiently acidic to permit its carbethoxylation by reaction with sodium hydride and diethyl carbonate. Ethylation of the resulting ester gave **73** in 80% yield. The latter underwent simultaneous deketalization and Friedlander condensation with **12** in acidic media to afford the tetracyclic compound **74** (rings ABCD) in 80% yield. The Chinese group found that this Friedlander condensation was more suitably done in acidic media than in alkaline media. Following reductive acetylation of the cyano group of **74**, formation of the *N*-nitroso-acetamide of **75** *in situ* in the acid filtrate obtained from the hydrogenation reaction resulted in rearrangement to its acetoxy analog. The researchers then cyclized the latter intermediate to *dl*-**8** in dilute acid from which *dl*-**1** resulted by Winterfeldt's method. The overall yield of the Chinese chemists' ten-step synthesis was an impressive 18%.

Wall's group modified the Chinese group's camptothecin synthesis by first converting **73** to **76**, the complete CDE rings portion of **8**, from which they could prepare *dl*-**1** by Friedländer condensation and subsequent transformations (*41*). Their synthetic modification, which the Chinese workers had attempted but could not execute successfully, improved the overall yield of *dl*-**1** significantly.

7. Miscellaneous Syntheses

a. Wall Synthesis. Wall *et al.* published another total synthesis in 1972 (*42*). In this approach, 1,3-dihydro-2-methoxycarbonyl-2*H*-pyrrolo-[3,4-*b*]quinoline (**77**) was used as starting material. Compound **77** reacted with keto ester **78** to give **79**. The latter was treated with liquid HCN in the presence of KCN to yield cyanolactone **80** via the cyanohydrin, from which amide **81** was obtained by treatment with basic hydrogen peroxide. After selective removal of the *N*-carbomethoxy group, treatment with alkali, lactamization, and dehydrogenation with DDQ, pyridone **82** was obtained. Finally, **82** was transformed to *dl*-**1** through reduction with lithium borohydride and acidification.

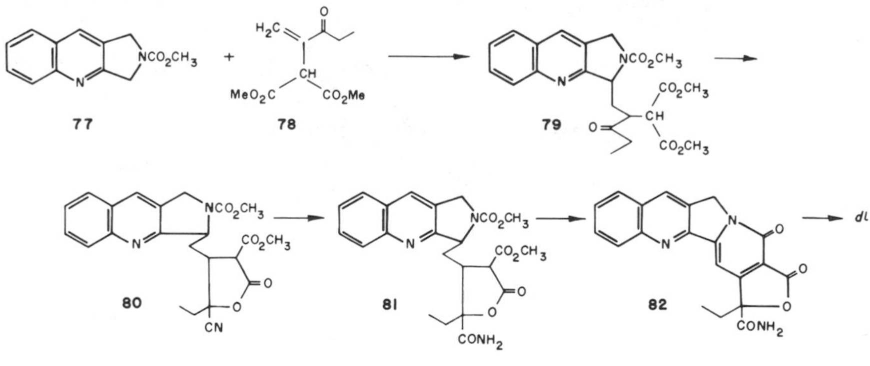

b. Sugasawa Synthesis. Sugasawa *et al.* utilized tricyclic amide **83** as starting material for the preparation of 7-methoxyindolizino[1,2-*b*]quinoline derivative **85** through condensation with diethyl acetonedicarboxylate, cyclization, and decarboxylation. The C-8 formylation of **85** occurred smoothly via Vilsmeier reaction, but the subsequent addition of malonate failed due to the instability of the reaction intermediate toward base. Consequently, **85** was reduced to tetrahydroquinoline **86**, then formylated to a Michael acceptor **87** which underwent addition with di-*tert*-butyl malonate to give **88**. 20-Deoxy-20-deethylcamptothecin (**48**) then was obtained by reduction with borohydride, hydrolysis, and dehydrogenation. Compound **48** was converted to *dl*-**1** by C-20 ethylation and oxidation (**43**).

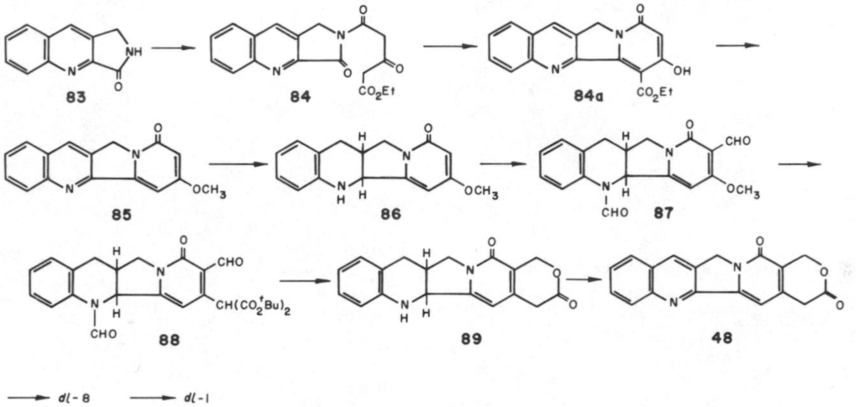

Sugasawa's group later reported a method for introducing the 20-ethyl and 20-hydroxy group in one step, thus improving the overall yield of camptothecin (*44*).

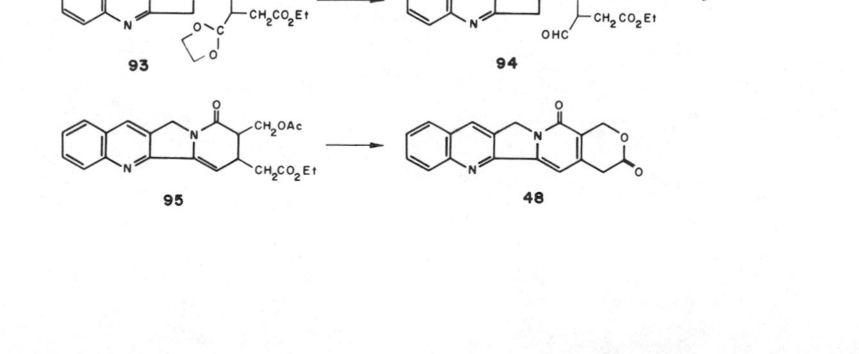

c. Meyers Synthesis. In this synthesis, oxazine amide **90**, prepared from tricyclic amine **35** with an oxazine ester, underwent Michael addition with the unsaturated ester **91** to give amide **92**. Sodium borohydride reduction of **92**, followed by cleavage gave aldehyde **93**. The latter, after reduction, was acylated and deketalized to afford **94**. Cyclization of **94** gave the dihydropyridone **95** that was aromatized with DDQ to the appropriate pyridone. Acid hydrolysis then produced the desired compound **48** (*45*).

d. Pandit Synthesis. Pandit and Walraven prepared lactone **96** from furfural by a six-step reaction sequence. Compound **96** reacted with tricyclic amide **35** to give amide **97**, that, after oxidation, was cyclized to pyridone **98**. Reduction of **98** by lithium borohydride followed by treatment with acid yielded **48** (*46*).

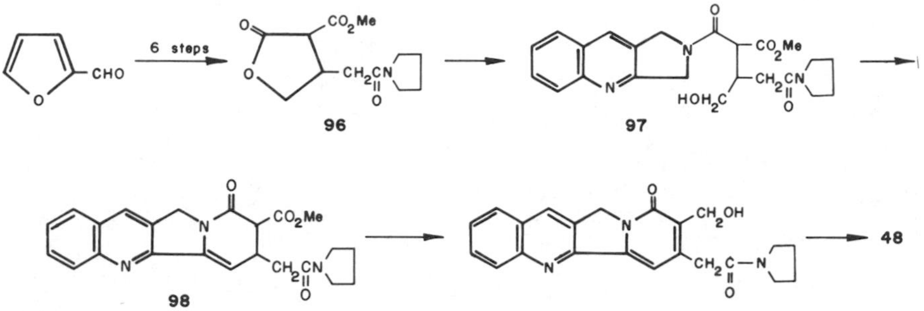

e. Shamma Synthesis. Pyridone **100**, obtained from urethane **99** through a nine-step sequence of reactions in 16% yield, was condensed with diethyl oxalate to give vinyl lactone **101** in 65% yield. The conversion of the latter to the δ-lactone **103** was carried out with sodium borohydride followed by HIO$_4$. Catalytic oxidation of the resulting hemiacetal **102** led to the desired δ-lactone **103**. 20-Deoxy-20-deethylcamptothecin (**48**) then was obtained by deketalization and Friedländer condensation (*47*).

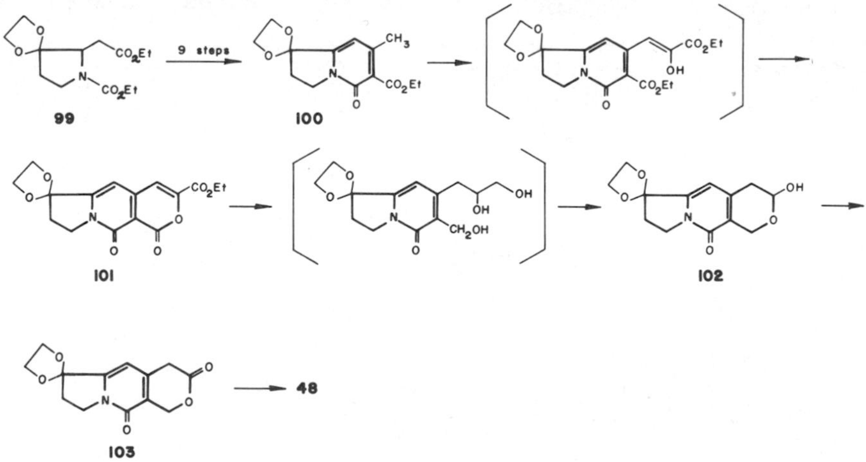

B. Total Synthesis of Camptothecin Alkaloids

1. Total Synthesis of Oxygenated Camptothecines

a. 10-Hydroxycamptothecin (3) and 10-Methoxycamptothecin (4). The Chinese scientists announced the first total synthesis of *dl*-**3** and *dl*-**4** in 1977 using their general synthetic method (Section II, A, 6). Recently, Cai *et al.* published the details of this work (*14*).

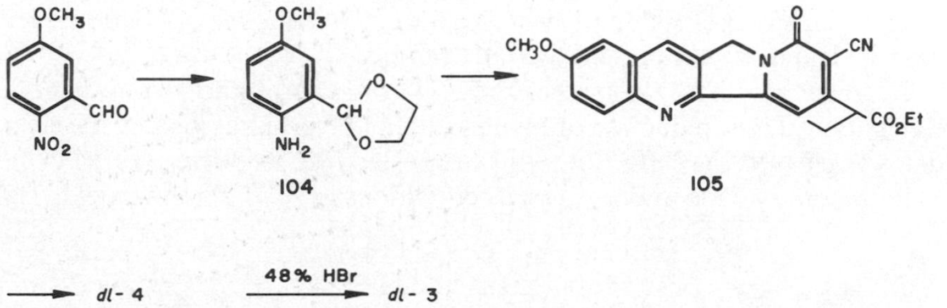

2-Nitro-5-methoxybenzaldehyde was protected with ethylene glycol followed by catalytic reduction to give **104**. This procedure improved the yield of reduction and made **104** suitable for long-period storage. Compound **104** was condensed with **73** to yield **105,** which was converted to *dl*-**4** by the method described in Section II, A, 6. Then *dl*-**4** was transformed to *dl*-**3** by HBr demethylation. The overall yields of *dl*-**3** and *dl*-**4** from **71** were 6.5 and 10%, respectively.

Wall *et al.* reported a yield of 10% and 21% for the synthesis of *dl*-**3** and *dl*-**4**, respectively, by their modification of the Chinese synthesis of *dl*-**1** (*41*).

Kametani *et al.* also synthesized *dl*-**3** and *dl*-**4** by using the methoxy derivative of **49** as starting material in their synthesis (Section II, A, 3, a).

b. 11-Hydroxycamptothecin (6) and 11-Methoxycamptothecin (7). Cai *et al.* synthesized these two alkaloids by their general method using 2-nitro-4-methoxybenzaldehyde as the starting material. The overall yield of *dl*-**6** and *dl*-**7** was lower than that of *dl*-**3** and *dl*-**4** (*15*).

Up to the present, all the oxygenated camptothecinoid alkaloids isolated from natural sources have been synthesized except for 9-methoxycamptothecin (**5**).

2. Total Synthesis of Mappicine (9)

Kametani *et al.* reported the total synthesis of *dl*-mappicine (*dl*-**9**) in 1975. Tetracyclic ester **106** was converted to **107** by treatment with diazomethane, from which dl-**9** was obtained through a four-step sequence of reactions. The overall yield of dl-**9** from **106** was 11% (*48*).

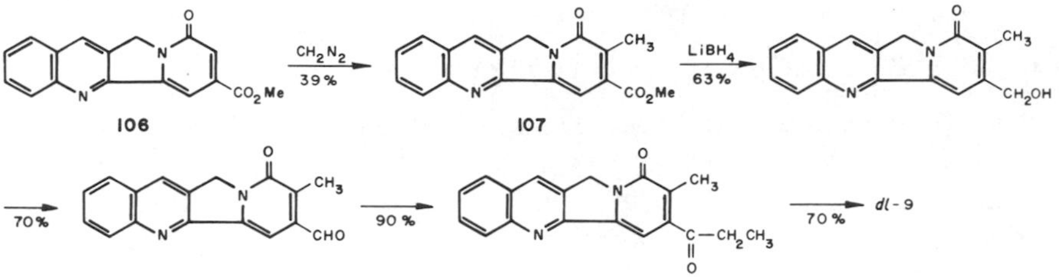

Govindachari *et al.* (*9*) synthesized dl-**9** from camptothecin for structure elucidation. Thus, reduction of **1** with sodium borohydride yielded the lactol that was cleaved by Pb(OAc)$_4$ to the keto ester **108**. Reduction of **108** with sodium borohydride under vigorous conditions gave dl-**9** in low yield.

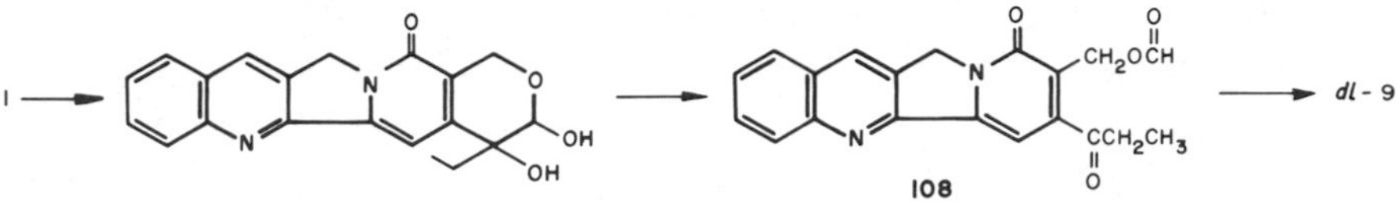

III. Biogenesis

A. Biogenetic Speculations

Although it is not immediately obvious from the structure of camptothecin that it could be derived biosynthetically from tryptophan and a monoterpene, Wenkert *et al.* speculated in 1967 (*49*) that **1** in fact might be a monoterpene indole alkaloid and used plausible chemical transformations of isositsirikine (**109**) as the basis from which to formulate a biogenetic scheme for **1**. Winterfeldt (*50*) later expanded on this idea based on his own finding that **110** underwent facile autoxidation to **111** *in vitro,* and proposed that geissochizine (**112**) was a plausible biogenetic precursor of **1**.

Hutchinson and co-workers took these two ideas into account, but also recognized the clear structural relationship between **1** and strictosamide (**114**). The latter neutral glucoside was known as a transformation product of strictosidine (**113**) under basic conditions (*51*). Thus, transformation of **114** into **1** was considered by them to be possible via three basic transformations: ring BC oxidation–recyclization, ring D oxidation, and removal of the C-21 glucose moiety followed by ring E oxidation (*52*). This biogenetic hypothesis also was proposed independently by Cordell (*53*).

B. PRELIMINARY BIOSYNTHETIC STUDIES

Hutchinson *et al.* established in initial trial-feeding experiments that radioactively labeled tryptophan was incorporated into **1** in apical cuttings of young seedlings of *C. acuminata* to the extent of $2.6 \times 10^{-2} - 4 \times 10^{-4}\%$. Similarly, radioactively labeled mevalonic acid and [6-³H]secologanin (**115**) were found to give rise to radioactive **1** *in vivo* (*52*). These incorporations of radioactivity into **1** were rather low and could not be confirmed as regiospecific because of a lack of suitable degradative chemistry for radioactive label localization. Sheriha and Rapoport (*54*) subsequently confirmed these initial observations using 8-month-old *C. acuminata* seedlings. They found that singly and doubly labeled radioactive precursors gave the following total incorporations into **1**: tryptophan (1.9%), tryptamine (0.02%), mevalonate (0.2%), and geraniol/nerol isomeric mixture (0.08%). Again, the lack of a suitable degradative chemistry prevented rigorous validation of these radioactivity incorporations.

Since the results of the initial experiments strongly indicated that **1** was a monoterpene indole alkaloid, the C-3 epimeric mixture of strictosidine/vincoside [(3R)-**113**], tritium labeled at C-5 by synthesis from [1-³H]tryptamine, was then tested as a precursor of **1**. The observed total radioactivity incorporation into **1** of 0.24% supported the implications of the initial data.

At this juncture three possibilities for the conversion of **113** to **1** *in vivo* were considered: via **113**, via **109**, or via **112**. Since radiochemically labeled **114** and **112** were available, two of these possibilities could be tested. It was

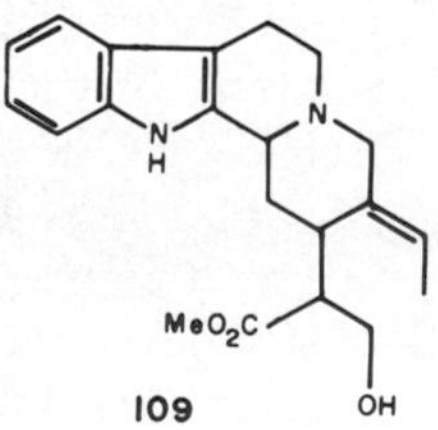

109

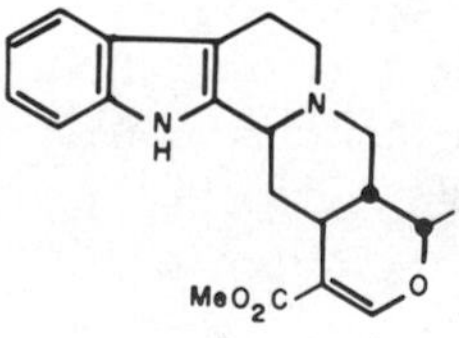

110

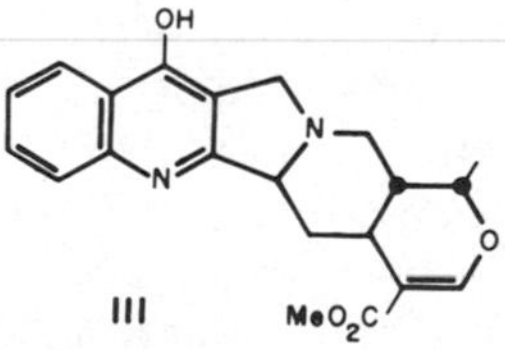

111

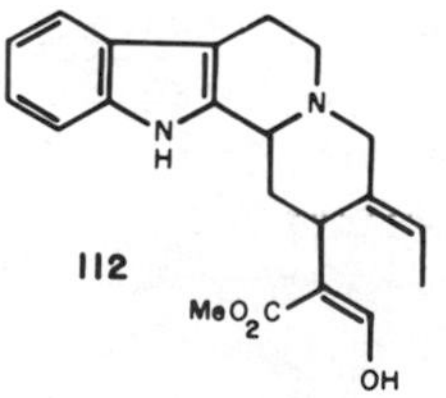

112

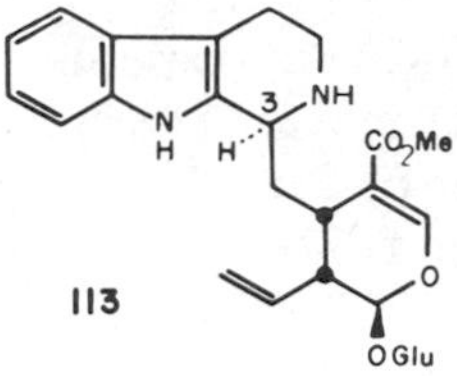

immediately clear from its efficient incorporation into **1** (1 – 4%) that only **114** needed to be considered for further experimentation in the elucidation of camptothecin's biosynthetic pathway (*52*).

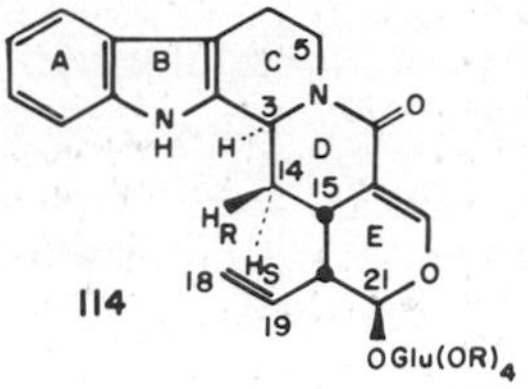

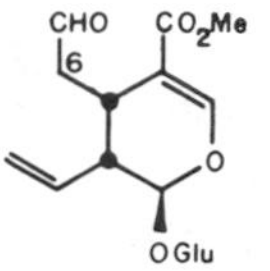

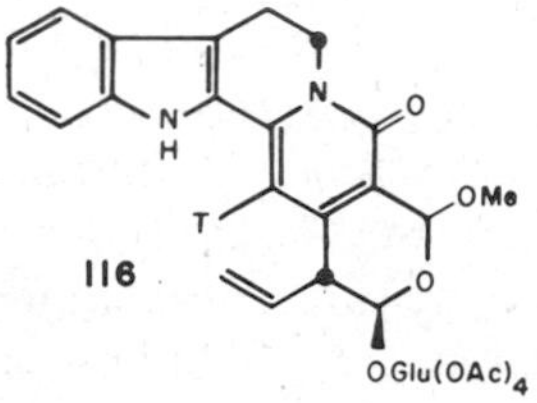

C. STRICTOSAMIDE: THE PENULTIMATE BIOSYNTHETIC PRECURSOR

Although the results of the feeding experiments summarized above strongly supported the role of strictosamide (**114**) as the key biosynthetic precursor of camptothecin, it had to be ascertained that it was regiospecifically incorporated into **1**. In spite of the lack of suitable degradative chemistry, the efficient incorporation of radioactive **114** into **1** suggested that the labeling regiochemistry could be determined directly by ^{13}C-NMR spectroscopic analysis. A suitable quantity of [5-^{13}C]-**114** containing 84 mol % ^{13}C was synthesized and fed to *C. acuminata* plants. The proton noise-decoupled ^{13}C-NMR spectrum of the resulting labeled **1** showed that only the resonance corresponding to C-5 had been significantly enhanced (55%) using the height of the C-17 methylene signal as the internal reference (*52*). This observed enhancement corresponds to a specific ^{13}C incorporation of $\sim 0.9\%$, in good agreement with the specific incorporations observed for [5-^{14}C]-**114** in separate radioactive feeding experiments. Consequently, the requisite certification of the role **114** plays as a specific biosynthetic precursor of **1** was firmly established.

Earlier incorporations of [5-^{14}C, 14-^{3}H]-**114** into **1** had been attended by only a 5 – 9% decrease in the ^{3}H : ^{14}C ratio. This was a surprising finding, since **114** was expected to lose about one-half of its ^{3}H labeling at C-14 during oxidative formation of the pyridone ring of **1**. Hutchinson *et al.* (*55*) considered three explanations for this low percentage ^{3}H loss: (1) **114** fortuitously might have been stereospecifically labeled with ^{3}H at C-14, and oxidation to **1** might remove hydrogen from only the unlabeled diastereotopic position; (2) conversion of **114** into **1** might involve an intramolecular migration of one of the two C-14 tritium atoms to another site in some biosynthetic intermediate leading to **1**, resulting in retention of both labels; or (3) loss of hydrogen from C-14 during oxidation of the D ring of **114** to the pyridone ring of **1** might be both nonstereospecific (and therefore nonenzymatic) and subject to a significant kinetic isotope effect discriminating against tritium removal.

They examined each of the above alternatives in turn. Analysis of the ^{1}H-, ^{2}H-, and ^{13}C-NMR spectra of samples of [14-^{2}H]-**114**, which had been prepared in the same manner as for [14-^{3}H]-**114**, clearly showed that the C-14 diastereotopic positions of **114** were equally ^{2}H labeled. Furthermore, these NMR spectra showed that ^{2}H was not incorporated into any other position in **114** during the isotopic labeling reactions. It was concluded that the samples of [5-^{14}C, 14-^{3}H]-**114** used in the biosynthetic feeding experiments were equally ^{3}H labeled intermolecularly at the diastereotopic hydrogens attached to C-14, assuming that the distribution of ^{3}H label at these two positions paralleled the established ^{2}H-labeling stereoselectivity. They next examined the possibility that the high retention of ^{3}H in the conversion of **114** to **1** *in vivo* was due to an intramolecular ^{3}H migration. For example, one likely possibility was a [1,4] migration of ^{3}H from C-14 to C-17 via some ionically charged intermediate. Chemical degradation of a sample of radioactive **1**, labeled by the incorporation of [5-^{14}C, 14-^{3}H]-**114**, to a C-17 lactone (analogous to **19**, p. 106) eliminated this possibility since the lactone degradation product contained 97% of the molar ^{3}H content of **1**. A second and more conclusive result (than the latter) was the finding that ^{2}H-NMR analysis of the 20-methylthiomethylene derivative of **1** labeled by [14-^{2}H, ^{3}H]-**114** showed ^{2}H to reside only at C-14 of **1**. The specific incorporation of ^{2}H (0.53% by NMR analysis) agreed closely with the same value calculated from the ^{3}H radioactivity (0.57%). These data established that [14-^{2}H, ^{3}H]strictosamide labels only H-14 of camptothecin *in vivo*.

It was established, therefore, that the precursor [14-^{2}H or ^{3}H]-**114** was nonstereospecifically labeled at C-14 and at no other position, and that the product **1** was labeled only at H-14. The results of an independent feeding experiment with [6,8-^{3}H]loganin further corroborated these observations, and also revealed indirectly that the mechanism of D ring oxidation of the unknown biosynthetic intermediate lying between strictosamide and camptothecin does not involve significant stereospecific loss of hydrogen (as ^{3}H) from the C-14 diastereotopic positions (*55*). Consequently, it appeared that the third presumption of the three possible explanations (*vide supra*) for the low ^{3}H loss attending the biosynthetic incorporation of [5-^{14}C, 14-^{3}H]-**114** into **1** is correct.

Verification that this presumption explains the observations will be difficult until it is possible to examine the D ring oxidation of poststrictosamide biosynthetic intermediates as a discrete event, i.e., by cell-free or purified enzyme experiments. Since experiments of this type are not yet feasible, Hutchinson *et al.* presented data from four other literature sources that support the sensibility of their rationalization (*55*). Furthermore, they observed that the D ring oxidation of tetraacetyl [5-^{14}C, 14-^{3}H]-**114** to -**116** with DDQ *in vitro* was attended by a strikingly high retention of ^{3}H (115%) relative to the intermolecular ^{14}C reference label.

D. POSTSTRICTOSAMIDE BIOSYNTHETIC EVENTS

The biogenetic hypothesis for camptothecin thus is valid overall as far as the above results support it. However, the exact sequence of biosynthetic transformations between strictosamide (**114**) and **1** remains unclear. Removal of glucose from **114** is, intuitively, likely to be the step immediately following the formation of **114** by analogy with the biosynthetic fate of strictosidine (**113**) in other higher plants (*52, 56, 57*). We also believe that formation of the pyrrolo[3,4-*b*]quinoline ring system should precede the oxidation of the D ring to a pyridone. This presumption is supported by the fact that **116** and related model compounds are inert to laboratory reagents and conditions known to transform **114** and other tetrahydro-β-carbolines smoothly into 12-hydroxy-9-oxo-9,11-dihydroindolizino[1,2-*b*]quinolines (J. L. Straughn and C. R. Hutchinson, unpublished observations). However, the observation that neither radioactive strictosamide aglucon (**117**) nor **118**, the quinolone analog of **114**, were incorporated into **1** in *C. acuminata* cuttings does not support our presumptions (*58*)..

In our current biosynthetic study of camptothecin, we are investigating the sequential relationships of the pathway intermediates and the mechanism of quinoline and pyridone ring formation. For example, we are testing the idea that the mechanism of formation of the pyrrolo[3,4-*b*]quinoline ring system *in vivo* could proceed by reduction of the ketolactam **119**, derived from strictosamide, to **119a**, followed by ring closure to quinoline **120** via stepwise ionic or concerted electrocyclic processes. The thermal cyclization of **121** to the corresponding analog of **120** supports the latter biosynthetic concept (J. L. Straughn and C. R. Hutchinson, unpublished observations). Reductive removal of the C-7 hydroxy of **117**, which also could form from **119** *in vivo* as it does *in vitro* (*59*), of course is an alternate biosynthetic possibility.

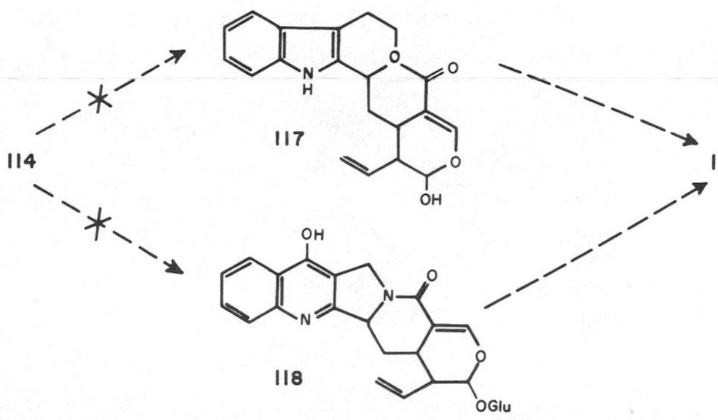

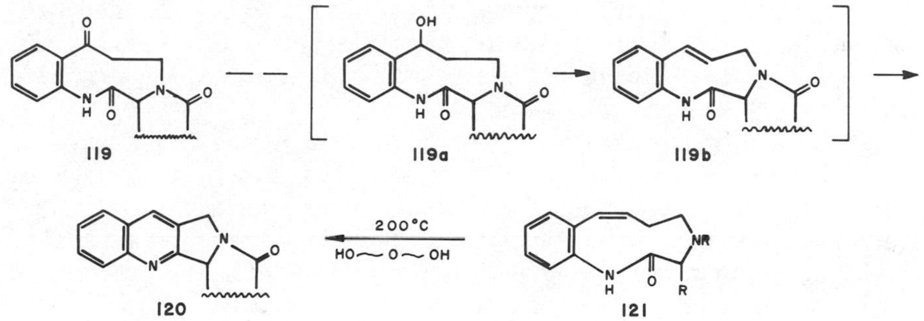

IV. Medicinal Chemistry

A. Physiological Properties

1. Antitumor Activity

a. *In Vivo.* Camptothecin exhibits a broad spectrum of antitumor activity in animal tests. The sodium salt of camptothecin (**122a**) has received more extensive study than the other forms of camptothecin. It showed fair-to-good activity against leukemia L1210, good-to-marked activity against leukemia P388, and was also quite active against P388 sublines resistant to either vincrisitine or adriamycin (*60, 61, 62, 63*). In mice having leukemia L5178 and K1964, **122a** treatment resulted in 70–90% survivors 6 months after death of all controls. It also showed good activity against solid tumors such as Walker 256 carcinoma and Yoshida sarcoma (*64*). It was also evaluated against several other tumor lines not commonly used for routine screening. Therein, it showed fair activity against Colon 38, CD8F$_1$ mammary tumor,

M5 ovarian carcinoma, and ependymoblastoma. However, **122a** was inactive against leukemia L1210 by intravenous administration, whereas camptothecin was active (*65*).

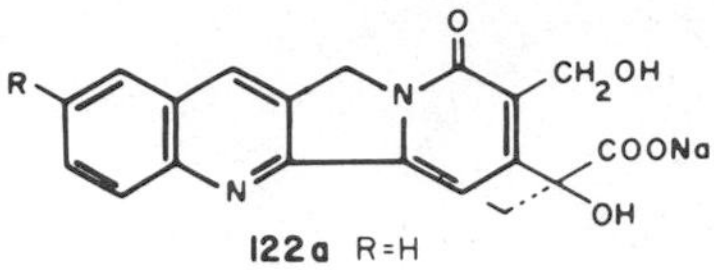

122a R = H
122b R = ONa

The following camptothecin analogs also have been tested for their antitumor activity. 9-Methoxycamptothecin (**5**) showed marked activity against leukemia P388, but only fair activity in L1210. 10-Methoxycamptothecin (**4**) showed marked activity against L1210, Walker 256, and B16 melanoma, but only fair activity in P388. 10-Hydroxycamptothecin (**3**) showed marked activity against L1210, P388, and good activity against B16 melanoma, Walker 256 carcinoma.

At the Shanghai Institute of Materia Medica, it was found that the disodium salt of 10-hydroxycamptothecin (**122b**) exhibited an obvious inhibitory action on both ascites and solid tumors, such as Ehrlich ascites carcinoma, ascetic reticule cell sarcoma, Yoshida sarcoma, sarcoma S37, and Walker carcinoma 256. Oral administration of **122b** also produced a significant inhibition (57.6%) on sarcoma S180 (*64*).

Gordon at Bristol Laboratories found that camptothecin, 10-hydroxycamptothecin, and 10-methoxycamptothecin were active against intraperitoneally implanted Madison M109 lung tumor. Among these compounds, 10-hydroxycamptothecin had the best activity. The activity of 10-hydroxycamptothecin makes it the second best drug ever tested against M109 (adriamycin is the first) and only the third drug (mitomycin C is the other) to produce any cures. Against subcutaneously implanted M109, 10-methoxycamptothecin was slightly better than 10-hydroxycamptothecin, whereas camptothecin was inactive (*67*).

b. *In Vitro.* Camptothecin is a potent inhibitor of the growth of leukemia cells, HeLa cells, and KB cells *in vitro*. Wall *et al.* summarized the results of the inhibition of the growth of 9KB cells by camptothecin and its analogs, and concluded that the 9KB cell cytotoxicity test has excellent predictive value for *in vivo* antitumor activity of camptothecines (*63, 65*).

2. Antiviral Activity

It was reported that camptothecin inhibited the replication of DNA viruses such as adenovirus, vaccinia virus, and herpesvirus, and had no effect on the replication of poliovirus, an RNA virus. Therefore, it would seem that camptothecin is active only when DNA serves as a template for nucleic acid synthesis (*63, 68*).

Tafur *et al.* found that 10-methoxycamptothecin is about eight times more potent than camptothecin as an inhibitor of herpesvirus as measured by plaque reduction method (*10*).

3. Plant Growth-Regulating Activity

Buta and Worley reported in 1976 that camptothecin exhibited selective plant growth inhibition. The growth of tobacco and corn were retarded, whereas no effect was noted on beans and sorghum when a 1×10^{-4} ppm emulsion of camptothecin was applied as a spray. Growth inhibition appeared to be confined to the meristematic portions of the plants (*69*). Later, it was pointed out that camptothecin completely controlled auxiliary bud growth in tobacco plants in the greenhouse. No phototoxicity was observed, and the treated buds on senescing plants remained green (*70*).

4. Insect Chemosterilant Activity

In 1974 Demilo reported that camptothecin was an effective chemosterilant for the housefly. In China, camptothecin was also studied for control of *Dendrolimus punctatus*. It was found that the mortality of larvae, pupae, and adults of *D. punctatus* increased after treatment with camptothecin. The hatchability of eggs decreased after treatment with 0.05% camptothecin (*71*).

B. STRUCTURE AND ANTITUMOR ACTIVITY RELATIONSHIPS

1. Optical Isomer and Antitumor Activity

It was found at the Shanghai Institute of Materia Medica that the disodium salt of *dl*-10-hydroxycamptothecin (*dl*-**122b**) showed antitumor activity similar to that of the disodium salt of (*S*)-10-hydroxycamptothecin (**122b**) on Ehrlich ascites carcinoma, ascitic reticule cell sarcoma, or lymphosarcoma in mice at a dosage twice that of **122a**. The subacute LD_{50} of *dl*-**122b** was twice that of **122**. Therefore it could be considered that the 20-(*R*) isomer is an inactive but less toxic component (*66, 14*). Wall *et al.* also reported that *dl*-**1** showed one-half the activity of natural **1** in the P388 assay (**65**). Thus, it is evident that the 20-(*S*) configuration of camptothecin is an absolute requirement for the antitumor activity.

2. Modifications in Ring A

Hartwell and Abbott established early that nuclear substitution in ring A of camptothecin has little effect on antitumor activity based on the activity of **3** and **4** (*72*). Other oxygenated camptothecines **5**, **6**, and **7** showed more activity than **1** in some animal tests.

Pan *et al.* reported in 1975 the partial synthesis of eleven 12-substituted camptothecin analogs, **123–126**, from **1** via nitration, reduction, diazotization, and suitable transformations. It was found that the 12-chloro analog **123** was more active than **1** in the leukemia L615 assay, and compounds **124** and **125** were more active against Ehrlich ascites carcinoma than **1** (*73*).

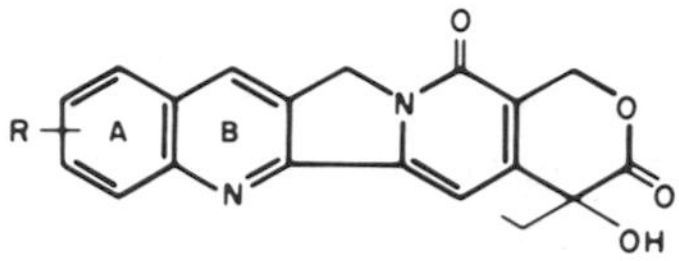

123	R=12-Cl	**127**	R=10-Cl
124	R=12-OH	**128**	R=10-OCH₂COOH
125	R=12-OCH₃	**129**	R=10-F
126	R=12-NO₂, NH₂, NHCOCH₃,	**130**	R=11-F
	Br, SH, CN, COOH, COOCH₃	**131**	R=10-OCH₂CH₂NEt₂ · HCl

Cai *et al.* prepared the ring A analogs dl-**127**–*dl*-**130** by total synthesis. Preliminary pharmacological tests revealed that the sodium salt of compounds *dl*-**127**, *dl*-**128**, and *dl*-**129** were more active than the sodium salt of **1** in Ehrlich ascites carcinoma assay, but compound *dl*-**130** was inactive (*74, 75*).

Wall *et al.* synthesized ring A analogs **128**, **131**, *dl*-**132**, *dl*-**133**, and *dl*-**134**. Compounds **128** and **131** were water soluble, whereas in compounds **132** and **133** ring A was replaced by a heterocyclic ring and **134** was a compound with an additional benzene ring fused onto ring A. Contrary to Cai's result, compound **128** when tested in the P388 system was completely inactive and was also inactive in 9KB cell assay. Another water soluble compound **131** showed good activity in the P388 and L1210 test and had a better therapeutic index than **1** and **3**. Wall *et al.* believe that the activity of **131** may be due to its conversion to **3** *in vivo*. The 12-aza analog **132** was less active and much less potent than **1** in the P388 assay. Compound **134** had activity comparable to

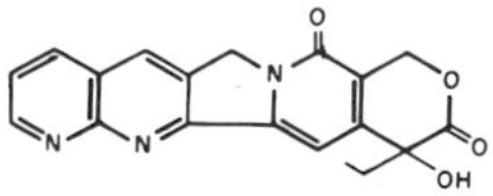

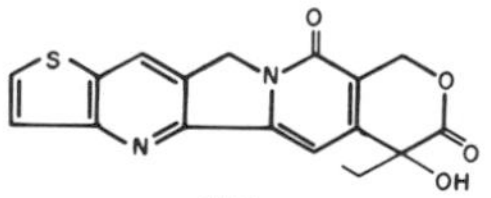

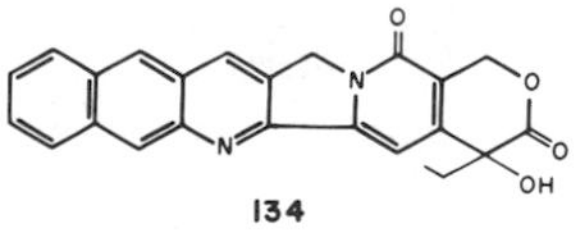

1. The biological activity of the thiophene analog **133** is not available at present (*41*).

Wall *et al.* thus have proposed that substitution in ring A by a small group is permissible, whereas a bulky group substituted in ring A leads to inactivation, probably because of steric or electronic inhibition of binding to DNA or RNA (*65*). However, based on the results mentioned above, i.e., the inactivity of compounds **126** and **130** and the different results of testing of compound **128**, this rationale needs further investigation.

3. Modifications in Ring B

Wall *et al.* prepared N_1 oxide **135** from camptothecin during the structure elucidation of camptothecin and found that the antitumor activity of this compound was decreased (*3*).

Winterfeldt *et al.* prepared *dl*-7-chlorocamptothecin (**136**) and *dl*-7-methoxycamptothecin (**137**) during their development of the total synthesis of *dl*-**1**. They also synthesized the *dl*-7-acetoxy analog **138** (*76*). It was found that **136** exhibited more activity than *dl*-**1**, but the *dl*-7-methoxy analog **137** was inactive (*77*).

Sugasawa *et al.* synthesized the tetrahydro analog **139** and found that each epimer of **139** showed no activity in the *in vivo* L1210 assay (*44*).

Recently, Miyasaka *et al.* reported the synthesis of 7-hydroxymethylcamptothecin, 7-acetoxymethylcamptothecin (**140**), and some related analogs (**141**) from camptothecin. Unfortunately, the pharmacological data are not yet available (*78*).

Thus it can be deduced that the basicity at N-1 and/or a flat molecular shape may play an important role in the biological activity of camptothecin.

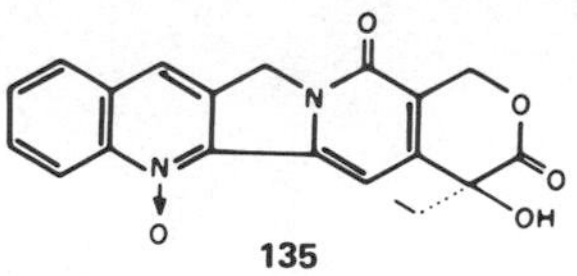

135

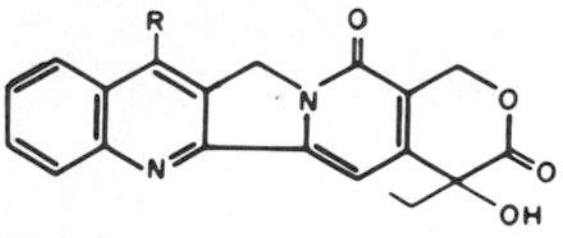

136 R=Cl
137 R=OCH₃
138 R=OCOCH₃
140 R=CH₂OH, CH₂OCOCH₃
141 R=CHO, COOH, CH₂OC(CH₂)₂COOH

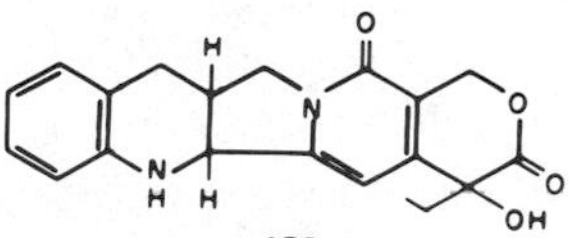

139

4. Modifications in Ring C

Hutchinson and Adamovics found that 5-hydroxy- and 5-acetoxycamp-tothecin (**142**) were completely inactive in the animal antitumor assays (*79*). Winterfeldt *et al.* reported the synthesis of *dl*-5-ethylcamptothecin (**143**) and found that this compound also was inactive *in vivo* (*77*). Miyasaka *et al.* also prepared some ring C analogs (**144**) but did not report on their pharamaco-logical activity (*78*).

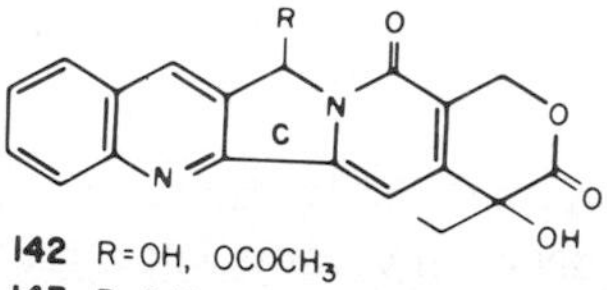

142 R=OH, OCOCH$_3$
143 R=C$_2$H$_5$
144 R=OCH$_3$, OC$_4$H$_9$(*n*), OCOC$_6$H$_5$

5. Modifications in Ring D

Danishefsky *et al.* prepared isocamptothecin (**145**) and isohomocamp-tothecin (**146**) during their total synthesis of *dl*-**1**. Compound **145** showed a slight activity *in vitro* in the inhibition of nucleic acid synthesis. Compound **146**, in which the hydroxy group of **145** was replaced by an hydroxymethyl group, was essentially inactive in the same bioassay (*27*).

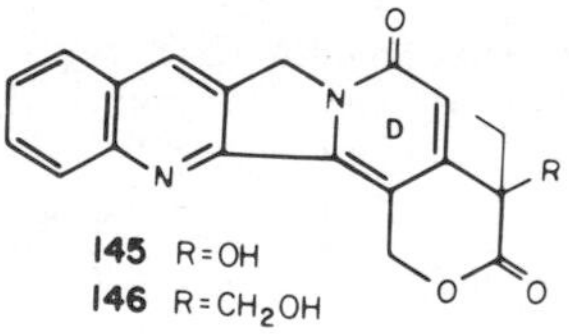

145 R=OH
146 R=CH$_2$OH

6. Modifications in Ring E

a. 20-Hydroxy Group. Camptothecin acetate (**147**) exhibited markedly reduced activity in the L1210 animal test. 20-Deoxy-(**8**), 20-chloro-(**148**), 20-allyl-(**149**), 20-ethyl-(**150**), and 20-hydroxymethylcamptothecin (**151**) were inactive in the animal tests (*3, 27, 44*). It seems that a hydroxy group at C-20 with the (*S*) configuration is required for the antitumor activity (Section IV, B, 1).

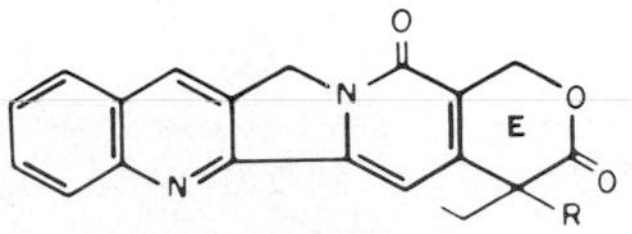

147 R = OCOCH₃
148 R = Cl
149 R = CH₂CH = CH₂
150 R = CH₂CH₃
151 R = CH₂OH

b. 20-Ethyl Group. Sugasawa *et al.* reported that when the 20-ethyl group of camptothecin was replaced by allyl (**152**), propargyl (**153**), or benzyl (**154**), this resulted in retention of activity in the *in vivo* L1210 test. Compound **152** was even more active than *dl*-1 (*44*). However, the antitumor activity was decreased when the 20-ethyl group was replaced by a phenacyl (**156**) (*44*).

Wall *et al.* prepared 18-methoxycamptothecin (**155**) and found that this compound exhibited activity comparable to that of *dl*-1 (*41*).

Thus, it appears that the 20-ethyl group can be replaced by a suitable substituent and still retain or even increase the antitumor activity.

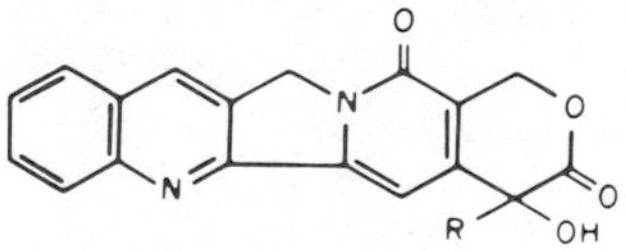

152   R = CH₂CH = CH
153   R = CH₂C ≡ CH
154   R = CH₂C₆H₅
155   R = CH₂CH₂OCH₃
156   R = CH₂COC₆H₅

c. Lactone Ring. Camptothecin, when dissolved in sodium hydroxide solution, results in opening of the lactone ring to form its sodium salt **122a.** This salt was often used in the pharmacological studies and clinical trials. Although **122a** showed marked activity against a variety of animal tumors (Section IV, A, 1), it was found by Wall *et al.* that the antitumor activity of this compound was only one-tenth that of camptothecin when administered by the intravenous route. Wall *et al.* explained that the pH of blood is 7.2 and at this pH the sodium salt of **1** cannot regenerate the α-hydroxy lactone ring required for antitumor activity (*54*).

The lactol **157** that was obtained by reduction of **1** with sodium borohydride was essentially inactive *in vivo* (*3*).

Wall *et al.* prepared the C-21 methylamide **158** by reaction of camptothecin with methyl amine. Compound **158** showed about 60% the activity of **1** in the animal L1210 assay. Adamovics and Hutchinson converted camptothecin to its C-21 isopropylamide (**159**) and five C-17 substituted analogs (**160 – 164**). Compounds **160** and **162** showed activity about 85% that of **1** in the P388 test *in vivo*, whereas the basic analogs **163** and **164** were inactive (*80*). It is noteworthy that a compound **165** prepared by Sugasawa *et al.* retained activity comparable to *dl*-**1** in spite of the lack of the lactone ring (*44*).

Therefore, it seems that the lactone ring is not an absolute requirement for the antitumor activity.

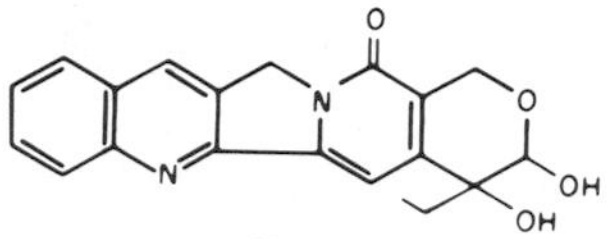

157

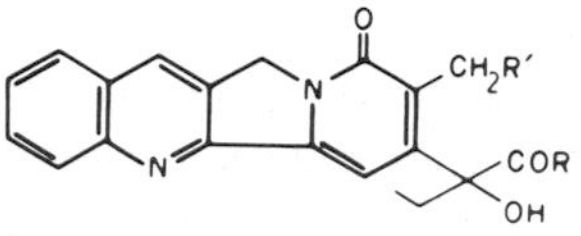

165

158	R = NHCH$_3$	R' = OH
159	R = NHCH(CH$_3$)$_2$	R' = OH
160	R = NHCH(CH$_3$)$_2$	R' = OCOCH$_3$
161	R = NHCH(CH$_3$)$_2$	R' = OCOC$_5$H$_{11}$
162	R = NHCH(CH$_3$)$_2$	R' = OCOCH=CH$_2$
163	R = NHCH(CH$_3$)$_2$	R' = c-C$_4$H$_8$N
164	R = NHCH(CH$_3$)$_2$	R' = —

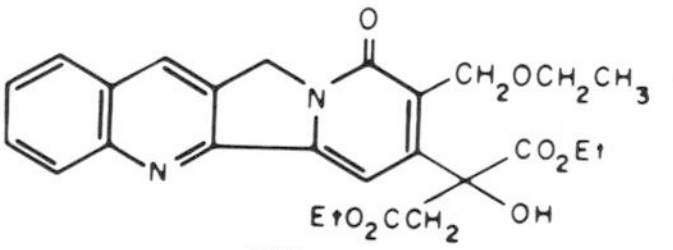

7. Modifications in Both Ring A (or B) and E

Cai *et al.* prepared *dl*-10-methoxy-20-deethyl-20-benzyl camptothecin (**166**) and found that this compound was more active than **1** in the Ehrlich ascites carcinoma assay (*74*).

Winterfeldt *et al.* reported the synthesis of *dl*-7-methoxy-20-deethyl-20-substituted analogs **167** and **168** and found that these compounds exhibited moderate antitumor activity in the animal P388 test. Thus it seems that 20-ethyl group substituted by carbethoxymethyl or morpholinocarbomethyl

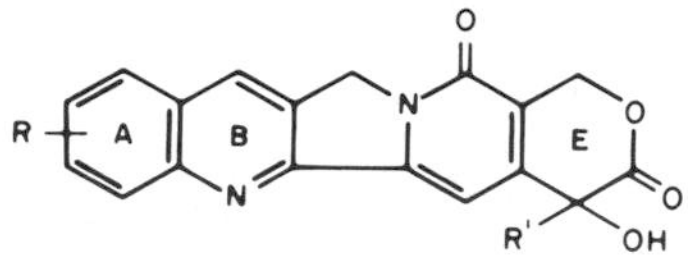

166	R = 10-OCH$_3$	R' = CH$_2$C$_6$H$_5$
167	R = 7-OCH$_3$	R' = CH$_2$COOC$_2$H$_5$
168	R = 7-OCH$_3$	R' = CH$_2$CON O

group would increase the antitumor activity as 7-methoxycamptothecin itself (**137**) was inactive (*77, 81*).

8. Miscellaneous

Since the presence of an α-hydroxy-δ-lactone in camptothecin was thought to be responsible for the antitumor activity of camptothecin, a number of rings DE and rings CDE analogs of **1** have been synthesized. Unfortunately, the pharmacological test of only a few of these analogs has been reported. The bicyclic ring DE analogs **169–173** were inactive in the L1210 animal test. The tricyclic ring CDE analog **174** was also inactive (*82–84*).

169 R = CH₃   R' = H
170 R = n-C₄H₉   R' = CH₃
171 R = CH₂C₆H₅ R' = H
172 R = CH₂

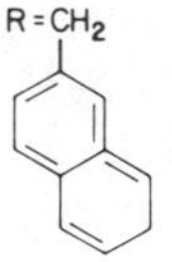

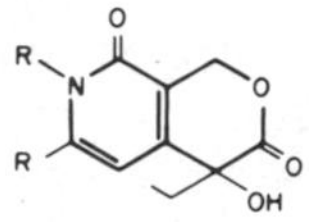

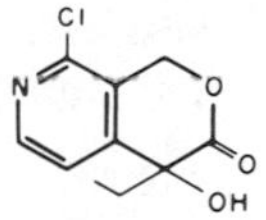

173 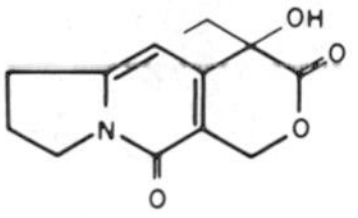   174

On the basis of these results, it is evident that ring A and ring B might play an important role for the antitumor activity of camptothecin. Flurry and Howland predicted that a tetracyclic system containing rings BCDE would be the minimum-size structure compatible with biological activity in the camptothecin series; but such a compound has not been synthesized as yet (*65*).

C. MECHANISM OF ACTION

The novel structure and significant antitumor activity of camptothecin stimulated several groups of researchers to investigate its effect on whole animals and on isolated mammalian cells in an attempt to understand its molecular mechanism of action (*63*).

The principal effect of camptothecin on cultured mammalian cells is the potent inhibition of polynucleic acid biosynthesis. This apparently is not due to the inhibition of nucleotide biosynthesis or the enzymatic activity of DNA and RNA polymerases, and the effect on RNA formation is easily reversible when the drug is removed. The drug affects the biosynthesis of ribosomal RNA more than other types of cellular RNA. It does not inhibit significantly protein biosynthesis.

Camptothecin induces single-strand breaks in cellular DNA in intact HeLa cells as viewed by alkaline sucrose density gradient analysis; this effect is reversible. Since camptothecin does not affect the enzymes involved in DNA biosynthesis, its inhibition of DNA formation appears to be the result of some effect on the template function of DNA. Most investigators have concluded that the latter event is the primary determinant of camptothecin's cytotoxicity.

Camptothecin is an effective inhibitor of the replication of viruses containing DNA, but not those containing primarily RNA. Its effect here again is on DNA biosynthesis and is reversible on the drug's removal. The observations with viral systems corroborate the conclusion that camptothecin is cytotoxic because of a disruption of the normal function of DNA in cellular ontogenesis. Consequently, the molecular mechanism of action of camptothecin should include a DNA-binding component and a mechanism for covalent bond breakage in polydeoxyribonucleotides.

Wall *et al.* suggested that camptothecin binds to the polynucleic acid in a manner similar if not identical to intercalation. They proposed that the lactone moiety in ring E may be located in a favorable orientation to form a covalent bond with a nucleophilic group appropriately located on the polynucleic acid (*65*).

Since several camptothecin analogs that are capable of inhibiting polynucleic acid biosynthesis and causing the fragmentation of DNA *in vitro* are completely inactive as antitumor agents in the animal assays, it may be that camptothecin must be "activated" *in vivo* to become cytotoxic. Moore proposed a bioreductive alkylation mechanism. The key features of the mechanism are the initial *in vivo* reduction of the quinoline nucleus to the dihydro form **176**. Subsequent eliminative ring opening of the lactone nucleus would give **177**. This compound can lose water to give **178** which has

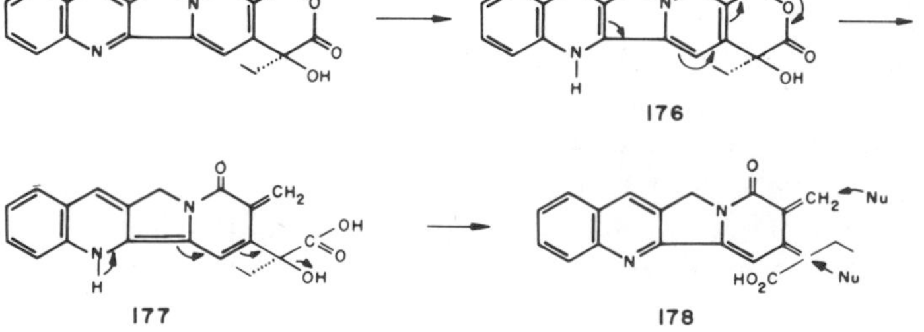

176

177   178

two reactive sites for alkylation by a nucleophile (Nu), the α-methylene lactam and the extended π system in conjugation with the quinoline ring (85).

Lown et al. (86) proposed a free-radical mechanism. Photosensitization of camptothecin may generate radicals to attack DNA or, in the presence of oxygen, generate hydroperoxy radicals. The photolytic reaction of campto thecin itself proceeded via formation of racemized photolabile campto- thecin hemiacetal 181 which suggested two alternative free-radical mecha nisms. In the anaerobic pathway, photodecarboxylation of the alkaloid may generate a diradical that can collapse to 181 or abstract an H atom from DNA

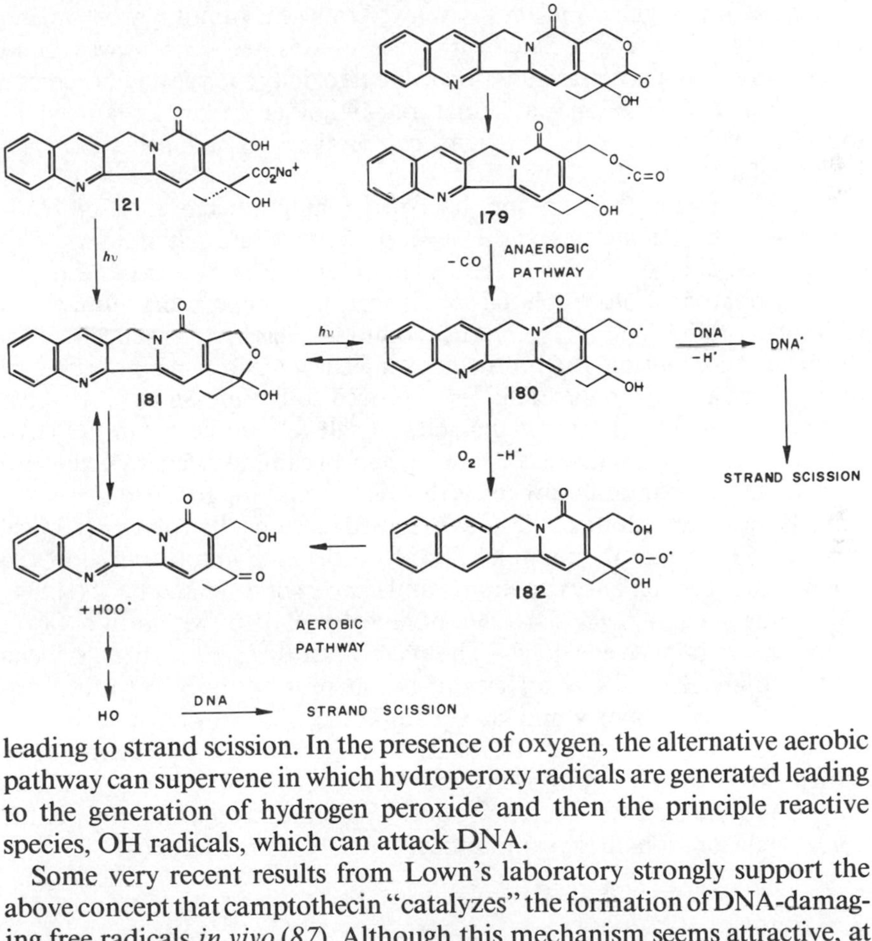

leading to strand scission. In the presence of oxygen, the alternative aerobic pathway can supervene in which hydroperoxy radicals are generated leading to the generation of hydrogen peroxide and then the principle reactive species, OH radicals, which can attack DNA.

Some very recent results from Lown's laboratory strongly support the above concept that camptothecin "catalyzes" the formation of DNA-damag- ing free radicals in vivo (87). Although this mechanism seems attractive, at this moment it is not certain that the antitumor activity of camptothecin is due to its potential ability to act as an alkylating agent or as a source of DNA-damaging free radicals.

D. Clinical Use

1. Antitumor Agent

The first report of the clinical use of camptothecin appeared in 1970. The sodium salt of camptothecin (**122a**) was administered experimentally to 18 patients with various advanced cancers of the gastrointestinal (GI) tract. Improvement was noted in 11 cases (*88*). Unfortunately, subsequent evaluation of this drug led to discouraging results. Only two of sixty-one patients suffering from advanced GI adenocarcinoma showed partial objective responses (*89*). In another group, only three of fifteen patients with advanced disseminated malignant melanoma showed transient tumor regression and no clinical benefit was associated with these responses (*90*). It also was found that camptothecin possessed dose-dependent toxicity such as hematopoietic depression, diarrhea, alopecia, hematuria, and other urinary irritant symptoms (*91*). Because of its high toxicity, camptothecin is no longer of interest in clinical testing in the United States.

At present, camptothecin and 10-hydroxycamptothecin are used clinically only in the People's Republic of China. The sodium salt of camptothecin was found to be effective in treatment of GI cancer, cancer of head and neck, lymphosarcoma, trophoblastic cancer, and some other tumors. In China camptothecin itself, but not its sodium salt, has been used in clinics in the form of suspensions via intravenous injection for treatment of primary liver cancer and leukemia. This form proved to be safe and effective. Its toxicity was less than that of the sodium salt of camptothecin (*92*). An experimental study for decreasing the toxicity of camptothecin by coadministration of monoammonium glycyrrhetate was also reported (*64*).

10-Hydroxycamptothecin (**3**) was first used in China. In a phase II clinical study on 63 patients, the sodium salt of 10-hydroxycamptothecin showed a therapeutic effect on liver carcinoma and cancers of head and neck (chiefly carcinoma of salivary glands). Eight of nineteen patients with primary liver carcinoma responded effectively. The effective rate in case of cancers of head and neck was 39.8%. The toxicity of the sodium salt of **3** was much less than that caused by the sodium salt of camptothecin, in particular the irritant action on the urinary tract. Researchers suggested that the difference of toxicity caused by camptothecin and 10-hydroxycamptothecin probably was due to the different excretion route, for camptothecin mainly via the urine while for 10-hydroxycamptothecin mainly via the feces (*66*).

2. Treatment of Psoriasis

Camptothecin was also used in China for treatment of psoriasis, a skin disease. A solution of camptothecin in dimethyl sulfoxide was used topically for 30 cases of psoriasis vulgaris and this therapy proved effective in all 30 cases. All cases except two had had previous treatment by various other methods (93).

REFERENCES

1. R. E. Perdue, M. E. Wall, J. L. Hartwell and B. J. Abbott, *Lloydia* **31**, 229 (1968).
2. M. E. Wall, M. C. Wani, C. E. Cook, K. H. Palmer, A. T. McPhail and G. A. Sim, *J. Am. Chem. Soc.* **88**, 3888 (1966).
3. M. E. Wall, *Biochem. Physiol. Alkaloide Int. Symp., 4th,* 77 (1969, 1972).
4. A. T. McPhail and G. A. Sim, *J. Chem. Soc. B* 923 (1968).
5. M. Shamma, *Experientia* **24**, 107 (1968).
6. A. G. Schultz, *Chem. Rev.* **73**, 385 (1973).
7. M. G. Wani and M. E. Wall, *J. Org. Chem.* **34**, 1364 (1969).
8a. T. R. Govindachari and N. Viswanathan, *Indian J. Chem.* **10**, 453 (1972).
8b. J. S. Agarwal and R. P. Rastogi, *Indian J. Chem.* **11**, 969 (1973).
9. T. R. Govindachari, K. R. Ravindranath, and N. Viswanathan, *J. Chem. Soc. Perkin Trans 1,* 1215 (1974).
10. S. Tafur, J. D. Nelson, D. C. DeLong, and G. H. Svoboda, *Lloydia* **39**, 261 (1976).
11. S. P. Gunsekera, M. M. Badaw, G. A. Cordell, N. R. Farnsworth, and M. Chitnis, *J. Nat. Prod.* **42**, 475 (1979).
12. J.-S. Hsu, T.-Y. Chao, L.-T. Lin, and C.-F. Hsu, *Hua Hsueh Hsueh Pao* **35**, 193 (1977); CA **90**, 28930k (1979).
13. L.-T. Lin, C.-C. Sung, and J.-S. Hsu, *K'o Hsueh Túng Pao* **24**, 478 (1979).
14. J.-C. Cai, M.-G. Yin, A.-Z. Min, D.-W. Feng, and X.-X. Zhang, *Hua Hsueh Hsueh Pao* **39**, 171 (1981); *K'o Hsueh Túng Pao* **22**, 269 (1977).
15. J.-C. Cai, C.-F. Tang, and Y.-S. Gao, *Yao Hsueh Hsueh Pao,* submitted.
16. L.-T. Lin, T.-Y. Chao, and J.-S. Hsu, *Hua Hsueh Hsueh Pao* **35**, 227 (1977); CA **89**, 22078s (1978).
17. J. G. Buta and M. J. Novak, *Ind. Eng. Chem. Prod. Res. Dev.* **17**, 160 (1978).
18. K.-P. Chu, L.-T. Lin, W.-C. Pan, T.-C. Chou, Y.-C. Huang, R.-Y. Hsieh, and S.-F. Liang, *K'o Hsueh Túng Pao* **23**, 761 (1978); CA **90**, 119733q (1979).
19. K. Sakato, H. Tanaka, N. Mukai, and M. Misawa, *Agr. Biol. Chem.* **38**, 217 (1974).
20. K. Sakato and M. Misawa, *Agr. Biol. Chem.* **38**, 491 (974).
21. J. A. Adamovics, J. A. Cina, and C. R. Hutchinson, *Phytochemistry* **18**, 1085 (1979).
22. M. Shamma and V. St. Georgiev, *J. Pharm. Sci.* **63**, 163 (1974).
23. C. R. Hutchinson, *Tetrahedron* **37**, 1047 (1981).
24. G. Stork and A. G. Schultz, *J. Am. Chem. Soc.* **93**, 4074 (1971).
25. R. Volkmann, S. Danishefsky, J. Eggler, and D. M. Solomon, *J. Am. Chem. Soc.* **93**, 5576 (1971).
26. S. Danishefsky, S. J. Etheredge, R. Volkmann, J. Eggler, and J. Quick, *J. Am. Chem. Soc.* **93**, 5576 (1971).
27. S. Danishefsky, R. Volkmann, and S. B. Horwitz, *Tetrahedron Lett.* 2521 (1973).
28. J. Quick, *Tetrahedron Lett.* 327 (1977).
29. J. C. Bradley and G. Büchi, *J. Org. Chem.* **41**, 699 (1976).
30. A. S. Kende, T. J. Bentley, R. W. Draper, J. K. Jenkins, M. Joyeux, and I. Kubo, *Tetrahedron Lett.* 1307 (1973).
31. E. Winterfeldt, *Liebigs Ann. Chem.* **745**, 23 (1971).

32. E. Winterfeldt, T. Korth, D. Pike, and M. Boch, *Angew. Chem. Int. Ed. Engl.* **11**, 289 (1972). M. Boch, T. Korth, D. Pike, H. Radunz, and E. Winterfeldt, *Chem. Ber.* **105**, 2126 (1976).
33. J. Warneke and E. Winterfeldt, *Chem. Ber.* **105**, 2120 (1972).
34. K. Krohn and E. Winterfeldt, *Chem. Ber.* **108**, 3030 (1975).
35. K. Krohn, H. W. Ohlendorf, and E. Winterfeldt, *Chem. Ber.* **109**, 1389 (1976).
36. T. Kametani, T. Ohsawa, and M. Ihara, *Heterocycles* **14**, 951 (1980); *J. Chem. Soc. Perkin Trans. 1,* 1563 (1981).
37. C. S. F. Tang and H. Rapoport, *J. Am. Chem. Soc.* **94**, 8615 (1972); C. S. F. Tang, C. J. Morrow, and H. Rapoport, *J. Am. Chem. Soc.* **97**, 159 (1975).
38. M. L. Rueppel and H. Rapoport, *J. Am. Chem. Soc.* **94**, 3877 (1972); D. L. Lee, C. J. Morrow, and H. Rapoport, *J. Org. Chem.* **39**, 893 (1974).
39. E. J. Corey, D. E. Crouse, and J. E. Anderson, *J. Org. Chem.* **40**, 2140 (1975).
40. Shanghai No. 5 and No. 12 Pharmaceutical Plant, Shanghai Institute of Pharmaceutical Industrial Research, and Shanghai Institute of Materia Medica, *K'o Hsueh Túng Pao* **21**, 40 (1976); CA **84** 122100n (1976); *Scientia Sinica* **21**, 87 (1978).
41. M. C. Wani, P. E. Ronman, J. T. Lindley, and M. E. Wall, *J. Med. Chem.* **23**, 554 (1980).
42. M. C. Wani, H. F. Campbell, G. A. Brine, J. A. Kepler, M. E. Wall, and S. G. Levine, *J. Am. Chem. Soc.* **94**, 3631 (1972).
43. T. Sugasawa, T. Toyoda, and K. Sasakura, *Tetrahedron Lett.* 5109 (1972); T. Sugasawa, K. Sasakura, and T. Toyoda, *Chem. Pharm. Bull.* **22**, 763 (1974).
44. T. Sugasawa, T. Toyoda, N. Uchida, and K. Yamaguchi, *J. Med. Chem.* **19**, 575 (1976).
45. A. I. Meyers, R. L. Nolen, E. W. Collington, T. A. Narwid, R. C. Strickland, *J. Org. Chem.* **38**, 1974 (1973).
46. H. G. M. Walraven, and U. K. Pandit, *Tetrahedron Lett.* 4507 (1975); *Tetrahedron* **36**, 321 (1980).
47. M. Shamma, D. A. Smithers, and V. St. Georgiev, *Tetrahedron* **29**, 1949 (1973).
48. T. Kametani, H. Takeda, H. Nemoto, and K. Fukumoto, *J. Chem. Soc. Perkin Trans. 1,* 1825 (1975).
49. E. Wenkert, K. G. Dave, R. G. Lewis, and P. W. Sprague, *J. Am. Chem. Soc.* **89**, 6741 (1967).
50. E. Winterfeldt, *Liebigs Ann. Chem.* **745**, 23 (1971); J. Warneke and E. Winterfeldt, *Chem. Ber.* **105**, 2120 (1972).
51. A. R. Battersby, A. R. Burnett, and P. G. Parsons, *J. Chem. Soc. C.* 1193 (1969).
52. C. R. Hutchinson, A. M. Heckendorf, P. E. Daddona, E. Hagaman, and E. Wenkert, *J. Am. Chem. Soc.* **96**, 5609 (1974).
53. G. A. Cordell, *Lloydia* **37**, 219 (1974).
54. G. M. Sheriha and H. Rapoport, *Phytochemistry* **15**, 505 (1976).
55. C. R. Hutchinson, A. H. Heckendorf, J. L. Straughn, P. E. Daddona, and D. E. Cane, *J. Am. Chem. Soc.* **101**, 3358 (1979).
56. J. Stockigt and M. H. Zenk, *FEBS Lett.* **79**, 233 (1977); *J. Chem. Soc. Chem. Commun.* 646 (1977); A. I. Scott, S. L. Lee de Capite, M. G. Culver, and C. R. Hutchinson, *Heterocycles* **7**, 979 (1977).
57. M. Rueffer, N. Nagakura, and M. H. Zenk, *Tetrahedron Lett.* 1593 (1978); R. T. Brown, J. Leonard, and S. K. Sleigh, *Phytochemistry* **17**, 899 (1978).
58. A. H. Heckendorf and C. R. Hutchinson, *Tetrahedron Lett.* 4153 (1977).
59. C. R. Hutchinson, G. J. O'Loughlin, R. T. Brown, and S. B. Fraser, *J. Chem. Soc., Chem. Commun.* 928 (1974).
60. R. C. Gallo, P. Whang-Pen, R. H. Adamson, *J. Nat. Cancer Inst.* **46**, 789 (1971).
61. B. J. Abbott, *Cancer Treat. Rep.* **60**, 1007 (1976).

62. Dr. M. Suffness, personal communication (1978).
63. S. B. Horwitz, *in* "Antibiotics III Mechanism of Action of Antimicrobial and Antitumor Agents" (J. W. Corcoran and F. E. Hahn, eds.), p. 48. Springer, New York (1975).
64. Shanghai Institute of Materia Medica, *Chinese Med. J.* **55**, 274 (1975).
65. M. E. Wall and M. C. Wani, *in* "Anticancer Agents Based on Natural Product Models" (J. M. Cassady and J. D. Douros, eds.), p. 417. Academic Press, New York (1980).
66. Shanghai Institute of Materia Medica, *in* "Adv. Med. Oncol. Res. Educ., Proc. Int. Cancer Congr., 12th, 1978, vol. 5" (B. W. Fox, ed.), p. 105. Pergamon, Oxford and New York (1979); *Chinese Med. J.* **58**, 598 (1978).
67. Dr. M. Gordon, personal communication (1979).
68. Y. Becker and U. Olshevsky, *Isr. J. Med. Sci.* **9**, 1578 (1973).
69. J. G. Buta and J. F. Worley, *J. Agric. Food Chem.* **24**, 1085 (1976).
70. J. F. Worley, D. N. Spaulding, and J. G. Buta, *Tobacco* **181**(8), 26 (1979); CA **91**, 50993a (1979).
71. Honan Institute of Forestry, *K'un C'hung Hsueh Pao* **21**, 108 (1978); CA **89**, 1665r (1978).
72. J. L. Hartwell and B. Abbott, *Adv. Pharmacol. Chemotherapy* **7**, 137 (1969).
73. P.-C. Pan, S.-Y. Pan, Y.-H. Tu, S.-Y. Wang, and T.-Y. Owen, *Hua Hsueh Hsueh Pao* **33**, 73 (1975).
74. J.-C. Cai, L.-L. Chen, and Y.-S. Gao, Papers at the 4th National Congress of the Chinese Pharmaceutical Society, Nanjing, China (1979).
75. J.-C. Cai, Z.-K. Leng, Y.-F. Ren, and Y.-S. Gao, Abs. 182nd Am. Chem. Soc. National Meeting, New York, NY, SORT 16 (1981).
76. H. W. Ohlendorf, R. Stranghóner, and E. Winterfeldt, *Synthesis* 741 (1976).
77. Prof. E. Winterfeldt, personal communication (1979).
78. T. Miyasaka, S. Sawada, and K. Nokata, *Heterocycles* **16**, 1713, 1719 (1981).
79. J. A. Adamovics and C. R. Hutchinson, unpublished observations.
80. J. A. Adamovics and C. R. Hutchinson, *J. Med. Chem.* **22**, 310 (1979).
81. E. Baxmann and E. Winterfeldt, *Chem. Ber.* **111**, 3403 (1978).
82. J. A. Bristol, D. L. Comins, R. W. Davenport, M. J. Kane, R. E. Lyle, J. R. Maloney, D. E. Portlock, and S. B. Horwitz, *J. Med. Chem.* **18**, 535 (1975).
83. M. E. Wall, H. F. Campbell, M. C. Wani, and S. G. Levine, *J. Am. Chem. Soc.* **94**, 3632 (1972).
84. S. Danishefsky and S. J. Etheredge, *J. Org. Chem.* **39**, 3430 (1974).
85. H. W. Moore, *Science* **197**, 527 (1977).
86. J. W. Lown and H.-H. Chen, *Biochem. Pharm.* **29**, 905 (1980).
87. J. W. Lown, H.-H. Chen, and J. A. Plambeck, *Biochem. Pharm.,* submitted (1980).
88. J. A. Gottlieb, A. M. Guarino, J. B. Call, V. T. Oliverio, and J. B. Block, *Cancer Chemother. Rep.* **54**, 461 (1970).
89. C. G. Moertel, A. J. Schutt, R. J. Reitmeier, and R. G. Hahn, *Cancer Chemother. Rep.* **56**, 95 (1972).
90. J. A. Gottlieb and J. K. Luce, *Cancer Chemother. Rep.* **56**, 103 (1972).
91. U. Schaeppi, R. W. Fleischman, D. A. Cooney, *Cancer Chemother. Rep., Part 3* **5**, 25 (1974).
92. Shanghai Institute of Materia Medica, *K'o Hsueh Tung Pao* **22**, 552 (1977).
93. C.-Y. Chiao and H.-S. Li, *Chinese Med. J.* (*Peking, Chinese ed.*) **4**, 208 (1974); *Biol. Abstr.* **60**, 39404 (1975).

—CHAPTER 5—

AMPHIBIAN ALKALOIDS

BERNHARD WITKOP

National Institutes of Health, Bethesda, Maryland

AND

EDDA GÖSSINGER

Institut für Organische Chemie der Universität Wien, Austria

I. Introduction

The introduction and epilogue of this chapter share an important concept: the phenomenon of *receptors* that selectively bind to specific agents; in this case the novel amphibian toxins whose isolation, structure, and syntheses will be described. In the end, this means that to some readers the expansion of

THE ALKALOIDS, VOL. XXI

ISBN 0-12-469521-3

our knowledge of the chemistry and syntheses of novel natural products, exciting as they may be to the organic chemist or for the art or skill of organic synthesis, is only a prologue. The true aim is summed up in the part dealing with pharmacology as an advance in the biomedical sciences, not an entirely fortuitous event, since all research not dealing with synthesis was initiated and expatiated on at the National Institutes of Health, an institution supported by public funds.

The important principle of freedom of research, even in a government institution, is here exemplified; a too narrow interpretation of scientific relevance and insistence on a fast solution of problems related to public health might have terminated the project after the first few exploratory expeditions into the impenetrable jungle of western Colombia (1).

EARLY RESEARCH ON TOXINS AND RECEPTORS

Not long after the centennial of Claude Bernard (1813–1878), the great French physiologist, we are reminded of his use of the South American arrow poison curare (2) for the more recent clarification of the mechanisms of neural excitability (3), ion flux (4, 5), and synaptic transmission (6). In this intervening century biochemical and physiological techniques have been refined. Two important ideas were the lodestars that guided the investigator: One is the concept of *receptor* first postulated by Paul Ehrlich (1854–1915) as an atom group, side chain, or receptive substance that has a specific combining property for a particular toxin, antigen, substrate, or drug (7, 8). This concept has made possible the use of neurotoxins as tools in neurobiology (9), the isolation of the acetylcholine receptor as an allosteric membrane protein, the formulation of changes of ionic permeability, and generation of action potentials in the precise terms of molecular biology and electrophysiology.

The second concept is that of *specificity,* again going back to Paul Ehrlich's role in the foundation of immunology and his development of antitoxins against the plant poisons ricin or abrin or against bacterial toxins. Nervous functions are now approached by the use of antibodies that have one major characteristic in common with neurotoxins: specificity. Both receptors and specificity of toxins and antibodies have an evolutionary basis, an aspect that was not overlooked by the early pioneers such as Claude Bernard or John Newport Langley. The same evolutionary pattern, in a manner yet mysterious, enters into the phenomenon of "chemical convergence" but fails to explain the simultaneous occurrence of tetrodotoxin in pufferfish, Californian newts, and South American frogs and marine mollusks (*vide infra*).

John Newport Langley (1852–1925) referred to Claude Bernard's classic experiment when he carried the receptor concept into synaptic studies with nicotine and curare, the beginning of the terms agonists, antagonists, and later desensitization as well as allosteric interaction (*10*). It is safe to say that of all membrane-bound receptors, our knowledge of the nicotinic acetylcholine receptor (*11*) has progressed farthest because of the use and discovery of selective neurotoxins ranging from curare to amphibian toxins (*9*).

In the laboratory of Heinrich Wieland (1877–1957) (*12*) both toad venoms and arrow poisons (*13*) were explored. The combination of these two subjects established a trail to western Colombia, where more than 20 years ago the active principle of the poison arrow frog of the Indians of the Choco jungle was discovered (*14*). How all the data on toxins and activities eventually led to new insights and the gradation from information to knowledge (*15*) is the aim of this chapter.

II. Location and Function of Amphibian Venoms: Steroidal Alkaloids, Tetrodotoxin

As the early investigators recognized (*16, 17*), amphibian venoms may offer only some passive protection. The venoms are located in skin glands, in some cases in the eggs of the species, and very rarely in the bloodstream (*18*). A mechanism for active release does not exist in general. The chemical diversity of the physiologically active ingredients (*19*) has been interpreted not only as protection against vertebrate predators, but also as an antibacterial or antimycotic device (*20*). A function as physiological regulators for sodium transport or for enzymes involved in ion flux, such as K^+-dependent adenosinetriphosphatase (*21*), cannot be excluded. There is no comparable occurrence of such highly diversified and concentrated active endogenous agents and toxins known elsewhere in the animal kingdom (*22*).

We shall not attempt here to review the unusual oligopeptides from Caudata (Urodela) or Salientia (Anura), such as the hypotensive undecapeptide physalaemin from *Physalaemus fuscumaculatus (23)*, or the potent opiate-like heptapeptide dermorphin from *Phyllomedusa sauvagei* (*24*) which Erspamer discovered. (*25–27*).

A. TOAD VENOMS

A thoroughly investigated group of physiologically active substances are the bufodienolides, isolated mainly from the parotid glands of bufonidae (*19, 28–31*). As early as 1922, Heinrich Wieland showed that bufotoxin, isolated from the common European toad *Bufo vulgaris,* is the ester of bufotalin with suberylarginine. His student Munio Kotake isolated gamabufotalin with an 11α-hydroxyl group from the Japanese toad whose dried skin forms the basis of a cardiotonic drug Ch'an Su or Senso in oriental medicine (*19, 28–31*).

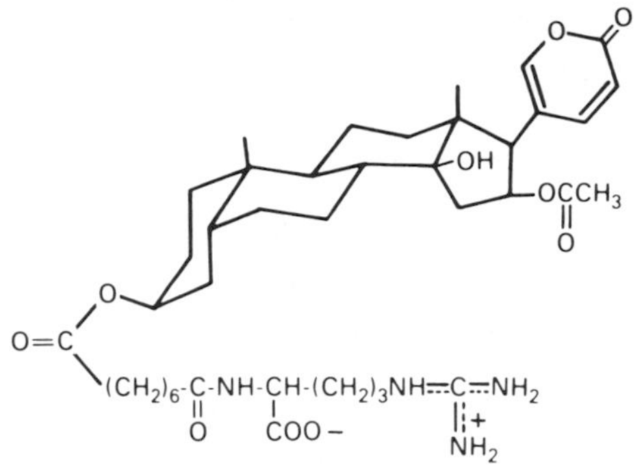

1

SCHEME 1. *Not only is bufotoxin (1) the first steroidal animal alkaloid, isolated in 1922 by H. Wieland and R. Alles (33), but its relation to the bile acids was surmised by Wieland as early as 1913.*

Bufogenins and bufotoxins are steroids and have the following structural features in common: A/B-*cis*-B/C-*trans*-C/D-*cis* (or *cis-anti-cis*) junctions of the tetracyclic system, a 14β-hydroxyl function, a 3β-hydroxyl group, and a six-membered doubly unsaturated lactone in position C-17 (bufogenins). If, instead of the 3β-hydroxyl, the 3β-O-suberylarginine residue (**30**) is present as in the bufotoxins, toxicity is markedly increased. Scheme 1 presents only bufotoxin from *B. bufo bufo* (*Bufo vulgaris*) as a representative example. Cholesterol serves as precursor for the biosynthesis of the toad venoms (*32*), and digitoxigenin (**32a**) and testosterone as precursors for the synthesis of bufalin (**32b**).

B. BIOGENIC AMINES

Biogenic amines is the term used by M. Guggenheim for basic substances related to the parent amino acids *pro forma* or *de facto*, by decarboxylation. They have been located in high concentration in the skins of more than 500 different amphibian species and genera (*19, 22, 34–38*). Their physiological activity is often more that of a local irritant than a general toxin. Thin-layer chromatography has helped to identify well-known biogenic amines (Schemes 2 and 3) in addition to amines that still have to be characterized and elucidated (*22, 39*). The spinaceamines from Leptodactylidae fall within the definition of true alkaloids.

The condensation products of endogenous biogenic amines with α-keto acids or their decarboxylation products, metabolites of the parent amino acids, are sometimes referred to as "mammalian alkaloids." Their medical significance is still an object of active investigation (*40, 41*).

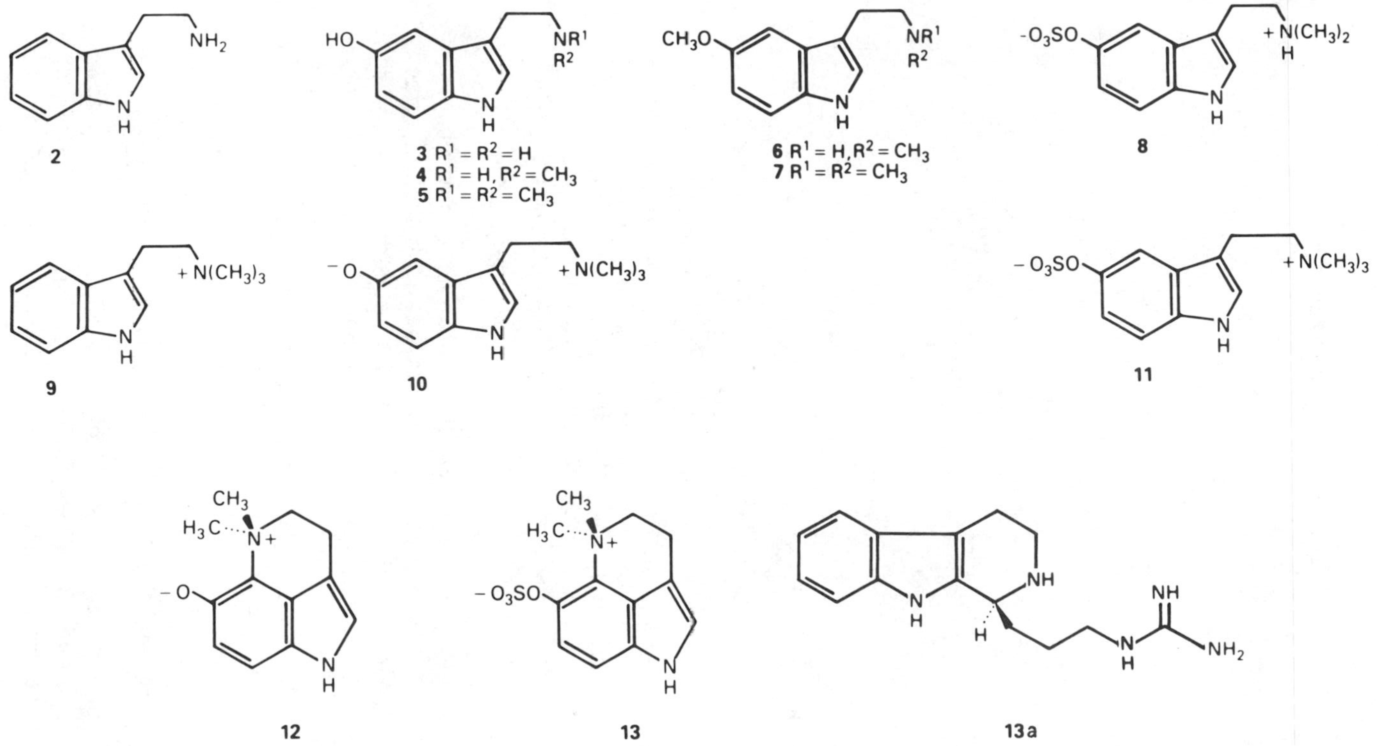

SCHEME 2. Biogenic amines derived from tryptophan. 2 Tryptamine (Salamandridae), 3 serotonin (Discoglossidae, Pipidae, Bufonidae, Leptodactylidae, Hylidae, Ranidae, Salamandridae, Atelopodidae, Hylidae, Ranidae), 4 N-methylserotonin (Bufonidae, Leptodactylidae, Atelopodidae, Hylidae, Ranidae), 5 Bufotenine (Bufonidae, Atelopodidae, Hylidae, Ranidae), 6 N-methyl-5-methoxytryptamine (Bufonidae), 7 O-methylbufotenine (Bufonidae), 8 bufotenine (Bufonidae, Atelopodidae, Hylidae, Ranidae), 6 N-methyl-5-methoxytryptamine (Bufonidae), 7 O-methylbufotenine (Bufonidae), 8 bufoteninesulfate (Bufonidae, Leptodactylidae, Nictimystes species), 9 trimethyltriptammonium salt (Leptodactylidae), 10 bufoteninidine (Pipidae, Bufonidae, Leptodactylidae), 11 Bufoviridine (Bufonidae, Nictimystes species), 12 Dehydrobufotenine (Bufonidae), 13 Bufothionine (Bufonidae, Hylidae), 13a (1S)(−)-tryptargine (Kassina senegalensis).

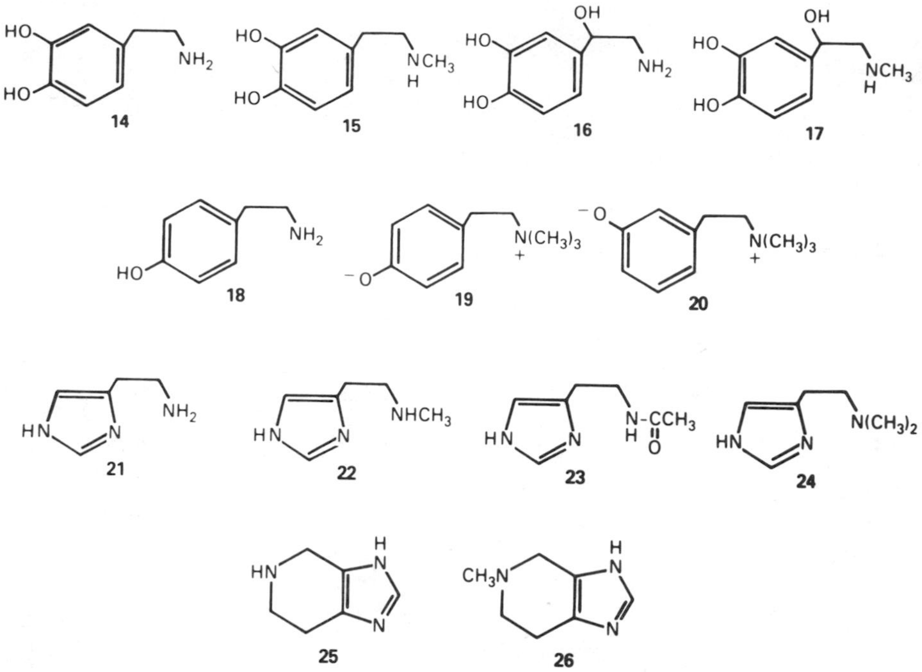

SCHEME 3. *14 Dopamine (Bufonidae), 15 epinine (Bufonidae), 16 (−)-noradrenaline (Bufonidae), 17 (−)-adrenaline (Bufonidae) 18 tyramine (Leptodactylidae), 19 candicine (Leptodactylidae), 20 leptodactyline (Leptodactylidae), 21 histamine (Hylidae, Leptodactylidae) 22 N-methylhistamine (Leptodactylidae), 23 N-acetylhistamine (Leptodactylidae) 24 N,N-dimethylhistamine (Leptodactylidae, Nictimystes species), 25 spinaceamine (Leptodactylidae), 26 6-methylspinaceamine (Leptodactylidae).*

Tetrahydrocarbolines, either as mammalian alkaloids or amphibian principles, have recently been reported not only in mammalian brain, urine, or platelets (*41, 42*), but also in the skins of an African frog *Kassina senegalensis* (*27*). This novel heterocycle is named (−)-trypargine (**13a**) to signify its origin formally, as a product of condensation of tryptamine with a metabolite of arginine in a manner that leads to retention of optical activity. The synthesis and absolute (1S)-configuration have just been reported and follow a similar route (*44*).

C. TETRODOTOXIN, A PROBLEM OF CHEMICAL CONVERGENCE

The spawn and skin of Californian newt (salamander) *Taricha torosa* of the order Caudata, and skin and eggs of the South American frog *Atelopus chiriquensis* of the order Anura, both contain tetrodotoxin (**27**) identical with the major toxin of the globe or puffer fish (Tetroadontidae). Some authors (45) do not consider this highly hydroxylated guanidino derivative an alkaloid whose biogenesis is still a mystery. There is no precedent for such a structure, a heteroadamantane skeleton containing the anion of a hemiorthoester neutralizing the charge of a guanidinium cation that is part of a five-membered ring. Its lethal effect (LD_{50} = 8 μg/kg mouse) is due to the reversible blockage of sodium-specific, voltage-sensitive channels.

Tetrodotoxin is also found in the blue-ringed octopus *Hapalochlaena maculosa*, a cephalopod (46) and in two marine snails *Babylonia japonica*, Japanese ivory shell (47) from which surugatoxin (48) was isolated, as well as in *Charonia sauliae*, trumpet shell (49). Fish of two different orders Tetraodontidae, such as puffer fish, as well as Gobiidae such as gobi fish *Gobius criniger (43)*, contain tetrodotoxin. It is difficult to imagine that such a substance arose several times during evolution (153).

It is no surprise, therefore, that isolation, structure elucidation (50–55, 45), synthesis (56–58, 45, 53, 55), synthetic approaches (59–65), pharmacological activity (32, 34–39), and application as a tool for the clarification of neurotransmission (8, 9, 66–68) have been the subject of repeated reviews.

The localization of tetrodotoxin varies. In *T. torosa*, *T. rivularis*, and *T. granulosa* tetrodotoxin occurs both in the skin and in the spawn (69–71), whereas Tetraodontidae contain it in roe, liver, and skin; the latter are capable of secreting the toxin from the skin (72). In mollusks the intestinal gland corresponds to the liver and contains the toxin.

Judging from the physiological effects of crude extracts there is reason to assume the presence of tetrodotoxin in other Salamandridae (newts and salamanders), such as *Dicmictylus viridescens (Triturus viridescens)*, *Triturus (Cynops) pyrroghaster*, *Cynops ensicauda*, and *Triturus marmorata* (69). Several species of Atelopodidae contain tetrodotoxin in their skin: *Atelopus ambulatorius*, *A. varius varius*, *A. senex*, and *A. chiriquensis* contain free tetrodotoxin in the skin and a bound form of the toxin in the eggs. In addition, tetrodotoxin is accompanied by chiriquitoxin **28**, the only known natural congener of tetrodotoxin (Scheme 4; 73–75). The presumable structure is a tetrodotoxin carrying a somewhat puzzling substituent $C_3H_6NO_3$ or $C_2H_6N_3O_4$ at C-6. Like tetrodotoxin, chiriquitoxin blocks sodium-specific channels, but, unlike tetrodotoxin, blocks potassium-specific channels as well (76, 77).

Another *Atelopus* species *A. zeteki* contains a low molecular weight, water-soluble toxin atelopidtoxin (78, 79). It is not related to tetrodotoxin or to chiriquitoxin and does not belong to the carbohydrates or to the steroids. Atelopitoxin recently was resolved into two components, the less toxic zetekitoxin C (LD_{50} = 80 μg/kg mouse) and the more active zetekitoxin AB (LD_{50} = 11 μg/kg mouse) (80). The paucity of material so far has prevented

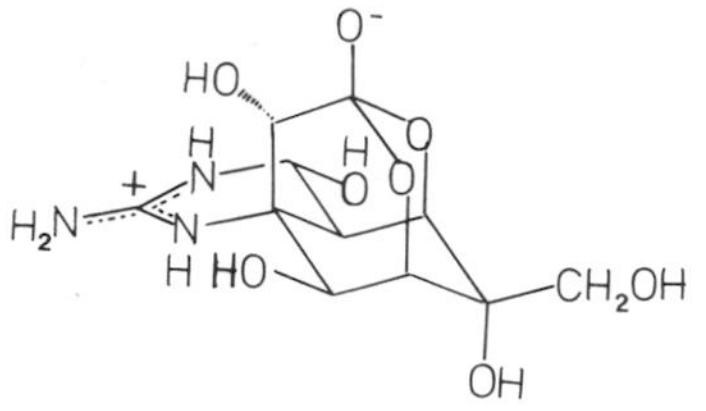

27

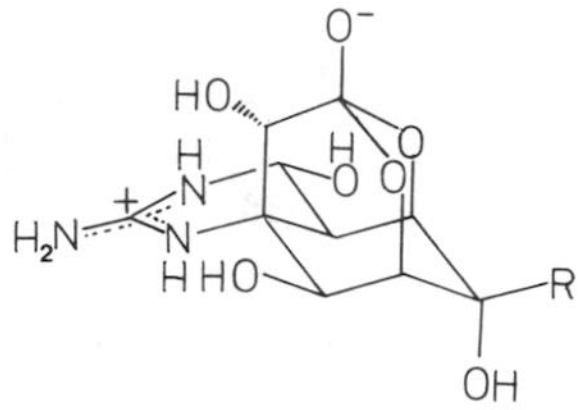

28

$$R = C_3H_6NO_3, C_2H_6N_3O_2 \text{ or } CH_2N_3O_3$$

SCHEME 4. **27** *Tetrodotoxin* (Taricha torosa, T. granulosa, T. rivularis (*Salamandridae*), Atelopus chiriquensis, A. varius varius, A. varius ambulatorius, A. senex (*Atelopodidae*), **28** *Chiriquitoxin* (Atelopus chiriquensis).

structural elucidation. *Atelopus planispina* contains another toxin which is awaiting characterization (*73*).

D. SALAMANDER ALKALOIDS

Wieland's pioneering work on toad venoms prompted his disciple Clemens Schöpf to investigate the secretion from the parotid glands of the European fire salamander *Salamandra maculosa taeniata* indigenous to western Europe, as well as *Salamandra maculosa maculosa* of southeastern Europe. The salamander toxins turned out to be steroidal alkaloids, all distinguished by a ring A expanded to a seven-membered azaheterocycle cis-fused to ring B. With the exception of **36, 37,** and **38,** ring A contains an oxygen bridge as part of a carbinolamine ether. All salamander alkaloids carry an oxygen function at C-16 (Scheme 5).

Samandarine (**29**) is the major ingredient of *Salamander maculosa taeniata* in addition to *O*-acetylsamandarine (**30**), samandaridine (**31**), samandarone (**33**), dehydrosamandrone (**34**), samanine (**36**), without oxygen bridge in ring A, cycloneosamandione (**37**), and cycloneosamandaridine (**38**). In addition, *S. m. maculosa* contains samandinine (**32**) with samandarone (**33**) the major ingredient, but no samandarine (**29**) and no samanine (**36**). Alkaloid **35** was isolated from *Cryptobranchus maximus* (*81*). The skin of the Australian toad species *Pseudophryne corroboree* contains an alkaloid with chromatographic properties analogous to samandarine (*82*).

Several reviews deal with isolation and structure elucidation of the salamander alkaloids (*83–87, 19;* Vol. IX, this treatise, *84*). More recent developments are the structure of samandinine (*88*), the amended structure of cycloneosamandione (*89, 90*), description of alkaloid **35** (*81*), and the

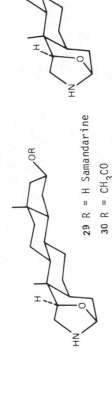

29 R = H Samandarine
30 R = CH₃CO
O-Acetylsamandarine

31

Samandaridine

32

Samandinine

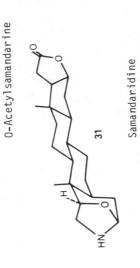

33

Samandarone

34

Dehydrosamandarone

35

from *Cryptobranchus maximus*

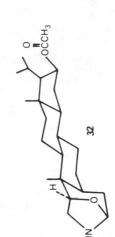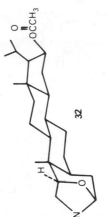

36

Samanine

37

Cycloneosamandione

38

Cycloneosamandaridine

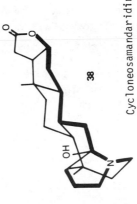

SCHEME 5. *Salamander alkaloids.*

suggestion of structure **38** for cycloneosamandaridine (*90, 91*). The biosynthetic pathway was examined again. Cholesterol, both in its free form and esterified with fatty acids, occurring in skin secretions (*19*) serves as intermediate (*92*).

Structures and absolute configurations were confirmed by partial syntheses for samandarine (*93, 94*), samandarone (*93*), samandaridine via samandarone (*95*), *O*-acetylsamandarine (*96*), cycloneosamandione (*89, 90*), samanine (*97–99*), and alkaloid **35** (*93, 100–103*). These synthetic investigations have been reviewed (*104, 105*).

E. The Batrachotoxins

1. Occurrence, Isolation, and Structure

Three groups of amphibians contain toxic steroids: toads secrete bufotoxins in their parotid glands, salamanders the samandarins both in their parotid glands and in their skin, but the most potent steroidal toxins belong to the small neotropical frogs of the genus *Phyllobates,* part of the family Dendrobatidae (superfamily: Bufonideae), in the order of Anura. There are morphological reasons as well as chemical characteristics which justify the assignment of the *Phyllobates* as an independent genus of Dendrobatidae that also contains *Dendrobates,* a species distinguished by a plethora of diverse alkaloids, and *Colostethus* a species whose skin secretions contain no alkaloids (*106, 107*).* The potent batrachotoxins occur only in *Phyllobates* and have never been encountered in *Dendrobates,* while, to a lesser extent, certain pumiliotoxins characteristic of *Dendrobates* seem to occur in *Phyllobates.* In addition, *Phyllobates terribilis,* besides large amounts of batrachotoxin, contains *d*-chimonanthine, *l*-calycanthine, i.e., both antipodes of the plant alkaloids as well as noranabasamine, thus far never encountered in any animal (*108*).

Skin secretions of *Phyllobates* are used by the Cholo Indians of the Choco rain forests in western Colombia for poisoning the tips of their blow darts for hunting with blow guns. The Indians designate the poison arrow frog as "Kokoi," an onomatopoetic expression. Its exact zoological name is now *Phyllobates aurotaenia* (*14*) after earlier explorers used other names such as *Ph. bicolor* (*14*), *Ph. chocoensis* (*109*), *Ph. melanorrhinus* (*110*), and the erroneous classification as *Dendrobates tinctorius* (*111*). *Phyllobates bicolor* is almost as toxic as *Ph. aurotaenia. Phyllobates terribilis* is larger, about 4 cm in length, contains about 1 mg of batrachotoxin, and both species are used by the Indians (*106*). Batrachotoxins were also identified in *Ph. vittatus* of

---

* There is one exception to this statement: *C. inquinalis* contains a water soluble toxin of unknown structure in its skin secretion (*107*).

southwestern Costa Rica and in some populations of *Ph. lugubris* of Panama by the use of thin-layer chromatography and bioassays (*106*). However, the latter two species are not used for poisoning blow darts.

Isolation and structural elucidation of the toxic compounds contained in the skin secretions of these frogs met with extraordinary difficulties, such as the problem of access to the peculiar habitats of these frogs as well as the lability of these toxins. The successful isolation of the active principles depended on the immediate methanol extraction of the fresh skins of the frogs right after their capture. In this way, five separate expeditions netted more than 8000 frogs, namely, *Ph. aurotaenia* whose extracts led to the isolation of three major alkaloids (*112*).

Historically, the nature of the toxin and the mode of isolation was first reported in 1963 (*14*). The skins of 330 frogs were extracted with methanol. A bioassay was needed to monitor the enrichment of the physiologically active ingredients. Initially, the method of distribution between water and chloroform was chosen and inactive ingredients were removed by chromatography on a column of silica gel. This was followed by two consecutive countercurrent distributions and thin-layer chromatography on silica gel. As soon as the lipophilic and basic character of the toxin was recognized, the method of isolation was simplified to a procedure which, with minor modifications, has since been adopted for all subsequent isolations (*113*).

The methanolic extracts are concentrated, diluted with water, and extracted with chloroform from which 0.2 N HCl removes the basic fraction. Dilute ammonia then liberates the bases which are again extracted into chloroform. In this way a hundred frog skins furnish about 5 mg of a crude base fraction. Repeated chromatography on silica gel leads to the isolation of four bases: homobatrachotoxin (HBTX) **39**, batrachotoxin (BTX) **40**, batrachotoxinin A (BTX-A) **41**, and pseudobatrachotoxin (PBTX) in a ratio of 28:11:47:1 (Scheme 6). Both HBTX and BTX give positive Ehrlich reactions with *p*-aminobenz- or cinnamaldehyde, while neither BTX-A nor the unstable PBTX, convertible to BTX-A by the addition of water, gives this color reaction. An evaluation of the fragmentation of the batrachotoxins in the mass spectrophotometer permitted the conclusion that these novel animal toxins are derived from steroids (*114*). Roentgen-ray crystallographic analysis of a very small crystal of the *p*-bromobenzoate of batrachotoxinin A (*113, 115, 116*) permitted the establishment of the relative and, in 1970, of the absolute configuration of BTX-A (*117*). The mass spectra allowed the localization of the difference between BTX-A on the one hand, and HBTX and BTX on the other hand, and pointed to C-17 as the point of attachment of a unit easily separated by the conditions of mass spectrometry. The nature of this unit became apparent through the positive Ehrlich reaction and the fragments in the mass spectrum of BTX whose molecular ion shows only a

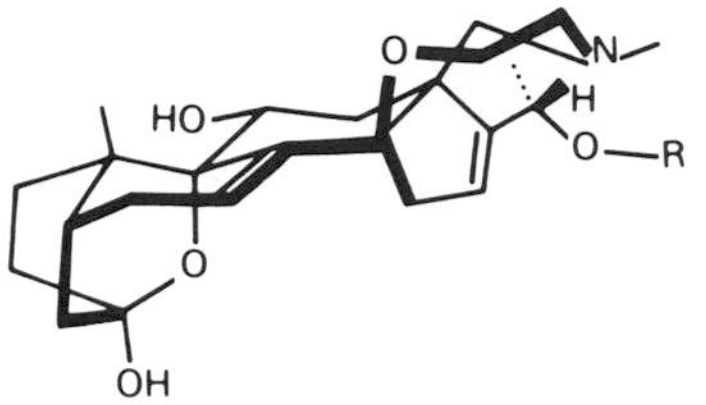

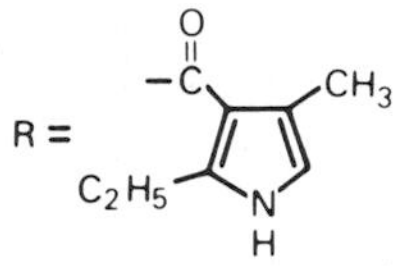

R =

Homobatrachotoxin (HBTX) **39**

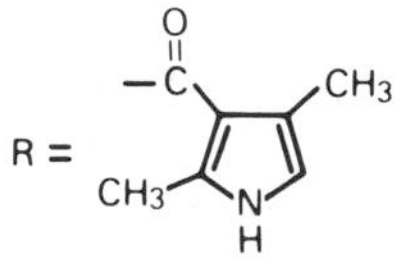

R =

Batrachotoxin (BTX) **40**

R = H

Batrachotoxinin A (BTX-A) **41**

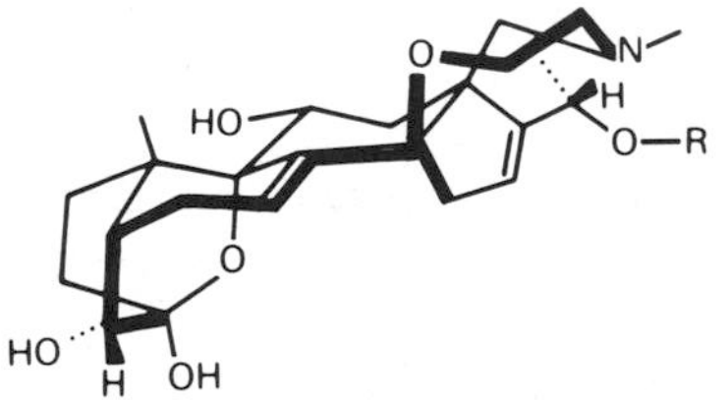

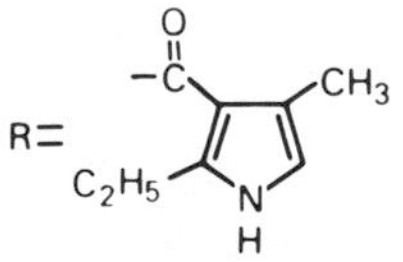

R=

4 β -Hydroxyhomobatrachotoxin **42**

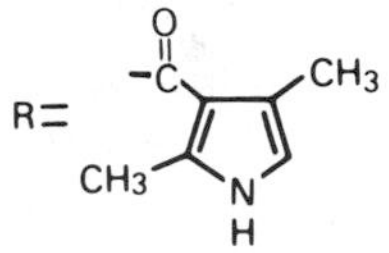

R=

4 β -Hydroxybatrachotoxin **43**

SCHEME 6. *The hexacyclic steroidal skeleton of the batrachotoxins forces the six rings, especially ring A, into unusual conformations, probably a contributory reason for their powerful and specific pharmacological activity* (116, 117).

very weak peak at m/z 538 ($C_{31}H_{42}N_2O_6$) but a strong fragment $C_7H_9NO_2$. The NMR signals present in BTX and HBTX but not in BTX-A and the differences in the IR and UV spectra all indicate that the secondary hydroxyl group at C-20 (which posseses the S-configuration) of BTX-A is esterified with 2,4-dimethylpyrrole-3-carboxylic acid in BTX and with 2-ethyl-4-methylpyrrole-3-carboxylic acid in HBTX. Both BTX and HBTX were convertible to BTX-A by strong alkaline hydrolysis. Vice versa, BTX-A under the conditions of the Schotten–Baumann reaction was esterified with the mixed anhydrides of 2,4-dimethyl- or 2-ethyl-4-methylpyrrole-3-carboxylic acid and monoethyl carbonic acid to BTX and HBTX, respectively. These interconversions prove the structures of the three major steroidal frog alkaloids. The structure of the elusive pseudobatrachotoxin is not known.

Variations of the acid component led to new esters of BTX-A and relationships between structure and activity. One representative of these semisynthetic toxins, namely, the ester of BTX-A with 2,4,5-trimethylpyrrole-3-carboxylic acid, proved to be twice as toxic ($LD_{50} = 1$ μg/kg mice) as BTX ($LD_{50} = 2$ μg/kg mice). The third methyl group in this ester occupies the α position which is free in BTX and may lead to a more stable toxin with a longer physiological half-life. The ester group is essential because BTX-A possesses only 1/500 of the potency of BTX. While the p-bromobenzoate is inactive, the benzoate, usually with tritium in the para position, is as active as BTX and therefore has become a valuable tool in pharmacology (*vide infra*).

The steroidal component is equally sensitive to minor alterations: reduction of the C-3—C-9 hemiketal to a secondary alcohol in C-3 reduces the toxicity to 1/200 of BTX. 4β-Hydroxyhomobatrachotoxin has only 1/100 of the potency of homobatrachotoxin.

The batrachotoxins are *sui generis* and share few structural features with the other amphibian steroids. The cis junction of rings A and B is characteristic of the samandarins (*85*) as well as of the bufotoxins and bufogenins (*28*). The steroids from toads possess the 14β-hydroxyl function and some representatives, such as gamabufotoxin and arenobufotoxin, have an 11α-hydroxyl. Bufotoxin, cinobufagin, and most salamander alkaloids carry an oxygen function at C-16 whose biochemical transformation to the Δ^{16}-double bond may be possible. Δ^{16}-Steroids are known only as metabolites of pregnenolones (*118*). The Δ^7-unsaturation, although not present in amphibian steroids, is common in steroids of animal or plant origin. In contrast, the hemiketal ether bridge between 3α and 9α which forces ring A into the boat form, as well as the N-methylethanolamine group which bridges C-18 and C-14 in a propellane-like arrangement have no precedent in naturally occurring steroids.

Incorporation or utilization of radioactive acetic acid or cholesterol have been useful for biosynthetic considerations for salamander alkaloids (*92*) and bufogenins (*32*). Such experiments were not successful with poisonous frogs (*119*). The frogs, analogous to puffer fish, lose much of their toxic strength in captivity as well as under stress, yet they retain enough of it to rule out the possibility that the toxin is of nutritional origin (*106*).

Recently the basic ingredients of 426 skins of *Ph. terribilis* were carefully screened for new and known alkaloids (*108*). The method of isolation as described above led to 780 mg of basic material. Reversed-phase silica gel chromatography then produced eight fractions that were resolved further on ℓ-Calycanthine, $[\alpha]_D^{25°} = -570°$

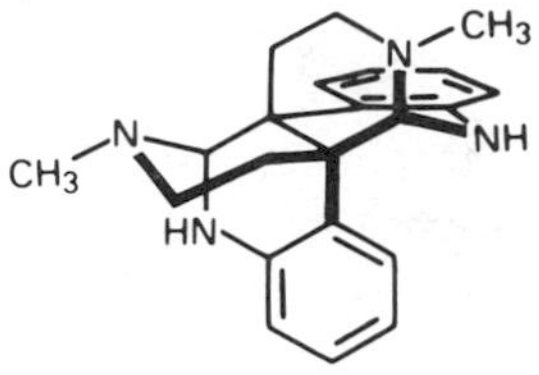

44   (animal)

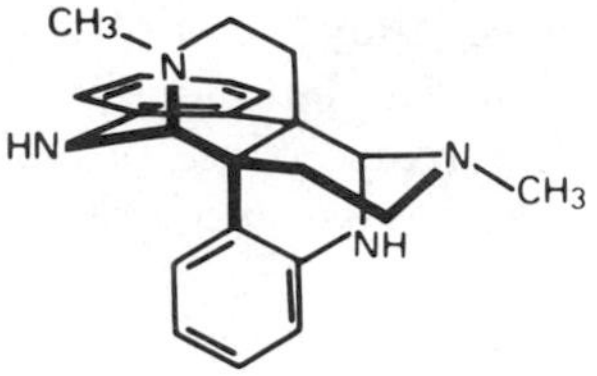

d-Calycanthine, $[\alpha]_D^{25°} = +550°$

45   (plant)

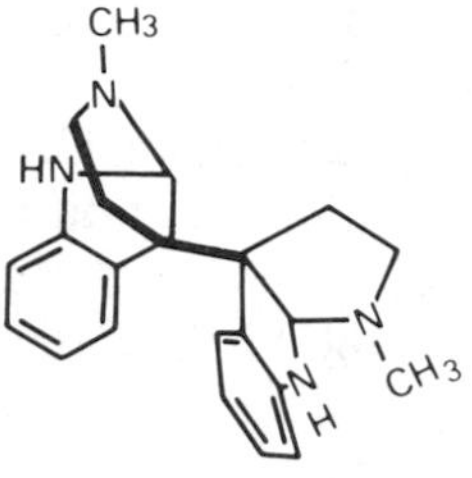

d-Chimonanthine, $[\alpha]_D^{25°} = +280°$

46   (animal)

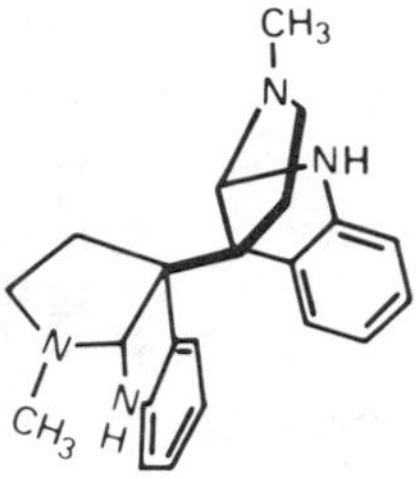

ℓ-Chimonanthine, $[\alpha]_D^{25°} = -329°$

47   (plant)

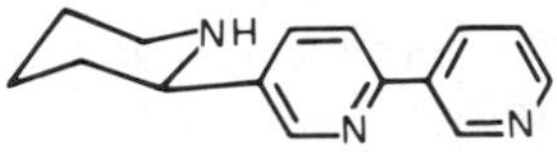

Noranabasamine, $[\alpha]_D^{25°} = -82°$

48

SCHEME 7. Phyllobates terribilis *contains the optical antipodes of dimeric indole alkaloids previously isolated from several plant families* (108).

silica gel and, with increasing elution time, yielded the following alkaloids (Schemes 6 and 7): noranabasamine **48** (15 mg), batrachotoxinin A (152 mg), 4β-hydroxybatrachotoxin **43** (2 mg), 4β-hydroxyhomobatrachotoxin **42** (3 mg), (+)-chimonanthine **46** (8 mg), (−)-calycanthine **44** (11 mg), batrachotoxin (175 mg), homobatrachotoxin (113 mg), 3-*O*-methylhomo-batrachotoxin (16 mg), and 3-*O*-methylbatrachotoxin (38 mg). Thus, totally unexpected minor alkaloids and congeners have been isolated and identified. The structures were elucidated by mass, ¹H-, and ¹³C-NMR spectrometry.

Noranabasamine **48** has been found in nature for the first time. Related alkaloids occur in Chenopodiaceae (*120*). *d*-Chimonanthine **46** and *l*-calycanthine **44** are interconvertible by the action of strong acid (*121*). Their antipodes **47** and **45** occur in several plant families. These plant alkaloids, their chemical properties as well as the syntheses of (±)-chimonanthine and (±)-calycanthine have been reviewed in this series (*121 – 124*).

4β-Hydroxybatrachotoxin **43** and 4β-hydroxyhomobatrachotoxin **42** are two novel members of the batrachotoxin family. Their structures rest on the comparison of their spectra with BTX and HBTX, respectively. In this way the location of the additional hydroxyl group in ring A became apparent from the mass spectra. ^{13}C- and ^{1}H-NMR data, especially the small width of the geminal proton signals next to the hydroxyl group, revealed its 4β position. The two *O*-methyl batrachotoxins are probably artifacts, since batrachotoxins are easily converted into their methyl ethers on brief exposure to acidic methanol.

Pseudobatrachotoxin was no longer found in this isolation. The earlier statement (*106*) that pumiliotoxins are present in *Phyllobates* could not be corroborated.

2. Partial Synthesis of Batrachotoxinin A

The unusual structural features of the batrachotoxins were accepted as a challenge to synthesize by several groups. Model studies may often detract from the decisive "frontal onslaught," but the successful partial synthesis of BTX-A (*125 – 126*) justified this approach as the final reward for a number of carefully planned and executed model studies (*127 – 133*). Besides this partial synthesis, three more studies (*134 – 137*) have been reported with the aim of a partial synthesis of BTX-A. So far there exists no total synthesis.

3. Construction of the Heterocyclic Propellane Moiety

Two groups reported successful approaches to the tricyclic propellane structure with the common C-13—C-14 axis (*128, 134*).

Starting with easily accessible 17β,19-diacetoxy-Δ⁴-androsten-3-one **49** (*138*) (Scheme 8) methods for the construction of the seven-membered homomorpholine ring were elaborated (*127*). The double bond in **49** was epoxidized with alkaline hydrogen peroxide to the 4α,5α-epoxide and, after selective acetylation, converted to the allylic alcohol **50** with hydrazine in acidic medium by the method of Wharton. After saturation of the Δ³-double bond, the C-19 acetate was selectively saponified and the primary alcohol so obtained converted to the aldehyde **51** by the method of Fetizon. Reaction with methylamine in a sealed tube yielded the imine reducible to the amine

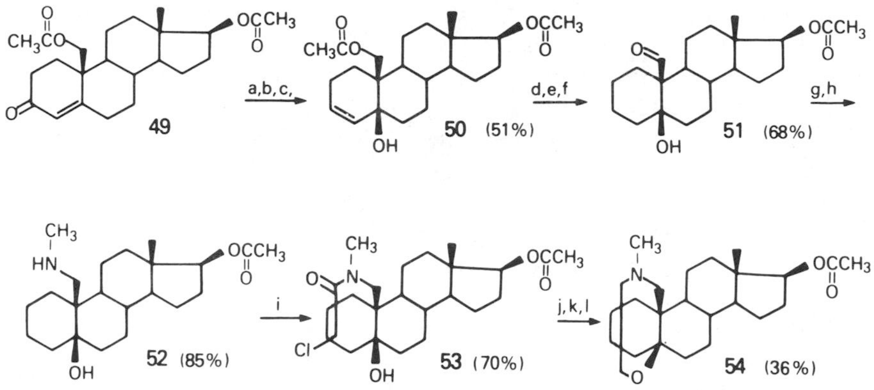

SCHEME 8. *First model synthesis of the seven-membered homomorpholine ring as part of a propellane involving rings A and B of an androstenone (127). Reaction conditions: (a) H_2O_2, 10% NaOH, $CH_3OH:CH_2Cl_2(2:1)$, $-11°C$; 3 days, 4°C; (b) $(CH_3CO)_2O$:pyridine $(1:1)$, 15 hr; (c) NH_2NH_2, H_2O, $HOCOCH_3$, 30 min; (d) PtO_2, H_2, C_2H_5OH; (e) 1% $NaHCO_3 \cdot H_2O$, C_2H_5OH ↥ 25 min; (f) Ag_2CO_3–celite, abs. ϕH, ↥; (g) CH_3NH_2, abs. ϕH, 120°C 15 hr; (h) $NaBH_4$, CH_3OH, 30 min; (i) $ClCH_2COCl$, NaOH, H_2O, $CHCl_3$, 10 min; (j) NaH, C_2H_5OH (cat.), THF, ϕH; (k) LAH, $(C_2H_5)_2O$; (l) $(CH_3CO)_2O$:pyridine $(1:1)$, 15 hr.*

52 that with chloroacetyl chloride under the conditions of the Schotten–Baumann reaction, furnished the amide 53. Sodium hydride formed the C-5 alcoholate that internally displaced the chloride by cyclization to the tricyclic propellane having C-5 and C-10 of the steroid system as a common axis. Reduction with lithium aluminum hydride and acetylation led to the desired homomorpholine 54. This reaction sequence was applied to the construction of the propellane system (128) with C-13 and C-14 as common axis (Scheme 9).

The C-18 methyl group of 20R-3β-acetoxy-20-hydroxy-5α-pregnane 55 (139) was functionalized in a nonclassical way by the hypoiodite reaction (140). When lead tetraacetate is used in the presence of half a mol equivalent of iodine, the reaction can be arrested at the level of the C-18 iodide. The acetate of the hemiketal 56 is then obtained by subsequent oxidation of the C-20 hydroxyl group followed by treatment with silver acetate. The acetyl group is easily saponified on chromatography on deactivated silica gel. The diacetate 57 is obtained by acetylation at elevated temperature. In order to introduce the oxygen function at C-14, the following method is used: bromination and dehydrobromination, when carried out twice consecutively, lead to the dienone 58 which with p-nitroperbenzoic acid yields the β-epoxide 59. Hydrogenolysis of the allylic C-15 epoxide bond, saturation of the Δ^{16}-double bond, followed by reduction of the keto group, leads to the

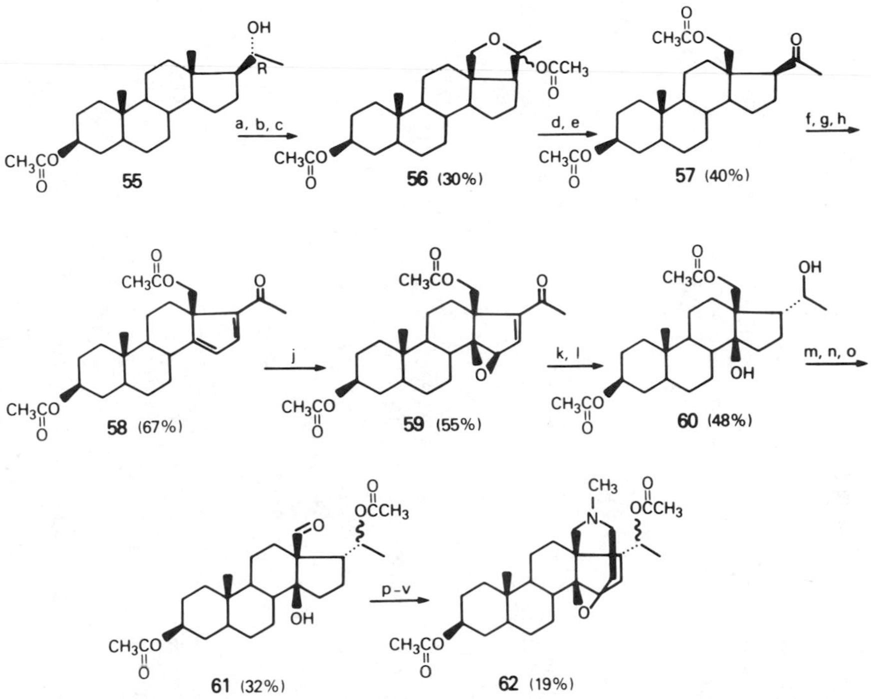

SCHEME 9. *Construction of the homomorpholine propellane system involving the common axis C-13, C-14 of rings C and D of a pregnane system (128). Reaction conditions: (a) Pb(OCOCH₃)₄, CaCO₃, cyclohexane, 1ↄ, 1 hr; ( ), 0.5 Eq I₂, hv 30 min 1ↄ; (b) Jones reagent; (c) AgOCOCH₃, CH₃OH, 1ↄ, 3 hr; (d) chromatography on deactivated silica gel; (e) (CH₃CO)₂O:pyridine (1:1), 10 hr, 95°C; (f) pyridine HBr · Br₂, CH₂Cl₂; (g) DMF, 1ↄ; (h) NBS, AIBN, CCl₄, 1ↄ, 1 hr; (i) NaI, (CH₃)₂CO, 1ↄ, 3 hr; (j) p-NO₂ØCOOOH, CHCl₃, 19 hr; (k) Pd/BaSO₄ H₂, C₂H₅OH; (l) LiAl[(CH₃)₃CO]₃H, THF, 5 hr; (m) (CH₃CO)₂O:pyridine (1:1), 15 hr; (n) NaHCO₃, H₂O:CH₃OH (1:10), 8 min 1ↄ; (o) Jones reagent; (p) CH₃NH₂,ØH, 115°C, 16 hr; (q) pyridine:(CH₃CO)₂O (1:1), 5 hr; (r) NaBH₄, CH₃OH/H₂O (10:1) 20 min; (s) ClCOCH₂Cl, NaOH, H₂O, CHCl₃, 20 min; (t) NaH, C₂H₅OH (cat.), THF, ØH, 1ↄ, 7 hr; (u) LAH, C₂H₅OH:THF (3:1); (v) (CH₃CO)₂O:pyridine (1:1).*

14-hydroxy-17α-pregnane derivative **60**, a mixture of the alcohols epimeric at C-20. Acetylation to the triacetate is followed by selective deacetylation of the C-18 acetate and separation into the C-20 epimeric acetates (ratio 4:1). The epimer obtained in a larger amount is oxidized to the aldehyde **61** by the method of Jones. The subsequent steps are repetitions of the reactions studied with the above model: methylamine forms the imine which (after reacetylation) is reduced to the amine, treated with chloroacetyl chloride, cyclized to the seven-membered ring and reduced to the homomorpholine. The diacetate **62** is the end product of this model study.

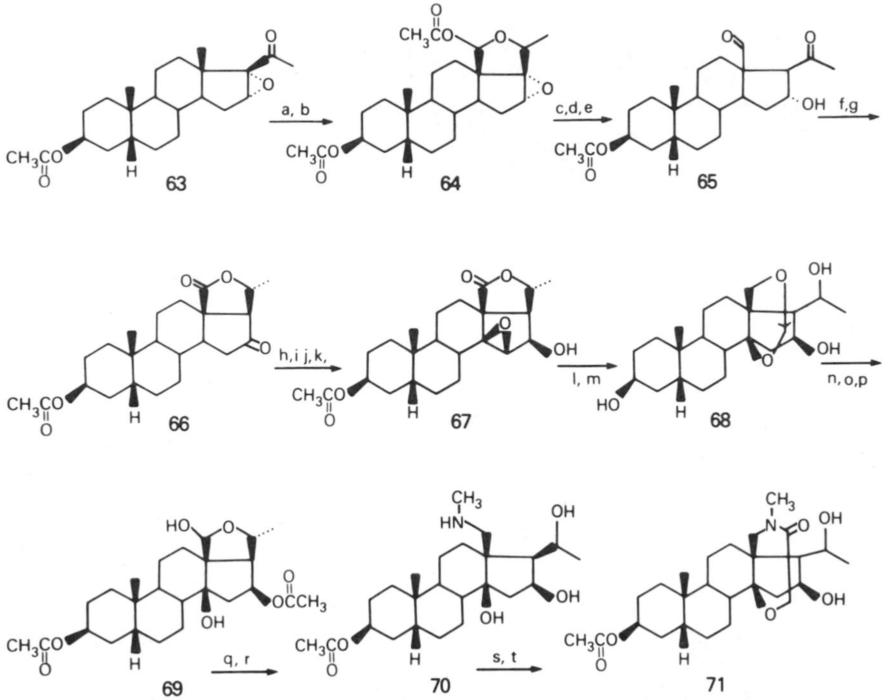

SCHEME 10. *Another approach to the propellane (71) system of batrachotoxinin A starting with the pregnanone 63 (134). Reaction conditions: (a) NaBH₄; (b) Pb(OCOCH₃)₄, I₂; (c) mild base; (d) DMSO, pyridine · SO₃; (e) Cr(OCOCH₃)₂, pyridine; (f) LiAl[(CH₃)₃CO]₃H; (g) CrO₃, H⁺; (h) bromination; (i) dehydrobromination; (j) LiAl[(CH₃)₃CO]₃H; (k) CH₃COOOH; (l) LiAlH₄; (m)(CH₃)₂CO, BF₃ · O(C₂H₅)₂, THF; (n)(CH₃CO)₂O; (o) H⁺, H₂O; (p) DMSO, pyridine · SO₃; (q) CH₃NH₂, C₂H₅OH, 120°C; (r) H₂, PtO₂, HOCOCH₃; (s) ClCH₂COCl, THF; (t) KOC(CH₃)₃, −20°C, THF.*

Another successful approach to the construction of the propellane system with the common C-14–C-13 axis (Scheme 10) (*134*) permits the introduction of the Δ¹⁶- unsaturation at a later stage.

3β-Acetoxy-16α,17α-epoxy-5β-pregnan-20-one **63** (*141*) is reduced to the C-20 epimeric alcohols. Their separation is unnecessary because of the subsequent functionalization of the C-18 methyl group by the use of the hypoiodite reaction: Only one of the epimeric alcohols leads to the acetate of the hemiketal **64** which, after partial saponification and oxidation of the C-20 hydroxyl group, yields the ketoaldehyde. The C-17 oxygen bond is activated by the adjacent carbonyl group and thus reductively cleaved by chromous acetate to the ketoaldehyde **65** whose keto group is selectively reduced by lithium tri-*tert*-butoxyaluminum hydride. The hemiacetal so formed is oxidized to the lactone with simultaneous oxidation of the C-16 alcohol to the keto lactone **66**. Bromination and dehydrobromination lead to the conjugated enone. Selective reduction with lithium tri-*tert*-butoxylaluminum hydride leads to the 16β-alcohol that directs the subsequent epoxidation stereoselectively to the 14β,15β position. The epoxide **67** is converted to the pentol with lithium aluminum hydride. Acetone in the presence of catalytic amounts of a Lewis acid selectively forms the acetonide **68**. Partial acetylation involves only the hydroxyl groups at C-3 and C-16, and cleavage of the acetonide yields a triol. The hemiacetal **69** is obtained by selective oxidation by the method of Doering. Methylamine converts it to the hemiaminal reducible to the amine **70** with platinum and hydrogen. The construction of the homomorpholine ring essentially follows the above-mentioned route: acetylation with chloroacetyl chloride and subsequent treatment with strong base at low temperature yields the cylic amide **71**. The omission of yields does not permit an evaluation of the efficiency of this approach.

4. Construction of the 3β,11α-Dihydroxy-3α,9α-oxido Moiety

The hemiketal grouping with an ether bridge between C-3 and C-9, though not known among natural steroids, has been explored previously (Scheme 11) (*142, 143*). Three different groups reported on the construction of the hemiketal (*129, 135–137*); yet all approaches are based on the experience (*129*) that contrary to earlier reports (*144*), the $\Delta^{9,11}$-unsaturation of 5β-pregnone derivatives reacts with osmium tetroxide.

The lactone **72** (Scheme 12) is selected (*129*) as a suitable starting material since this compound permits the introduction of all the structural details

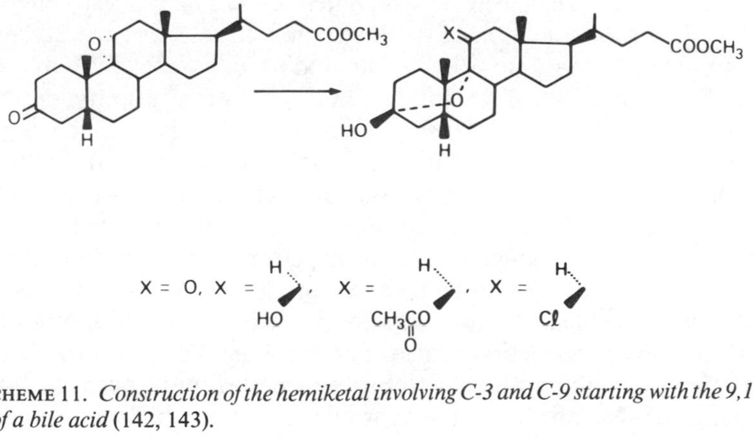

SCHEME 11. *Construction of the hemiketal involving C-3 and C-9 starting with the 9,11-epoxide of a bile acid* (142, 143).

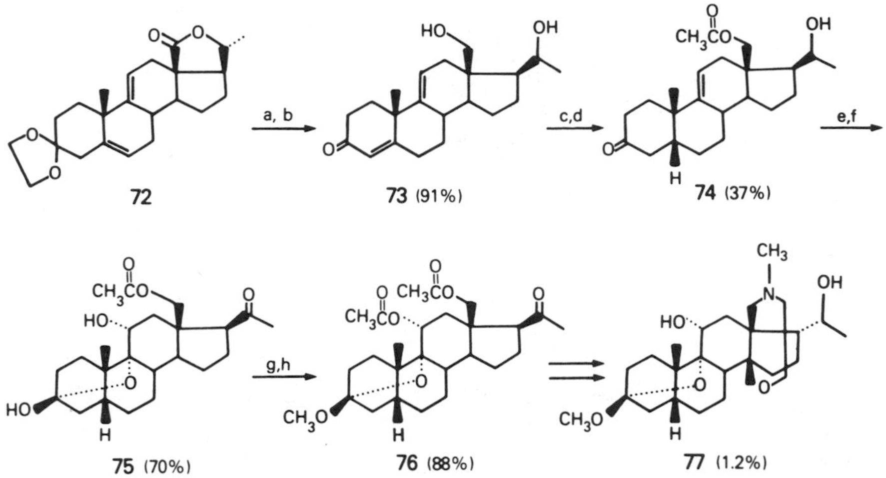

SCHEME 12. *Synthesis of 3-O-methyl-17α,20ζ-tetrahydrobatrachotoxinin A, 77 (129). Reaction conditions: (a) LAH, THF, 1ℓ, 2 hr; (b) TosOH, (CH₃)₂CO:H₂O (15:1), 1ℓ, 2 hr; (c) Pd/C, 1 atm H₂, 0.1 N KOH, C₂H₅OH; (d) (CH₃CO)₂O:pyridine (2:3), 1 hr; (e) Jones reagent; (f) OsO₄, pyridine, 7 days; dioxane, NH₄Cl, H₂O, H₂S; (g) 0.1 N HCl, CH₃OH, 20 min; (h) (CH₃CO)₂O:pyridine (1:1), 4 hr, 70°C.*

characteristic of batrachotoxinin A. This lactone (**72**) has been prepared from 11α-progesterone in a sequence of ten steps (*140*).

Lactone **72** is reduced to the 18,20-dialcohol and then converted to enone **73** by acid hydrolysis. For steric reasons the double bond in ring A is more accessible than the $\Delta^{9,11}$-double bond to catalytic reduction under basic conditions with hydrogen approaching from the β face. The 5β-pregnene derivative so obtained is selectively acetylated to **74**. The modest yield of **74** is improved by saponification of the diacetate and selective reacetylation. Oxidation to the methyl ketone by the method of Jones is followed by a very slow reaction of the C-9–C-11 double bond with osmium tetroxide. After cleavage of the 9α,11α-osmium ester, the hemiketal **75** is obtained in good yield. In order to protect the labile hemiketal grouping the mixed methyl ketal is formed by the action of acidic methanol. There was still some apprehension as to whether this conformationally and configurationally altered steroid would be amenable to construction of the homomorpholine ring by the method developed with simple models (*128*). Fortunately, this method was applicable with minor modifications in spite of the sensitivity of the ketal function to the action of acid. The diacetate **76** is now brominated, not as previously with pyridine hydrobromide perbromide, but with *N*-bromosuccinimide. Repetition of the bromination and dehydrobromination leads to the dienone, whose epoxidation involves the formation of the strongly acidic p-nitrobenzoic acid and cleavage of the ketal. This cleavage is suppressed by the addition of methanol to the epoxidation mixture. Catalytic reduction and construction of the homomorpholine ring as described above eventually furnish 3-O-methyl-17α,20ξ-tetrahydrobatrachotoxinin A, **77**.

A doctoral thesis of 1980 describes an attempt to synthesize 7,8-dihydro-batrachotoxinin A (*137*). The starting material (Scheme 13) is commercial 11α-hydroxyprogesterone **78** which, by catalytic hydrogenation in pyridine, is selectively converted to the 5β-pregnane derivative. Dehydration with sulfuryl chloride introduces the $\Delta^{9,11}$-unsaturation. Selective ketalization of the C-3 carbonyl group is possible with glycol in the presence of selenium dioxide in acidic medium. After reduction of the C-20 ketone, a mixture of the acetate of the hemiacetal and of the lactone is obtained by the application

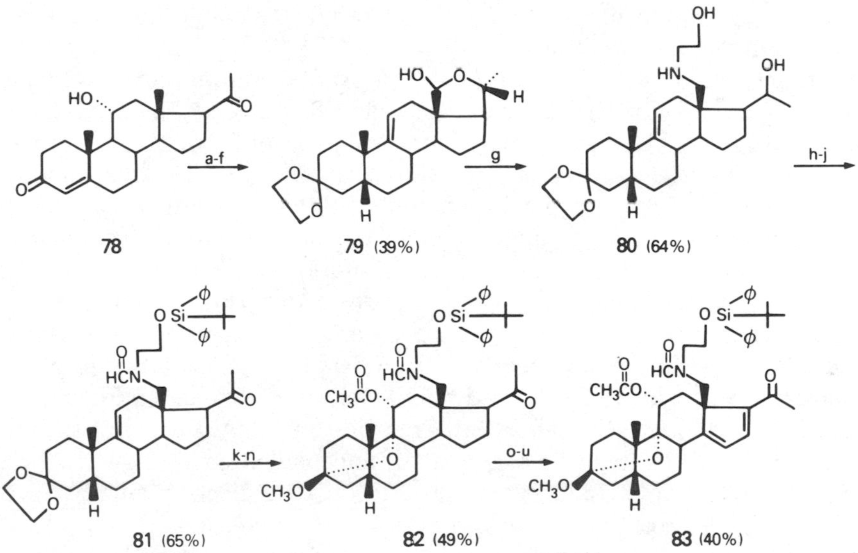

SCHEME 13. *Incomplete approach to 7,8-dihydrobatrachotoxinin A* (137). *Reaction conditions:* (a) H_2,Pd/C, pyridine, 1 atm, 9 hr; (b) SO_2Cl_2, pyridine/CHCl$_3$, $-78°C$, 1 hr; 0°C, 1.5 hr; (c) TosOH (cat.), (CH$_2$OH)$_2$, SeO$_2$, CH$_2$Cl$_2$, 15 hr; (d) LiAl(CH$_3$O)$_3$H, THF, $-78°C$, 3 hr; (e) Pb(OCOCH$_3$)$_4$, I$_2$, cyclohexane, hv, ↑↓, 1.5 hr; (f) DBAH, ØCH$_3$, $-78°C$, 30 min; (g) NaBCNH$_3$, 6 Eq HOCH$_2$CH$_2$N̄H$_3$ Cl$^-$, CH$_3$OH, 6 days; (h) HCO$-$O$-$COCH$_3$, CH$_2$Cl$_2$, 0°C, 2 hr; (i) (CH$_3$)$_3$CSiØ$_2$Cl, imidazole, DMF, 2.5 hr; (j) CrO$_3$·2 pyridine, CH$_2$Cl$_2$, 0°C, 1 hr; (k) HOOCCH$_3$:THF:H$_2$O, 50°C, 18 hr; (l) OsO$_4$, pyridine, 9 days; pyridine/H$_2$O, NaHSO$_3$, 45 min; (m) TosOH, CH$_3$OH, 2.5 hr; (n) (CH$_3$CO)$_2$O, pyridine, 15 hr; (o) [(CH$_3$)$_3$Si]$_2$NH, (CH$_3$)$_3$SiI, CH$_2$Cl$_2$, $-20°C$, 15 min; 25°C, 30 min; (p) ØSCl, CH$_2$Cl$_2$, $-78°C$, 20 min; (r) m-ClØCOOOH, CH$_2$Cl$_2$, 0°C, 40 min; (s) CaCO$_3$, CH$_2$Cl$_2$, ↑↓, 45 min; (t) NBS, AIPC, CCl$_4$ hv, ↑↓, 20 min; (u) LiBr, Li$_2$CO$_3$, DMF, 130°C, 45 min.

of the hypoiodite reaction under special conditions. The hemiacetal **79** is obtained in ~ 40% yield by reduction of this mixture with diisobutylaluminum hydride. Ethanolamine is attached to C-18 by reductive amination **(80)**. A number of transformations are necessary before glycolization of the $\Delta^{9,11}$-double bond becomes possible: The amino group is protected as the formamide, and the primary alcohol by formation of the stable *tert*-butyldiphenylsilyl ether; the secondary alcohol at C-20 is oxidized to the ketone **81** and the C-3 ketal hydrolyzed. As expected, the reaction of this compound with osmium tetroxide is exceedingly slow, but, given enough time, proceeds in good yield. In this way all the necessary functions in rings A, B, and C are present. The hemiketal is now stabilized by formation of the mixed methyl ketal and **82** is obtained by protection of the 11α-hydroxyl group through acetylation. Now the stage is set for the construction of the dienone **83** in six steps. But the attempted intramolecular nucleophilic 1,6-addition of the alkoxide generated by removal of the protective silyl group by fluoride ion and the generation of the anion of the primary alcohol failed to form 7,8-dihydrobatrachotoxinin A.

Another attempt (Scheme 14) (*135, 136*) starts with the easily available cholic acid **84** which already has rings A and B in the cis junction and in which the 7α-hydroxyl group serves as a precursor for the desired $\Delta^{7,8}$-unsaturation. The 12α-hydroxyl group likewise is useful for the elaboration of the $\Delta^{9,11}$-double bond, a route for which there are precedents (*145*).

The transformation of cholic acid **84** to the enone **85** follows classic methods (*145*). Thioketalization and desulfuration of **85** lead to the isolated $\Delta^{9,11}$-olefin. Selective saponification liberates the 3α-hydroxyl group which is oxidized to the ketone **86**. Analagous to the procedure shown in Scheme 11, epoxidation of **86** produces the 9α,11α-epoxide. However, all attempts to convert this epoxide to the desired hemiketal with an oxygen bridging C-3 and C-9 failed (*136*). Therefore, the reaction of the olefin with osmium tetroxide had to be tried. This turned out to be an extremely slow reaction with the formation of undesired by-products, such as the $\Delta^{9,11}$-en-12-one in 15% yield. Cleavage of the osmium ester furnishes the 9α,11α-diol whose tendency to form the hemiketal is slight, to judge from ¹H-NMR spectroscopic data. However, the 9α,11α,7α-triol, available by transesterification with methoxide in methanol, forms a mixture of ketone and hemiketal both of which are converted to the mixed methylketal **87** by treatment with acidic methanol. However, when the olefin **86** is first converted to the 7α-alcohol by transesterification, the reaction with osmium tetroxide is complete within 1 hr, proceeds quantitatively, and forms no undesired by-products. After cleavage of the osmium ester and conversion to the mixed ketal, the diol **87** is obtained in very good yield. Selective acetylation of the 11α-hydroxyl group makes possible the introduction of the Δ^7-unsaturation by dehydration with

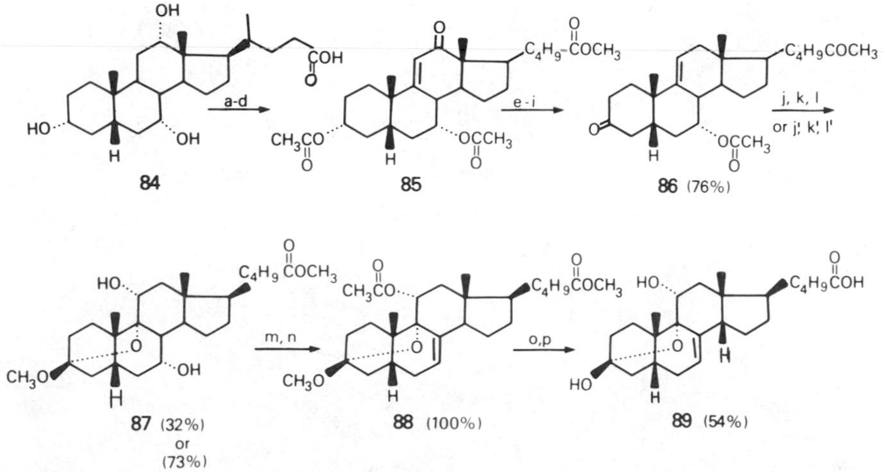

SCHEME 14. *Elaboration of the $\Delta^{7,8}$-unsaturation starting with cholic acid* (135, 136). *Reaction conditions: (a) CH_2N_2, $(C_2H_5)_2O$; (b) $(CH_3CO)_2O$, dioxane, pyridine, 20 hr; (c) CrO_3, $HOCOCH_3$, 100°C, 1.5 hr; (d) SeO_2, CH_3COOH, 1l, 18 hr; (e) $(CH_2SH)_2$, HCl gas, $CHCl_3$, $-15°C$; 0°C, 20 hr; (f) Ra/Ni, abs. C_2H_5OH, 1l, 8 hr; (g) K_2CO_3, CH_3OH/H_2O, 45°C, 6 hr; (h) CH_2N_2, $(C_2H_5)_2O$, dioxane; (i) $NaCr_2O_7 \cdot 2H_2O$, $HOCOCH_3$, 10 hr; (j) OsO_4, pyridine, 8 days; H_2S, NH_4Cl, $H_2O/dioxane$; (k) CH_3ONa, CH_3OH, 1l, 2 hr; (l) CH_3OH, HBr (cat.) .5 hr; (j') NaONa, CH_3OH, 1l, 3 hr; (k') OsO_4, pyridine, 1 hr; (l') H_2S, NH_4Cl, $H_2O/dioxane$; (m) $(CH_3CO)_2O$, pyridine, 40°C, 6 hr; (n) $POCl_3$, pyri, 15 hr; (o) $HOCOCH_3/H_2O$, $HClO_4$, 12 hr; (p) 0.2 N NaOH, C_2H_5OH.*

phosphorus trichloride in pyridine. Acidic hydrolysis followed by alkaline saponification convert **88** into the acid **89** in which rings A, B, and C are structurally and functionally related to batrachotoxinin A.

Two other variations (*130*) describe the synthesis of two compounds related to batrachotoxinin A with regard to the configuration and functions within the area of rings A, B, and C. In both cases the lactone **26** serves as starting material.

The first approach (Scheme 15) starts with deketalization of **72** to the enone **90**. The electronically more susceptible $\Delta^{9,11}$-olefin is attacked by osmium tetroxide from the α face. After cleavage of the osmium ester and acetylation of the secondary alcohol function, the conjugated dienone **91** is obtained by dehydrogenation with *p*-dichlorodicyanoquinone in acidic medium. Epoxidation leads selectively to the $6\alpha,7\alpha$-epoxide which is opened to the diol **92** by selective cleavage of the allylic oxygen function. The subsequent hydrogenation proceeds with addition of hydrogen from the β face. The hemiketal forms spontaneously and is converted to the mixed ketal **93**. Reductive opening of the lactone ring with lithium aluminum hydride leads

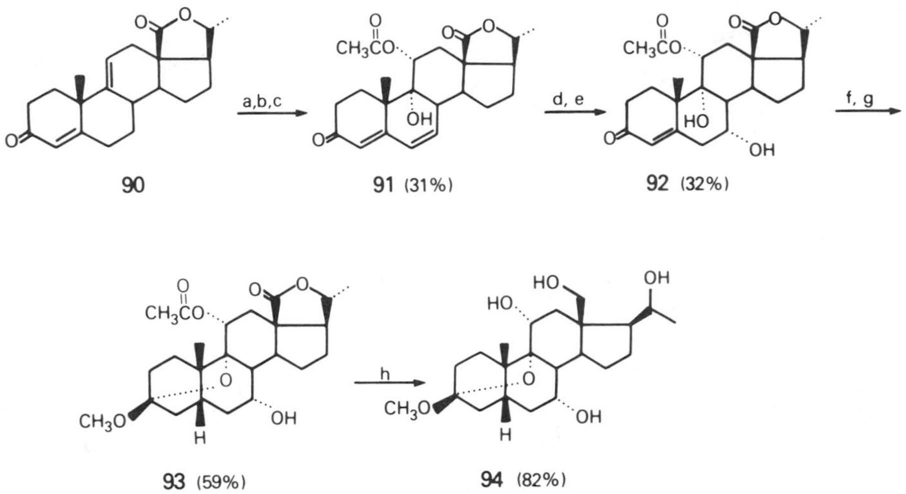

90   **91** (31%)   **92** (32%)

93 (59%)   **94** (82%)

SCHEME 15. *Approach to an analog of batrachotoxinin A **94** starting with the unsaturated lactone **90** (130). Reaction condition: (a) OsO₄, pyridine, 14 hr, dioxane, NH₄Cl, H₂O, H₂S; (b) (CH₃CO)₂O:pyridine (1:1), 15 hr; (c) 0.5 N HCl, dioxane, DDQ, 6 hr; (d) p-NO₂φ COOH, CHCl₃, 4 hr; (e) Pd/BaSO₄, cyclohexene, abs. CH₃OH, 1ᴸ, 15 min; (f) Pd/C, 1 atm H₂, CH₃OH; (g) 0.01 N HCl, CH₃OH; (h) LAH, (C₂H₅)₂O, 1ᴸ, 2 hr.*

to a tetrol **94** which poses a problem with regard to differential introduction of protective groups. Therefore, a second approach was explored.

First, the lactone ring of **72** is reductively opened, the diol so obtained acetylated, and the ketal protective group removed. The opening of the lactone notably changes the conformation and ring strain and is reflected in different reactivity toward osmium tetroxide, leading to a mixture of osmium esters both in rings A and C in a ratio of 4:1, respectively. Accordingly the following route (Scheme 16) was selected.

Again, lactone **72** is reductively opened with subsequent deketalization. The enone **73** so obtained is epoxidized with alkaline hydrogen peroxide to the diastereoisomeric epoxides **95** ($\alpha:\beta$ = 1:2). The primary alcohol at C-18 is selectively acetylated and the secondary alcohol at C-20 is converted to the ketone. Osmium tetroxide now reacts within 14 hr to give the osmium ester which is opened to the 9α,11α-diol **96**. After acetylation of the 11α-hydroxyl, zinc in acetic acid regenerates the Δ⁴-double bond. The enone so-formed is dehydrogenated (*vide supra*) to the dienone and converted to the 7α,8α-epoxide **97**. When the allylic C—O bond of the epoxide is hydrogenolyzed at C-6 the enone **98** is obtained whose Δ⁴-double bond is hydrogenated from the β-side and the hemiketal is formed spontaneously and stabilized as the mixed methyl ketal **99**. The Δ⁷,⁸-double bond is introduced by dehydration of the

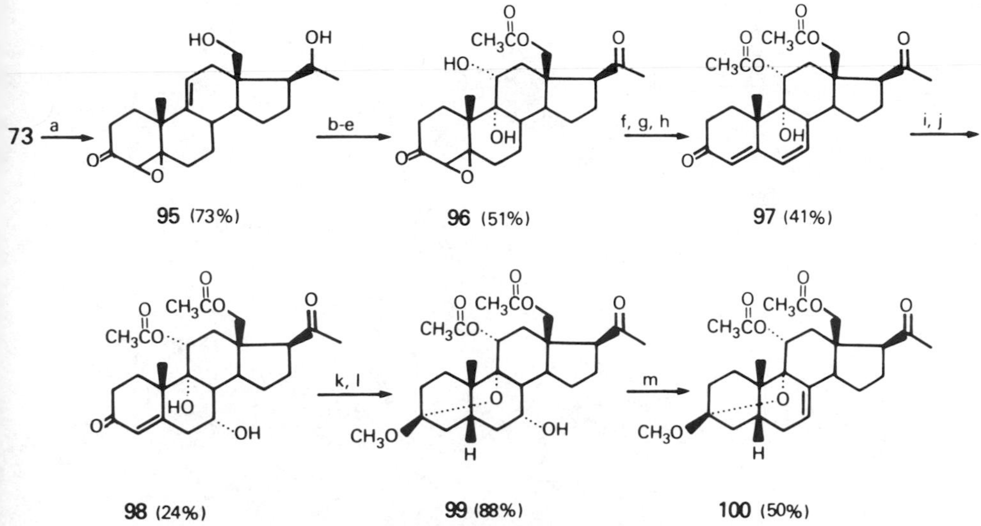

95 (73%)   96 (51%)   97 (41%)

98 (24%)   99 (88%)   100 (50%)

SCHEME 16. *Lactone 72, enone 73, and epoxyketone 95 mixture of one part α- and two parts β-epoxyketones, serve as starting materials for the synthesis of the Δ⁷,⁸-methylketal 100 (130). Reaction conditions: (a) 30% H₂O₂, 4 N KOH, CH₃OH:CH₂Cl₂(2:1), −20°C; 4°C, 60 hr; (b) (CH₃CO)₂O, pyridine, 20 min; (d) Jones reagent; (e) OsO₄, pyridine, 14 hr; dioxane, NH₄Cl, H₂O, H₂S; (f) (CH₃CO)₂O:pyridine (1:1), 15 hr; (g) Zn, HOCOCH₃, 1ʰ; (h) 0.45 N HCl, dioxane, DDQ; (i) p-NO₂ØCOOOH, CHCl₃, 4 hr; (j) Pd/BaSO₄, cyclohexene, dioxane, 1ʰ, 20 min; (k) Pd/C, 1 atm H₂, CH₃OH; (l) 0.01 N HCl, CH₃OH; (m) POCl₃, pyridine, 15 hr.*

7α-hydroxyl with phosphorus oxychloride in pyridine to yield **100.** This step completes the model sequence.

5. Synthesis of (20R)-3-O-Methyl-7,8-dihydrobatrachotoxinin A

The next step in the elaboration of a synthesis of batrachotoxinin A was the introduction of the Δ¹⁶-unsaturation. Three publications (*131–133*) expatiate on the most feasible route to 7,8-dihydrobatrochotoxinin A. The starting material (Scheme 17) is again the lactone 72, which in several steps as described, is converted to **101,** an intermediate in the synthesis of tetrahydrobatrochotoxinin A. By catalytic hydrogen transfer from cyclohexene it is possible to open the epoxide at the allylic C-15–oxygen bond with conservation of the Δ¹⁶-unsaturation. At this point one has the option to reduce this double bond in order to arrive at tetrahydrobatrochotoxinin A in higher yield than before (*131*).

However, in order to arrive at 7,8-dihydrobatrachotoxinin A, the Δ¹⁶-double bond must be retained and this retention requires a number of modifica-

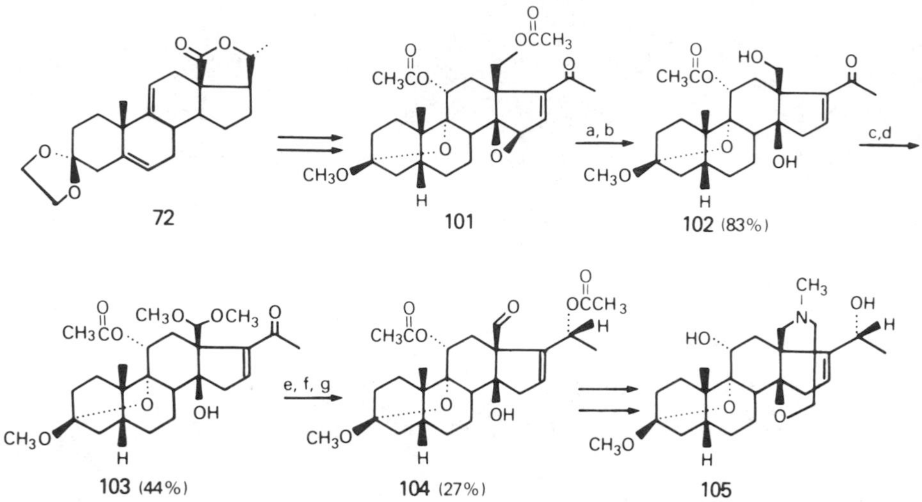

SCHEME 17. *Synthesis of* (20R)-3-O-methyl-7,8-dihydrobatrachotoxinin A **105** (131, 132). *Reaction conditions: (a) Pd/BaSO₄, cyclohexene, CH₃OH,* 1ʟ, *25 min; (b) 1 N NaOH, dioxane/H₂O, 10 min; (c) CrO₃, CH₂Cl₂, abs. pyridine, 15 min; (d) 0.1 N HCl, CH₃OH, 2 hr; (e) DBAH, ϕCH₃, − 78°C, 10 min; (f) (CH₃CO)₂O:pyridine (1:1), 17 hr, 45°C; (g) (CH₃)₂CO, TosOH, 3.5 hr.*

tions (*132*), especially in the construction of the homomorpholine ring which in the case of tetrahydrobatrachotoxinin A no longer posed a problem. A differential reactivity between the primary and allylic hydroxyl group is no longer observed. For this reason the C-18 acetate group, after opening of the epoxide ring in **101** is selectively saponified to yield the diol **102**. The primary alcohol is then oxidized to the aldehyde. Here again the presence of the additional Δ^{16}-double bond has a labilizing influence: both the β-hydroxyal-dehyde and the vinylogous β-hydroxyketo groupings are prone to fragmen-tations under the catalytic influence of acid or base. One of these disadvan-tages is the low yield of the C-18 aldehyde. When the aldehyde group is protected as the dimethyl acetal **103**, the C-20 keto group cannot be reduced safely with sodium borohydride because the Δ^{16}-double bond is partially reduced. If however diisobutylaluminum hydride is utilized, the allyl alcohol with *R*-configuration at C-20 is selectively formed. After acetylation of this alcohol, the aldehyde **104** is liberated. The subsequent construction of the imine, because of the lability of **104,** requires cautious and mild conditions. The reduction of the imine to the amine requires lithium cyanoborohydride because it exerts its reducing power even in a weakly acidic medium. The remaining steps copy the earlier procedures and lead to a 7,8-dihydrobatra-

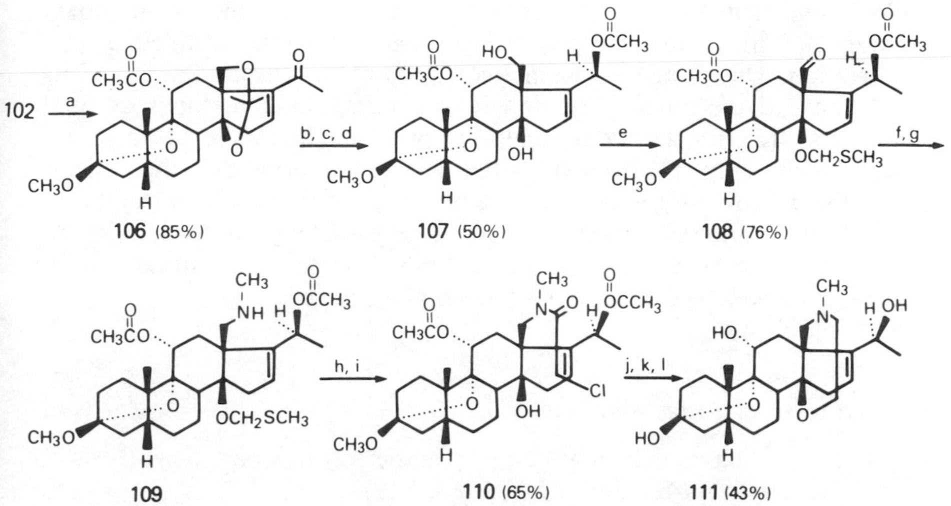

SCHEME 18. *Synthesis of 7,8-dihydrobatrachotoxinin A 111* (133). *Reaction conditions:* (a) $(CH_3)_2C(OCH_3)_2$, *TosOH, 20 min;* (b) *NaBH$_4$, abs. CH$_3$OH, $-30°C$, 19 hr;* (c) $(CH_3CO)_2O$: *pyridine* (1:1), *15 hr;* (d) *TosOH, CH$_3$OH, 15 min;* (e) $(CH_3)_2SO:(CH_3CO)_2O$ (1:1), *17 hr;* (f) *CH$_3$NH$_2$, ØH, 80°C, 7 hr;* (g) *NaBH$_4$, CH$_3$OH:H$_2$O* (15:1), *15°C, 10 min;* (h) *ClCH$_2$COCl, NaOH, H$_2$O, CHCl$_3$, 0°C, 15 min;* (i) *0.05 N HCl, CH$_3$OH;* (j) *NaH, ØH:THF* (1:1), *C$_2$H$_5$OH (cat.), ⇅, 6 hr;* (k) *LAH, (C$_2$H$_5$)$_2$O, ⇅, 5 hr;* (l) *TosOH, (CH$_3$)$_2$O:H$_2$O* (10:1), *⇅, 1 hr.*

chotoxinin A **105** which is epimeric with regard to the hydroxyl group at C-20. The structure of **105** was corroborated by Roentgen-ray analysis (*146*).

6. Synthesis of 7,8-Dihydrobatrachotoxinin-A

A markedly improved route both to epimeric and authentic 7,8-dihydro-batrachotoxinin A was then elaborated (Scheme 18) (*133*).

The intermediate **102** (Scheme 17) is converted to the acetonide **106**. With the 14β-hydroxyl function masked in this way the reduction of the enone succeeds. If the voluminous lithium tri-*tert*-amyloxyaluminum hydride is used, more of the C-20 alcohol with the *R*-configuration ($R:S = 3:1$) is obtained. If sodium borohydride is employed, no reduction of the Δ^{16}-unsaturation is observed and, at low temperatures, mostly the *S*-configuration of the C-20 allylalcohol ($R:S = 1:4$) is observed. The route of choice to 7,8-dihydrobatrachotoxinin A, therefore, made use of the sodium borohydride reduction. After acetylation of the allyl alcohol, the acetonide is opened to the diol **107** by the action of acid. The conversion to aldehyde **108** makes use of dimethyl sulfoxide in acetic anhydride, a method developed by

Goldmann and Albright, that simultaneously protects the 14β-hydroxyl group by a methylthiomethoxy group, thus suppressing competing fragmentation. This protection facilitated the formation of the imine and its reduction to the amine **109** by essentially the same methods as used previously (*129*). Chloroacetyl chloride provides the amide **110** after removal of the methylthiomethyl residue. The homomorpholine ring is closed with the help of sodium hydride. The reduction of the lactam with lithium aluminum hydride simultaneously liberates the hydroxyl groups at C-20 and C-11. The free hemiketal **111**, i.e., 7,8-dihydrobatrachotoxinin A, is obtained after acid treatment of the *O*-methyl precursor.

7. Synthesis of Batrachotoxinin A

The preceding model studies now open the way to a partial synthesis of batrachotoxinin A (**41**) (*125, 126*) (Scheme 19).

Starting with lactone **72**, the epoxide **112**, possessing all the required functionalities in rings A, B, and C, is obtained by the above-mentioned methods (*130*) (Scheme 16). The opening of the epoxide **112** (Scheme 19) with chromous acetate now leads to higher yields. After hydrogenation of the Δ⁴-double bond, the 3,9-hemiketal is formed and protected as mixed methylketal. Since the presence of the 7α-hydroxyl group makes this ketal very sensitive to acid, acetylation is achieved under drastic conditions before the dienone **113** is formed by two consecutive brominations and dehydrobrominations. Epoxidation and cleavage of the oxiran at the allylic C-15—O bond lead to the 14β-hydroxy derivative **114**. Selective saponification and ketalization lead to the 14,18-acetonide. Reduction of the C-20 keto group with sodium borohydride at low temperature furnishes a mixture of the allylic alcohols epimeric at C-20 ($S:R = 4:1$). To improve the yield of the desired *S*-alcohol, the allylic alcohol with *R*-configuration at C-20 is reoxidized on activated manganese dioxide and reduced again. After acetylation the triacetate **115** is obtained. Hydrolysis of the acetonide and oxidation with dimethylsulfoxide in acetic anhydride yield the C-18 aldehyde **116** in which the 14β-hydroxyl group is protected. The construction of the homomorpholine ring follows steps described previously: the aldehyde is converted to the methylimine, reduced to the amine, and with chloroacetyl chloride transformed to the amide **117**. The 14β-hydroxyl group is liberated by the action of weak acid and cyclization to the homomorpholine ring achieved by the action of sodium hydride. Excess sodium hydride is converted to methoxide by the addition of methanol and used for transesterification on heating. The triol **118** obtained is acetylated without involvement of the 7α-hydroxyl group, since it is sterically less accessible. The Δ⁷-double bond is now

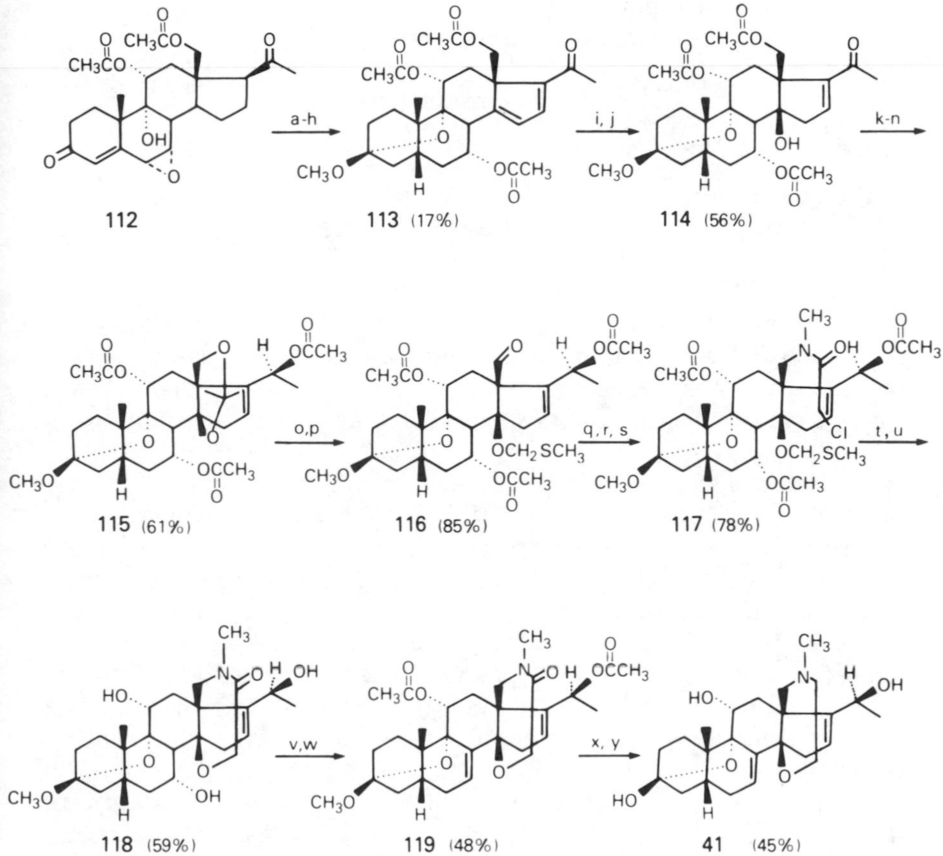

SCHEME 19. *Synthesis of batrachotoxinin A* (126). *Reaction conditions:* (a) $Cr(OCOCH_3)_2$, *pyridine 30 min, Ar;* (b) *Pd/C, 1 atm H_2, CH_3OH;* (c) *0.01 N HCl, CH_3OH;* (d) *$(CH_3CO)_2O$:pyridine (1:1), 105°C, 20 hr;* (e) *NBS, AIBN, hv, CCl_4 ↑↓ 45 min;* (f) *LiBr, Li_2CO_3, DMF, 130°C, 2.5 hr;* (g) *NBS, AIBN, hv, CCl_4 ↑↓, 20 min;* (h) *LiBr, Li_2CO_3, DMF, 130°C, 15 min;* (i) *p-$NO_2\phi COOOH$, $CH_3OH/CHCl_3$ ↑↓, 15 hr;* (j) *Pd/BaSO_4, cyclohexene, CH_3OH ↑↓, 30 min;* (k) *$NaHCO_3$, CH_3OH:H_2O (9:1), ↑↓, 15 min;* (l) *$(CH_3)_2$, TosOH;* (m) *$NaBH_4$, CH_3OH, −30°C, 20 hr;* (n) *$(CH_3CO)_2$:pyridine (1:1), 3 hr;* (o) *TosOH, CH_3OH;* (p) *DMSO abs., $(CH_3CO)_2O$, 15 hr;* (q) *CH_3NH_2, ϕH, 85°C, 9 hr;* (r) *$NaBH_4$, CH_3OH;* (s) *$ClCH_2COCl$, NaOH, H_2O, $CHCl_2$, 0°C, 15 min;* (t) *0.05 N HCl, CH_3OH;* (u) *NaH, ϕH:THF, C_2H_5OH cat., ↑↓, 2 hr; CH_3OH;* (v) *$(CH_3CO)_2O$:pyridine (1:1), 15 hr;* (w) *$SOCl_2$, pyridine, 2 hr;* (x) *LAH, $(C_2H_5)_2O$, ↑↓, 5 hr;* (y) *TosOH, $(CH_3)_2CO$:H_2O (9.1).*

introduced in high yield by dehydration with thionyl chloride, a process that produces 10% of the Δ^6-double bond. The dienone **119** is now converted to batrachotoxinin A **41** by reduction with lithium aluminum hydride and acid hydrolysis.

III. Spiropiperidines, cis-Decahydroquinolines, and Octahydroindolizidines

A. The Histrionicotoxins

1. Occurrence, Isolation, and Structure

The histrionicotoxins have all been isolated from extracts of the skins of the South American frog species *Dendrobates*. They form a group of ten alkaloids, three major (**120, 121, 123**) and seven minor ones (**122, 124–129**), (*147, 148*) (Table I).

Their common characteristic is the l-azaspiro[5.5]undecan-8-ol system observed here for the first time in a natural product. The variations among the ten alkaloids are located in the side chains in positions 2 and 7 which, without precedent, feature acetylenic as well as allenic unsaturations in animal alkaloids.

The first six alkaloids of this group, together with the gephyrotoxins, were isolated from the Colombian frog *Dendrobates histrionicus* (*147, 148*). Methanolic extracts of the skins of 1100 frogs gave a basic fraction (1.2 g) that, after purification with the help of silica gel and Sephadex LH-20 columns, yielded histrionicotoxin **120** (HTX, 226 mg), isodihydrohistrion-icotoxin **121** (320 mg), neodihydrohistrionicotoxin **122** (19 mg), tetrahy-drohistrionicotoxin **124** (2 mg), isotetrahydrohistrionicotoxin **125** (6 mg), octahydrohistrionicotoxin **127** (9 mg), gephyrotoxin **325** (47 mg), and significant amounts of mixed fractions. The structure and absolute configu-ration of two major alkaloids, HTX (*148*) and isodihydro-HTX (*148, 149*) were determined by Roentgen-ray crystallographic analysis of the hydro-chloride or hydrobromide, respectively. The structures of minor alkaloids were determined either chemically by catalytic hydrogenation to the identi-cal perhydrohistrionicotoxin **130** (PHTX) not encountered in nature, or by analysis of the ^{1}H-NMR and mass spectral data. Two further C_{19} alkaloids, allodihydro-HTX **123** whose presence had been suspected earlier (*147*) and allotetrahydro-HTX **126,** were isolated and characterized in the same way in a later and larger isolation procedure involving skins from 3200 *D. histrion-icus* frogs (*150*). Two additional congeners with the same ring skeleton but shorter side chains in positions 2 and 7, designated as "HTX-235A" **129,** and "HTX-259" **128,** were isolated and characterized in a careful analysis of skin extracts of 18 different *Dendrobates* subspecies with the help of adsorption and gas chromatography coupled with mass spectrometry, mass, and ^{1}H-NMR spectral data (*151*).

While there are so far no conclusive experiments concerning the biosyn-thesis of these toxins, two plausible hypotheses have been advanced: One (*152*) (Scheme 20) postulates intermediate polyketides, especially a δ-trike-

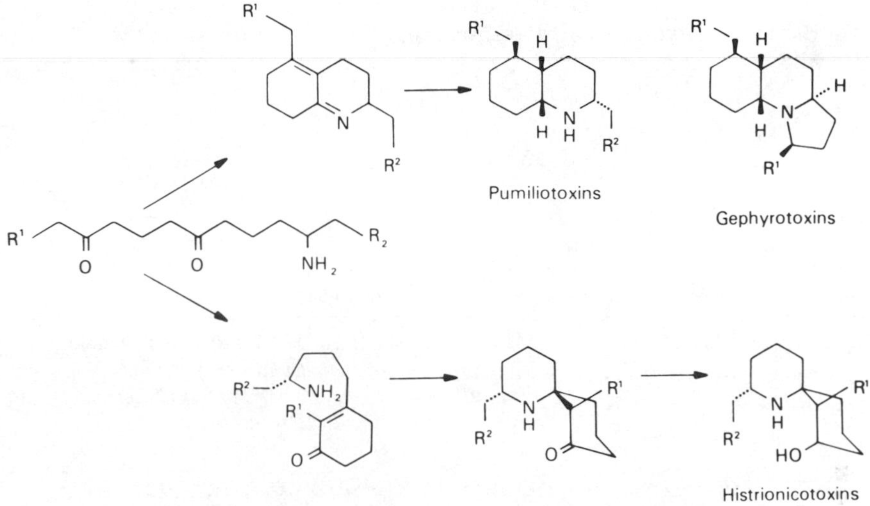

SCHEME 20. *Hypothetical scheme for the biosynthesis of the pumiliotoxins, histrionicotoxins, and gephyrotoxins via polyketide precursors* (152).

tone or a comparable long-chain 1-amino-5,10-diketone as the building stone for histrionicotoxins as well as pumiliotoxins and gephyrotoxins. The author suggests that the carbocycle of histrionicotoxin is formed first and then the heterocycle by intramolecular Michael addition. However, feeding experiments (*119*) with radioactive acetate suffer from the drawback of the slow formation of the toxins in frogs kept in captivity, thus leading to exchange of radioactive label with the general metabolic pool.

The second hypothesis (Scheme 21) assumes first the formation of the 2,6-disubstituted piperideine followed by intramolecular Mannich-type cyclization to the l-azaspiro[5.5]undecane derivative (*151*).

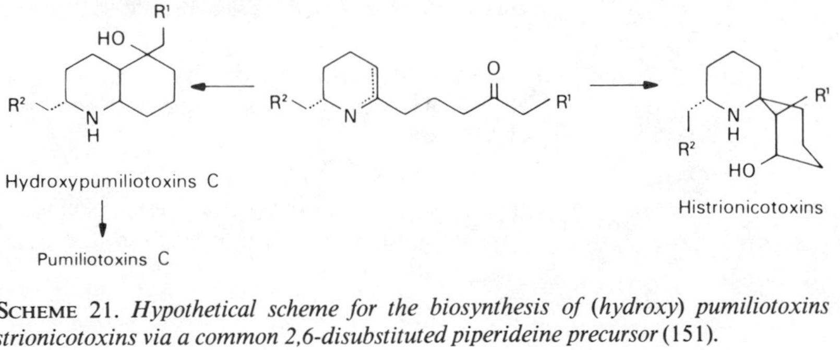

SCHEME 21. *Hypothetical scheme for the biosynthesis of (hydroxy) pumiliotoxins and histrionicotoxins via a common 2,6-disubstituted piperideine precursor* (151).

TABLE I

THE HISTRIONICOTOXINS: STRUCTURES, ABSOLUTE CONFIGURATIONS AND NOMENCLATURE

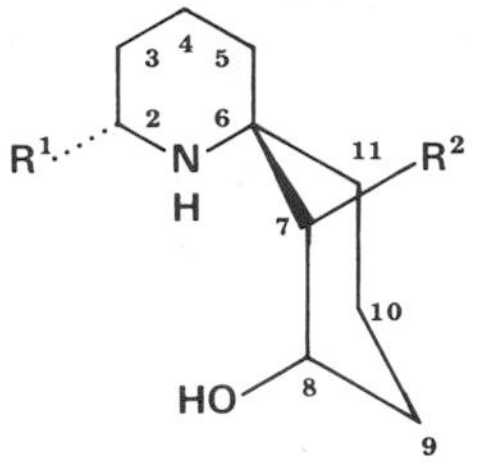

Structure[a]	Nomenclature, occurrence
$R^1 = CH_2CH\overset{Z}{=}CH-C\equiv CH$ $R^2 = CH\overset{Z}{=}CH-C\equiv CH$ (2S, 6R, 7S, 8S)	HISTRIONICOTOXIN (HTX), **120,** Chief alkaloid from all populations of *D. histrionicus, D. azureus, D. granuliferus, D. occultator, D. parvulus D. pictus, D. tinctorius, D. triviatus, D. truncatus.*
$R^1 = CH_2CH_2CH=C=CH_2$ $R^2 = CH\overset{Z}{=}CH-C\equiv CH$ (2S, 6R, 7S, 8S)	ISODIHYDROHISTRIONICOTOXIN, **121,** Chief alkaloid of all populations of *D. histrionicus, D. azureus, D. granuliferus, D. occultator, D. parvulus, D. pictus, D. tinctorius.*
$R^1 = CH_2CH\overset{Z}{=}CH-C\equiv CH$ $R^2 = CH\overset{Z}{=}CH-CH=CH_2$ (2S, 6R, 7S, 8S)	NEODIHYDROHISTRIONICOTOXIN, **122,** Minor alkaloids of populations of *D. histrionicus* from Playa de Oro, Quebrada Guangui, Guayacana, Rio Baba, *D. pictus, D. trivittatus, D. truncatus.*
$R^1 = CH_2CH_2CH_2C\equiv CH$ $R^2 = CH\overset{Z}{=}CH-C\equiv CH$ (2S, 6R, 7S, 8S)	ALLODIHYDROHISTRIONICOTOXIN, **123,** Chief alkaloid of all populations of *D. histrionicus, D. pictus, D. trivittatus, D. truncatus;* alkaloid of *D. granuliferus, D. tinctorius.*
$R^1CH_2CH\overset{Z}{=}CH-CH=CH_2$ $R^2 = CH\overset{Z}{=}CH-CH=CH_2$ (2S, 6R, 7S, 8S)	TETRAHYDROHISTRIONICOTOXIN, **124,** Minor alkaloids of populations of *D. histrionicotoxin* from Santa Cecilia, Quebrada Docordo, Quebrada Vicordo, Guayacana, Rio Baba, Rio Palenque, and of *D. parvulus.*
$R^1 = CH_2CH_2CH=C=CH_2$ $R^2 = CH\overset{Z}{=}CH-C=CH_2$ (2S, 6R, 7S, 8S)	ISOTETRAHYDROHISTRIONICOTOXIN, **125,** Minor alkaloid of all Colombian populations of *D. histrionicus,* and *D. pictus.*
$R^1 = CH_2CH_2CH_2C\equiv CH$ $R^2 = CH\overset{Z}{=}CH-CH=CH_2$ (2R, 6R, 7S, 8S)	ALLOTETRAHYDROHISTRIONICOTOXIN, **126,** trace of population of *D. histrionicus* from Guayacana.

TABLE I   (*Continued*)

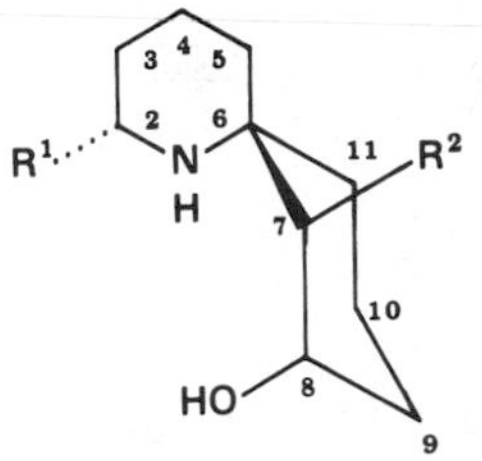

Structure[a]		Nomenclature, occurrence
$R^1 = CH_2CH_2CH_2CH=CH_2$ $R^2 = CH_2CH_2CH=CH_2$ (2R, 6R, 7S, 8S)		OCTAHYDROHISTRIONICOTOXIN, **127**, Main alkaloids of populations of *D. histrionicus* from Rio Guapi, Guayacana, Rio Palenque, of *D. granuliferus, D. parvulus, D. tinctorius, D. truncatus.*
$R^1 = CH_2CH=CH_2$ $R^2 = CH\overset{z}{=}CH-C\equiv CH$ (2S, 6R, 7S, 8S)		HISTRIONICOTOXIN-259, **128**, Main alkaloid of *D. tinctorius, D. trivittatus,* Minor alkaloid of *D. auratus, D. azureus, D. granuliferus,* of populations of *D. histrionicus* from Playa de Oro, Quebrada Vicordo, Quebrada Docordo, of *D. occulator, D. truncatus.*
$R^1 = CH_2CH=CH_2$ $R^2 = CH=CH_2$ (2S, 6R, 7S, 8S)		HISTRIONICOTOXIN-235 A, **129**, Minor alkaloid of *D. auratus,* trace alkaloid of *D. granuliferus, D. histrionicus* of populations from Quebrada Vicordo, *D. parvulus, D. trivittatus, D. truncatus.*

[a] The letters in parentheses denote absolute configuration.

2. Syntheses and Synthetic Approaches Starting with the Carbocyclic Ring

The histrionicotoxins have acquired an almost pivotal position for the study of the ion transport mechanism (ion conductance modulator) of the nicotinic acetylcholine receptor (*8*). The demand for samples from neurophysiological research centers all over the world soon exceeded supply by a wide margin. The synthetic task was simplified by the observation that perhydrohistrionicotoxin, **130**, the end product of catalytic hydrogenation of most histrionicotoxins, is tantamount in its pharmacological activity to the natural toxins. Octahydro-HTX, **127**, with two terminal olefinic bonds is especially active and is so far the only naturally occurring representative whose racemic form has been synthesized (*154*).

The majority of synthetic pathways first approaches the construction of a carbocyclic system followed by formation of the heterocyclic moiety. The first synthesis (*155*) (Scheme 22) starts with cyclopentanone **131** that is reductively dimerized to the diol and, after pinacolone rearrangement, yields

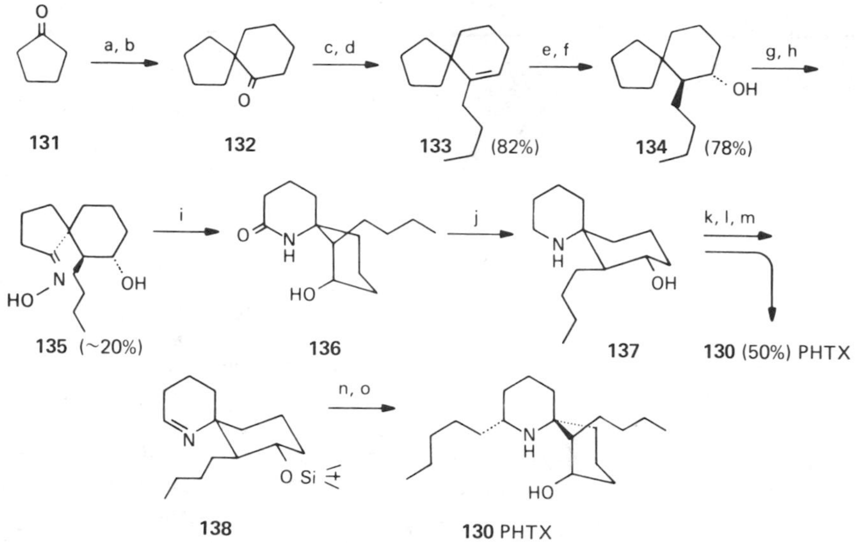

SCHEME 22. *Synthesis of racemic perhydrohistrionicotoxin 130 starting with cyclopentanone 131* (155). *Reaction conditions:* (a) *To Mg–Hg–THF, 0.45 Eq TiCl₄ cyclopentanone was added; 10% K₂CO₃, H₂O;* (b) *H⁺;* (c) *3 times:* [1] *C₄H₉Li, C₆H₁₄: (C₂H₅)₂O,* 1ℓ, *5 min* [2] *1 Eq CH₃OH;* (d) *SOCl₂, pyridine, −78°C, 6 hr; 4 Eq* ⟨N H⟩ ; (e) *R₂BH;* (f) *NaOH, H₂O₂;* (g) *1.2 Eq NOCl, 1.4 Eq pyridine, CH₂Cl₂ <0°C;* (h) *hv;* (i) *TosCl, pyridine, ØH 12 hr;* (j) *exc. LAH, THF,* 1ℓ, *36 hr;* (k) *1.3 Eq NaH, ClSi≶, THF;* (l) *NBS, THF, 0°C;* (m) *KOt-C₅H₁₁, −40°C;* (n) *n-C₅H₁₁Li, hexane, 12 hr;* (o) *(C₄H₉)₄N⁺F⁻, THF.*

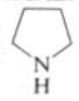

the spiroketone 132. Addition of butyllithium, under special conditions in order to counteract the competing formation of enolate, and dehydration of the resulting tertiary alcohol, yield the spiro olefin 133 which, after hydroboration and treatment with basic H₂O₂, furnishes the secondary alcohol 134. This completes the stereospecific construction of two of the four chiral centers of PHTX. Photolysis of the nitrous acid ester of the secondary alcohol introduces an oxime function on the five-membered ring at a cis position in relation to the original secondary hydroxyl group 135. The price for this desirable stereoselectivity has to be paid by low yield and by competing side reactions such as oximation of the butyl side chain. The Beckmann rearrangement regioselectively produces the lactam 136, a key intermediate in the majority of syntheses of PHTX.

The stereoselective introduction of the side chain of the heterocycle is solved as follows: The lactam 136 is reduced to the cyclic amine 137. Protection of the hydroxyl group as silyl ether is necessary in order to obtain, after bromination and elimination, the imine 138 with the substituents at the carbocycle in the equatorial position. Alkylation with pentyllithium then leads to (±)-PHTX **130.**

The attempt was made to introduce the desired side chain into depentyl-perhydrohistrionicotoxin **137** after N-nitrosation, metallation, and alkylation according to Seebach (*156*), but it failed because the PHTX obtained after denitrosation was the C-2 epimer.

Another synthesis (*157, 158*) (Scheme 23) starts with ethyl cyclohexene carboxylate **139** whose anion is alkylated with ethyl 4-bromobutyrate, followed by Dieckmann condensation, saponification, and decarboxylation to yield the unsaturated spiroketone **140.** The protected oxime **141** is converted to the bromohydrin **142** regio- and stereoselectively, with only 10% of the stereoisomeric bromohydrin and some bromo ketone formed. The path via an epoxide of **141** is not feasible, because its alkylative opening introduces the substituent exclusively in β position in relation to the quaternary carbon atom of the spirocyclic system. Therefore, a new method (*159*) was developed. Thus, oxidation and oximation yield the brominated dioxime monobenzyl ether **143** which with excess butynyllithium first forms the enenitroso intermediate followed by 1,4-addition of a second equivalent of butynyllithium in a stereoselective manner controlled by the proximity of the oxime ether grouping at the five-membered ring. Reduction of the newly introduced unsaturated side chain leads to **144.** Removal of the unprotected

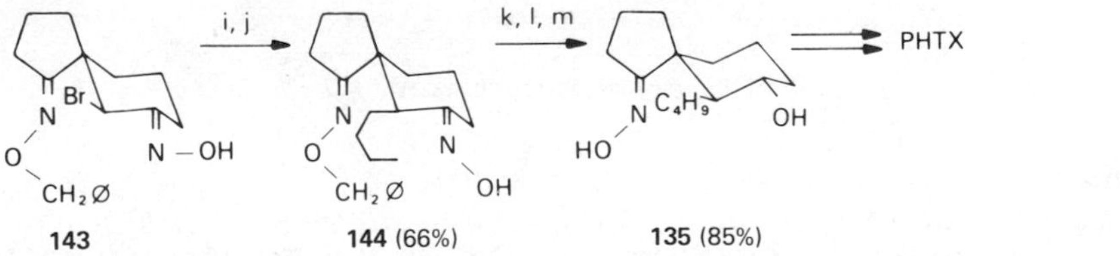

SCHEME 23. *Synthesis of racemic perhydrohistrionicotoxin starting with ethyl cyclohexene-carboxylate* **139** *(157, 158). Reaction conditions: (a) LDA, HMPT, THF, −78°C; Br(CH$_2$)$_3$COOC$_2$H$_5$, −20°C, 3 hr; (b) NaH, THF; (c) THF:H$_2$O:H$_2$SO$_4$ (8:2:1), 1ℓ, 4 hr; (d) N$^+$H$_3$OHCl$^-$, pyridine abs. C$_2$H$_5$OH, 18 hr; (e) KH; BrCH$_2$Ø, (CH$_3$OCH$_2$)$_2$; (f) NBS, DMF/H$_2$O, −20°; (g) Jones reagent; (h) N$^+$H$_3$OHCl$^-$, NaOCOCH$_3$, HOOCCH$_3$, 2 hr; (i) excess Li C≡CC$_2$H$_5$, −78 to −10°C; (j) H$_2$, Pd/C, C$_2$H$_5$OOCCH$_3$; (k) TiCl$_3$·H$_2$O, CH$_3$OH, N$^+$H$_4$OOCCH$_3$; (l) H$_2$, Pd/C, C$_2$H$_5$OH; (m) Na, NH$_3$, THF, (CH$_3$)$_2$CHOH.*

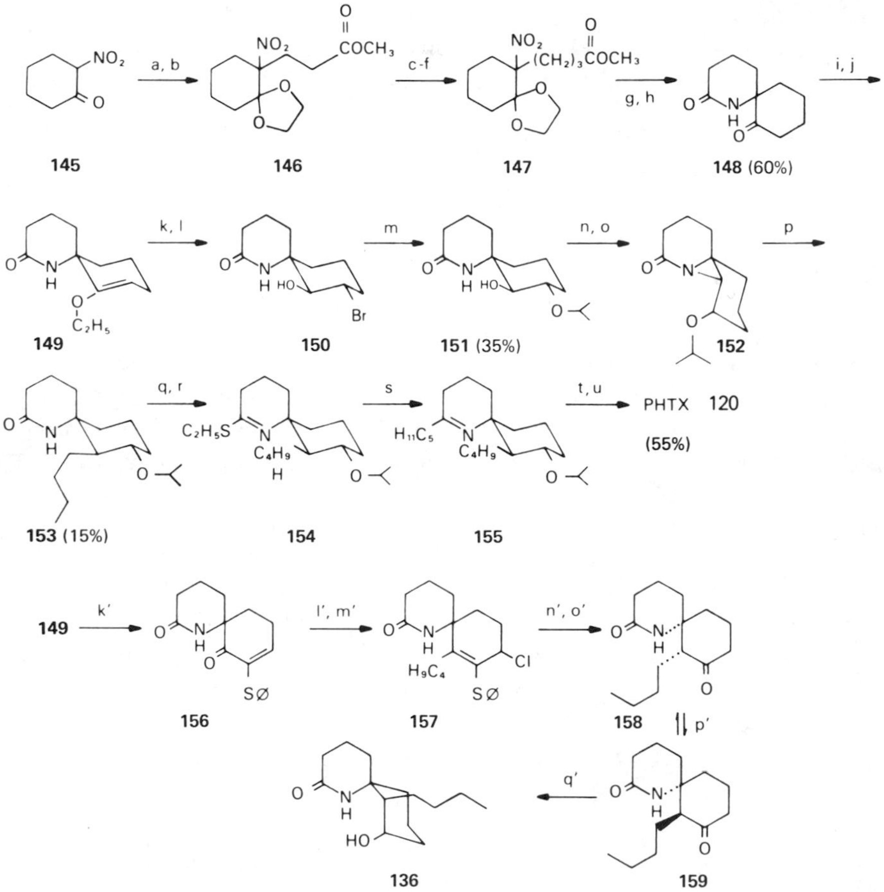

SCHEME 24. *Synthesis of racemic perhydrohistrionicotoxin* **130** *starting with* α-*nitrocyclo-*

hexanone **145** (160). *Reaction conditions:* (a) $\overset{\frown}{HO\,OH}$, H^+, ØH; (b) $\overset{O}{\overset{\|}{\diagdown\diagdown COCH_3}}$, *Triton B*, t-*BuOH;* (c) *NaOH*, H_2O/CH_3OH; (d) *SOCl*$_2$, ØH, 50°C; (e) CH_2N_2; (f) $AgBF_4$ $(C_2H_5)_3N$, CH_3OH, 0°C; (g) Ra/Ni, H_2, CH_3OH, 50°C; (h) CF_3COOH, H_2O, 75°C; (i) $(C_2H_5O)_3$ CH,H^+; (j) Δ; (k) Br_2; (l) $NaBH_4$; (m) $(CH_3)_2CHOH,(CH_3)_2CHONa$; (n) MesCl, pyridine; (o) NaH, wet ØH; (p) $(C_4H_9)_2CuLi$, THF; (q) P_2S_5, ØH; (r) $(C_2H_5)_3O^+B^-F_4$; (s) $C_5H_{11}Li$, C_6H_{14} $(C_2H_5)_2O$, DBAH; (t) BBr_3; (u) AlH_3, cyclohexane; (k') ØSCl, CH_2Cl_2; (l') C_4H_9MgCl, THF; (m') $SOCl_2$; (n') Zn, HCl; (o') conc. HBr; (p') CH_3ONa,CH_2Cl_2; (q') Li or Ca; liq NH_3, −78°C.

oxime, hydrogenolysis of the *O*-benzyloxime, and stereoselective reduction of the carbonyl group leads to **135,** mentioned before (Scheme 22), as precursor of PHTX as described above (*155*).

Another synthesis of PHTX (Scheme 24) (*160*) starts with α-nitrocyclohexanone **145** (*161*) which, after ketalization, undergoes Michael addition with methyl acrylate to yield **146,** convertible to the homologous **147** by an Arndt–Eistert reaction. Reduction of the nitro group and deketalization furnishes the ketolactam **148.** Its enol ether **149** is converted to the bromo ketone that undergoes stereoselective reduction with sodium borohydride to **150** guided by the proximity of the amide nitrogen. Treatment with sodium isopropoxide, via an intermediate oxiran, yields the isopropyl ether **151** with retention of the original configuration at C-8. In a hitherto unknown sequence, the free secondary hydroxyl of the glycol half-ether **151** is mesylated to form, through intramolecular substitution by the amide anion, the acylaziridine **152,** which with dibutylcopperlithium opens stereoselectively to the lactam **153** possessing three of the four chiral centers of PHTX in the desired configuration. Introduction of the fourth and final chiral center proceeds via the thiolactam and thiolactim ethyl ether **154** which with pentyllithium and DBAH as a catalyst yields the imine **155.** Cleavage of the isopropyl ether and reduction of the imine with aluminum hydride leads preponderantly (9:1) to PHTX because of preferred attack from the side on which the secondary hydroxyl function is located. The ketolactam **148,** is also accessible by two other variants (*162, 163*, Schemes 25 and 26).

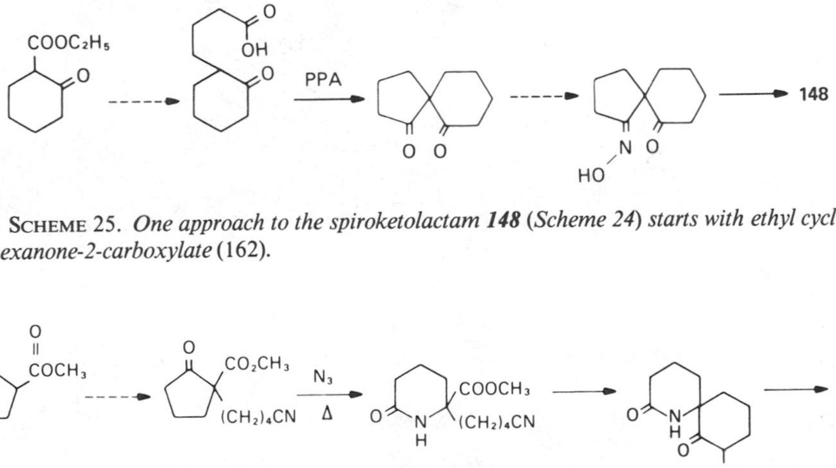

SCHEME 25. *One approach to the spiroketolactam **148** (Scheme 24) starts with ethyl cyclohexanone-2-carboxylate* (162).

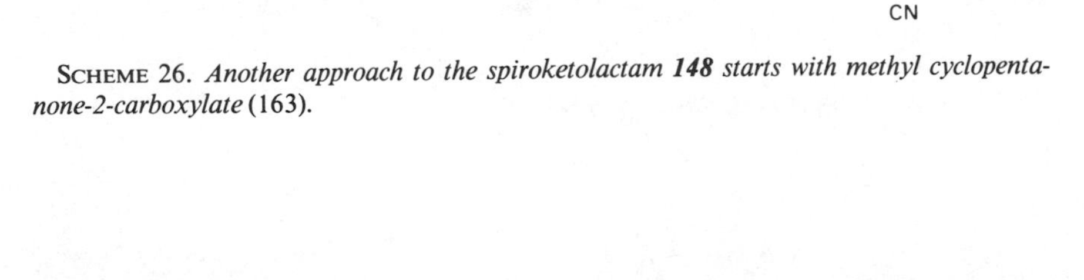

SCHEME 26. *Another approach to the spiroketolactam **148** starts with methyl cyclopentanone-2-carboxylate* (163).

Another modification (Scheme 24) of the above synthesis (*160*) originates with the enol ether **149** of the ketolactam **148**. An improvement in the overall yield of PHTX is achieved by the temporary sacrifice of stereoselective introduction of the butyl side chain. Thus, **149** is converted to the phenylthioenol ether **156** that undergoes 1,2-addition with butylmagnesium chloride to give the intermediate tertiary allyl alcohol yielding the allylic chloride **157** with thionyl chloride. Reduction with zinc and HCl and subsequent hydrolysis produces the mixture of ketones **158** and **159** in a ratio of 1:3. The desired ketone **159** is preferred in the equilibrium established under basic conditions. Hydroxylactam **136** is formed from ketone **159** stereoselectively by reduction with lithium or calcium, but *not* sodium, in liquid ammonia. Of the modes of stereoselective introduction of the pentyl side chain into the heterocycle described so far the following variation is the best: The secondary alcohol is protected by acetylation; the thiolactam and, after deacetylation, the tetrahydropyranyl ether are prepared. The thiolactim ether, after reaction with pentyllithium and DBAH, removal of the oxygen protective group, and reduction (AlH₃) yields PHTX **130** and its C-2 epimer.

The first and so far only synthesis of the racemic form of a naturally occurring histrionicotoxin (Scheme 27) was achieved as follows (*154*).

The starting material is 2-but-3′-enylcyclohexa-1,3-dione **160** accessible from the reaction of pent-4-enyl-cadmium with methyl-4-chloroformyl-

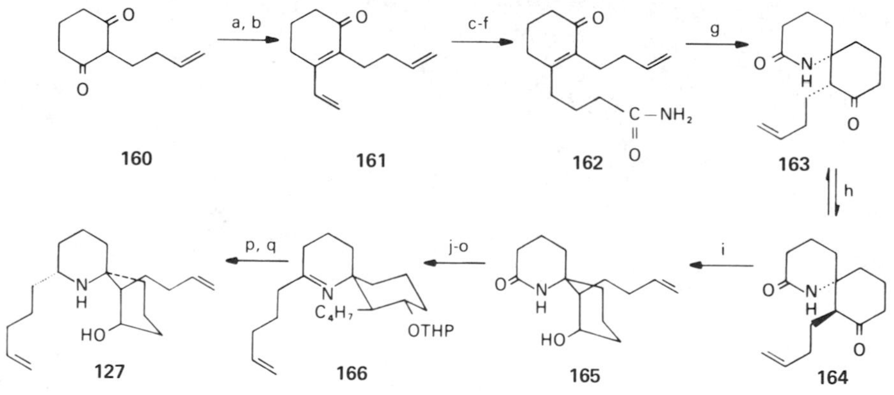

SCHEME 27. *Synthesis of racemic octahydrohistrionicotoxin **127** (154) in 14% overall yield.*

(*a*) C_2H_5OH, H^+; (*b*) $CH_2{=}CHMgBr$, THF; (*c*) $H_3COOCCH_2CNH_2$, $NaOCH_3$; (*d*) NaOH;

(*e*) HCl; (*f*) 100°C,  ; (*g*) $HC(OC_2H_5)_3$, H^+, C_2H_5OH; (*h*) CH_3ONa, CH_2Cl_2;

(*i*) Li, NH_3, $-78°C$; (*j*) $(CH_3C)_2O$, pyridine; (*k*) P_2S_5; (*l*) NaOH; (*m*) , H^+; (*n*) $(C_2H_5)_3O^+$ BF^-; (*o*) C_5H_9Li, DBAH cat; (*p*) H^+, H

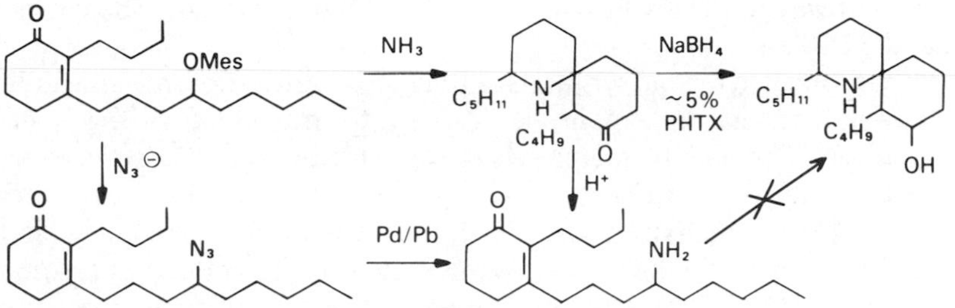

SCHEME 28. *Intramolecular Michael addition and its easy reversal in the approach to spirolactams via suitably substituted cyclohexenones* (164).

butyrate yielding methyl dec-9-en-5-one carboxylate and base-catalyzed cyclization. The monoethyl enol ether of **160,** after reaction with vinylmagnesium bromide, furnishes the trienone **161** to which the anion of methyl 3-amino-3-oxopropionate is added in 1,6-fashion to yield, after saponification and decarboxylation, the amide **162.** Its sodium salt undergoes intramolecular Michael addition to the spirocyclic diastereoisomeric systems **163** and **164** in a ratio of 2:1. As described above, equilibration in a basic medium shifts this ratio to 1:4. The desired ketone **164** is reduced to the correct alcohol **165.** The stereoselective introduction of the 5-carbon side chain proceeds as described with substitution of pentyllithium by pent-4-enyllithium. After this, a reductive step (AlH$_3$) leads from the imine **166** to racemic octahydrohistrionicotoxin **127** and its 2-epimer in a ratio of 6:1. The same sequence was used for a better and more practical approach to PHTX.

The idea of a cyclohexenone synthone and of an intramolecular Michael addition proved to be a theme with variations (*164*). A drawback of the spirocyclic ketones so formed (Scheme 28) is their tendency to undergo easily a retro-Michael reaction, especially when the amide group is replaced by a methylene amine moiety. Even unsubstituted 1-azaspiro[5.5]-undecan-8-ones were observed by several authors (*165, 166*) to behave in this way.* In the attempted cyclization to 7-butyl-2-pentyl-1-azaspiro[5.5]undecan-8-one (Scheme 28), retro-Michael addition supervenes.

An analogous pathway was also recorded in an undocumented abstract of a lecture (October 6–7, 1979) (*152*) in which the attempt was made to utilize biogenetic speculations as a lodestar for synthesis.

Variations on the use of the monoenol ether of cyclohexa-1,3-dione as synthon for the synthesis of suitable functionalized spirocyclic intermediates

---

* These authors observed the same type of retro-Michael reaction even with the lactam, 1-azaspiro[4.5]deca-2,7-dione (*167*).

on the pathway to PHTX have been recorded (*164, 168, 169*) (Schemes 29, 30, and 31).

In a complicated formal synthesis of PHTX, a sigmatropic rearrangement, the sulfilimine rearrangement, is utilized to introduce nitrogen (*170*) (Scheme 32). The starting material is the epoxide of cyclohexenone which is converted to the olefin **167** by a Wittig reaction. The anion of an appropriate allyl ether opens the oxiran in α-position to the double bond to yield the desired enol ether, which with acidic methanol furnishes the diastereomeric cyclic acetals **168** (2.5 : 1). The trans configuration of the bicyclic **168** permits only one chair conformation for the cyclohexane ring and thus makes it an appropriate compound for the stereoselective [2.3]-sigmatropic rearrangement known to occur at the exocyclic double bond of the cyclohexane ring stereoselectively with equatorial attack. For this reason **168** is ozonized to

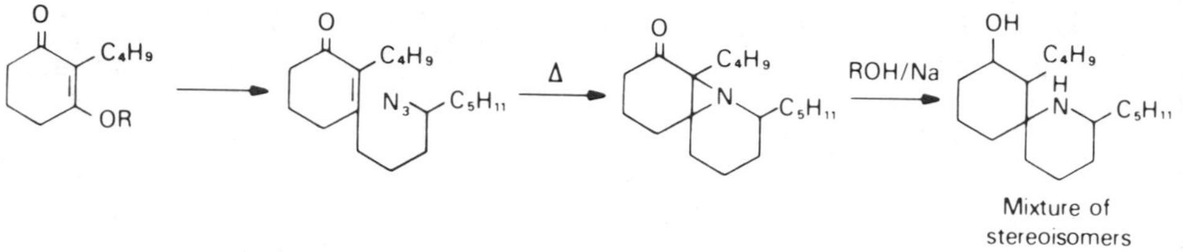

SCHEME 29. *Thermal decomposition of an azidocyclohexenone leads to a mixture of stereoisomeric spiropiperidines via an aziridine intermediate* (164).

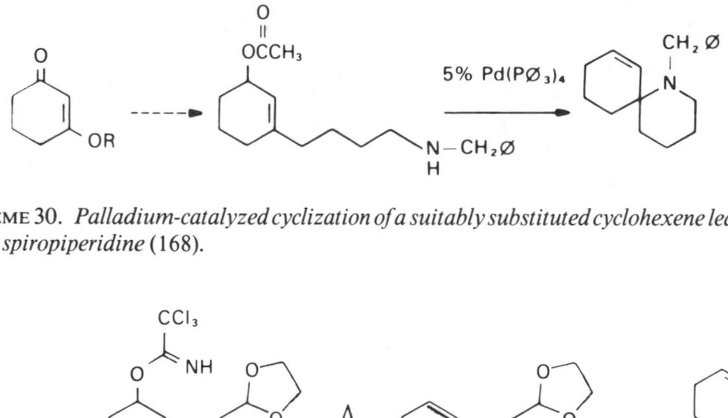

SCHEME 30. *Palladium-catalyzed cyclization of a suitably substituted cyclohexene leads to an olefinic spiropiperidine* (168).

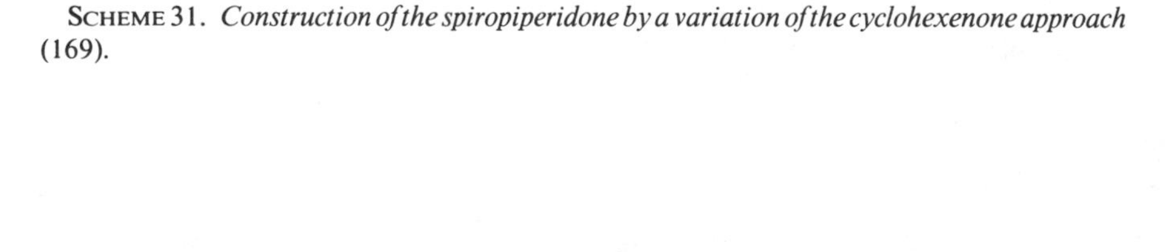

SCHEME 31. *Construction of the spiropiperidone by a variation of the cyclohexenone approach* (169).

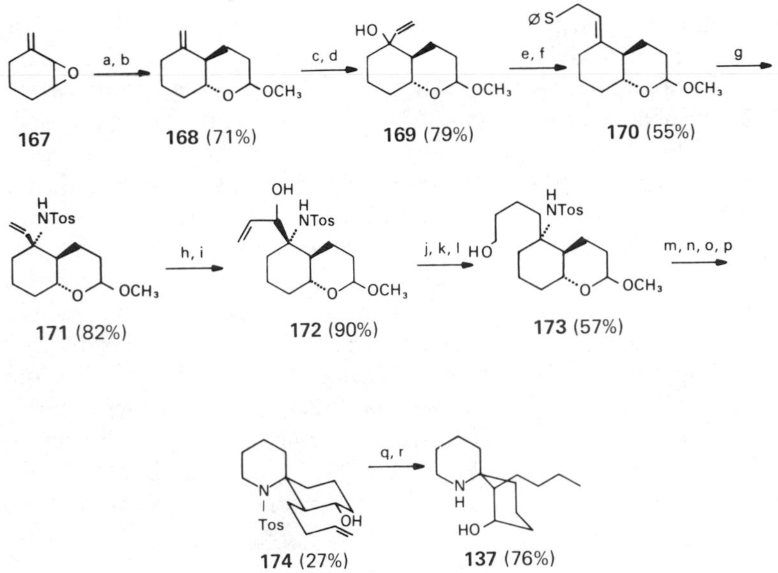

SCHEME 32. *Multistep synthesis of depentyl-PHTX 137 starting with cyclohexenone epoxide* (170). *Reaction conditions: (a)* $(CH_3)_2N-CH_2CH_2-O-CH_2 CH=CH_2$, *sec-BuLi, THF,* $-78°C$; (b) CH_3OH, *conc HCl, 2 hr;* (c) O_3, CH_2Cl_2; H_2CCOOH, *Zn,* $0°C$; (d) $\nearrow\!\!\!\sim$ MgBr, $ØH$, *3 hr,* $0°C$; (e) PBr_3, $N(C_2H_5)_3$ $60°C$; (f) $Ø-SNa$, *up to* $25°C$, *20 hr;* (g) *chloramine-T, 3* H_2O, *NaOH;* (h) O_3, CH_2Cl_2, $-78°C$; $(CH_3)_2S$; (i) $\nearrow\!\!\!\sim$ MgBr, $ØH$, $0°C$, *1 hr;* (j) KH; ICH_2 $Sn(C_4H_9)_3$, *THF;* (k) *Li* C_4H_9, *THF,* $-100°C$; (l) PtO_2, *4 atm* H_2, CH_3OH; (m) $ClSO_2CH_3$, $N(C_2H_5)_3$, CH_2Cl_2, $-23°C$; (n) *NaH, THF,* $1l$, *4 hr;* (o) *5% aq. HCl, THF,* $50°C$, *4 hr;* (p) $Ø_3P^+ CH_3 Br^-$, *Li* C_4H_9, *THF,* $1l$, *4 hr;* (q) PtO_2, *1 atm,* H_2, CH_3OH; (r) 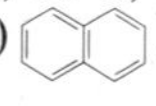 [naphthalene], *Na,* $(CH_3OCH_2)_2$.

yield the ketone which with vinylmagnesium bromide leads to the allylic alcohol **169.** Conversion to the allylic bromide and replacement by thiophenoxide regioselectively produces the thiophenylether **170,** that with chloramine-T leads to the sulfilimine. Rearrangement and immediate hydrolysis yield the allylic tosylamide **171.** Ozonization produces the aldehyde that with vinylmagnesiumbromide forms the secondary ally alcohol **172.** Side-chain extension is accomplished via the *O*-methylenetributyltin compound, destannylation by base, and rearrangement to the primary alcoholate **173** via [2,3]-sigmatropic rearrangement and reduction of the double bond. The hydroxyl group is mesylated and replaced by the anion of the tosylamide forming the spiro compound that on treatment with acids, followed by Wittig olefination yields **174,** convertible to **137,** an intermediate of an earlier PHTX synthesis (*155*), by hydrogenation and detosylation with sodium naphthalide.

Simultaneous construction of three chiral centers in the carbocyclic moiety can be achieved by use of a Diels–Alder reaction (*171*) (Scheme 33).

The dienophile in this case is ethyl 3-acetoxycyclohex-1-enylcarboxylate **175** that reacts with the diene **176** (*172*) to yield the octalinone **177**. Thioketalization, desulfuration with Raney nickel, and selective saponification of the acetate yield the octalinol **178** convertible by acid to the lactone **179** with migration of the double bond as a result of the prolonged reaction time. This olefin is converted to the ketoaldehyde **180** in two steps. A selective Wittig reaction involves only the aldehyde group, and reduction of the resulting olefin completes the construction of the butyl side chain in **181**. The ketonic side chain serves as precursor of the piperidine ring by being extended by methoxycarboxylation to a β-keto ester and removal of the keto function in four steps to yield the lactone **182**. Dieckmann condensation and demethoxycarboxylation with DABCO produce the spirocyclic ketone **183** from which oximation and Beckmann rearrangement lead to the useful lactam intermediate **136**. If, in the above sequence, instead of the acetate of

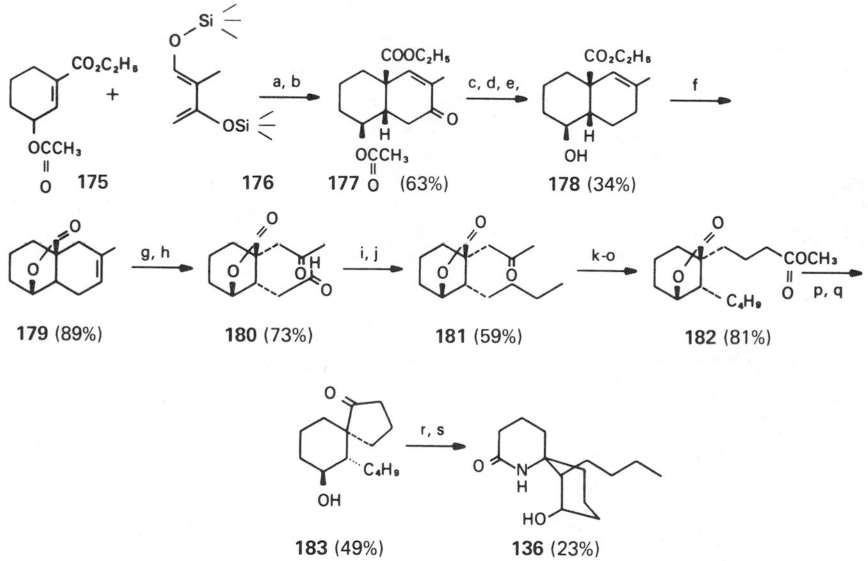

SCHEME 33. *Simultaneous introduction of three chiral centers by the use of the Diels–Alder reaction* (171, 172). *Reaction conditions:* (a) *170° C, 48 hr, 190° C, 72 hr, mesitylene;* (b) *5% HCl;* (c) *(CH₂SH)₂, H⁺;* (d) *Ra/Ni, THF;* (e) *KOH:H₂O:C₂H₅OH (1:6:30), 50°C;* (f) *ØCH₃,*

TosOH, 8 hr, ⇅; (g) *OsO₄,* *; (h) H₃IO₅, THF:H₂O (2:3), − 50°C, 25°C;* (i) *Ø₃P-*

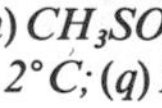

C̄HCH₃, THF:DMSO(15:3.5); (j) *H₂, PtO₂, CH₃OH;* (k) *CH₃OMgOCO₂CH₃, DMF, 135°C,*

24 hr; (l) *NaBH₄, CH₃OH;* (m) *CH₃SO₂Cl-C₂H₅OH, ØH, 4°C;* (n) *DBU, (C₂N₅)₃N, 7°C;* (o) *H₂,*

PtO₂, CH₃OH; (p) *KH, THF, 2°C;* (q) *DABCO, xylene, ⇅;* (r) *N̄H₃OH Cl⁻;* (s) *TosCl, pyridine.*

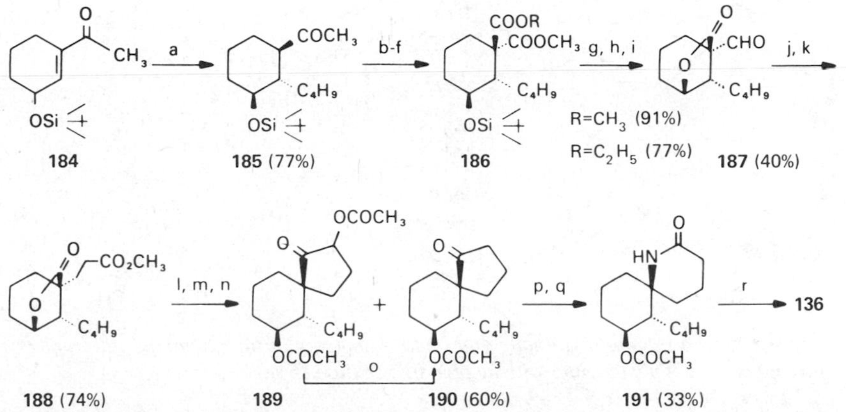

O
CH₃
OSi
184

a →

COCH₃
C₄H₉
OSi
185 (77%)

b-f →

COOR
COOCH₃
C₄H₉
OSi
186
R=CH₃ (91%)
R=C₂H₅ (77%)

g, h, i →

O
CHO
O
C₄H₉
187 (40%)

j, k →

O
CO₂CH₃
O
C₄H₉
188 (74%)

l, m, n →

OCOCH₃
O
C₄H₉
OCOCH₃
189

+

O
C₄H₉
OCOCH₃
190 (60%)

o

p, q →

O
HN
C₄H₉
OCOCH₃
191 (33%)

r → **136**

SCHEME 34. *Synthesis of the O-acetylketolactam 191 (and racemic perhydrohistrionicotoxin 136) in a largely stereoselective manner* (173, 174). *Reaction conditions:* (a) C₄H₉ Cu. AlCl₃; (b) LDA, − 70 to − 40° C; (CH₃)₃SiCl, (C₂H₅)₃N, − 70 to − 10° C; (c) O₃, − 70° C; (d) CH₂N₂; (e) LDA, CO₂; (f) CH₂N₂ or CH₃CHN₂; (g) dil. HCl, 50° C; (h) DBAH, ØCH₃:hexane (1:4), − 70° C; (i) PCC; (j) (CH₂O)₂P(O)CHCO₂CH₃, ØH:(C₂H₅)₂O (1:1), 0° C; (k) H₂, PtO₂, CH₃OH; (l) Na, (CH₃)₃SiCl, ØCH₃; (m) 5% HCl, 0° C, 15 min; (n) (CH₃CO)₂O, pyridine, 4-dimethylamino-pyridine; (o) Zn, HOCOCH₃; (p) N⁺H₃OH Cl⁻; (q) TosCl, pyridine; (r) NaOMe, CH₃OH.

the dienophile **175** the corresponding silyl ether is used for the Diels–Alder reaction with the diene **176,** no cyclization is observed.

However, the authors have developed a method (*173*) that solves the problem of stereoselective preparation of the cyclohexane derivative with the three vicinal substituents in equatorial position (*174*) by alkylation of the 1-acetylcyclohexene derivative **184** with a butylcopper–aluminum chloride complex yielding the desired butyl derivative **185** (Scheme 34). Kinetic enolization of the methyl ketone **185,** formation of the silylenol ether, ozonization, esterification with diazomethane, carboxylation of the carbanion, and esterification produce the diester **186,** again in a stereoselective manner. Acid treatment effects lactonization, and the remaining ester group is reduced to the alcohol and dehydrogenated to the aldehyde **187.** Horner–Emmons olefination and catalytic hydrogenation lead to the lactone **188.** This lactone undergoes acyloin condensation with sodium and trimethyl silyl chloride. Treatment with acid and subsequent acetylation yield a mixture of the spirocyclic α-acetoxyketone **189** and the ketone **190.** The main product **189** is easily convertible to the desired **190** by reduction with zinc in acetic acid. Oximation and Beckmann rearrangement lead to **191** and after saponification to the lactam **136.** The same authors report a variant (*174, 175*) that leads to the lactam epimeric at C-7.

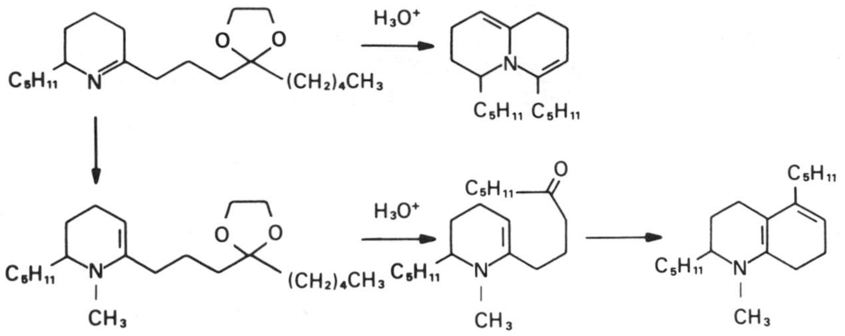

SCHEME 35. *Formation of quinolizidines and quinolines in the attempted intramolecular Mannich reaction of piperideines with ketonic side chains* (164).

This completes all those synthetic approaches that build up PHTX starting with the carbocyclic part.

3. Synthesis and Synthetic Approaches Starting with the Heterocyclic Ring

This concept corresponds with one of the hypotheses of the biogenic pathways (151). However (Scheme 35), instead of an intramolecular Mannich reaction under acidic conditions, ring closure occurs either by formation of an intermediary iminium salt furnishing a disubstituted quinolizidine or by starting with N-methylpiperideine by reaction of the enamine with the ketone of the side chain leading to 1-methyl-2,6-dipentyl-1,2,3,4,7,8-hexahydroquinoline, a compound closely related to pumiliotoxin C and its congeners (164).

The reason for this result may be found in the instability of the 1-azaspiro[5.5]undecan-8-one system in acidic medium, caused by Grob fragmentation (retro-Mannich reaction) as shown in Scheme 36 (176). Again a hydrogenated quinoline derivative is formed. These results emphasize the possibility of a common biogenetic precursor for both histrionicotoxins and pumiliotoxins.

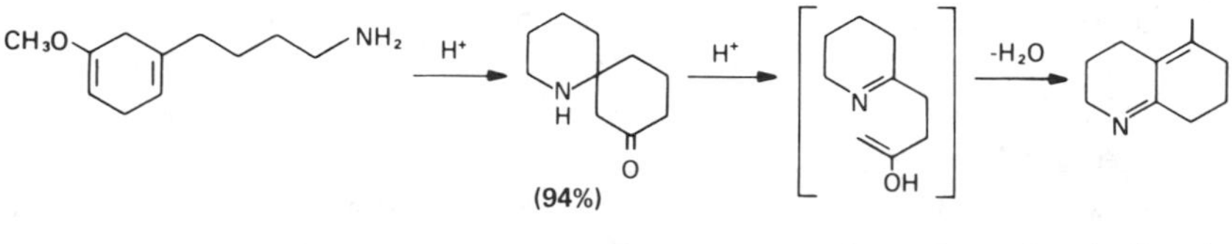

SCHEME 36. *Grob fragmentation of ketonic histrionicotoxin derivatives to hexahydroquinolines* (176).

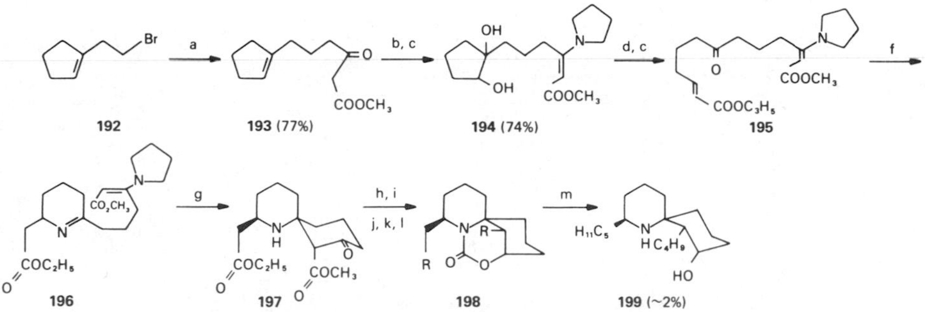

SCHEME 37. *Epimeric perhydrohistrionicotoxins via modified intramolecular Mannich reactions* (177). *Reaction conditions:* (a) $\bar{C}H_2-CO-\bar{C}H-COOCH_3$, *HMPT/THF;* (b) ⟨N-H pyrrolidine⟩, *HOOCCH$_3$, ØH, 3 hr;* (c) OsO_4, *pyridine-$(C_2H_5)_2O$, -40 to $-15°C$, 50 min; NaHSO$_3$, H$_2$O, $-10°C$, 0.5 hr;* (d) Ag_2CO_3/celite, ØH, ↑; (e) $(C_2H_5O)_2POCH_2COOC_2H_5$, NaH, THF, -30 to $-20°C$; (f) liq NH$_3$, 25°C, 24 hr; (g) TosOH · H$_2$O; (h) NaBH$_4$, 1% KOH, C$_2$H$_5$OH, $-20°C$; (i) COCl$_2$, CH$_2$Cl$_2$, 0°C; (j) DBAH, $-78°C$, CH$_2$Cl$_2$:R = CHO; (k) Ø $(CH_3)_2P^+CH_2HC=CH_2$, K$^+$ CH$_2$−SO CH$_3$, THF, DMSO, 0°C:R = CH=CH−CH=CH$_2$; (l) H$_2$, Pd/C, THF:R = C$_4$H$_9$; (m) Li, CH$_3$NH$_2$, $-78°C$.

Another attempt to utilize a modified intramolecular Mannich reaction led to a PHTX epimeric both in positions C-2 and C-7 (*177*) (Scheme 37).

The bromoethylcyclopentene **192** reacts with the dianion of methyl acetoacetate to give the β-ketoester **193**. Glycolization of **193** leads to the diol **194** which with silver carbonate on Celite produces a ketoaldehyde convertible to the unsaturated ester **195** by a selective Horner–Emmons reaction. Treatment with ammonia in a sealed tube leads to the piperideine **196**, which in several steps via **197**, gives the carbamate **198** and a stereoisomer of PHTX **199**, with the C-2 and C-7 side chains in configurations epimeric to the natural product (*177*).

In analogy with the biomimetic olefin cyclization, Speckamp *et al.* (*178–180*) developed the method of the α-aciliminium–olefin cyclization. This method was successfully applied by several groups (*166, 178, 181–186*) leading to the shortest route to PHTX (Scheme 38). Thanks to continuous improvements the lactam **136** became easily accessible (*178, 179, 182, 183–185*).

One of the hurdles in this steeple chase was the reluctance of the ketoamide **203** to cyclize. It was therefore considered not advisable (*181–183*) to isolate the carbinolamide **202** formed from glutarimide **200** by reaction with a large excess of *E*-non-4-enylmagnesium bromide **201** (*187*). Instead, the reaction mixture, after removal of solvent, is treated with concentrated formic acid to yield via **202** *O*-formylketolactam **206** in addition to traces of the alternate spirocycle **207**. Tolerable yields (∼ 30%) are achieved only by higher temper-

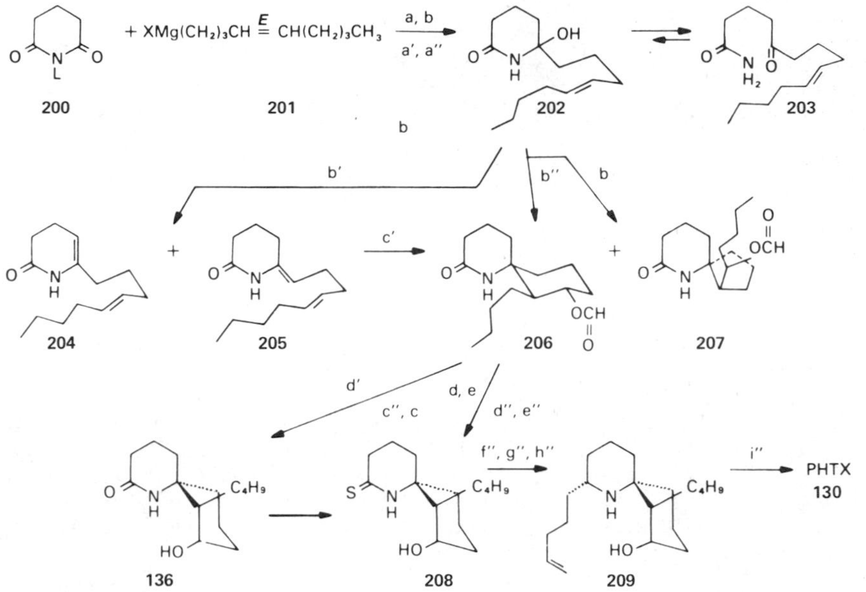

SCHEME 38.   *The method of α-acyliminium – olefin cyclization constitutes the shortest route to racemic perhydrohistrionicotoxin **130** (183 – 185). Reaction conditions: when L = H, X = Br: (a) THF, 18 hr; THF↑ ; (b) HCOOH 42°C, 14 days, 30% yield; (c) KOH, $C_2H_5OH:H_2O$ (1:1), 95 – 100% yield; (d) P_2S_5, ØH, ⇅, 1 hr; (e) NaOH, CH_3OH, 88% yield (183). When L = MgI, X = Cl: (a') $(C_2H_5)_2O$, ⇅, 1 hr; NH_4Cl, H_2O, 66% yield; (b') TosOH (cat.), $ØCH_3:DMF$ (50:1), ⇅, 48 hr, 75% yield; (c') HCOOH, 38 hr, 40% yield; (d') $NaOCH_3$, CH_3OH, 15 hr (184). When L = MgBr, X = Br; (a") CH_2Cl_2, ⇅, 18 hr; (b") HCOOH, 25°C, 48 hr, 33% to ~25% yield; (c") $NaOCH_3$ CH_3OH, 25°C, .5 hr, 95% yield; (d") P_2S_5, Ø H, ⇅, 1.5 hr, (e") 1 N NaOH, CH_3OH, 3 hr; (f") CH_3I, CH_2Cl_2, 18 hr; (g") $MgCl_2$ anh. CH_2Cl_2; ⟋⟍⟍⟍ $MgCl$, ⇅, 24 hr; (h") AlH_3, $ØCH_3$, – 72°C, 5 hr; 25°C, 14 hr (i") Pd/C, 1 atm H_2, THF, 4 hr (185).*

atures and protracted reaction times. Deformylation of **206** gives the keto-lactam **136.** On the other hand, **206** may be converted to the thiolactam and to **208,** an intermediate that helped Kishi (*154*) to arrive at PHTX in good yield (Scheme 24). This route to **209** (or **208**) constitutes, therefore, a new formal synthesis of PHTX **130** in an impressive overall yield of 19%.

The unfavorable equilibrium **202** ⇌ **203** poses a problem that was solved by Evans in two ways (*184, 185*). One method (Scheme 39) starts with the magnesium salt of glutarimide **200** that reacts with E-non-4-enylmagnesium chloride **201** to yield **202** and **203** as a 1:1 mixture which on dehydration becomes a mixture of the enamides **204** and **205** in a ratio of 9:1. The improvement is that this mixture cyclizes at room temperature with formic acid within 36 hr to yield the requisite O-formylketolactam **206** in 40% yield, in addition to 30% of the alternate undersirable spirocyclic **207.**

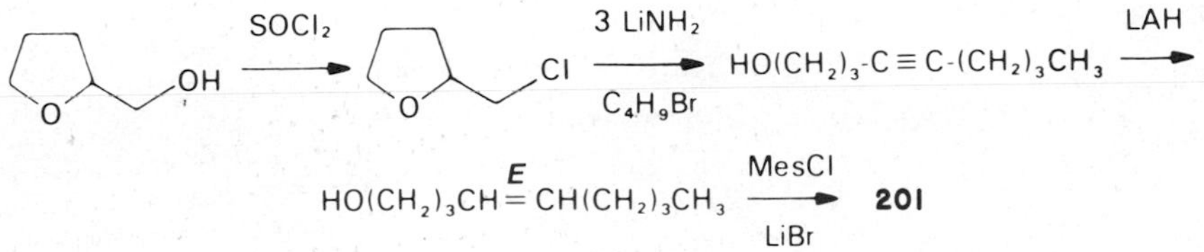

SCHEME 39. *Preparation of* E-*non-4-enylmagnesium chloride* **201** *starting with 2-hydroxy-methyltetrahydrofurane* (187).

The second improvement (*185*) involves preparation of the magnesium salt of glutarimide anion **200** to react with the Grignard complex of **201** prepared according to Scheme 40 in dichloromethane to form exclusively the carbinol **202** which directly is cyclized to **206** with formic acid within 48 hr in 33% yield. The *O*-formylketolactam **206** is converted to 7-butyl-2-pent-4′-enyl-1-azaspiro[5.5]undecan-8-ol **209** in a variation of Kishi's method (*154*): The intermediate thiolactam **208** was prepared with phosphorus pentasulfide, methylated to the thiolactim ether, which reacted with magnesium chloride in dichloromethane to a magnesium complex that with pent-4-enylmagnesium chloride yielded the imine reducible stereoselectively with aluminum hydride to the decahydrohistrionicotoxin **209** (93 parts and 7 parts of the C-2 epimer). The terminal olefinic bond of **209** lends itself to hydroboration and oxidation for the introduction of a terminal primary alcohol function, convertible to the 2-nitro-4-azidophenyl ether, a photoaffinity marker for binding studies with electroplax membrane fractions of *Torpedo californica* (*185*).

If in the sequence pictured in Scheme 38 the *E*-non-4-enylmagnesium chloride is replaced by *Z*-non-4-enylmagnesium bromide, the lactam so obtained is epimeric at C-7 in comparison with the *O*-formylketolactam **206** (*183*).

If morpholine-3,5-dione **210** (Scheme 41) instead of glutarimide **200** is employed, the analogous oxaspirolactam **211** comparable to **207** is obtained (*183*), without a trace of the oxa analog of **206**. This is indeed an unexpected result that illustrates the precarious sensitivity of the principles of regioselec-

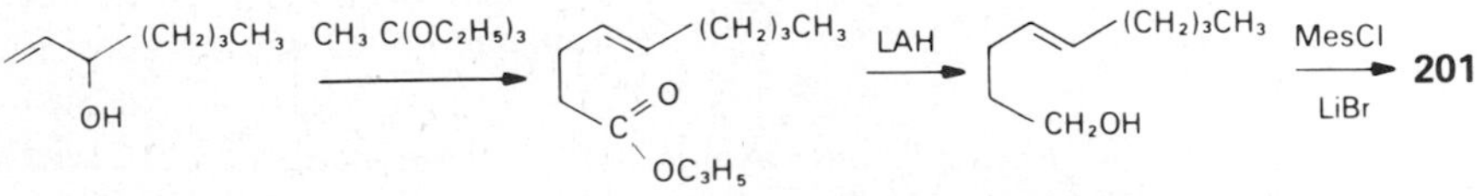

SCHEME 40. *Preparation of* E-*non-4-enylmagnesium chloride* **201** *from* n-*hept-1-en-3-ol* (184).

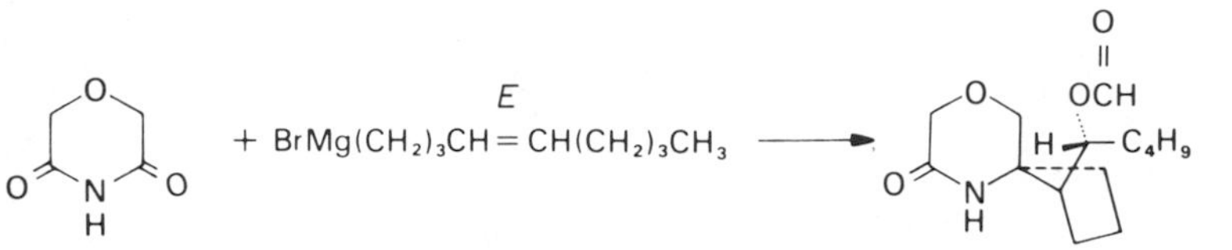

SCHEME 41. *Use of morpholine-3,5-dione 210 instead of glutarimide 200 leads exclusively to the oxaspirolactam 211, comparable to 207 without a trace of the oxa analog of 206 (183).*

tivity of the acyliminium–olefin cyclization to reaction conditions and minor structural alterations.

The advances described above prompted a group at NIH to improve on the preparation of the lactam for the preparation of 2-deamyl-PHTX **137** on a large scale (*186*) in order to demonstrate the usefulness of this simplified HTX for neurophysiological research (*166*).

The same workers for the first time synthesized both optical antipodes of PHTX (*166*). For this synthesis, the lactam **136** (Scheme 22) is treated with 3-(+)-α-methylbenzyl isocyanate and the resulting diastereoisomeric carbamates are separated by preparative HPLC. After hydrolysis the two optical antipodes of the lactam are obtained. Optically active PHTX is prepared by the two known methods (*154, 185*) with the C-2 epimer being the minor product (ratio: 7:3). The levoratory antipode (−)-(2R,6R,7S,8S)-7-butyl-2-pentyl-1-azaspiro[5.5]undecan-8-ol* is identical with the product of perhydrogenation of natural histrionicotoxin and its congeners.

Another method (Scheme 42) utilizing an intramolecular olefin cycliza-

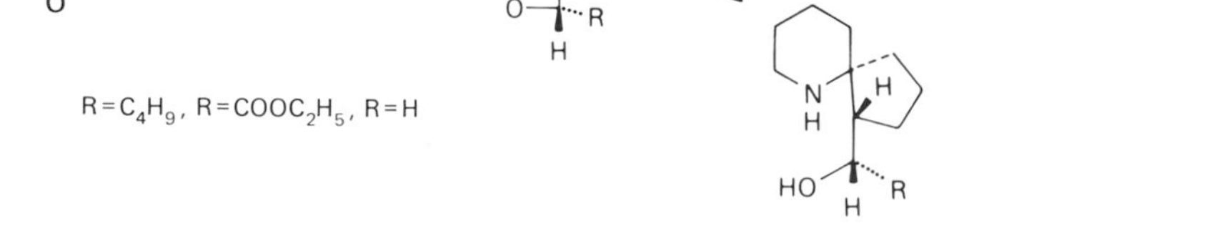

SCHEME 42. *Intramolecular 1,3-dipolar addition, only in the simplest case (R = H), leads to the basic skeleton of histrionicotoxin (188, 189).*

* The nomenclature (*145*) used for isodihydrohistrionicotoxin would designate a diastereomer. In an analogous fashion, the nomenclature for PHTX required correction as indicated.

tion, now based on 1,3-dipolar addition, was attempted (*188, 189*) in order to construct three of the chiral centers of PHTX simultaneously. However, intramolecular cyclization of a suitably substituted 3,4,5,6-tetrahydropyridine 1-oxide resulted in the unwelcome 1-azaspiro[5.4]decane derivative. Only if R equaled H, was thermal equilibration to the 1-azaspiro[5.5]undecane derivative possible and then subsequent reductive cleavage to the basic skeleton of the histrionicotoxins (*189*).

4. Models for the Synthesis of the Z-Enyne Side Chains

Although there exists so far no synthesis for histrionicotoxin itself, a number of model studies explore the construction of the side chains (*190–195*). Methylcyclohexane carboxylate **212** serves as model (Scheme 43) and by reaction with the anion of dimethylacetamide is converted to the β-ketoamide **213**. With hydrazine this forms a pyrazoline which thallic nitrate converts to the acetylenic ester **214**. Reduction with diisobutylaluminum hydride to the alcohol and partial hydrogenation furnishes the Z-allyl alcohol **215**. Oxidation with manganese dioxide yields the aldehyde, which is transformed to a mixture of Z- and E-chloroalkadiene **216** by Wittig olefination with chloromethylenetriphenylphosphoniumylide. Elimination of hydrogen chloride with excess methyllithium leads to the lithium acetylide that on treatment with trimethylsilyl chloride is isolated as trimethylsilyl-protected Z-enyne **217**. Treatment with fluoride anion furnishes the free Z-enyne **218** (*190*).

Wittig olefination with trimethylsilylpropargylphosphoniumylide **220** itself can be utilized with good stereoselectivity, provided the temperature is kept low (– 40°C) as shown with the acetonide of glyceraldehyde **219** (*191*) (**Scheme 44,** upper). As the temperature is raised (– 40°C) the original ratio of Z-enyne **221** to E-enyne **222** decreases from 5 : 1 to 1.25 : 1.

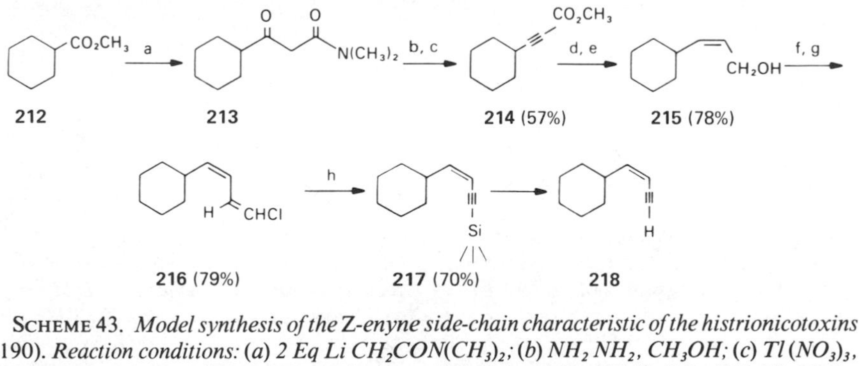

SCHEME 43. *Model synthesis of the Z-enyne side-chain characteristic of the histrionicotoxins* (190). *Reaction conditions: (a) 2 Eq Li CH₂CON(CH₃)₂; (b) NH₂ NH₂, CH₃OH; (c) Tl (NO₃)₃, CH₃OH,* 1ℓ, *0.25 hr; (d) 3 Eq DBAH, hexane, 0°C, .5 hr; (e) H₂, Pd/CaCO₃; ØH; (f) MnO₂, CH₂Cl₂; (g) ClCHPØ₃, THF, 3 hr; (h) CH₃Li, THF, 12 hr; (CH₃)₃SiCl.*

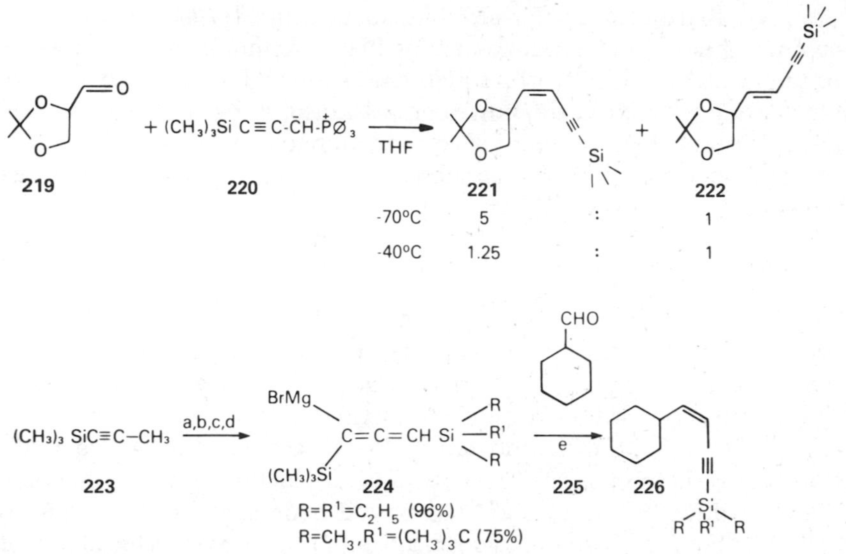

219 + **220** (CH₃)₃Si C≡C-ĊH-ṖØ₃ → **221** + **222**

	221	:	222
-70°C	5	:	1
-40°C	1.25	:	1

(CH₃)₃ SiC≡C−CH₃ →(a,b,c,d) **224** →(e) **226**

223

R=R¹=C₂H₅ (96%)
R=CH₃,R¹=(CH₃)₃C (75%)

225

SCHEME 44. *Z- and E-enynes* **221** *and* **222** *are obtained by Wittig olefination. The ratio is temperature dependant* (191). *The trialkylsilylated Z-enyne* **226** *(lower right) is obtained by Peterson olefination of cyclohexane carbaldehyde* **225** (195). *(a) (CH₃)₃CLi, (C₂H₅)₂O, −5°C, 1 hr; (b) (C₂H₅)₃SiCl or (CH₃)₃CSi(CH₃)₂Cl, 12 hr; (c) (CH₃)₃cLi, THF, −78°C, 15′; (e) −78°C, 5′; 50°C, 3 hr.*

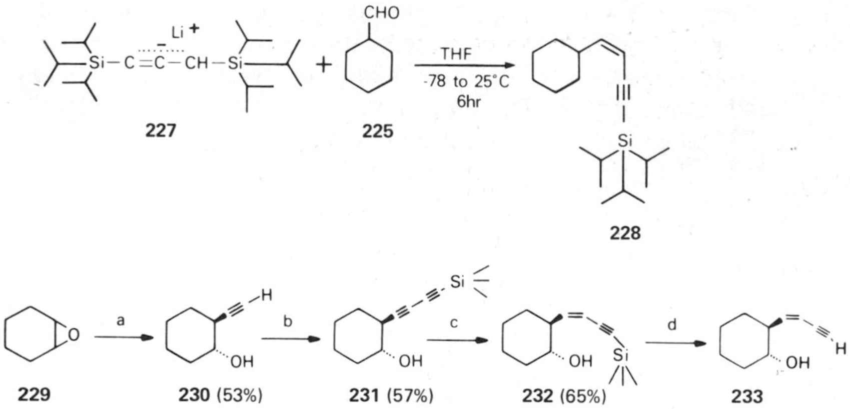

227 + **225** →(THF, -78 to 25°C, 6hr) **228**

229 →(a) **230** (53%) →(b) **231** (57%) →(c) **232** (65%) →(d) **233**

SCHEME 45. *A voluminous trialkylsilylated propynyllithium complex* **227** *stereoselectively forms the Z-enyne* **228** *with cyclohexane carbaldehyde* **225** (194). *Another stereoselective approach to the Z-enyne* **233** *utilized partial reduction of an intermediate protected diyne* **231** (192). *Reaction conditions: (a) LiC̄≡CH·(NH₂CH₂)₂, HMPTA, 80°C; (b) CuCl·[(CH₃)₂NCH₂]₂, (CH₂)₃SiC≡CH, acetone, air; (c) H₂, Pd/BaSO₄, quinoline, hexane:methanol (20:1); (d) N⁺(C₄H₉)₄F⁻.*

Recently two groups (*194, 195*) succeeded in preparing the conjugated terminal trialkylsilylated Z-enyne **228** in high stereoselectivity by Peterson olefination (Schemes 44, lower and 45, upper). 1-Trimethylsilylpropyne **223** (Scheme 44, upper) (*196*) reacts after lithiation with triethylsilyl chloride (or *tert*-butyldimethylsilyl chloride) furnishing the 1,3-bistrialkylsilylpropyne (*195*). Lithiation with *tert*-butyllithium and exchange of lithium for magnesium yields the allenylmagnesium bromide **224,** which by reaction with cyclohexanecarboxaldehyde **225,** leads to 95% (or 75%) yield with high stereoselectivity to the protected Z-enyne **226** ($Z:E = 23:1$ or $Z:E = 30:1$). The stereoselectivity decreases if lithium is used instead of magnesium and the less voluminous trimethylsilyl group (*195*).

Likewise, in the second approach (*194*) a voluminous trialkylsilyl group is utilized to attain high stereoselectivity (Scheme 45, upper). With 1,3-bistri-isopropylsilylpropynyllithium **227,** cyclohexanecarbaldehyde **225** is converted to the protected Z-enyne **228** with high stereoselectively ($Z:E = 20:1$). These authors also demonstrated the influence of the counter ion: If the same Peterson olefination is conducted in the presence of HMPT, the good solvating properties for lithium ions change the stereoselectivity to furnish the conjugated E-enyne.

Another route (*192*) starts with cyclohexene oxide **229** (Scheme 45, lower), which opens with the acetylide to give *trans*-1-hydroxy-2-ethynylcyclohexane **230.** The key step in this sequence is the modified Glaser reaction, namely, oxidative coupling with trimethylsilylethyne with the help of the Hay catalyst to give the 2-trimethylsilylbutadiynylcyclohexanol **208.** Partial reduction exclusively yields the Z-enyne **232** and desilylation the desired 1,2-trans-disubstituted cyclohexane derivative **233.**

There is another method by which a protective group for the terminal alkyne becomes expendable (*193*). Z-1-Iodohex-1-ene **234** (Scheme 46) reacts with the more stable ethynylzinc chloride **235,** easily accessible by reaction of ethynyllithium with anhydrous zinc chloride, in the presence of tetrakis(triphenylphosphine)palladium as catalyst to yield stereospecifically the Z-enyne **236.**

After completion of this section (see also addendum p. 248) the stereoselective synthesis of perhydro- and octahydrohistrionicotoxins were summarized in a somewhat different manner in the "Festband" of Heterocycles

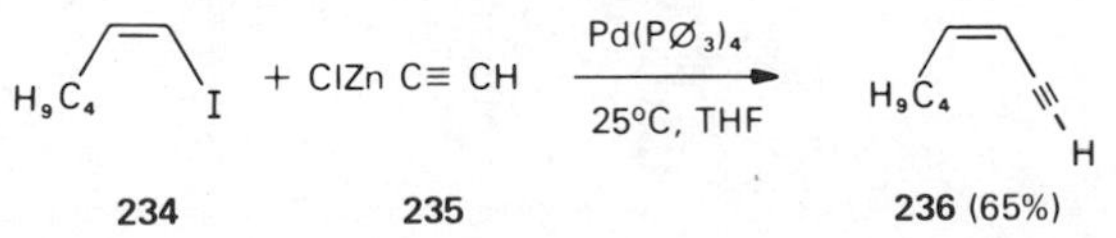

SCHEME 46. *Stereospecific construction of the Z-enyne **236**, without the use of a protective group, starts with Z-1-iodo-1-hexene **234** and reaction with ethynyl zinc chloride **235** (193).*

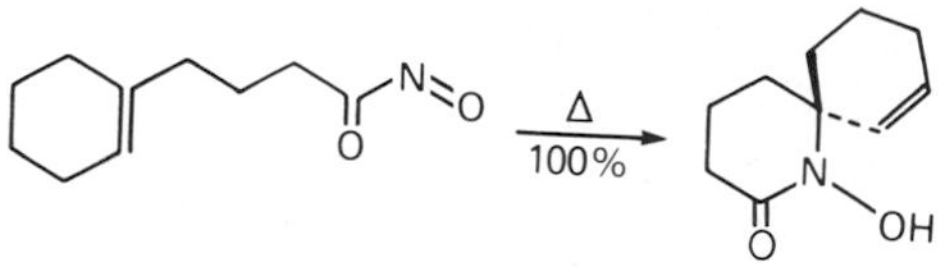

SCHEME 47. *A kinetically controlled, thermal regioselective, intramolecular ene-reaction of an acylnitrosocyclohexene produces an olefinic hydroxamic acid with the skeleton of histrionicotoxin* (198).

which appeared in honor of Prof. Kyosuke Tsuda's 75th birthday (*197*). There Pearson's series of spirocyclizations of tricarbonyldienyliumiron hexafluorophosphate complexes are reviewed. Recently suitable acylnitroso derivatives have been found to undergo kinetically controlled, thermal, regioselective, intramolecular ene reactions (Scheme 47) which approach a hydroxamic acid precursor of histrionicotoxin in quantitative yield (*198*).

B. PUMILIOTOXIN C AND CONGENERS

1. Occurrence, Isolation, and Structures

Among the *Dendrobates* alkaloids the pumiliotoxins with 66 congeners form the largest group (*151*). On the basis of structure they are divided into two distinct subgroups.

Pumiliotoxins-C. This is a group of 42 alkaloids containing a decahydroquinoline ring system, a side chain in the 2 position and, in many cases, another side chain in the 5 position. This group of 42 alkaloids is subdivided into the pumiliotoxins C, a group of 26 alkaloids and into congeners containing an additional hydroxyl group either in one of the side chains or in the ring system, the hydroxypumiliotoxins C.

Pumiliotoxin C (PTX-C) **237** one of the three major alkaloids from *Dendrobates pumilio,* an extremely diversified and variable species from Panama (*199*), was the first representative whose structure was elucidated by Roentgen-ray crystallographic analysis (*200*) as 2-*S*-propyl-4a-*S*-4a-*H*-5*R*-methyl-9a-*H*-9a-*R*-decahydroquinoline **237** (Scheme 48) and confirmed by ,

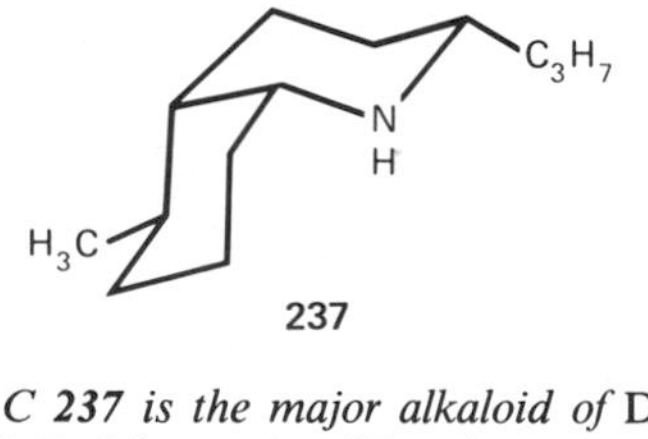

237

SCHEME 48. *Pumiliotoxin C 237 is the major alkaloid of* Dendrobates auratus *and* D. pumilio, *and a trace alkaloid in* D. lehmanni *and* D. minutus.

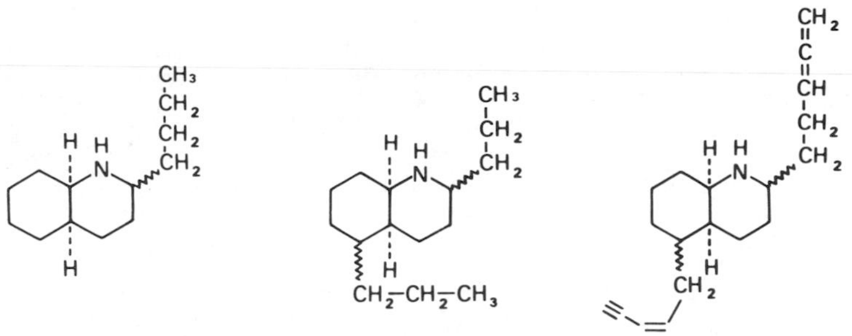

238: Alkaloid 195B   239: Alkaloid 223AB   240: Alkaloid 269A

SCHEME 49. *Three cognate alkaloids from* Dendrobates histrionicus *designated as Alkaloid 195B **238**, Alkaloid 223AB **239**, and Alkaloid 269A **240**.

asymmetric synthesis (*216*). The isolation procedure (*200*) follows the one described for histrionicotoxins (*150*). The same alkaloid was also isolated from *Dendrobates auratus* (*200, 201, 151*).

Three cognate alkaloids **238, 239, 240** (Scheme 49) were then isolated from *Dendrobates histrionicus* (*150*). Their structural assignments rest on NMR and mass spectroscopic data and their comparison with those of the known PTX-C and histrionicotoxins. In this way an alkaloid originally considered to be a deoxyhistrionicotoxin turned out to be related to PTX-C. This was confirmed by the preparation of and comparison with deoxyperhydrohistrionicotoxin (*150*) from PHTX through removal of the secondary hydroxyl group and catalytic hydrogenation of the intermediate olefin. The enyne and allene side chains, so characteristic of the histrionicotoxins, are therefore also peculiar to pumiliotoxin C congeners (Scheme 49).

The majority of pumiliotoxin C alkaloids was discovered within the framework of the comprehensive investigation of 18 different species of *Dendrobates* which also led to new histrionicotoxins (*151*). While the structural assignment rest more or less securely on mass spectral fragmentation data and some chemical transformations, no binding stereochemical conclusions are possible (Table II).

Five independent groups accepted the challenge of synthesizing the relatively simple pumiliotoxin C by widely divergent schemes. A synopsis of these activities up to 1977 was presented in a review (*202*).

2. Syntheses via Beckmann Rearrangement

A plausible starting material is the *cis*-tetrahydroindan-1-one **241** (*203, 204*) (Scheme 50). The Beckmann rearrangement of its oxime with retention of configuration leads to *cis*-octahydro-2-quinolone, convertible to the

TABLE II
CONGENERS OF PUMILIOTOXIN C^a

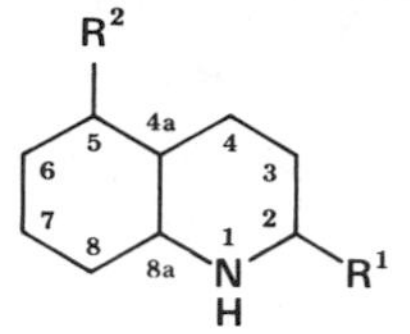

R^2	R^1
H	C_2H_5, C_3H_7, C_4H_9, $CH_2CH=CH-C\equiv CH$
	$CH_2CH=CH-CH=CH_2$, $(CH_2)_2CH=CH_2$, C_6H_{13} (b),
	C_7H_{13} (a) (b), C_7H_{15}
CH_3	C_2H_5, C_3H_7, $CH_2CH=CH-CH=CH_2$
	C_5H_{11} (a) (b), C_6H_7, C_7H_{15} (a) (b)
C_2H_5	C_4H_9, $CH_2CH=CH-C\equiv CH$
$CH_2-CH=CH_2$	$CH_2-CH=CH_2$
C_3H_7	C_3H_7
$CH_2-CH=CH-C\equiv CH$	C_3H_7, $CH_2CH_2CH=C=CH_2$
$CH_2CH=CH-CH=CH_2$	$CH_2-CH=CH_2$
C_9H_{15}	$CH_2-CH=CH_2$

^{a} The small letters in parentheses signify ring hydroxylated (a) or side chain-hydroxylated (b) alkaloids.

N-benzyl derivative **242**. Attack of the 6,7-double bond of the octahydroquinolone **242** by m-chloroperbenzoic acid forms the oxiran from the less hindered side, and its opening by bromide ion proceeds in an axial approach from the less hindered side to provide the desired bromohydrin. Subsequent

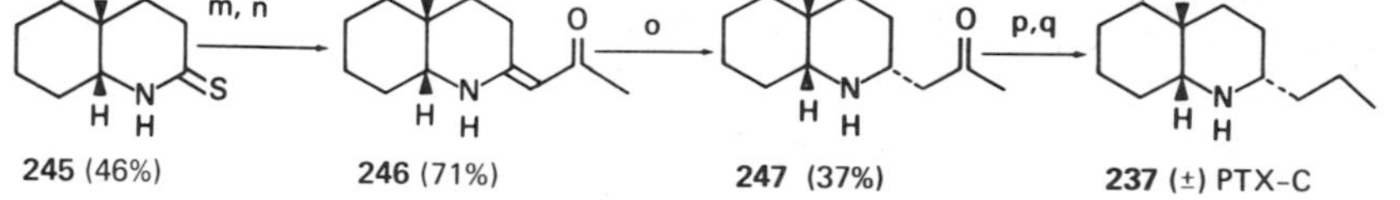

SCHEME 50. *Synthesis of racemic pumiliotoxin C* **237** *starting with cis-tetrahydroindanone* (203, 204). *Reaction conditions: (a)* $\overset{+}{N}H_3OH$ Cl^-, $NaOCOCH_3$, CH_3OH, 13 hr; *(b)* TosCl, *pyridine, 10 hr; (c)* NaH, ϕH, $ClCH_2\phi$; *(d)* m-ClϕCOOOH, CH_2Cl_2, 20 hr; *(e)* 48% HBr, $CHCl_3$; *(f)* Jones reagent; *(g)* LiBr, Li_2CO_3 DMF, 110° C, 3 hr; *(h)* $(CH_3)_2CuLi$, $(\bar{C}_2H_5)_2O:THF(1:1)$; *(i)* [$^{SH}_{SH}$, $BF_3 \cdot O(C_2H_5)_2$, $CHCl_3$, 5 hr; *(j)* Ra/Ni, C_2H_5OH, 1↓; *(k)* liq. NH_3, Na, $O(C_2H_5)_2$; *(l)* P_2S_5, ϕH, 1↓; *(m)* $BrCH_2COCH_3$, CH_2Cl_2; *(n)* ϕ_3P, $KOt-C_4H_9$, $HOt-C_4H_9$, ϕH; *(o)* PtO_2, 1 atmH$_2$; *(p)* [$^{SH}_{SH}$, $BF_3O(C_2H_5)_2$, $CHCl_3$, 40 hr; *(q)* Ra/Ni, CH_3OH, 1↓.

oxidation of the secondary hydroxyl group and dehydrobromination complete the sequence to the olefinic ketolactam **243**. This enone accepts the methyl group from the dimethylcopper-lithium complex through axial attack by 1,4-addition to the conformation with less steric interaction to yield the methylated ketolactam **244** with high stereoselectivity. The carbonyl function in position 7 has now become expendable and is removed by desulfurization of the thioketal with Raney nickel. The C-2 side chain is introduced into the debenzylated thiolactam **245** by the sulfide contraction method of Eschenmoser, providing the vinylogous amide **246**. Catalytic hydrogen adds from the less hindered side to furnish the amino ketone **247** in modest yield. Thioketalization and desulfurization lead to racemic pumiliotoxin C, **237**.

Two variations on the indanone theme start with 4-methyl-*cis*-hydrindan-1-one **9** (Schemes 51 and 52). In the first one (*205*) the desired relative configurations at C-4a and C-5 are provided by the stereospecificity of an intramolecular Diels–Alder reaction (*205*). To achieve this goal, *E*-hex-4-enal **248** prepared by Claisen–Cope rearrangement from 3-but-1-enevinyl

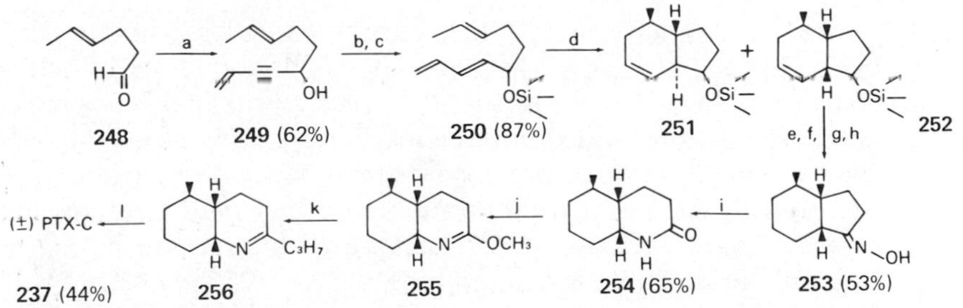

SCHEME 51. *Construction of four chiral centers by stereoselective intramolecular Diels–Alder reaction* (205). *Reaction conditions:* (a) $\diagup\!\!\!\!\equiv$ $- MgBr, (C_2H_5)_2O, 0°C, 2 hr;$ (b) 1.5 Eq LAH, 3 Eq NaOCH_3, THF, 2 hr; (c) (CH_3)_3SiHNCOCH_3, C_6H_{14}, $1\!\!\downarrow$, 1 hr; 5°C, 16 hr; (d) ϕCH_3, 245°C, 16 hr; (e) KF, CH_3OH, 4 hr; (f) PtO_2, H_2, 3 hr; (g) Jones reagent; (h) $\overset{+}{N}H_2OHCl^-$, NaOAc, CH_3OH, 45 min; (i) TosCl, NaOH, dioxane/H_2O (3:4), 15 hr; (j) (CH_3)_3$\overset{+}{O}$ BF_4^-, N(C_2H_5)- (iC_3H_7)_2(cat.), CH_2Cl_2, 1 hr; (k) C_3H_7MgBr, ϕH, $1\!\!\downarrow$, 3 hr; (l) 10% Pt/C, CH_3OH, 3 hr.

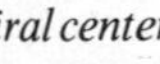

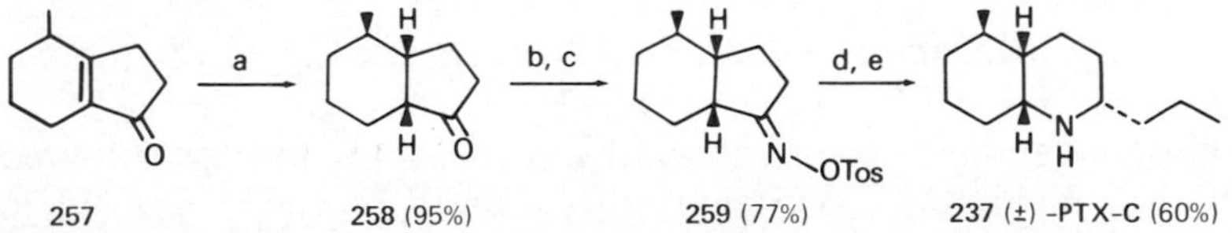

SCHEME 52. *Synthesis of racemic pumiliotoxin C,* **237,** *starting with 4-methyl-4,5,6,7-tetrahydroindanone* **257** (206). *Reaction conditions:* (a) Pd (black), 1 atm H_2, dioxane (12% CH_3CH_2COOH), 12 hr; (b) NH_2OH · HCl, NaOCOCH_3, CH_3OH, 5 hr; (c) TosCl, pyridine, −20°C, 1 hr; 0°C, 5 hr; (d) 3 Eq (n-C_3H_7)_3Al, CH_2Cl_2, 30 min; (e) 4 Eq DBAH, 1 hr.

ether yields the decadienynol **249** by reaction with butenynemagnesium bromide. Model experiments showed that Z,E-5-trimethylsilyloxydeca-1,3,8-triene, obtainable from **249** by reduction with zinc and KCN and formation of the silyl ether, cyclizes exclusively to the hydrindenol **252**. However, the yield is depressed because of a competing sigmatropic 1,5H-shift. This experience prompted a switch to the isomeric triene **250**, available from the acetylenic precursor **249** by reduction with lithium aluminum hydride in the presence of sodium methoxide and formation of the trimethylsilyl ether. The thermal intramolecular cyclization of **250** does not proceed until the temperature reaches 245 °C and then produces the two hydrindene epimers **251** and **252** in a ratio of 1 : 2 in 51% yield. Desilylation, catalytic reduction of the unsaturation, and oxidation leads to the *trans*- and *cis*-hydrindanones in a ratio of 1 : 2. Conversion to the oxime is faster with the *cis*-hydrindanone thus permitting an easy separation from the undesired trans isomer by crystallization. The oxime **253** is rearranged to the lactam **254** which, by Meerwein's method, forms the lactim ether **255**. Propylmagnesium bromide introduces the side chain in the imine **256** which accepts catalytic hydrogen from the less hindered side to yield racemic pumiliotoxin C, **237**.

In the most recent approach (*206*, Scheme 52), 4-methyl-*cis*-hydrindan-1-one **258** is prepared by catalytic hydrogenation of 4-methyl-4,5,6,7-tetrahydroindan-1-one **257** (*207*) in a highly stereoselective manner. The perhydroindanone **258** is transformed into the tosyl oxime **259** and then to the imine **256** (Scheme 51) in one step by a method newly developed by the authors (*206*) with the aid of tripropylaluminum that both catalyzes the Beckmann rearrangement as well as alkylates the reactive intermediate. The imine **256,** without isolation, is reduced to (±)-PTX-C, **237**, with DBAH.

3. Synthesis via $\Delta^{1,8a}$-Octahydroquinolines

Another short synthesis (*208, 209*) starts with the mixture of diastereo-isomers of $\Delta^{1,8a}$-octahydroquinoline **262**, accessible through a modification of an earlier procedure (*210*) by reaction of the enamine **260** with 1-bromo-3-aminohexane **261** (Scheme 53). The diastereomeric imines **263** are then obtained by acidic saponification of the ester **262** and by decarboxylation. Catalytic hydrogenation provides mainly the *trans*-decahydroquinoline **264** in addition to (±)-PTX-C (25%), (±)-**237**.

The same authors report the synthesis of levorotatory pumiliotoxin C (*209; 211*, footnote 2, correction): *S*-3-amino-1-bromohexane (+)-**261** is prepared (*211*) by starting with ethyl 3-oxohexanoate **265** which with benzylamine is transformed into the enamine **267** (Scheme 54). By reduction with sodium in isopropanol, both the unsaturation as well as the ester group

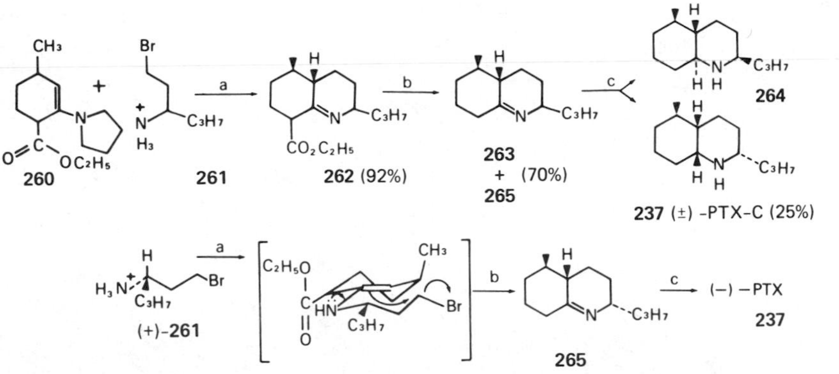

SCHEME 53. *Synthesis of racemic (±)-237 and levorotatory natural pumiliotoxin C 237 by the use of racemic or S-3-amino-1-bromohexane (±)-261 (208, 209). Reaction conditions: (a) DMF, 100°C, 6 hr; (b) 20% HCl ↑↓; (c) Pd/C, C₂H₅OH, H₂.*

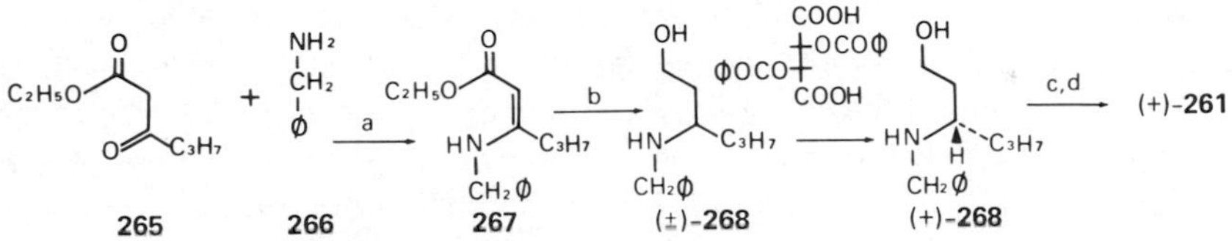

SCHEME 54. *Synthesis of S-3-amino-1-bromohexane, (+)-261 (209, 211). Reaction conditions: (a) ØCH₃, HCOOH (cat.), H₂O↓; (b) Na, (CH₃)₂CHOH, xylene; (c) (COOH)₂, (CH₃)₂CO, Pd/C, 20 atm H₂, H₂O:C₂H₅OH (1:3); (d) 48% HBr.*

are reduced; the amino alcohol **268** is resolved into optical antipodes with the help of dibenzoyl-(+)-tartaric acid. The dextrorotatory alcohol is converted to the S-3-amino-1-bromohexane (+)-**261** by hydrogen bromide. At this point the previous condensation (Scheme 53) with the enamine **260** is repeated to provide via **265** levorotatory pumiliotoxin C, (−)-**238,** identical with the natural toxin.

7,8-Dihydro-5(6H)quinolones have been explored as potential intermediates for the synthesis of decahydroquinolines, but only with limited success and difficulties as soon as 2-substituents were introduced (212).

4. Syntheses via Intramolecular Michael Addition

The key step in another synthetic approach (Scheme 55) is an intramolecular Michael addition of an amide to a suitably substituted cyclohexenone derivative (213, 214), accessible through Diels–Alder reactions. 1,3-Bistrimethylsilyloxydienes are introduced as new synthons. The diene **269** avail-

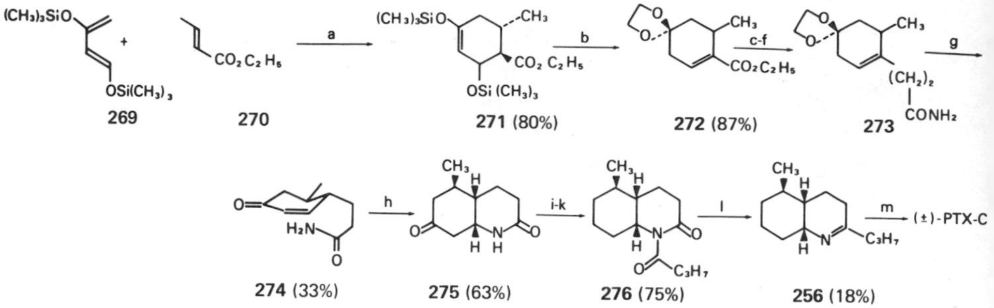

SCHEME 55. *Intramolecular Michael addition as a key step in the synthesis of racemic pumiliotoxin C (213, 214). Reaction conditions: (a) xylene, 170°C, 48 hr; (b) ϕH, [$^{OH}_{OH}$, TosOH ⇂; (c) LAH, $(C_2H_5)_2O$, 0°C, 4 hr; (d) LiC_4H_9, HMPT, TosCl, THF, $-78°C$; 0°C, 12 hr; (e) C_4H_9Li, CH_3CN, THF, $-78°C$; CuI; THF$-30°C$; (f) 5 M NaOH, H_2O_2, CH_3OH, 0°C; (g) 1% HCl, CH_3COCH_3 12 hr; (h) $NaOCH_3$, $HOCH_3$ ⇂, 40 min; (i) $BF_3 \cdot O(C_2H_5)_2$, [$^{SH}_{SH}$, CH_2Cl_2, 42 hr; (j) Ra/Ni, C_2H_5OH ⇂, 24 hr; (k) NaH, THF; C_3H_7COCl, HMPT, 3 hr; (l) CaO, 250°–300°C, 2 hr; (m) 2 M HCl, PtO_2, H_2, 7 hr.*

able from the sodium salt of acetoacetaldehyde with zinc chloride–triethylamine and trimethylchlorosilane, with ethyl crotonate **270** undergoes thermal regioselective cyclization to the cyclohexene **271** which is converted to the ketal **272**. In order to lengthen the side chain, the ester group is first reduced to the alcohol, converted to the unstable chloride by the method of Stork, and transformed to the nitrile with the help of cyanomethylcopper. The nitrile with hydrogen peroxide in alkaline medium furnishes the amide **273**. Deketalization to **274** and intramolecular Michael addition in basic medium yield the ketolactam **275**. The carbonyl group in position 7 is removed via the thioketal and desulfurization. The side chain in position 2 is introduced first as the butyryllactam **276** which on pyrolysis in the presence of calcium oxide, rearranges to imine **256** reducible to racemic pumiliotoxin C (±)-**237**.

A second synthesis (*214*) with the aim of reaching the ketolactam **275** by intramolecular Michael addition (Scheme 56) starts with 1,3-bistrimethylsilyloxy-5-methylcyclohexa-1,3-diene **277** which with acrylonitrile cyclizes to give preponderantly the exo adduct [ratio of exo and endo = 3:1 (*215*)]. This exo adduct by treatment with acid yields the methylbornanone **279** after separation from the undesired diastereoisomers. α-Bromination, reduction of the ketone, and treatment of the bromohydrin with zinc leads to the bornene derivative **280** that in strong acid undergoes a type of retroaldol reaction to form the unstable cyclohexenone **281** easily saponifiable to the amide **274**. As mentioned before, intramolecular Michael addition provides the ketolactam **275**. Its butyryl derivative **276** is thermally rearranged to **256** and then reduced to racemic pumiliotoxin C (±)-**237**.

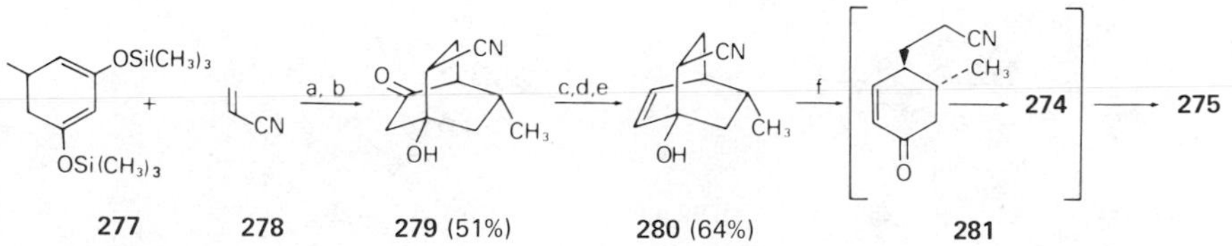

277   278   279 (51%)   280 (64%)   281

SCHEME 56. *A modified procedure leads,* via *the methylbornanone* **279** *and intramolecular Michael addition to the ketolactam* **275** *(213–215). Reaction conditions: (a) xylene, 170°C, 96 hr; (b) 10% HCl, 0°C, 12 hr; chromatography; (c) pyridine, Hbr·Br$_2$, HOCOCH$_3$, 2 hr; (d) NaBH$_4$, C$_2$H$_5$OH, 0°C, 2 hr; (e) Zn, HOCOCH$_3$, 30 min; (f) 70% HClO$_4$, HOCOCH$_3$, 100°C, 5 hr.*

5. Synthesis via *Intra*molecular Cyclization of a Dieneamide to an Olefin

Racemic, dextrorotatory (unnatural), and levorotatory (natural) pumilio-toxin C have all become available by a synthesis in which the key step is an intramolecular Diels–Alder reaction (*216, 217*). The advantage of this method is based on the directing influence of the initial chiral center in α position to the nitrogen that controls the construction of the remaining three chiral centers with the correct relative steric arrangement in a single step. The potential of this approach was first probed in model reactions (*218, 219*). It soon became obvious that the selective formation of the *cis*-decahydroquin-oline occurs with an open chain diene amide with an additional isolated double bond, provided that the *N*-acyl residue does not become part of the ring to be formed, or in other words, when one of the carbon chains involved in the cyclization possesses a tetragonal (*sp³*-hybridized) carbon atom in α-position to the nitrogen (*218*). This required the elaboration of a new approach to *N*-acyl-*N*-alkyl-1-amino-1,3-butadienes (*219*) in the following manner (Scheme 57).

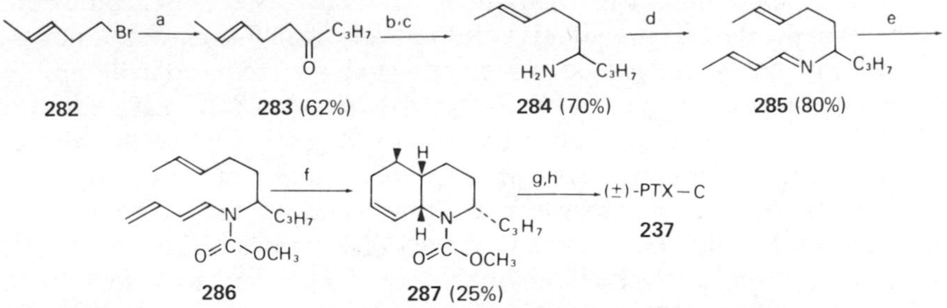

282   283 (62%)   284 (70%)   285 (80%)

286   287 (25%)

SCHEME 57. *Intramolecular Diels–Alder reaction as a key step to racemic pumiliotoxin C (217). Reaction conditions: (a) Mg, (C$_2$H$_5$)$_2$O; C$_3$H$_7$CN ⇅; (b) NH$_3$OHCl⁻, H$_2$O, Na$_2$CO$_3$; (c) LAH, (C$_2$H$_5$)$_2$O ⇅; (d) crotonaldehyde, molecular sieve, ϕH; (e) [(CH$_3$)$_3$Si]$_2$NNa, ClCOOCH$_3$, ØCH$_3$, −40°C; (f) ØCH$_3$, 215°C, 20 hr; (g) H$_2$Pd/C, CH$_3$OH, 5 hr; (h) conc HCl/HOCOCH$_3$/ H$_2$O (1:1:1), 30 hr.*

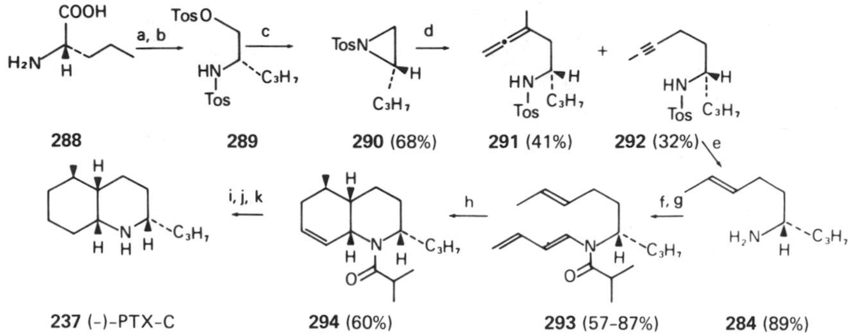

SCHEME 58. *Synthesis of levorotatory (natural) pumiliotoxin C starting with optically active norvaline (288)* (216). *Reaction conditions: (a) LAH; (b) TosCl, pyridine; (c) KOH, CH₃OH; (d) H₃C—C≡C—CH₂—MgBr; (e) Na, liq NH₃; (f) ⌇⌇⌇CHO , molecular sieve; (g) NaH, (CH₃OCH₂)₂, −30°C; ClCOCH(CH₃)₂; (h) ØCH₃ [(CH₃)₃Si]₂NCOCH₃ (cat.), 230°C, 16 hr; (i) H₂/Pd, CH₃OH; (j) DBAH; (k) HCl, CH₃OH.*

Pent-3-enylmagnesium bromide, prepared from **282,** reacts with butyronitrile to yield *E*-non-2-en-6-one **283** which, via the oxime, is converted to the amine **284** and with crotonaldehyde to the Schiff base **285.** With strong base this azomethine forms the delocalized anion and with methyl chlorocarbonate, the carbamate of the triene **286.*** Intramolecular Diels–Alder cyclization at 215°C provides the desired *cis*-octahydroquinoline **287** in addition to products resulting from elimination reactions.

The 7,8-olefinic bond is reduced catalytically and after acidic saponification of the carbamate provides racemic pumiliotoxin C (±)-**237.**

The same concept is utilized for the synthesis of the optically active antipodes of pumiliotoxin and for final clarification of the absolute configuration of the natural toxin (*216*) about which there existed some confusion. In this case the starting material is commercially available natural L-norvaline **288** with the *S*-configuration (Scheme 58) which is reduced to the amino alcohol and transformed to the *N,O*-ditosyl derivative **289.** Treatment with alkali effects intramolecular substitution to the *N*-tosylaziridine **290.** But-2-ynylmagnesium bromide reacts with its primary carbon atom with opening of **290** to the acetylenic *N*-tosylamine **292** whereas the major product, the allenic *N*-tosylamine **291,** is formed by attack of the vinylic C-3 of but-2-ynyl magnesium bromide. Reduction of the acetylenic bond with sodium in ammonia provides *E*-6-amino-non-2-ene **294** which is condensed to the Schiff base with crotonaldehyde and converted to the trieneamide **293** by reaction with base followed by isobutyryl chloride. Competing elimination

* A milder method of N-acylation of dieneamides is described in reference (*220*).

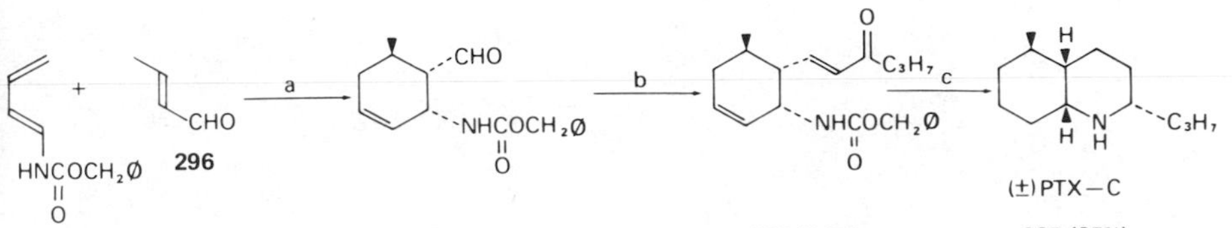

SCHEME 59. *Synthesis of racemic pumiliotoxin C starting with an* intermolecular *Diels-Alder reaction* (226). *Reaction conditions:* (a) *4-t-butyl-1,2-dihydroxybenzene (cat.), 110°C, 160 min;* (b) (CH₃O)₂P(O)CH₂COC₃H₇, *NaH, THF, – 10°C, 20 min; 1 hr, 1↓;* (c) *0.2 N HCl, C₂H₅OH, 1 atm H₂, Pd/C.*

reactions in the subsequent thermal intramolecular cyclization to **294** were suppressed to some extent by the addition of catalytic amounts of the silylating agent bis(trimethylsilyl)acetamide. Catalytic reduction of the 7,8-double bond, partial reduction of the amide to the iminium stage, followed by acid hydrolysis, provided levorotatory (−)-pumiliotoxin C.

6. Synthesis via *Inter*molecular Cyclization of a Dieneamide to an Olefin

The conjugated dieneamide, in this instance prepared from the dienoic acyl azides by Curtius rearrangement, was also used as synthon in an approach involving an intermolecular Diels–Alder reaction as the key step. This approach was tested in model reactions (*221–223*), as well as in its theoretical implications (*224*). When unsaturated carbonyl compounds were used as dienophiles, the reaction proceeded with high regioselectivity to the desired trisubstituted cyclohexenes. Model experiments (*224*) had shown that stereoselectivity, i.e., the *endo–exo* ratio as well as the reaction rate increase if the *N*-acyl group possesses an additional donor of electrons. This observation led to the use of the *O*-ethyl-(*225*) or better *O*-benzyl-carbamate (*226*).

Crotonaldehyde **296** reacts with diene amide **295** (*223, 227b*) at 110°C to yield the cyclohexene **297** with high regioselectivity and good stereoselectivity (*endo:exo* = 76:24) (*227a*)* (Scheme 59). The chiral center in α-position to the aldehyde group easily undergoes thermal epimerization, possibly aided by participation from the neighboring carbamate group. For this reason, the subsequent lengthening of the aldehyde side chain through the Horner–Emmons reaction to the enone **298** has to be carried out at low temperature (– 10°C). When the conjugated and unconjugated double

* Footnote 25 in reference (*227a*) corrects the ratio of *exo* to *endo* products from 1:9 (*225, 226*) to 24:76.

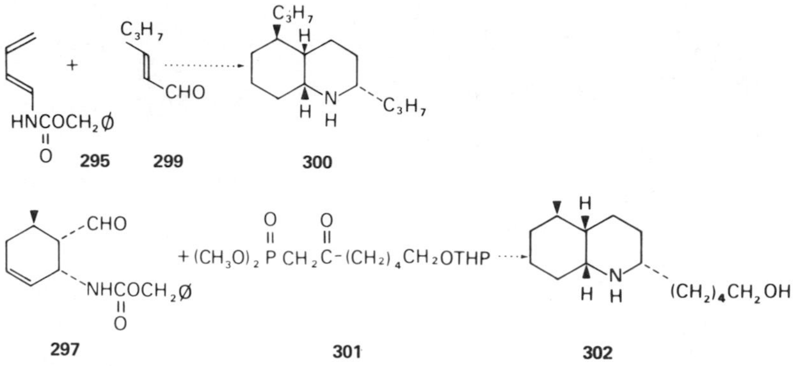

C_3H_7

+

C_3H_7

CHO

HNCOCH$_2\emptyset$

‖
O   295   299

C_3H_7
H

N
H H C_3H_7

300

CHO

NHCOCH$_2\emptyset$
‖
O

297

$$+ (CH_3O)_2 \, P \, CH_2 C-(CH_2)_4 CH_2 OTHP$$

301

H

N
H H $(CH_2)_4 CH_2 OH$

302

Scheme 60. *Variations of the preceding synthesis with regard to the aldehyde or the phosphonate components* (226).

bonds are reduced, debenzylation and decarboxylation occur in the same step with liberation of the amino group and formation of the cyclic Schiff base. The same step lets hydrogen approach from the less hindered side in a highly stereoselective manner to give racemic pumiliotoxin C **237** in good yield.

7. Synthesis of Homologs and Hydroxylated Analogs

The above scheme is capable of useful extensions and variations (*226*). For instance, the aldehyde component may be modified. If crotonaldehyde is substituted by *E*-hex-2-enal **299**, another congener of pumiliotoxin C **300** can be synthesized, namely, alkaloid 237A ($R_1 = C_3H_7$, Scheme 60) isolated from *Dendrobates auratus* as well *D. histrionicus* (*151*).

Variation of the phosphonate component, for instance the derivative **301** by reaction with the cyclohexene **297**, leads to the tetrahydropyranyl ether of **302** from which acid liberates the hydroxypumiliotoxin C **302**.

The methodology of the aforementioned syntheses (*217, 226*) was used to prepare the C-5-epimeric *cis*-decahydroquinoline-5-carboxylic acids in order to test their inhibitory action on GABA receptors (*228*).

C. PUMILIOTOXIN A, PUMILIOTOXIN B, AND THEIR ALLOPUMILIOTOXIN EQUIVALENTS

This group of 24 alkaloids differs from pumiliotoxin C by having a common indolizidine ring system that carries side chains at positions C-6 and C-8. The side chain at C-6 is connected to the ring by an exocyclic double bond (Schemes 61–63).

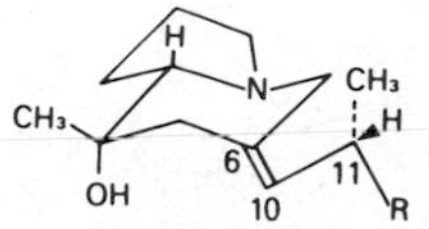

303, 304, 305

SCHEME 61. *Alkaloid 237A: 303, R = CH₂CH₂CH₃; major alkaloid of* D. abditus, *trace*

alkaloid of D. histrionicus; *Pumiliotoxin A: 304, R =* $-CH_2$ E CH_3 ... *; the con-*

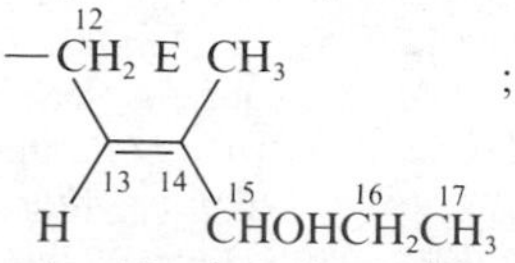

H $CHOHCH_2CH_3$

figuration of the C atom carrying the hydroxyl group in the side chain is not known; major alkaloid of D. pumilio; *minor or trace alkaloid of* D. auratus, D. granuliferus, D. lehmanni, D. minutus, D. occultator, D. viridis, *and* D. bombetes. (230). *Pumiliotoxin B: 305,*

$R =$ $-CH_2$ CH_3 ... H ... OH ... H HO H CH₃ ; *major alkaloid of* D. auratus, D. granuliferus, D. lehmanni,

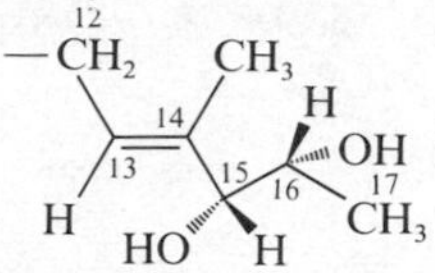

D. leucomelas, D. occultator, D. pumilio, D. viridis; *minor or trace alkaloid of* D. abditus, D. histrionicus, D. minutus, *and* D. bombetes.

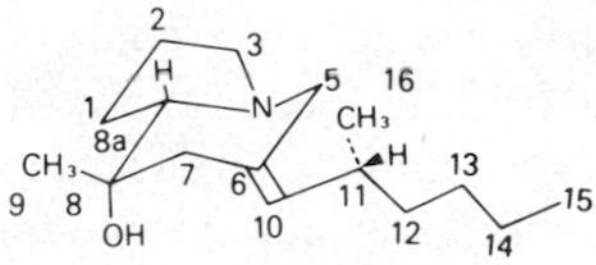

306

SCHEME 62. *Alkaloid 251 D 306 is the main alkaloid of* D. tricolor, D. silverstonei (229), *and* D. bombetes (230), *and a minor or trace alkaloid of* D. auratus, D. histrionicus, D. lehmanni, *and* D. minutus.

1. Occurrence, Isolation, and Structure

Two unstable representatives of this structural type, namely, pumiliotoxin A and pumiliotoxin B, have been mentioned as early as 1967 (*199*). Their structural elucidation became a recent achievement (*231, 232–234, 380*) after a more simple member of this group was isolated in sufficient quantity from *Dendrobates tricolor* as the major alkaloid. 750 specimens of this frog were collected in Ecuador and processed as described for the histrionicotoxins (*150*) to give 80 mg of alkaloid 251D (Scheme 62). Its crystalline hydrochloride by Roentgen-ray analysis yielded the structure and the absolute configuration. The fragmentation pattern in the mass spectrum shows that alkaloid 251D produces the pyrrolinium ion, one of the charac-

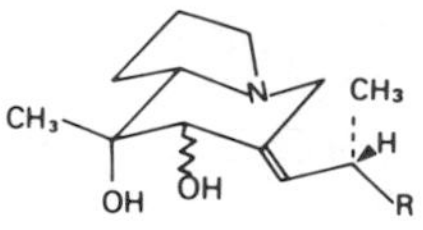

307, 308, 309

SCHEME 63. *Alkaloid 253 307*, $R = CH_2CH_2CH_3$; *major alkaloid of* D. abditus, *minor alkaloid of* D. lehmanni *and* D. auratus. *Alkaloid 267 A,* **308**, $R = CH_2CH_2CH_2CH_3$; *major alkaloid of* D. auratus, D. azureus, D. fulguratus, D. granuliferus, D. lehmanni, D. leucomelas, D. minutus, *and* D. tinctorius; *trace alkaloid of* D. auratus (Rio Campana), D. histrionicus, *and* D. pumilio. *Allopumiliotoxin B* **309**, *recently has been separated into two isomers differing only in the configuration at C-7: Allopumiliotoxin B', with the hydroxy group at C-7 in equatorial position and allopumiliotoxin B" with the hydroxy group in C-7 in axial position (371).* R =

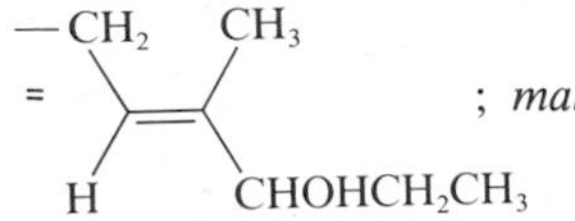

; *main alkaloid of* D. tricolor, D. silverstonei (229), *and minor alkaloid of* D. abditus, D. bombetes (230), D. leucomelas, *and* D. pumilio.

teristic peaks of pumiliotoxins A and B and also of the allopumiliotoxins. Additional information from ^{1}H-NMR and ^{13}C-NMR spectral data permitted structural assignments for several more congeners of this group: Alkaloid 237, pumiliotoxin A and B (*231*) (Scheme 61) that share the indolizidine system, the methyl substituent, and the tertiary alcohol both at C-8, as well as the side chain attached by an exocyclic double bond to C-6 and an allylic methyl substituent in position 11 of the side chain. The individual differences among the congeners rest in the length and substitution of the side chain. The structure and configuration of the side chain of pumiliotoxin B was further elucidated very recently (*232, 233, 373*). With the help of the nuclear Overhauser effect measured with pumiliotoxin B itself and with the ^{1}H-NMR data of the α,β-unsaturated aldehyde obtained by cleavage of the vicinal diol at positions 15 and 16 of the side chain with manganese dioxide, and the corresponding unsaturated acid obtained by oxidation of the aldehyde, it could be shown that the double bond at C-13 and C-14 posseses *E*-configuration (*232*), which is contrary to earlier tentative assignments (*231*). The threo configuration of the vicinal diol of pumiliotoxin B could be demonstrated by comparison of the nuclear Overhauser effect of the acetonide-d_6 (*233*) and of the phenylboronate (*232*), respectively, of natural pumiliotoxin B and model substances with the allylic vicinal diol with threo and with erythro configuration.

The absolute configuration of the chiral centers at C-15 and C-16 of the side chain was established by comparison of the ozonolysis product of pumiliotoxin-B-diacetate with the synthetic 3,4-diacetoxy-2-pentanone, derived from (+)-tartaric acid (*371*). Since the synthetic product proved to be the optical antipode of the ozonolysis product, the chiral centers at C-15 and C-16 of pumiliotoxin B possess *R* configuration. Pumiliotoxin A appears to possess also *E*-configuration of the 13,14 double bond (*234*). The absolute configuration at C-15 of pumiliotoxin A remains to be verified. The structure of three additional congeners with a hydroxyl group at C-7 of unknown configuration, belonging to the allopumiliotoxins, was also assigned: Alkaloid 253, alkaloid 267A, and allopumiliotoxins B' and B" (Scheme 63) (*231, 232, 382*) are now represented as **307, 308**, and **309**.

2. Syntheses of Congeners of Pumiliotoxin A

The aim of sophisticated synthesis is convergency rather than linearity (*235*). A convergent asymmetric synthesis in ten steps (*236*) utilizing as the key reaction the newly developed stereospecific intramoleclar cyclization of a vinylsilaneiminium precursor (Scheme 64) led to alkaloid 251D (**306**).

The scheme was tested first in the synthesis of the 11-nor analog of the simplest representative of the pumiliotoxin A group, noralkaloid 237A (**310**) that possessed no chiral center in the side chain.

L-Proline served as optically active starting material. In its protected form as methyl *N*-carbobenzoxyprolinate **311** reacts with Grignard reagent to form the tertiary alcohol, easily dehydrated to the olefin **312**. Its epoxidation yields the diastereoisomeric oxirans **313** and **314** in a ratio of 1:1.

Improvement of this ratio could be achieved by using hexane (**313**:**314** = 1:2) instead of CH_2Cl_2 as solvent (*381*). Very recently a highly stereoselective preparation of **314** in 60% overall yield was developed (*381*) by bromolactonization with participation of the carbamate unit of **312** and consecutive epoxide formation via basic saponification. Since their configuration could not be determined, the isomers are separated and reacted separately with lithium silylvinylalanate **315**, available from hex-1-yne by silylation. Reduction with DBAH to the silylvinylalane and reaction with methyllithium gives **315**. Attack of **315** on the isomeric oxirans opens the three-membered ring at the primary carbon atom through substitution by the vinyl residue. The tertiary alcohol so formed displaces the benzyl residue with formation of the cyclic carbamate **316**. At this stage, NMR data permitted steric assignments and identification of the two diasteroisomers. The final confirmation, however, was not in hand until the synthesis of alkaloid 251D was complete. In the carbamate **316** two chiral centers of alkaloid 251D are already present; saponification to the secondary amine **317**, reaction with paraformaldehyde and acid to the iminium intermediate **318**, and intramolecular electrophilic attack on the vinylsilane group with removal of the silyl group provide the 11-noralkaloid 237A **310**.

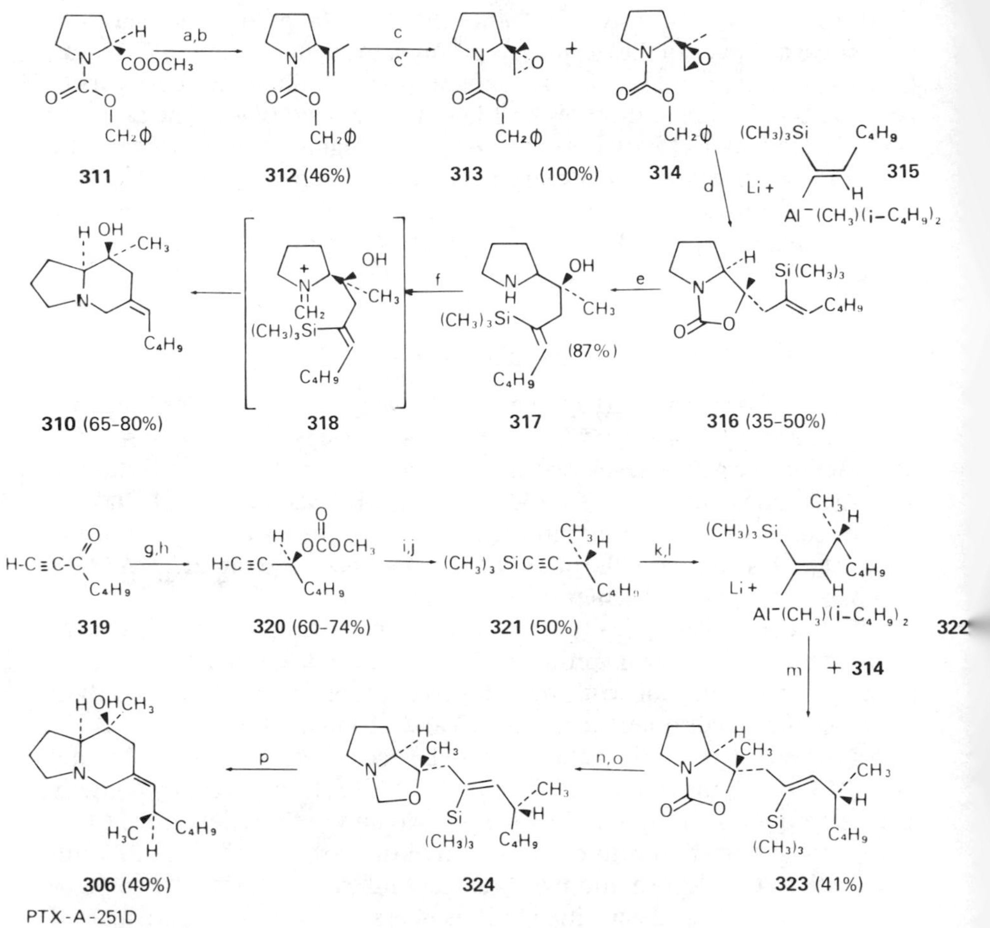

PTX-A-251D

SCHEME 64. *L-Proline as a synthon for the synthesis of 11-demethyl-Alkaloid 237A and Alkaloid 251D, 306, both congeners of pumiliotoxin A (236, 381). Reaction conditions: (a) 2.2 Eq CH₃MgI; (b) SOCl₂, pyridine THF, −45° C; (c) m-ClØCOOOH, CH₂Cl₂; chromatography; or (c′) NBS,DMSO:H₂O (50:1), rt; 0.3 M NaOH, THF:H₂O (5:2), rt, 2 hr; (d) (C₂H₅)₂O, ↿⇂; (e) 20% KOH, CH₃OH/H₂O, ↿⇂; (f) 2 Eq (CH₂O)ₓ, 1 Eq d-camphorsulfonic acid, C₂H₅OH, ↿⇂; (g) β-3-pinanyl-9-borabicyclo[3.3.1]nonane; 82 ± 5% ee; (h) ClCOOCH₃; (i) (CH₃)₃ SiCl; (j) 4 Eq CH₃MgBr, 2 Eq CuI, THF; (k) 1 Eq DIBAH; (l) CH₃Li; (m) (C₂H₅)₂O, ↿⇂; (n) 20% KOH, CH₃OH/H₂O ↿⇂; (o) 1 Eq(CH₂O)ₓ, C₂H₅OH, 80° C; (p) 1 Eq d-camphorsulfonic acid, C₂H₅OH, ↿⇂.*

The synthesis of alkaloid 251D **306** (*236*) differs from the above sequence only in the use of chiral silylvinylalanate **322** instead of the optically inactive lower homolog **315** (Scheme 64, lower). For its preparation, hept-1-yne-3-one **319** with the help of an optically active borane is reduced to *S*-heptynol in high optical yield and converted to its methyl carbonate **320**. The acetylene group is converted to the silylalkyne and the methyl carbonate residue exchanged with Walden inversion for a methyl group **321** with the help of a methylcopper complex. Treatment of **321** with first DBAH, then with methyllithium provides the alanate **322** which opens the epoxide **314** to the cyclic carbamate **323**. After saponification, reaction with paraformaldehyde yields the oxazolidine **324** undergoing acid-catalyzed intramolecular stereospecific cyclization to yield the optically active indolizidine alkaloid PTX-A-251D **306**.

D. The Gephyrotoxins

1. Occurrence, Isolation, and Structure

The gephyrotoxins form a group of at least six alkaloids isolated from the skin of *Dendrobates* species (*151*). Their common feature is the perhydropyrrolo[1,2-*a*]pyridine or indolizidine moiety. They are either perhydropyrrolo[1,2-*a*]quinolines with unsaturated side chains typical of histrionicotoxins or indolizidines resembling pumiliotoxins A and B (*231*) (Scheme 65). Since they *bridge* the gap between pumiliotoxins and histrionicotoxins, they appropriately are named bridge (gephyra) alkaloids or gephyrotoxins.

Gephyrotoxin (older code names: 287C = HTX-D) 325. This base has been discovered among the alkaloids from *Dendrobates histrionicus*. Its isolation was first described in 1974 under the code name of HTX-D (*237*). At that time, the methanolic extracts of the skins of 1100 frogs yielded 47 mg of the new base. Chemical and mass spectrometric data suggested the formula $C_{19}H_{29}NO$, the presence of a pent-2-en-4-inyl side chain, and the easy formation of an acetate. On catalytic hydrogenation three molar equivalents of hydrogen were consumed with the formation of a saturated perhydrogephyrotoxin. A second isolation process, starting with skins from 3200 frogs and involving repeated purification by column chromatography on Sephadex LH-20 yielded 15 mg of crystalline gephyrotoxin. A single crystal of the hydrobromide was analyzed by Roentgen-ray analysis to yield the relative as well as absolute configuration as expressed in **325** (*150*). The presence of a *cis*-decahydroquinoline system, substituted in positions 2 and 5 (i.e., 3a and 6, using pyrroloquinoline numbering) shows the relationship to pumiliotoxin C with the noteworthy difference that the configuration at C-2 (or 3a) is reversed. The pent-2-en-4-inyl side chain is, of course, the hallmark of the histrionicotoxins, again reminding us of gephyrotoxin bridging a gap between two groups of toxins. In the work-up of large numbers of *Dendrobates histrionicus* populations from the Guayacana region, small amounts of another congener were isolated.

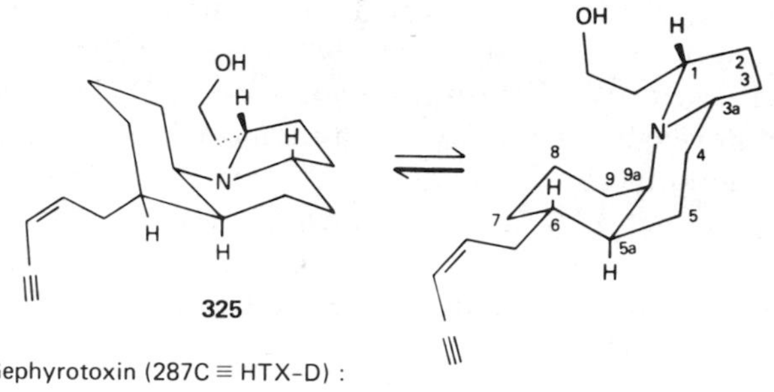

325

Gephyrotoxin (287C ≡ HTX–D) :

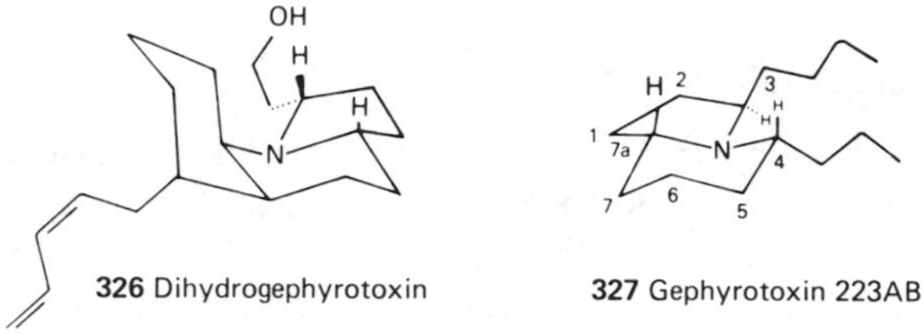

326 Dihydrogephyrotoxin   **327** Gephyrotoxin 223AB

Gephyrotoxin 257 A : $C_{18}H_{25}N$
(structure unknown)

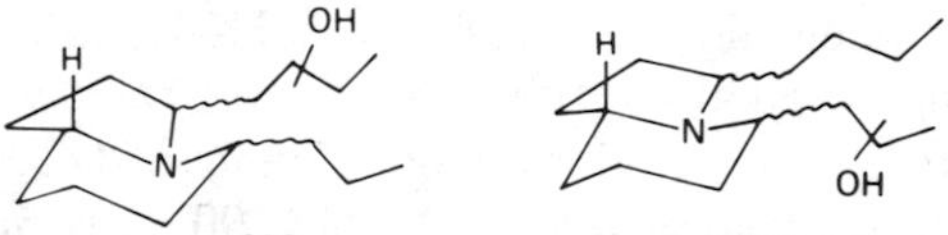

328   **239**

Gephyrotoxin 239 CD and Gephyrotoxin 239 AB

SCHEME 65. *Structures of gephyrotoxins.*

Dihydrogephyrotoxin 326. This alkaloid yielded the same perhydrogephyrotoxin as gephyrotoxin on catalytic hydrogenation. According to spectroscopic data, the difference rests in the unsaturated side chain where instead of a terminal triple bond, we find an olefinic bond (*150*).

Dendrobates trivittatus contains a base designated alkaloid 257A, which cannot be acetylated but consumes four molar equivalents of hydrogen and probably possesses the composition $C_{18}H_{25}N$ and may share the basic ring system with gephyrotoxin (*151*).

Three further congeners are indolizidine alkaloids. Thus, an alkaloid, code name 223AB **327,** was isolated from *Dendrobates histrionicus* but, as the comparison by the combination of gas chromatography with mass spectrometry showed, was present also in other *Dendrobates* species (*231*). Its structure **327** was proved by synthesis of the four possible diastereoisomers and

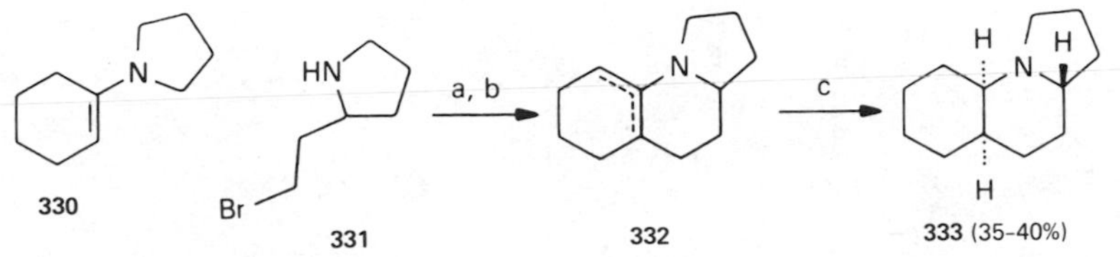

SCHEME 66. *Synthesis of gephyran, the basic tricyclic ring system of the gephyrotoxins* (241). *Reaction conditions:* (a) *DMF, 95°C;* (b) *OH⁻;* (c) *H₂/Pd/C.*

subsequent comparison with the natural base by gas chromatography (*238, 239*). This base, therefore, contains the 2,6-*cis*-disubstituted piperidine moiety of pumiliotoxin C, rather than the 2,6-*trans* disubstituted piperidine moiety of gephyrotoxin. There is also a resemblance with the trail phero-mones of the Pharao ant *Monomorium pharaonis* (*240*).

Skin extracts of *D. histrionicus* and *D. occultator* furnished two more congeners with equal molecular weight for which structures **328** and **329** have been considered (*151*). They cannot be hydrogenated but are easily acetylated.

2. Synthetic Approaches to the Gephyrotoxins: Gephyran **333**

Gephyrotoxin **325**, structurally related both to the pumiliotoxins as well as to the histrionicotoxins, was approached in an attempt to arrive at the basic bicyclic skeleton gephyran (*241*) by the same key steps that permitted the synthesis of pumiliotoxin (*208, 209*), namely, by annelation of a cyclohex-enylpyrrolidine **330** with an appropriately substituted 1-halo-3-amino com-ponent **331** (Scheme 66). Catalytic reduction of the resulting tricyclic enamine leads to a mixture of gephyran **333** and lesser amounts of the diastereoisomeric *rac*-3a-*cis*-9a-*trans*-5a-perhydropyrrolo[1,2-*a*]quinoline.

3. Syntheses of (±)-Perhydrogephyrotoxin

Likewise in the successful syntheses of perhydrogephyrotoxin, three of the five chiral centers are built up by utilizing methods that permit the construc-tion of the disubstituted *cis*-decahydroquinoline system of pumiliotoxin C (*225, 226*). The key step in this sequence, a stereoselective dieneamide Diels–Alder reaction, leads to a bifurcation from which two different routes (*242, 243*) branch out to perhydrogephyrotoxin. The main difference be-tween the two approaches is the construction of the fourth chiral center at C-2 of the quinoline system. The first synthesis (*242*) involves a pericyclic reaction, a cationic aza-Cope rearrangement (*244–248*), to effect the stereo-selective introduction at C-2 of the quinoline ring (Scheme 67).

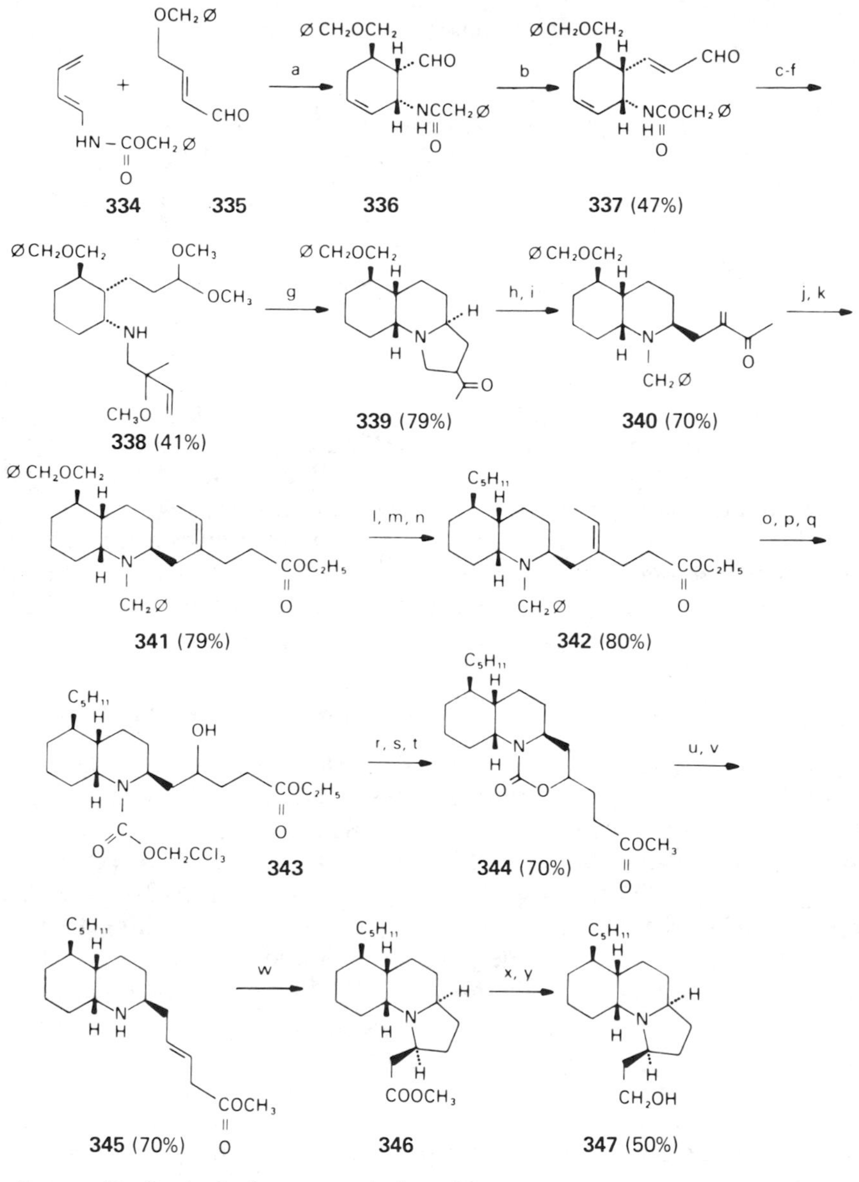

SCHEME 67. *Synthesis of racemic perhydrogephyrotoxin,* **347** *(242), by taking advantage of steps utilized in the construction of the* cis-*decahydroquinoline system of the pumiliotoxins* (225, 226). *Reaction conditions:* (a) 110°C; (b) Ø₃P⁺−CH⁻CHO; (c) pyridine. TosOH/CH₃OH; (d) Pd/C,H₂, CH₃OH; (e) ![structure](CHO / OCH₃); (f) NaBH₄; (g) ØH, 0.9 Eq TosOH · H₂O; (h) BrCH₂Ø, CHCl₃, 1↓; (i) 2% NaOH; (j) NaBH₄; (k) CH₃ C(OCH₃)₃, CH₃CH₂COOH,Δ; (l) BF₃·O(C₂H₅)₂, C₂H₅SH; (m) TosCl; (n) LiCu(C₄H₉)₂, −20°C; (o) CCl₃CH₂OCOCl; (p) O₃; (q) NaBH₄ (r) LiOH; (s) NaH, (t) CH₂N₂; (u) SeO₂; (v) Zn/HOCOCH₃; (w) NaOCH₃/HOCH₃ (cat.); (x) LAH; (y) chromatography.

The synthesis starts with the Diels–Alder reaction of the α,β-unsaturated aldehyde 335 (*242*) and dieneamide 334 (*223*) that produces mainly the endo adduct 336 besides 10% of the exo adduct. A Wittig olefination furnishes the unsaturated aldehyde 337. After acetalization and catalytic reduction of the double bond, the secondary amine is liberated with retention of the benzyl ether. Reductive amination gives 338 as a mixture of diastereoisomers, which has the N-substituent required for the subsequent sigmatropic rearrangement. When 338 is heated in benzene with *p*-toluenesulfonic acid monohydrate, a sequence of reactions occurs: The liberated aldehyde cyclizes to the *cis*-octahydroquinolinium salt. The subsequent sigmatropic rearrangement proceeds stereoselectively across the convex face of the bicyclic iminium ion. The 2-azonia sigmatropic [3.3] rearrangement is followed by a Mannich reaction rendering the rearrangement irreversible. The tricyclic 339 possesses the correct β-oriented C-3–C-3a bond, but is a 3:2 mixture of C-2 epimeric isomers carrying an acetyl group. Hofmann elimination of the benzylammonium base of 339 affords sterically pure enone 340, which after reduction ($NaBH_4$), conversion to the mixed ortho ester, and ortho ester Claisen rearrangement gives 341. Selective acid-catalyzed O-debenzylation, O-tosylation, and coupling with lithium dibutyl cuprate results in 342 possessing the requisite pentyl side chain in position 5 of the *cis*-decahydroquinoline. N-Debenzylation with chlorocarbonate trichloroethyl ester, removal of the ethylidene group by ozonolysis, reduction of the resulting ketone to the alcohol 343, saponification of the ethyl ester with lithium oxide, ring closure to cyclic carbamate by treatment with sodium hydride, and subsequent reesterification with diazomethane gives the cyclic carbamate 344. By elimination with selenium dioxide and vinylogous reductive cleavage with zinc, 344 is transformed to the α,β-unsaturated ester 345. On treatment with base intramolecular Michael addition yields 346 besides the C-1 epimer in a ratio of 8:1. Reduction of the ester 346 to the primary alcohol and separation from the unwanted diastereomer gives (±)-perhydrogephyrotoxin 347 in an overall yield of 2.75%.

The second synthesis (*243*) of perhydrogephyrotoxin 297 is more convergent in the sense explained by Veluz (*249*) and elaborated by Hendrickson (*235*). The construction of the side chains is advantageously transferred to the reactants of the Diels–Alder reaction and the Horner–Emmons reaction, respectively. The fourth chiral center is obtained by hydride reduction of the imine (Scheme 68).

Thus, the aldehyde 348 which is already equipped with the pentyl side chain at C-5 of the future *cis*-decahydroquinoline, reacts with the diene 334 mainly to give the endo adduct 349 that epimerizes more easily than the analog 336. For this reason, 349 without further purification is converted to the enone 350 by reaction with dimethyl 2-oxo-7-(ethylidenedioxy)heptyl-

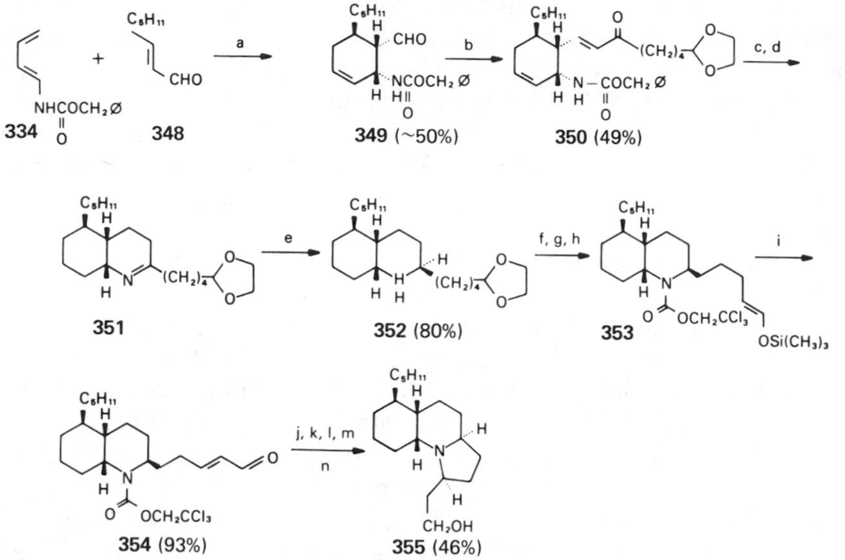

SCHEME 68. *A more convergent approach to (±)-perhydrogephyrotoxin 347, proceeding in an overall yield of 15% (243). Reaction conditions: (a) 110°C, 3 hr; (b) $(C_2H_5O)_2 P(O)(CH_2)_4$* �: *(c) 10% Pd/C,H_2, $C_2H_5OCOCH_3$, CF_3COOH excess; (d) hexane-1 N NaOH; (e) 25 Eq LAH, −20°C; (f) CCl_3CH_2OCCl, 1,2,2,4,4-pentamethylpiperidine,CCl_4; (g) THF: HO-COCH_3: 1 N HCl (4:3:2), rt, 5 hr; (h) $(CH_3)_3SiO-S(O)_2CF_3$, i-$Pr_2N C_2H_5$, $\emptyset CH_3$; $HN(C_2H_5)_2$; (i) Pd(OAc)_2, CH_3CN/DMF; (j) CH_3OH, pyridine, TosOH; (k) 10 Eq Zn/Pb, THF:1 M NH_4OAc (4:1); (l) HCl; (m) 20 Eq 1% NaOMe; (n) $NaBH_4$.*

phosphonate (*226*) a procedure which at this early stage introduces the second side chain. Hydrogenation in strongly acidic medium not only saturates the double bond but also removes the N-protecting group. Base-catalyzed cyclization leads to the octahydroquinoline **351**. The author was led to use hydride addition because of his observation that Δ^1-*cis*-octahydroquinolines add organolithium and Grignard reagents from the more congested α face (*242, 250*). Direction of addition of hydrides depends on the conditions of the reduction. Only with lithium aluminum hydride at − 15°C is the desired *cis*-decahydroquinoline **352** (besides 10% of the C-2 epimer) obtained. The secondary amino group is then protected, the aldehyde liberated, and its enolate protected by a trimethylsilyl group yielding **353**. Oxidation with palladium acetate furnishes the unsaturated aldehyde **354** that is again acetalized before the N-protecting group is removed with a zinc–lead couple. The aldehyde is liberated again and intramolecular Michael addition effected by the action of base. Reduction of the aldehyde and removal of the undesired C-1 epimer yields (±)-perhydrogephyrotoxin **347** in an overall yield of 15%, a considerable improvement by comparison with the preceding synthesis (*242*).

4. (±)-Dihydrogephyrotoxin and (±)-Gephyrotoxin via *N*-Acyliminum–Olefin Cyclization

This synthetic pathway uses the acyliminium–olefin cyclization (*178*) as the key step, a reaction type also utilized in the approach to perhydrohistrionicotoxins (*181, 183–185*) as described in four publications (*251–253, 254*).

At first (*251*), the stereochemistry of the key reaction was tested (Scheme 69). Displacement with inversion of the hydroxyl group of *trans*-2-vinyl-cyclohexanol **355** is effected utilizing succinimide, diethyl azodicarboxylate, and triphenylphosphan according to Mitsunobu. Reduction with diisobutyl-aluminum hydride leads to the diastereomeric carbinolamine **356** that on treatment with formic acid cyclizes to 5-formyloxygephyrone **357**. The high stereoselectivity is caused by the fact that the conformation with the acyliminium moiety in equatorial position is energetically more favorable and that ring closure of the energetically less favorable conformation would lead to strong $A^{1,3}$ strain during the formation of the second six-membered ring. Extending the preceding synthesis, the substituent at C-1 is introduced to yield 6-depentylperhydrogephyrotoxin **362** (*252*) (Scheme 70).

trans-2-Ethinylcylclohexanol **358** is treated with succinimide according to Mitsunobu, the triple bond reduced to the olefin, and one of the imide carbonyls then reduced to the carbinolamine, which with formic acid undergoes cyclization to 5-formyloxygephyrone **357**. After saponification of

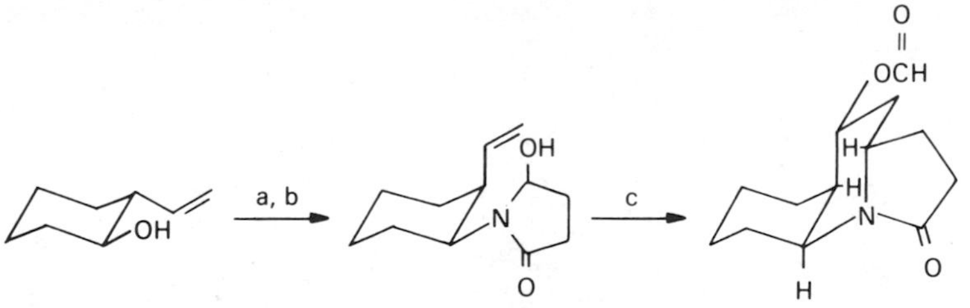

SCHEME 69. *Model synthesis of 5-formylgephyrone **357** by stereoselective acyliminium–olefin cyclization (251). Reaction conditions: (a) succinimide, DEAD, PØ₃; (b) DBAH; (c) HCOOH.*

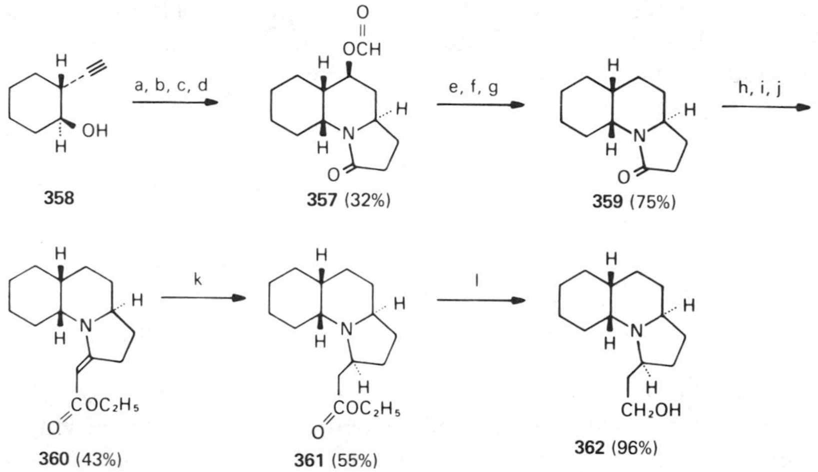

SCHEME 70. *Synthesis of racemic 6-depentylperhydrogephyrotoxin 362* (252). *Reaction conditions:* (a) *succinimide, DEAD,PØ₃;* (b) *H₂, Pd/BaSO₄, pyridine;* (c) *DBAH, ØCH₃* (d) 97% *HCOOH;* (e) *NaOH;* (f) *NaH, CS₂, CH₃I;* (g) *(C₄H₉)₃ SnH;* (h) *P₂S₅;* (i) *BrCH₂CO₂C₂H₅;* (j) *Ø₃P, CHCl₃, (C₂H₅)₃N;* (k) *NaBH₃CN;* (l) *LAH.*

the formate, the 5-hydroxy function is removed via the xanthate by treatment with tri-*n*-butylstannan, according to Barton, to yield the gephyrone **359**. The two-carbon side chain is introduced via the thioamide and Eschenmoser's "sulfide contraction procedure." The resulting enamine **360** is reduced with sodium cyanoborohydride to a 4:1 mixture of C-1 epimeric esters from which the desired tricyclic ester **361** is separated and further reduced to (±)-6-desamylperhydrogephyrotoxin **362** (*252*).

A variation of this synthesis (*252*) (Scheme 71) introduces the two-carbon

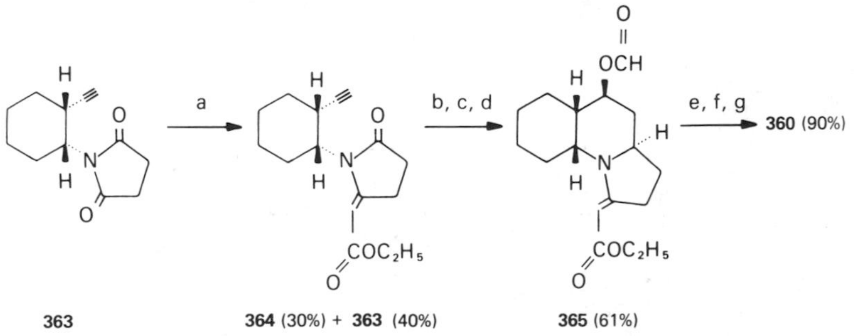

SCHEME 71. *An improved approach to the enamine intermediate **360**, a precursor of racemic 6-depentylperhydrogephyrotoxin 362* (252). *Reaction conditions:* (a) *Ø₃P⁺—C⁻HCO₂C₂H₅;* (b) *H₂, Pd/BaSO₄, pyridine;* (c) *DBAH, −65°C;* (d) *HCOOH;* (e) *NaOH;* (f) *NaH, CS₂, CH₃I;* (g) *(C₄H₉)₃ SnH.*

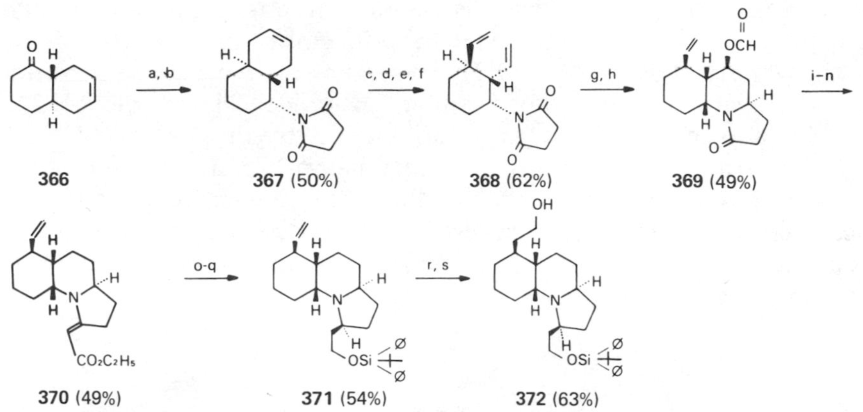

366 **367** (50%) **368** (62%) **369** (49%)

370 (49%) **371** (54%) **372** (63%)

SCHEME 72. *A formal synthesis of racemic gephyrotoxin (253). (a) LAH, −70°C; (b) succinimide, DEAD, PØ₃; (c) O₃; (d) NaBH₄; (e) o-NO₂Ø—SeCN, (C₄H₉)₃P; (f) H₂O₂; (g) DBAH, −65°C; (h) HCOOH; (i) NaOH; (j) NaH, imidazole, 60°C, CS₂, CH₃I; (k) (C₄H₉)₃SnH; (l) (p-CH₃OØ-PS₂)₂, 100°C; (m) BrCH₂COOC₂H₅; (n) Ø₃P, (C₂H₅)₃N, CH₂Cl₂ or (m) CH₃I; (n) C₂H₅O₂CCH⁻COO⁻ Mg²⁺, DMF; (o) NaBH₃CN; (p) LAH; (q) t-BuØ₂SiCl, DMF, imidazole; (r) THF; (s) NaOH, H₂O₂.*

side chain at C-1 into the succinimide **363** before cyclization of the vinylogous carbinolamide of **364** to **365**. The reason for the improvement in yield is discussed.

A formal synthesis of (±)-gephyrotoxin **325** (Scheme 72) starts with the reduction of *trans*-octalone **366** by lithium aluminum hydride at −70°C, by axial attack of the hydride. After substitution by succinimide, according to Mitsunobu, the imide **367** is obtained. Ozonolysis of **367** and reduction leads to the diol that is converted to the diene **368** via the diselenide, oxidation to the corresponding selenoxides, and elimination. The diene possesses the desired configuration of the trisubstituted cyclohexane and adopts, according to NMR data, a chair conformation in which the two vinyl groups occupy axial sites. Therefore, the subsequent acyliminium–olefin cyclization follows the same stereochemical course as described before, yielding the tricyclic lactam **369**. Deformylation and dehydroxylation to **370***, according to Barton, is followed by the introduction of the two-carbon side chain by Eschenmoser's method or by a method developed by the author (*256*), i.e., conversion of the lactam to the methylmercaptoiminium salt and reaction with the dibasic magnesium salt of monoethyl malonate leading to **370**. Reduction of the enamine double bond with cyanoborohydride yields a 3:2

---

* Another (but not superior) route from **368** to **370** involves formation of the acylaminoradical and cyclization with the help of tributylstannane (*382*).

mixture of diastereomers. Finally reduction of the ester groups with lithium aluminum hydride and protection of the primary hydroxyl group yields the diphenyl-*tert*-butylsilyl ether **371**. Hydroboration of the olefinic side chain at C-6 and treatment with alkaline hydrogen peroxide yields the alcohol **372** which is also an intermediate in an earlier gephyrotoxin synthesis (*255*).

The intermediate **370** serves as starting material for a different route to (±)-gephyrotoxin **325** (*254*) in which the enyne side chain is constructed according to Yamamoto (*195*, Scheme 73). The isolated double bond of **370** is hydrated to a primary hydroxyl group by hydroboration and successive treatment with alkaline hydrogen peroxide. After protection as the di-

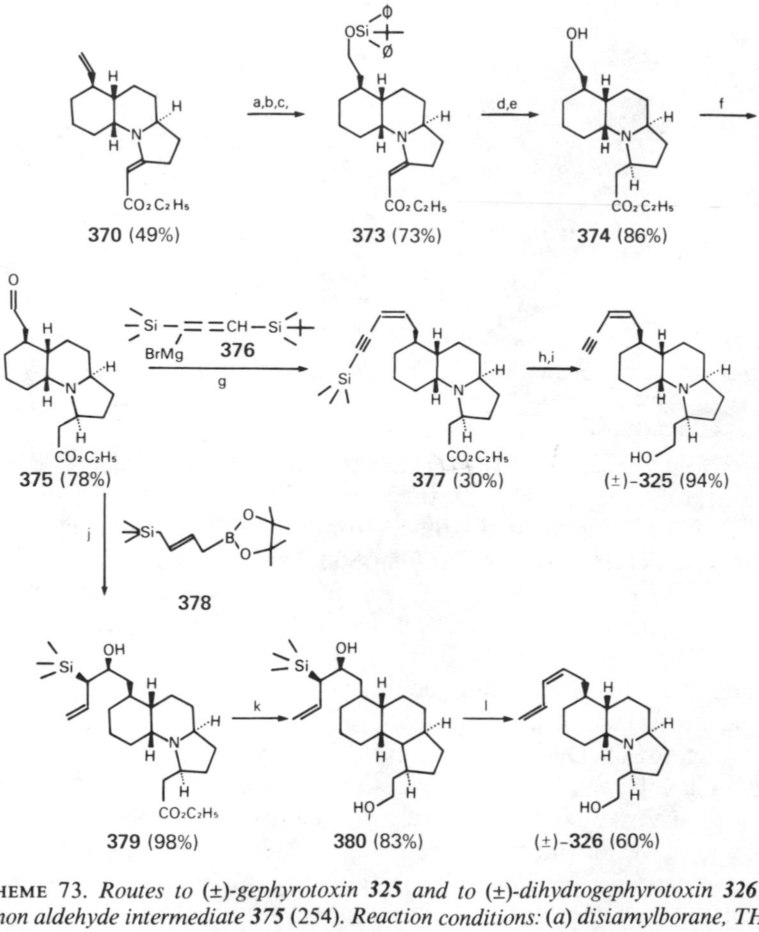

SCHEME 73. *Routes to* (±)-*gephyrotoxin* **325** *and to* (±)-*dihydrogephyrotoxin* **326** *via a common aldehyde intermediate* **375** (*254*). *Reaction conditions:* (*a*) *disiamylborane, THF;* (*b*) *NaOH, H₂O₂;* (*c*) *t-BuO₂SiCl, DMF, imidazole;* (*d*) *H₂, Pt/Al₂O₃;* (*e*) (*C₄H₉*)₄ *N⁺F⁻, DMF;* (*f*) (*COCl*)₂, *DMSO,* (*C₂H₅*)₂*N;* (*g*) (*C₂H₅*)₂ *O/THF,* − 78° *C; 50° C, 3 hr;* (*h*) *DBAH;* (*i*) (*C₄H₉*)₄*N⁺F⁻, DMF;* (*j*) *N*(*C₂H₄OH*)₃, *H₂O;* (*k*) *DBAH;* (*l*) *KH, THF.*

phenyl-*tert*-butylsilyl ether **373,** selective hydrogenation of the enamine double bond with platinum on alumina as catalyst and desilylation lead to the saturated alcohol **374,** which is then converted to the aldehyde **375** and transformed by Peterson olefination with 1,3-bistrialkylsilylallenylmagnesium bromide **376** to the protected enyne **377.** Reduction of the ester group and deprotection of the enyne side chain completes the synthesis of (±)-gephyrotoxin **325** in an overall yield of 1.3%.

Branching off from the aldehyde **375,** the route to (±)-dihydrogephyrotoxin **326** (*254*) (Scheme 73) leads to a 2-hydroxy-3-trimethylsilylpent-4-enyl side chain at C-6 of **379** by the use of *E*-1-trimethylsilyl-1-propene-3-boronate **378** (*257*). Reduction of the ester group and Peterson elimination yields (±)-dihydrogephyrotoxin, **326** (*146,* footnote 16).

5. (±)-Gephyrotoxin via Stereodirected Catalytic Hydrogenations (*255*)

The stereoselective elaboration of the five chiral centers in this synthesis rests on stereocontrolled catalytic hydrogenation (Scheme 74). *N*-Benzylsuccinimide **381** is converted to the 2,5-disubstituted exoolefinic pyrrole derivative **382** by the repeated use of ethoxyethynylmagesium halide and acidic work-up. Catalytic hydrogenation with palladium on charcoal in methanol under acid conditions with simultaneous debenzylation yields mainly the *cis*-2,5-disubstituted pyrrolidine **383,** besides 8% of the trans isomer. The pyrrolidine **383** is converted to the bicyclic carbamate by reaction with carbobenzoxychloride, reduction of the ester groups with lithiumborohydride, and ring closure under strong basic conditions. The remaining primary hydroxyl group is transformed to the benzyl ether and the cyclic carbamate saponified to yield the monobenzylated dialcohol **384** that forms the enamine **385** with cyclohexa-1,3-dione. The hydroxyl group is mesylated followed by treatment with lithium bromide in DMF leading to intramolecular α-alkylation. Hydrogenolysis of the benzyl ether yields the tricyclic enone **386.** Because of the directing effect of the hydroxyethyl substituent, hydrogenation of the enone **386** with platinum on alumina in dry ethyl acetate furnishes the gephyran **387** with high stereoselectivity. Selective acetylation of the primary alcohol and conversion of the secondary alcohol to the ketone **388** make possible the introduction of the two-carbon side chain by reaction with ethoxyethinylmagnesium chloride, followed by deacetylation to give the unsaturated esters **389,** as a 1 : 1 mixture of the *Z*-and *E*-olefin. To suppress the directing effect of the hydroxyl group and simultaneously increase the steric hindrance of attack from the side of the C-1 substituent, the bulky diphenyl-*tert*-butylsilyl ether is prepared so that hydrogenation with rhodium on alumina as catalyst and hexane as solvent yields the desired **390** besides 10% of the unwelcome C-6 epimer. Reduction

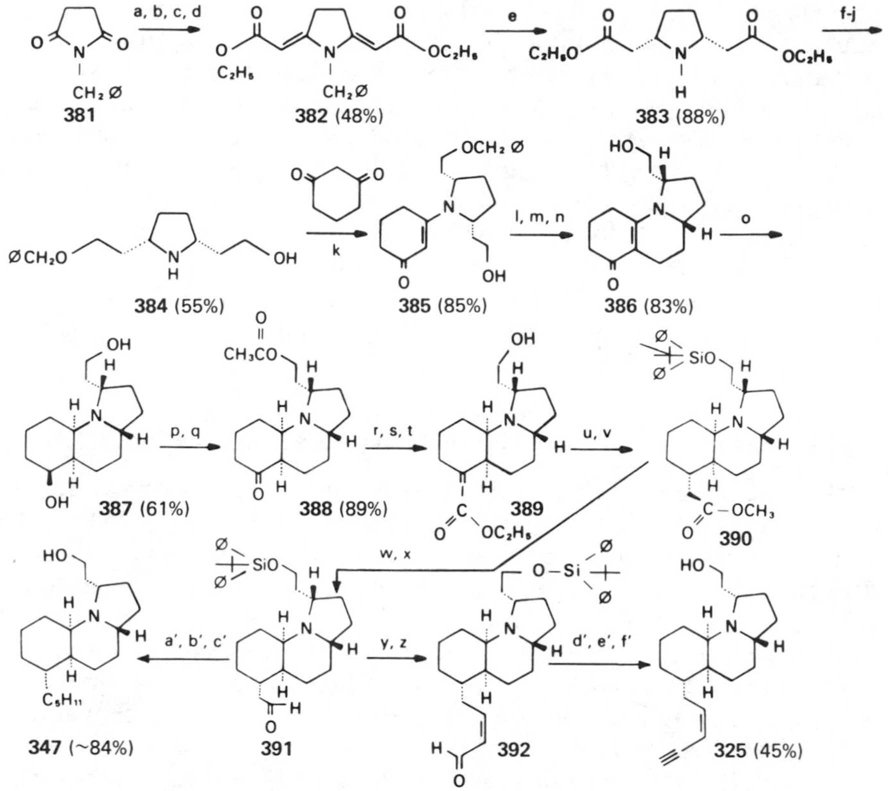

SCHEME 74. *Stereoselective stepwise elaboration of the five chiral centers in a total synthesis of racemic gephyrotoxin 325 (255). Reaction conditions: (a) $C_2H_5OC\equiv C-MgCl$; (b) 5% HCl, 0°C; (c) $C_2H_5C\equiv C-MgBr$; (d) 5% HCl; (e) 4 atm. H_2, 10% Pd/C, $HClO_4$, CH_3OH; (f) ØOCOCl, pyridine, CH_2Cl_2; (g) $LiBH_4$; (h) KH/THF; (i) $ØCH_2Br$, DMF; (j) $Ba(OH)_2$, H_2O, Δ; (k) pyridine, TosOH; (l) MesCl, $(C_2H_5)_3N$, CH_2Cl_2; (m) Li Br, DMF; (n) H_2, 10% Pd/C, $HClO_4$/ CH_3OH; (o) 4 atm H_2, 5% Pt/Al$_2O_3$, abs. $C_2H_5OCOCH_3$; (p) $(CH_3CO)_2O$, pyridine; (q) $(COCl)_2$, DMSO, $(C_2H_5)_3N$, −65°C to rt; (r) $C_2H_5OC\equiv CMgCl$; (s) CH_3MgBr; (t) 5% HCl, 0°C; (u) Ø$_2tBuSiCl$, imidazole, DMF; (v) 5% Rh/Al$_2O_3$, 1 atm H_2, hexane, −20°C; (w) LAH; (x) PCC, CH_2Cl_2; (y) C_2H_5O CH=CH−P^+Ø_3Br, $NaOC_2H_5$; (z) TosOH, CH_3COCH_3/H_2O, 0°C; (a') $CH_3CH_2CH−P̈Ø_3$; (b') 4 atm H_2, 5% Rh/Al_2O_3, C_2H_5OH; (c') $(C_4H_9)_4N^+F^-$, DMF; (d') $ClCH_2PØ_3$, C_4H_9Li; (e') CH_3Li, $(CH_3)_3SiCl$; (f') $(C_4H_9)_4N^+F^-$, DMF.*

of the ester group furnishes the alcohol which is then oxidized to the aldehyde **391**. Olefination of **391** with triphenylpropylphosphonium bromide and base, reduction of the newly formed double bond, and desilylation yield (±)-perhydrogephyrotoxin **347**. However, when the aldehyde **391** is reacted with 2-ethoxyethenyltriphenylphosphonium bromide and sodium ethoxide, the unstable Z-α,β-unsaturated aldehyde **392** is obtained and converted to (±)-gephyrotoxin **325** by a method elaborated by Corey and Ruden (*190*).

6. Asymmetric Synthesis of Gephyrotoxin (*258*)

Commercial L-pyroglutamic acid **393** (Scheme 75) (*258*) is converted to the nitrile **394** by the use of known methods (*259*). The two-carbon side chain is introduced by Eschenmoser's sulfide contraction method and by retro Claisen reaction to yield the olefinic ester **395**. A mixture of the epimeric pyrrolidines, in which **396** prevails in a ratio of 2.3 : 1, is obtained by catalytic hydrogenation with platinum in methanol under acidic conditions. The *cis*-2,5-disubstituted pyrrolidine **396** is converted to the cyclic carbamate **397**. The nitrile group is then reduced at − 105°C with diisobutylaluminum-hydride to the imine. Acidic work-up and reduction of the resulting aldehyde by sodium borohydride leads to the primary alcohol that is protected by a methoxymethyl group. Basic cleavage of the cyclic carbamate leads to the optically active pyrrolidine **398** then further processed as described in the synthesis of (±)-gephyrotoxin (*255*) (Scheme 74). The resulting optically active end product is (+)-gephyrotoxin, while the natural toxin is levorotatory. This result raises the question of the absolute configuration of natural gephyrotoxin.

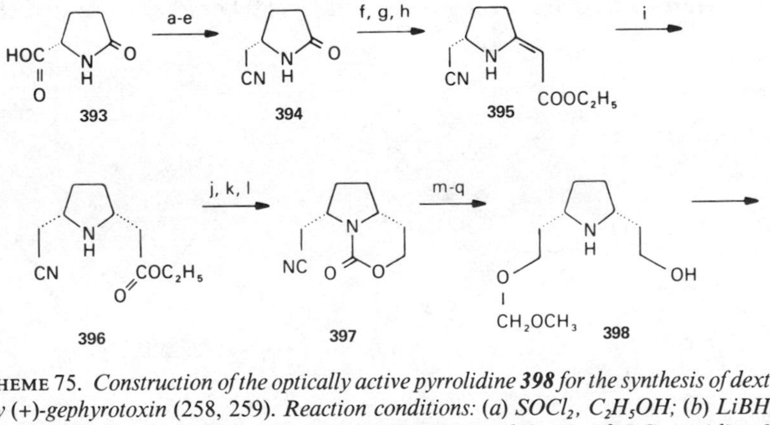

SCHEME 75. *Construction of the optically active pyrrolidine **398** for the synthesis of dextrorotatory (+)-gephyrotoxin (258, 259). Reaction conditions: (a) $SOCl_2$, C_2H_5OH; (b) $LiBH_4$; (c) H_2O/CH_3COOH; (d) $\emptyset_3P$, CBr_4, CH_3CN; (e) $\emptyset CH_3$, KCN/Al_2O_3, N_2; (f) P_2S_5, pyridine, 80°C; (g) $CH_3COCHBrCO_2C_2H_5$, $NaHCO_3$, rt to ⇅; (h) KOH, C_2H_5OH, 60°C; (i) 1 atm H_2, 5% Pt/C, $HClO_4/CH_3OH$; (j) $\emptyset-O-C-O-Cl$, pyridine, CH_2Cl_2; (k) $LiBH_4$; (l) KH; (m) DBAH, THF/ $\emptyset CH_3$, − 105°C; (n) 3 N HCl; (o) $NaBH_4$, DME; (p) CH_3OCH_2Br, i-$Pr_2NC_2H_5$; (q) $Ba(OH)_2/ H_2O$, ⇅.*

7. Synthesis of 3S*-3-Butyl-5S*-5-propyl-9S*-9H-perhydroindolizine 327 (formerly referred to as Gephyrotoxin 223AB)

This indolizidine alkaloid has been synthesized by two methods (238, 239). One procedure (239, 260) yields 327 (Scheme 76), the racemic equivalent of the natural base, only as a byproduct but permits comparison and identification with the natural product and the product of the second synthesis (238).

To obtain the dihydroindolizine derivative 404, an olefinic Friedel–Crafts reaction of valeroyl chloride 399 with allylchloride 400 leads to a mixture of 1,2-dichlorooctan-4-one and 1-chloroocten-4-one, that after treatment with methanol followed by potassium tert-butoxide yield exclusively the trans-substituted cyclopropyl derivative 401. Ring cleavage with methanol at elevated temperature and conversion of the keto group to an amino group furnish the acetal 402. Treatment of 402 with the cyclopropyl derivative 403 prepared in the same way with butyryl chloride as the starting material, in the presence of the complex of 2,6-dimethylpyridine with trifluoroacetic acid in xylene yields 404 (260). Hydrogenation of the dihydroindolizine 404 with rhodium on alumina yields mainly the indolizidines with cis-2,5 substituents 407 and 406 beside smaller amounts of the indolizidines with a trans-substituted pyrrolidine ring, 405 and 327. The synthesis of the isomer 406 was also achieved via acyliminium cyclization in thirteen steps (239, 383) besides 30% of 405 and in a highly stereoselective manner via alkylation of an intermediary $\Delta^{3a,4}$-didehydroindolizidinium salt (261).

In the stereoselective synthesis of 327 (238) (Scheme 77), the two chiral

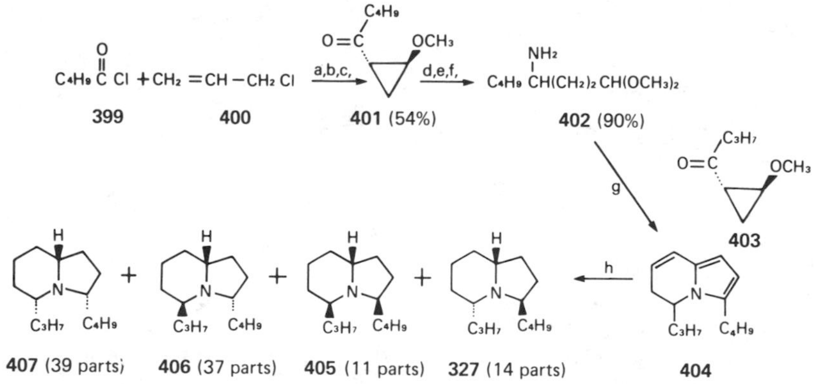

SCHEME 76. Synthesis of 3-S*-3-butyl-5S*-5-propyl-9-S*-H-perhydroindolizine, 327 (gephyrotoxin 223AB) (239, 260). Reaction conditions: (a) AlCl₃, CH₂Cl₂, −20°C; (b) Δ, CH₃OH; (c) t-BUOK, THF; (d) Δ, CH₃OH; (e) NH₃OHCl⁻, CH₃COO⁻Na⁺; (f) Na, n-C₄HgOH; (g) 2,6-dimethylpyridine, CF₃COOH, CF₃COOH, xylene, Δ; (h) H₂, RhAl₂O₃, C₂H₅OH.

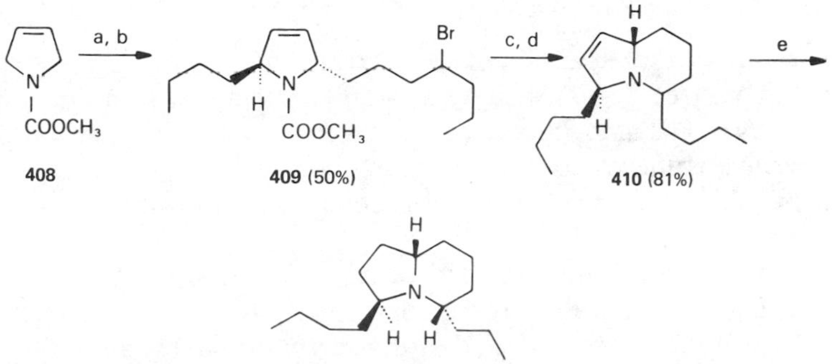

408 **409** (50%) **410** (81%)

327 (38%)

SCHEME 77. *Another approach to racemic gephyrotoxin congener 223AB,* **327** *(238). Reaction conditions:* (a) *LDA, BuBr,* $-40°C$; (b) *LDA, 1,4-dibromoheptane;* (c) $(CH_3)_3SiI$; (d) Na_2CO_3/CH_3OH, ↓; (e) PtO_2/CH_3COOH.

centers at the pyrrolidine ring are constructed by two-fold consecutive metallation and alkylation of 1-methoxycarbonyl-3-pyrroline **408** (*262*). Analogous to the alkylation of the kindred nitrosamines (*263*), the second alkyl substituent enters at the unsubstituted α position next to nitrogen. The advantage compared with the nitrosamine (*263*) is the higher stereoselectivity leading to 2,5-*trans*-dialkylpyrrolidines.

Thus metallation and alkylation with butyl bromide, followed by metallation and alkylation with 1,4-dibromoheptane lead to the 2,5-*trans*-dialkylpyrrolidine derivative **409** as a 1 : 1 mixture of C-4' epimeric diastereomers. Cleavage of the carbamate with trimethylsilyl iodide and base-catalyzed ring closure yield the unsaturated indolizidines **410**. Catalytic hydrogenation with platinum in acetic acid leads to only one racemic isomer **327** with the same relative configuration as the natural base by a mechanism so far unexplained.

IV. Neurotoxins, Receptors, and Ion Transport

A. BATRACHOTOXIN: A VALUABLE TOOL FOR THE STUDY OF VOLTAGE-SENSITIVE SODIUM CHANNELS

In the last decade batrachotoxin has been used in the laboratories of biochemists, biophysicists, molecular pharmacologists, and physiologists as one of the major tools for understanding the mechanism of passive ion transport in voltage-sensitive, sodium-specific channels in nerve and muscle. Historically there are three stages to this development.

1. Early Molecular Pharmacology

If we compare the expansion of knowledge produced by batrachotoxin to the growth of a tree (Fig. 1), we have a so-called cliogram (from *Clio,* muse of history, *264*). The first important observation was made by Edson X. Albuquerque; namely, the ability of batrachotoxin to cause persistent activation of the voltage-sensitive sodium channel that normally opens only during action potentials (*272*).

The concomitant sodium-dependent depolarization of excitable tissues has many effects, among them release of the neurotransmitter acetylcholine, contracture of skeletal muscle, and cardiac arrhythmia. It was recognized at an early stage that tetrodotoxin, the poison of the puffer fish, counteracted most effects of batrachotoxin (*272*), yet their specific active sites were different (*272–275*). This novel observation automatically led to further receptor studies.

2. Binding Studies and Cooperative Activation of Sodium Ionophores

W. A. Catterall showed that sodium channels have at least three separate receptor sites for neurotoxins (*276*): Tetrodotoxin and saxitoxin bind at receptor site 1 and block the ion transport activity of the sodium channel. The steroidal alkaloids batrachotoxin and, to a lesser degree, veratridine as well as the terpinoids aconitine and grayanotoxin, bind at receptor site 2, a common binding site probably in the sodium channel, belonging to a subunit exercising control over other subunits that must interact for the various functions of the sodium channel (*273*). Khodorov speaks of "negative modifiers for the inner channel receptor" (*277*).

The detailed binding studies have now been carried to a point at which the voltage-dependent sodium channel acquires the connotation of a (multiple) "drug receptor" in the sense of Paul Ehrlich (*278*).

The toxins of the scorpion *Leiurus quinquestriatus* and of sea anemones are polypeptides. They bind to receptor site 3 on the sodium channel, probably at the outside, and change the kinetics and steady state of the process of inactivation.

In addition to these well-characterized toxins, it now appears that additional classes of neurotoxins may also act at different receptor sites inside or outside the sodium channel whose structure and function they help to clarify by essentially biochemical operations (*279, 280*).

Fig. 1. A so-called cliogram, a dendroid presentation of the initial impact of special neurotoxins on the dynamics of voltage-sensitive channels (*267*).

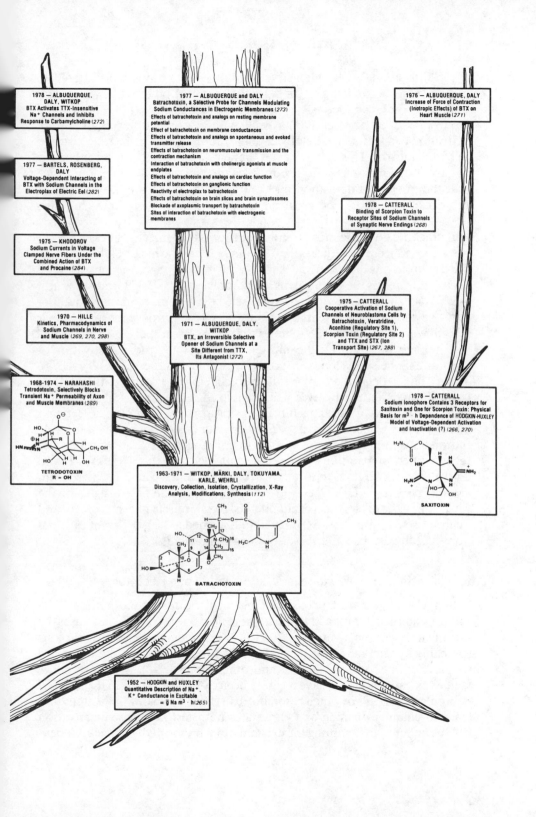

Ion flux studies with radioactive sodium (^{22}Na$^+$) prompted Catterall to postulate an allosteric model for batrachotoxin that binds with high affinity to the open, ion-conducting conformation of the channel (*281*). Both in the electroplax of electric eel (*282*) as well as in myelinated nerve (*283, 284*), activation of sodium channels by depolarization increases the rate of action of batrachotoxin. This demonstrates that batrachotoxin binds selectively to specific functional states of the sodium channel and provides clear evidence that the process of activation and inactivation of the sodium channel is describable in terms of conformational changes of protein subunits (*272–274*).

Batrachotoxin is the most useful probe for biochemical studies of neurotoxin receptor site 2 on the sodium channel. It has the highest affinity of the four neurotoxins that bind at site 2 and is a full agonist activating all channels, whereas the other toxins are partial agonists. [^{3}H]Batrachotoxinin A 20-α-benzoate (BTX-B), first synthesized by W. Burgermeister, was later (*285*) developed as a potential radioligand for neurotoxin receptor site 2. Recent work by Catterall (*286*) shows that BTX-B is a specific ligand at this receptor site and confirms the allosteric model of neurotoxin activation of sodium channels developed in previous ion flux studies (*281, 287–289*). These studies open the way to future biochemical experiments in which batrachotoxin and derivatives will be used to relate structural features of sodium channels to specific functional properties. One can expect that batrachotoxin will be used as a specific activator of sodium channels to study the ion transport properties of purified membrane fractions and purified and reconstituted preparations of sodium channels, and that [^{3}H]BTX-B and derivatives will be used to locate neurotoxin receptor site 2 in the sodium channel structure. This work will continue to provide new insight for the molecular basis of electrical excitability, skeletal muscle pharmacology, and neuromuscular disorders (*5*). What at first seemed to be a ceiling of expectations has, in part, become a floor of facts by now (*289*).

3. The Action of Batrachotoxin: The Assembly of a Receptor

Batrachotoxin simultaneously changes all the major properties of the sodium channel: it shifts activation to more negative potentials, decreases the rate of closing or inactivation (*290, 291, 291a*), alters the selectivity for ions (probably by interacting with the selectivity filter), and varies the sensibility to anesthetics, scorpion venoms, and other toxins (*292*). Khodorov's rate studies on gating kinetics have recently been extended to internally perfused neuroblastoma cells in which batrachotoxin slows down one of the steps that leads to channel opening and eliminates both fast and slow inactivation (*293*). Rather than assuming the existence of a preformed, voltage-independent BTX receptor that does not undergo conformational changes during channel gating (i.e., the capability to open and close in response to changes of membrane potential), Khodorov prefers the idea of a batrachotoxin receptor that acquires its active conformation or is formed and assembled in the process of transition of the channel to the open state. Such an assembly, *in situ* or *in statu nascendi,* of a receptor is reminiscent of Bertil-Hille's "modulated receptor hypothesis" (*294*).

4. Models of Ion Channels

The architecture of an ion-specific, transmembrane channel has so far been discussed in terms of models, such as the channels that are formed with the *N*-formyltetradecapeptide gramicidin A, B, or C (*295*) when inserted in an artifical membrane (*296, 297*).

However, there are no free carboxy or amino groups in the gramicidin channel, whereas the real sodium channels must possess ϵ-amino groups of lysine or β- and γ-carboxyls belonging to bound aspartic or glutamic residues, respectively (*298*).

It should be possible to show the effect of methylation of negatively charged carboxyl groups presumably present at the entry to the channel. Indeed, the sensitivity of the sodium channel of nerve nodes from *Rana pipiens* was abolished by methylation with trimethyloxonium ion (*299*). Likewise, the insensitivity to batrachotoxin (and reduced sensitivity to grayanotoxin and veratridine) of the sodium channels of the frogs that elaborate batrachotoxin is ascribed to a modification of the binding site which may prevent binding of the pyrrole moiety of the molecule. Batrachotoxinin A lacks the 2,4-dimethylpyrrole-3-carboxylic acid and is much less toxic, commensurate with grayanotoxin and veratridine (*300*).

The lack of sensitivity to their own venom is displayed by many animals. The nerve from the newt *Taricha torosa* is about 30,000 times less sensitive to tetrodotoxin than frog nerve; the nerve of the Atlantic puffer fish *Spheroides maculatus* is about 1000 times less sensitive. However, both nerves lose none of their sensitivity to saxitoxin (*301*).

With the occurrence of tetrodotoxin in Atelopid frogs of Costa Rica (*302*) and of chiriquitoxin in the harlequin frog *Atelopus chiriquensis* of Central America (*303*), the response of these frogs to the various endogenous toxins and their cross-over poses an interesting question. Subtle chemical changes capable of modifying membrane permeability need not be restricted to free carboxy or amino groups. Earlier experiments indicated a requirement for sulfhydryl groups for BTX to exert its effects (*275*). Modification of carboxyl groups at or near the tetrodotoxin receptor by a water-soluble carbodiimide is believed to abolish binding of tetrodotoxin to crab nerve fibers (*304*).

ε-Ammonium groups of bound lysine are assumed to be present at the entry to the sodium channel. An ammonium ion may be complexed efficiently by a triad of oxygens in an appropriate geometric arrangement (*305*). Kosower (*306*) suggests that such a triad of oxygens is present in batrachotoxin, namely, the two oxygen atoms of the 3,9-hemiketal as well as the secondary alcoholic hydroxyl at C-11 (Fig. 2). This triad is assumed to bind an ammonium ion more strongly than a nearby carboxyl group. In addition, there should be a vicinal site suitable for hydrogen bonding to an anion. Channel opening would require "single group rotation" of the crucial hypothetical ε-lysine group (Fig. 3). Proximity of the BTX-open channel complex to the entrance might inhibit the approach of ions, since Khodorov finds a low conductance for the BTX-opened channel.

Similar oxygen triad complexes with lysine ammonium ions may be pictured for grayanatoxin I (Fig. 4), aconitine (Fig. 5), or veratridine (Fig. 6). As hypothetical or tenuous as these pictures may be, they fulfill a useful function: In the sense of Paul Ehrlich they stimulate the eidetic approach in the same way as Ehrlich's hypothetical side-chain theory of the amboceptor or the receptors for drugs, toxins, and antigens eventually matured into models confirmed by X-ray crystallography (*7*).

Blockage of the voltage-dependent calcium channel by lipid-soluble toxins, such as batrachotoxin, veratridine, aconitine, and grayanotoxins has been observed quite recently in neuroblastoma cells (*307*). This finding, if it can be generalized, not only suggests common structural features for Na^+ and Ca^{2+} channels, but also the possibility that tritiated batrachotoxin might become the first tool available for the biochemical titration of the calcium channel.

B. HISTRIONICOTOXIN AND THE NICOTINIC ACETYLCHOLINE RECEPTOR

1. Acetylcholine Receptors

As long ago as 1907, J. N. Langley postulated (*308*) that a specific "receptive substance" was present at the contact between nerve and skeletal muscle, a substance that could bind the tobacco alkaloid nicotine to initiate muscle contraction or bind tubocurarine (from *Strychnos*) with the consequences that neither nicotine nor nerve stimulation produced muscle contraction. Since the experiments of Loewi, Dale, and Feldberg in the 1920s and 1930s, it is known that nicotine actually mimics the action of a substance released from the nerve, i.e., the neurotransmitter acetylcholine (AcCh) (for a historical review, see *309, 310*). The acetylcholine receptor (AcChR) itself is an integral component of the skeletal muscle plasma membrane, and the binding of AcCh or other agonists results in the opening of a transmembrane, cation-selective channel (*311*). Within the past 10 years the efforts of

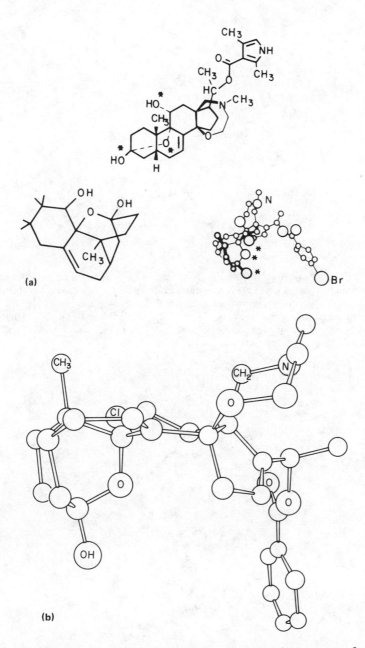

(a)

(b)

FIG. 2. (a) Molecular formula for batrachotoxin and structural arrangement from X-ray crystallography for batrachotoxinin A-20-p-bromobenzoate. The oxygen triad capable of interacting with an alkylammonium ion is starred. A partial structure which shows the three oxygens in the triad more clearly is given at the lower left (E. Kosower, unpublished). (b) Computer-drawn diagram of batrachotoxinin A p bromobenzoate showing the triad of oxygen atoms 3, 9, and 11 (*116, 117*).

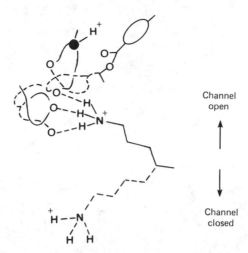

FIG. 3. Channel opening by batrachotoxin via lysine complex. Schematic formula for batrachotoxin complexed with the ε-ammonium group of the lysine at the channel entry. The negatively charged group that normally interacts with the ammonium ion on opening may be located near the protonated nitrogen (pK_a 6–7). SGR ("Single Group Rotation") of the lysine is inhibited by the relatively strong triple H bond complex to the oxygen triad (E. Kosower, unpublished).

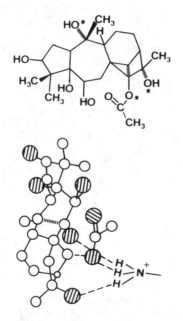

FIG. 4. Molecular formula for grayanotoxin I (GTX I), with oxygen atoms in the triad (see text) starred. The crystal structure for GTX I (based on that for GTX II) is shown below with a triple H bond from the ε-ammonium of lysine to the oxygen triad; the strong interaction maintains the channel in the open condition (E. Kosower, unpublished).

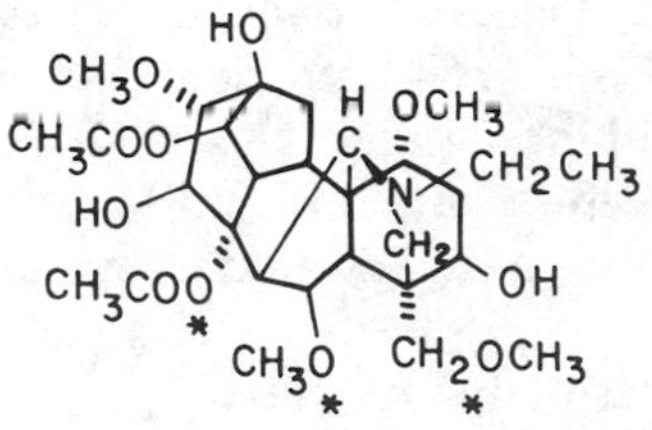

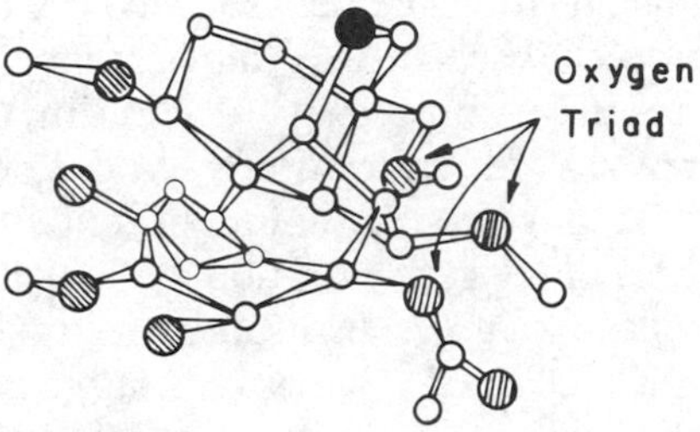

FIG. 5. Molecular formula for aconitine. The oxygen atoms in the triad are starred. The probable crystal structure is shown with an oxygen triad shown. Presumably, the triad may complex with the ε-ammonium ion of a lysine (E. Kosower, unpublished).

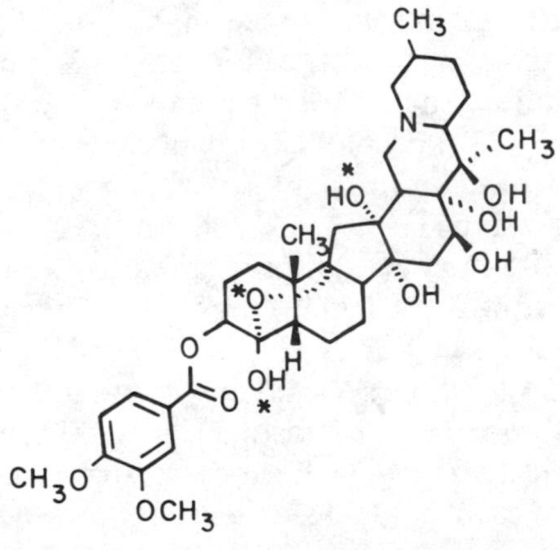

FIG. 6. Molecular formula for veratridine. The oxygen atoms thought to participate in the triad are starred. The general resemblance of the molecular arrangement in veratridine to that in batrachotoxin implies that the primary interaction is the same, with the oxygen triad forming a triple H bond to the ε-ammonium ion of lysine (E. Kosower, unpublished).

electrophysiologists and biochemists have added up to spectacular progress in defining the structure of the AcChR (reviewed in *312, 10*) and of the relationship between ligand binding and ion transport (*313, 314*). In our current understanding of receptor structure and function, histrionicotoxin (HTX) has served and will continue to function as an invaluable tool in those investigations (*315–317*).

Electrophysiological studies define functional properties of the AcChR. Within a fraction of a millisecond of its release from the nerve, AcCh binds to the AcChR and effects the opening of an ion channel permitting the passage of ~ 10,000 ions/millisecond. Individual channels remain open for several milliseconds (*318*). The relationship between receptor occupancy and channel activation is complex, because it is known that exposure to AcCh for seconds or longer results in progressive decline in the number of open channels [i.e., there is a desensitization of the response (*319, 320*)] a phenomenon observed more in microiontophoretic than natural release of acetylcholine. Analysis of the concentration dependence of desensitization and of recovery indicates that desensitization occurs because the receptor conformation binding AcCh with highest affinity is not the open channel conformation but an inactive "desensitized" conformation.

2. AcChR: Biochemistry and Ultrastructure

The pharmacological acetylcholine receptor includes both the AcCh binding site and the structure of the ion channel. Progress in the biochemical characterization of the structure of the AcChR has been possible because nature has provided an unusual tissue highly enriched in AcChR and molecules (toxins) that bind to the AcChR with high affinity and selectivity. The AcChR constitutes about 1% of the protein in the electric organs of the marine ray *Torpedo* (*321*), and kilogram quantities of tissue are available that easily yield 50 mg of pure AcChR. The principal polypeptide neurotoxins in the venom of elapid snakes, i.e., cobras, bind in a quasi-irreversible fashion to the AcCh binding site itself (*312*). For example, α-bungarotoxin (α-BgTx, MW ~ 8000) constitutes ~ 30% of the dry weight of the venom of *Bungarus multicinctus,* the Formosan krait, and has been used to quantify the distribution of AcChR in vertebrate skeletal muscle and in electric tissue. In both tissues receptors are present at a density of ~ $10,000/\mu m^2$ in the region of plasma membrane just underlying the nerve terminal (*322–324*). Since that density is equivalent to one AcChR for every 100 Å^2, the nicotinic postsynaptic membrane is a highly specialized structure.

The protein binding AcCh, tubocurarine, and α-BgTx is an integral membrane protein that can be purified from detergent extracts of *Torpedo* electric tissue. In detergent, the purified AcCh binding protein exists as a complex macromolecule (MW $\sim$ 250,000) containing two AcCh binding sites. That macromolecule is composed of four distinct polypeptides of MW 40,000 (α), 50,000 (β), 60,000 (γ), and 65,000 (δ), with a relative stoichiometry of $\alpha_2\beta\gamma\delta$ (for review see 312, 10). In this complex structure, the AcCh binding site is contained (at least in part) within the α subunit, since that is the subunit labeled by covalent affinity labels directed to the AcCh binding site (325, 326). It should be appreciated that these results define the AcCh binding protein and do not establish whether that protein also contains the ion channel. Nevertheless experiments to be described below establish that the AcCh binding protein also contains the ion channel.

Additional information about receptor structure and function is available because it is possible to isolate (327) the highly specialized nicotinic postsynaptic membrane from *Torpedo* electric tissue. Analysis of the polypeptide composition of those membranes reveals that they contain, besides the AcChR, only one additional polypeptide of MW 43,000, the "43 K protein" (328–330). When examined by negative electron microscopy, the isolated membranes appear as a population of vesicles and membrane fragments with dimensions ranging from 0.1 μm to 1 μm. The individual vesicles are covered (331) at a density of 10,000/μm^2 with particles 8 nm in diameter that appear as rosettes or "doughnuts." Those particles are the AcChR itself, since they are labeled by antibodies directed against the purified AcChR (332, 333). The available data provide a dramatic, low resolution definition of receptor structure: Each individual AcChR protrudes 55 Å beyond the lipid bilayer on the extracellular side and it contains a central aqueous pit (Fig. 7). Hopefully within the next years, it will be possible to relate the observed rosette structure to the packing of individual receptor subunits and to identify AcCh binding sites as well as the structures of the ion channel.

3. Histrionicotoxin and the AcChR Ionophore

An open ion channel is, of course, little more than a hole and as such it is difficult to identify biochemically. However, if ligands that bind to the ion channel could be identified, such ligands would provide a mechanism to identify the channel structure. Is the ion channel of the AcChR contained wholly within the complex protein ($\alpha_2\beta\gamma\delta$) visualized in the electron microscope? Does one subunit function as the ion channel? We do not have answers to these questions, but histrionicotoxins (HTX) have provided important clues.

HTX is a potent nicotinic noncompetitive antagonist, i.e., a ligand that antagonizes the response to AcCh without blocking the binding of AcCh. The suggestion to test histrionicotoxin on the nicotinic receptor came from the exact knowledge of the spatial arrangement of the key functional groups,

FIG. 7. A schematic representation of the subsynaptic membrane. The closely packed receptor molecules in the disulfide-bonded heavy form are cross-linked by the peripheral, histrionicotoxin-binding 43,000 dalton polypeptide on their internal cytoplasmic face and by components of the basal lamina on their cleft face (*351, 10*).

the secondary amine and secondary alcohol functions (Fig. 8, *149*). The distance between the two groups is 2.762 Å in HTX which, at micromolar concentrations, blocks the AcCh-induced depolarization of vertebrate skeletal muscle (*315–317, 334*) or of electric organ (*335*), but concentrations of HTX greater than 0.1 mM are necessary to prevent the binding of ^{125}I-labeled α-BgTx to skeletal muscle (*316*) or of ^{3}H-labeled AcCh to *Torpedo* membranes (*335, 336*).

Histrionicotoxins in frog nerve–muscle preparations indirectly block elicited twitch and depress both peak amplitudes and decay time constants of end-plate potentials. In addition, they decrease the rate of rise and prolong the falling phase of muscle action potential. The histrionicotoxins act at three membrane channels: the one associated with the acetylcholine receptor, the sodium channel, and the potassium channel. Blockade at all three channels is dependent on stimulus and concentration, but the rank order of potencies of histrionicotoxins for each of three channels differs: Hypothetically, on a scale

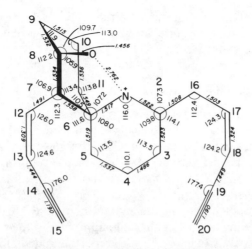

FIG. 8. The bond lengths and angles of histrionicotoxin (*149*) connect the spatial relationship of the secondary amino and alcohol groups with the agonist acetylcholine, an indication that led to the discovery of the inhibitory activity of HTX on ion flux in the acetylcholine receptor channel.

of potency with 10 at the top, the affinity for the potassium channel would rank 10, the sodium channel 4, and the AcChR ionophore about 8.

The fact that HTX is a nicotinic noncompetitive antagonist does not guarantee that it acts by binding to a specific site, let alone the ion channel itself. Many compounds, including aromatic amines, fatty acids, and detergents act as nicotinic noncompetitive antagonists. Those compounds have in common the fact that they act as general membrane stabilizers (reviewed in *337*), and it is possible that they act at the nicotinic synapse as elsewhere in this manner. In initial studies of [³H]HTX binding to skeletal muscle (*315*), no evidence of a specific binding site was found. However, the isolated *Torpedo* membranes provide a more favorable preparation for the analysis of ligand binding and with that preparation, Albuquerque and co-workers demonstrated (*338, 339,* see also *340*) that [³H]HTX does bind to a specific site distinct from the site of binding of AcCh and of α-BgTx. While [³H]HTX was displaced neither by tubocurarine nor by AcCh, it was displaced by a variety of aromatic, secondary, and tertiary amines, and this binding site is thus referred to as an *anesthetic binding site.* Binding to this site can be measured not only by the use of [³H]HTX, but also by synthetic aromatic amines, such as [³H]meproadifen (*341*) and [³H]phencyclidine (*342*), although their molecular target on the ionic channels appears to be different (*343–345*). Those ligands potentially provide a mechanism to identify the channel structure. Initial characterization of the HTX site has focused more on structure – activity relations than on the precise definition of the number of sites, but recent results indicate one HTX site per two AcCh sites, i.e., one HTX site per receptor (*346*).

4. Topography of the Receptor

Once it was established that there was a distinct HTX site, efforts were made to determine whether the site was contained within the AcChR itself. Initial studies analyzed the binding of HTX after the proteins of the *Torpedo* membranes had been fractionated in detergent solutions. Evidence was presented that the HTX binding site was separable from the AcChR (*339*) and was in fact the nonreceptor MW 43,000 protein (*329*). However, the use of detergents is problematic since detergents themselves are potent inhibitors of [³H]HTX binding (*347*). An alternate approach to the problem avoided the use of detergents. Cohen *et al.* (*348*) reported that alkaline pH could be used to extract the nonreceptor MW 43,000 protein from the *Torpedo* membranes. After alkaline extraction, the only protein in the *Torpedo* vesicles was the AcChR itself and the binding of [³H]meproadifen and HTX to those membranes was unaffected by removal of the MW 43,000 protein. In addition, the binding of agonists resulted in the normal permeability response. The conclusions based upon alkaline extraction have been confirmed in many laboratories (*349, 350*), and it is now recognized that the HTX site is associated with the AcChR itself. The nonreceptor MW 43,000 protein, though possibly coupled with the receptor, has recently been located, presumably on the *intracellular* surface of the postsynaptic membrane, while the AcChR polypeptides are exposed on the *extracellular* surface (*351*). A model (Fig. 7) for this topography is shown in the Harvey Lecture delivered by Changeux (*10*). Recently, Saitoh *et al.* (*352*) introduced an arylazide photoaffinity label for the anesthetic site. On photolysis, [³H]azidotrimethisoquin was incorporated into the δ subunit of the AcChR, a reaction that was blocked by HTX and other amine noncompetitive antagonists. A schematic representation of one possible arrangement of the peptide chains of *Torpedo* receptors in an on-face view is presented in Fig. 9 (*353*). The two monomers are cross-linked through a disulfide bridge between two δ chains. The chains are arranged around a central pit, possibly the entrance to the ion channel common for the transport of both sodium and potassium ions. Each α chain contains a 1-nm long acetylcholine binding site with a negatively charged group and disulfide susceptible to reduction. The toxin binding sites (arrows) are both in the α-chains in identical locations.

Figure 10 (*354*) shows the anchoring of the receptor in the lipid bilayer of the membranes. Such experiments establish the utility of HTX: it remains

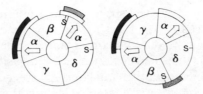

FIG. 9. Hypothetical arrangements of the receptor chains around the central pit (possibly identical with the entrance to the ion channel (*353*). The sulfur (S) involved in the $\delta-\delta$ disulfide, the sulfur involved in the $\beta-\beta$ disulfide, and the two bungaro or cobra toxin-binding sites (arrows) are located in their respective chains to satisfy as closely as possible the following conditions: (a) that the first toxin-binding site should fall within the range of 45–85° of the $\delta-\delta$ disulfide; (b) that the second toxin-binding site should fall in the range of 120–180° of the $\delta-\delta$ disulfide (these ranges are represented by unfilled arcs); (c) the angle between the two toxin sites should fall in the range of 80–140° (filled arc, measured from first toxin site); (d) the range of the angle between the $\delta-\delta$ disulfide and the $\beta-\beta$ disulfide ($\beta-\delta$ angle) is 50–80° (striped arc, measured from the $\delta-\delta$ disulfide); (e) in addition, the toxin sites are assumed to be in identical locations in the two α chains; hence, the angle between them is the sum of the angle subtended by an α chain and the angle subtended by the intervening chain. It has not been settled whether histrionicotoxin has its binding site in the δ or in the α subunit. Photoaffinity labeling with [³H]azodotrimethisoquin leads to incorporation in the δ subunit (*352*), while binding studies with tritiated anesthetics point to binding of the α site, a result that would not exclude the exciting possibility of an HTX-binding site at the interface between subunits (J. Cohen, private communication).

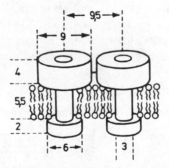

FIG. 10. Hypothetical Model of the dimeric acetylcholine receptor embedded in a membrane of 5.5-nm thickness. The center-to-center distance between the two receptor monomers is shown as 9.5 nm. Each monomer (250,000 daltons) is likely to contain a membrane-spanning, cation-conducting channel which is under the control of the monomer's two binding sites for acetylcholine. Histrionicotoxin has one binding site for two acetylcholine sites, i.e., one HTX site per receptor monomer (*346, 354*).

the ligand binding with highest specificity and affinity to the anesthetic binding site of the AcChR, and it is structurally distinct from the aromatic amine compounds that bind to the same site.

Now that it is clear that there is a specific binding site in the AcChR for HTX and amine noncompetitive antagonists, it is certainly important to determine whether that site is contained within the ion channel itself. Electrophysiological experiments provide results compatible with a channel-blocking site for HTX (317) as well as for aromatic amines (355, 356), but at this time prudent restraint is advisable. Analysis of receptor function in Torpedo vesicles (357) establishes that the receptor must be characterized by four different functional conformations, only one of which is ion transporting. Different functional states are characterized by different binding affinities for cholinergic agonists (358) and also for [³H]HTX (359). Further work will be necessary to determine whether HTX interacts with the structure of the ion channel itself or with a distinct regulatory site (344). In either case, nature has provided us with a powerful tool to probe the structure and function of the AcChR beyond the AcCh binding site itself.

C. PRELIMINARY PHARMACOLOGY

1. Gephyrotoxin, a Muscarinic Antagonist

The scarcity of the natural gephyrotoxins thus far has prevented systematic pharmacological evaluation. Judging from its antagonism to the action of acetylcholine on guinea pig ileum and its reversal of negative inotropic effects caused by acetycholine in guinea pig ilium, gephyrotoxin appears to be a muscarinic antagonist approximately ten times weaker than atropine (151). In frog sartorius muscle gephyrotoxin has been shown to suppress sodium and potassium conductances and to react with the ionic channel of the nicitonic receptor in its closed (activated but nonconducting) and open conformations (360).

2. Pumiliotoxin C, a Neuromuscular Blocking Agent

Pumiliotoxin C and various synthetic analogs inhibit indirectly elicited twitches in the rat phrenic nerve diaphragm preparation. At 40 μM, the most potent compound 2,5-di-n-propyl-cis-decahydroquinoline (PTX-C$_{II}$) blocks the indirectly elicited twitch by 50% in 45 min. The blockade is readily reversible. At 40 μM, PTX-C$_{II}$ has no effect on muscle membrane potential, but markedly depresses the amplitude of endplate potentials and the amplitude and frequency of miniature endplate potentials. In α-bungarotoxin-treated preparations, PTX-C$_{II}$ does not inhibit the directly elicited twitch in chronically (10 day) denervated rat soleus muscle. Binding of ^{125}I-labeled α-bungarotoxin to acetylcholine receptors in *Torpedo* electroplax membranes is inhibited by PTX-C$_{II}$ with a potency comparable to that of *d*-tubocurarine. The pumiliotoxin C class of alkaloids appears to block neuromuscular transmission primarily via interaction with the acetylcholine receptor rather than with the associated ionic channel (*361*).

3. Pumiliotoxins A and B: Cardioactive Inotropic Agents and Specific Inhibitors of Calcium-Dependent ATPase:

As was shown in the preceding section, pumilitoxins A, B, and 251D are all novel derivatives sharing the indolizidine skeleton. They have been studied in several systems. Pumiliotoxin B in skeletal muscle of rat, frog, and crayfish reversibly potentiates and prolongs the directly elicited muscle twitch up to 12-fold in a concentration- and frequency-dependent manner. In crayfish skeletal muscle, PTX-B increases the rate of rise of the "calcium-dependent" action potential and shortens its duration. PTX-B inhibits calcium-dependent adenosinetriphosphatase from sarcoplasmic reticulcum preparations. The potentiation and prolongation of the muscle twitch is assumed to be due to (1) a facilitation of release of calcium from storage sites within the sarcoplasmic reticulum, (2) mobilization of calcium from extracellular sites, and (3) blockage of the re-uptake of calcium by calcium-dependent adenosinetriphosphatase (*362, 363*).

Pumiliotoxin A is a much less potent inhibitor of Ca^{2+}-ATPase in sarcoplasmic reticulum vesicles from frog and rat hind-limb muscles than pumiliotoxin B. PTX-251D is not only a potent inhibitor of Ca^{2+}-ATPase, but also of Na$^+$-, K$^+$-, and Mg^{2+}-ATPases in rat brain synaptosome. Based on structure–activity profiles, inhibition of Ca^{2+}-ATPase by the indolizidine alkaloids presumably explains their prolongation of twitch in intact muscle.

4. Other Pharmacological Activities

Although the paucity of natural frog toxins has delayed pharmacological evaluation *in vivo,* synthetic racemic 2-depentyl perhydrohistrionicotoxin and its *N*-methyl homolog were tested in mice in which the LD$_{50}$ was found to be above 100 mg/kg (*166*). In the hot-plate assay neither compound possessed notable antinociceptive activity (*166*), had little effect on blood pressure and, in the Lipschitz assay, exhibited some diuretic effect with a concomitant inhibition of carbonic anhydrase (tests carried out at the Hoechst Research Laboratories, Frankfurt, Germany).

The Ecuadorian frog *Dendrobates tricolor* contains an alkaloid with powerful analgesic action. The Colombian frog *Dendrobates opistomelus* is notable for the contents of alkaloids that are much more toxic than any other congeners isolated so far from the genus *Dendrobates*. One particular population of *Dendrobates histrionicus* stands out for an agent capable of marked sympathetic activation (*151, 107*).

V. Epilogue

As mentioned in the introduction, the concept of receptors has been advanced considerably by the discovery of amphibian toxins whose unexpected and unprecedented structures could not have been foreseen.

Receptors that are membrane-bound, such as the nicotinic acetylcholine receptor and receptors or allosteric sites belonging to ion channels, are much harder to define than receptors for substrates such as glucose, oxygen, and hormonal peptides for which Roentgen-ray crystallography has provided precise expressions of the enzyme–substrate complex (e.g., with hexokinase, hemoglobin, or carboxypeptidase, respectively).

Hypothetical pictures of complicated membrane-bound receptors may be ahead of their time, but in this respect recall Ehrlich's side-chain models. He knew their limitation and, as a chemist, recognized their temporary and auxiliary nature. But as eidetic stepping stones, they were extremely useful in the long climb to exact presentations based primarily on Roentgen-ray crystallography.

"Why still pursue the study of natural products?" (*364*) is a question that has been asked and answered in several languages (*365–370*). This chapter covering 20 years of interdisciplinary investigations, should lay to rest any doubts about the value, timeliness, and significance of research on natural products.

Acknowledgements

The authors are grateful to the following colleagues for informative letters, unpublished manuscripts, résumés, and special points of view: Edson X. Albuquerque, Department of Pharmacology and Experimental Therapeutics, University of Maryland School of Medicine; William A. Catterall, Department of Pharmacology, University of Washington School of Medicine; Jonathan B. Cohen, Department of Pharmacology, Harvard Medical School; Stanley C. Froehner, Department of Biochemistry, Dartmouth Medical School; David J. Hart, Department of Chemistry, Ohio State University; Arthur Karlin, Department of Biochemistry and Neurology, College of Physicians and Surgeons, Columbia University; Boris I. Khodorov, Vishnevsky Surgery Institute, Moscow; Edward M. Kosower, Department of Chemistry, Tel-Aviv University; Harry Mosher, Stanford University, Larry Overman, Department of Chemistry, University of California, Irvine; Takashi Tokuyama, Osaka City University, Osaka.

REFERENCES

1. M. Latham, *National Geographic Magazine* **129**, 683 (1966).
2. Claude Bernard, "Leçon sur les effets des substances toxiques et médicamenteuses." Baillère, Paris, 1857.
3. N. Shaker, A. T. Eldefrawi, E. X. Albuquerque, and M. D. Eldefrawi, *J. Pharm. Exp. Therap.* **220**, 172 (1982).
4. E. X. Albuquerque, M. Adler, C. E. Spivak, and L. Aguayo, *Annals N. Y. Acad. Sci.* **358**, 204 (1980).
5. T. Narahashi, *Physiol. Reviews* **54**, 813–889 (1974). An extensive review with 671 references.
6. B. Katz, "Nerve, Muscle and Synapse." McGraw Hill, New York, 1966.
7. B. Witkop, *Naturwiss. Rundsch.* **34**, 361 (1981).
8. B. Witkop, *in* "Natural Products, Receptors, and Ligands; Natural Products as Medicinal Agents" (J. L. Beal, and E. Reinhard, eds.) pp. 151–184. Hippokrates Verlag, Stuttgart, 1981.
9. B. Ceccarelli, and F. Clementi, "Neurotoxins: Tools in Neurobiology Advances in Cytopharmacology," Vol. 3. Raven Press, New York, 1979.
10. J.-P. Changeux, *Harvey Lect.* **75**, 85–254. This review carefully explores the historical aspects and interconnections.
11. A. Karlin, "Molecular Properties of Nicotinic Acetylcholine Receptors, Cell Surface Reviews" (G. Poste, G. N. Nicolson, and C. E. Cotman, eds.) Vol. 6. Elsevier N. Holland, New York, 1980.
12. B. Witkop, *Angew. Chem. Int. Ed. Eng.* **16**, 559 (1977).
13. B. Witkop, *Angew. Chem.* **55**, 85 (1942). The classical monograph in this field is L. Lewin (1923) Die Pfeilgifte, Leipzig.
14. F. Märki, and B. Witkop, *Experientia* **19**, 329 (1963).
15. H. V. Wyatt, *Nature (London)* **235**, 86 (1972).
16. C. Phisalix, and G. Bertrand, *Compt. Rend.* **116**, 1080 (1893).
17. C. Phisalix, and G. Bertrand, *Compt. Rend.* **135**, 46 (1902).
18. B. Lutz, *in* "Venomous Animals and Their Venoms" (W. Bücherl, E. E. Buckley, eds.) Vol. II, p. 423. Academic Press, New York, 1971.
19. G. Habermehl, *Naturwissenschaften* **56**, 615 (1969).
20. H. J. Preusser, G. Habermehl, M. Sablofski, and D. Schmall-Haury, *Toxicon* **13**, 285 (1975); *in* "Animal, Plant and Microbial Toxins" (A. Ohsaka, K. Hayashi, and Y. Sawai eds.), Vol. I, p. 273. Plenum Press, N. Y., 1976.
21. J. Flier, M. W. Edwards, J. W. Daly, and C. W. Myers, *Science* **208**, 503 (1980).
22. M. Roseghini, V. Erspamer, and R. Endean, *Comp. Biochem. Physiol.* **54C**, 31 (1976).
23. V. Erspamer, *Experientia* **18**, 562 (1962).
24. P. C. Montecucchi, R. De Castiglione, S. Piani, L. Gozzini, and V. Erspamer, *J. Peptide Protein Res.* **17**, 275–283 (1981).
25. V. Erspamer, and P. Melchiorri, *in* "Growth Hormone and Other Biologically Active Peptides" (A. Pecile and E. E. Miller, eds.) pp. 185–200. Excerpta Medica, Amsterdam, 1980.
26. V. Erspamer, P. Melchiorri, *Trends Pharmacol. Sci.* **1**, 391 (1980) and references cited.
27. T. Akizawa, K. Yamasaki, T. Yasuhara, T. Nakajima, M. Roseghini, G. F. Erspamer, and V. Erspamer, *Biomed. Res.* **3**, 232 (1982).
28. K. Mayer, and H. Linde, *in* "Venomous Animals and Their Venoms" (W. Bücherl, E. E. Buckley, eds.), Vol. II, p. 521. Academic Press, New York, 1971.
29. P. G. Marshall, *in* "Rodds Chemistry of Carbon Compounds" (S. Coffey, eds.), Vol. IID, p. 412. Elsevier Amsterdam, 1970.

30. G. R. Pettit, and Y. Kamano, *J. Chem. Soc., Chem. Comm.* **45** (1972).
31. K. Nakanishi, *in* "Natural Products Chemistry" (K. Nakanishi, T. Goto, S. Ito, S. Natori, S. Nozoe, eds.), Vol. I, p. 469. Academic Press, New York (1972).
32. M. D. Siperstein, A. W. Murray, and E. Titus, *Arch. Biochem. Biophys.* **67,** 154 (1957).
32a. T. Y. R. Tsai and K. Wiesner, *Can. J. Chem.* **60** (16), 2161–2163 (1982).
32b. K. Wiesner, T. Y. R. Tsai, F. J. Jaggi, C. S. J. Tsai, and G. D. Gray, *Helv. Chim. Acta* **65** (7), 2049 (1982).
33. H. Wieland, and R. Alles, *Ber.* **55,** 1789 (1922).
34. V. Deulofeu, and E. A. Rùveda, *in* "Venomous Animals and Their Venoms," (W. Bücherl, E. E. Buckley, eds.), Vol. II, p. 475. Academic Press, New York (1971).
35. J. Daly, and B. Witkop, *in* "Venomous Animals and Their Venoms," (W. Bücherl, E. E. Buckley, eds.), Vol. II, p. 497. Academic Press, New York (1971).
36. M. Roseghini, R. Endean, and A. Temperilli, *Z. Naturforsch. C.: Biosci.* **31,** 118 (1976).
37. F. Märki, A. V. Robertson, and B. Witkop, *J. Am. Chem. Soc.* **83,** 3341–3342 (1961).
38. F. Märki, J. Axelrod, and B. Witkop, *Biochem. Biophys. Acta* **58,** 367–369 (1962).
39. J. W. Daly, and B. Witkop, *Angew. Chem. Int. Ed. Eng.* **2,** 421–440 (1963).
40. A. Brossi, *in* "Proceedings Workshop on Beta-Carbolines and Tetrahydro-isoquinolines" (F. Bloom, J. Barhas, M. Sander, E. Usdin, eds.). New York, 1982.
41. C. Melchior, and M. A. Collins, *CRC Crit. Rev. Toxicol.* **9,** 313–356 (1982).
42. N. S. Buckholtz, *Life Sci.* **27,** 893 (1980); H. Honecker, H. Rommelspacher, *Naunyn-Schmiedeberg's Arch. Pharmacol.* **305,** 135 (1978); H. Rommelspacher, H. Honecker, M. Barbay, and B. Meicke, Ibidem, **310,** 35 (1979); D. W. Shoemaker, J. T. Cummins, and T. G. Bidder, *Neuroscience* **3,** 233.
43. Y. Hashimoto and T. Noguchi, *Toxicon* **9,** 79 (1971); T. Noguchi and Y. Hashimoto, *Toxicon* **11,** 305 (1973).
44. M. Shimizu, M. Ishikawa, Y. Komoda, and T. Nakajima, *Chem. Pharm. Bull.* **30**(3), 909; M. Shimizu, M. Ishibawa, Y. Komoda, T. Nakajima, K. Yamaguchi, and S.-I. Sakai, *Chem. Pharm. Bull.* **30,** 3453 (1982).
45. T. Goto, *in* "Natural Products Chemistry" (K. Nakanishi, T. Goto, S. Ito, S. Natori, S. Nozoe, eds.), Vol. II, p. 457, 461. Academic Press, New York, 1975.
46. D. Scheumack, M. E. H. Howden, I. Spence, R. J. Quinn, *Science* **199,** 188 (1978).
47. T. Noguchi, J. Maruyama, Y. Ueda, K. Hashimoto, and T. Harata, *Nippon Suisan Gakkaishi* **47,** 909 (1981); *CA* **95,** 145045.
48. T. Kosuge, H. Zenda, A. Ochai, N. Masaki, M. Noguchi, S. Kimura, and H. Narita, *Tetrahedron Lett.* 2545 (1972).
49. H. Narita, T. Noguchi, J. Maruyama, Y. Ueda, K. Hashimoto, Y. Watanabe, K. Hida, *Nippon Suisan Gakkaishi* **47,** 935 (1981); *CA* **95,** 145046r.
50. K. Tsuda, *Naturwissenschaften* **53,** 171 (1966).
51. R. B. Woodward, *Pure and Applied Chem.* **9,** 49 (1964).
52. T. Goto, Y. Kishi, S. Takahashi, and Y. Hirata, *Tetrahedron* **21,** 2059 (1965).
53. P. J. Scheuer, "Chemistry of Marine Natural Products," p. 149. Academic Press, New York, 1973.
54. L. Chevolet, *in* "Marine Natural Products, Chemical and Biological Perspectives" (P. J. Scheuer, ed.), Vol. IV, p. 78. Academic Press, New York, 1981.
55. T. Goto, and Y. Kishi, "Tennenbutsu Yukikagaku" (H. Yoshimasa, ed.), p. 29. Iwanami, Shoten, Tokyo, 1981.
56. Y. Kishi, F. Nakatsubo, M. Aratani, T. Goto, S. Inoue, H. Kakoi, *Tetrahedron Lett.* 5127, 5129 (1970).
57. Y. Kishi, M. Aratani, T. Fukuyama, F. Nakatsubo. T. Goto, S. Inoue, H. Tanino, S. Sugiura, and H. Kakoi, *J. Am. Chem. Soc.* **94,** 9217 (1972).

58. Y. Kishi, T. Fukuyama, M. Aratani, F. Nakatsubo. T. Goto, S. Inoue, H. Tanino, S. Sugiura, and H. Kakoi, *J. Am. Chem. Soc.* **94,** 9129 (1972).
59. J. F. W. Keana, J. S. Bland, P. E. Eckler, V. Nelson, and J. Z. Gougoutas, *J. Org. Chem.* **41,** 2124 (1976).
60. J. F. W. Keana, and P. E. Eckler, *J. Org. Chem.* **41,** 2850 (1976).
61. E. Viera, Jr., Ph.D. Thesis, Harvard University, 1969.
62. R. D. Sitrin, Ph.D. Thesis, Harvard University, 1972.
63. J. Upeslacis, Ph.D. Thesis, Harvard University, 1975.
64. M. Funabashi, K. Kobagashi, J. Yoshimura, *J. Org. Chem.* **44,** 1618.
65. R. J. Nachman, Ph.D. Thesis, Stanford University, 1981.
66. B. W. Halstead, *in* "Poisonous and Venomous Marine Animals of the World," Vol. 2, U.S. Government Printing Office, Washington, D.C., 1967.
67. J. M. Ritchie, and R. B. Rogard, *Rev. Physiol. Biochem. Pharmacol.* **79,** 1 (1977).
68. J. M. Ritchie, *in* "Cell Membrane Receptors for Drugs and Hormones: A Multidisciplinary Approach" (R. W. Straub and L. Bolis, eds.), p. 227. Raven Press, New York, 1978.
69. M. S. Brown, and H. S. Mosher, *Science* **140,** 295.
70. H. D. Buchwald, L. Durham, H. G. Fisher, R. Harada, H. S. Mosher, C. Y. Kao, and F. A. Fuhrman, *Science* **143,** 474 (1964).
71. H. S. Mosher, F. A. Fuhrman, H. D. Buchwald, and H. G. Fischer, *Science* **144,** 1100 (1964).
72. R. J. Down, *in* "Animal, Plant and Microbial Toxins" (A. Ohsaka, K. Hayashi, Y. Sawai, eds.), Vol. II, p. 533. Plenum Press, New York, 1976.
73. Y. H. Kim, G. B. Brown, H. S. Mosher, and F. A. Fuhrman, *Toxicon* **15,** 135 (1975).
74. L. A. Pavelka, Y. H. Kim, and H. S. Mosher, *Toxicon* **15,** 135 (1977).
75. L. A. Pavelka, Ph.D. Thesis, Stanford University, Part I, 1980.
76. C. Y. Kao, P. N. Yeoh, F. A. Fuhrman, and H. S. Mosher, *J. Pharm. Exp. Therap.* **217,** 416 (1981).
77. C. Y. Kao, and P. N. Yeoh, *J. Physiol.* **272,** 54 (1977).
78. F. A. Fuhrman, and G. J. Fuhrman, *Science* **165,** 1376 (1969).
79. J. Shindelman, H. S. Mosher, and F. A. Fuhrman, *Toxicon* **7,** 315 (1969).
80. G. B. Brown, Y. H. Kim, H. Küntzel, H. S. Mosher, G. J. Fuhrman, and F. A. Fuhrman, *Toxicon* **15,** 115 (1977).
81. S. Hara, and K. Oka, Japan. Patent Nr. 7018, 663 (1970); *CA* .**3,** 56316g.
82. G. Habermehl, *Z. Naturforsch.* **20b,** 1129 (1965).
83. C. Schöpf, *Experientia* **17,** 285 (1961).
84. G. Habermehl, *Naturwissenschaften* **53,** 123 (1966).
85. G. Habermehl, *in* "The Alkaloids" (R. H. F. Manske, ed.), Vol. 9, p. 427. Academic Press, New York, 1967.
86. G. Habermehl, "Progress in Organic Chemistry," Vol. 7, p. 35. Butterworth, London, 1968.
87. G. Habermehl, *in* "Venomous Animals and Their Venoms" (W. Bücherl, E. E. Buckley, eds.), Vol. II, p. 569. Academic Press, New York, 1971.
88. G. Habermehl, and G. Vogel, *Toxicon* **7,** 163 (1969).
89. G. Habermehl, and A. Haaf, *Z. Naturforsch.* **23b,** 1551 (1968).
90. K. Oka, and S. Hara, *J. Am. Chem. Soc.* **99,** 3859 (1977).
91. K. Oka, and S. Hara, *J. Org. Chem.* **43,** 4408 (1978).
92. G. Habermehl, and A. Haaf, *Chem. Ber.* **101,** 198 (1968).
93. S. Hara, and K. Oka, *J. Am. Chem. Soc.* **89,** 1041 (1967).
94. S. Hara, and K. Oka, *Tetrahedron Lett.* 1987 (1969).
95. G. Habermehl, *Chem. Ber.* **96,** 2029 (1963).

96. G. Habermehl, *Chem. Ber.* **96**, 840 (1963).
97. G. Habermehl, and A. Haaf, *Ann. Chem.* **722**, 155 (1969).
98. R. B. Rao, and L. Weiler, *Tetrahedron Lett.* 4971 (1963).
99. K. Oka and S. Hara, *Tetrahedron Lett* 1193 (1969).
100. Y. Shimizu, *Tetrahedron Lett.* 2919 (1972).
101. M. H. Benn, and R. Shaw, *J. Chem. Soc., Chem. Comm.* 288 (1973).
102. M. H. Benn, and R. Shaw, *Canad. J. Chem.* **52**, 2936 (1974).
103. Y. Shimizu, *J. Org. Chem.* **41**, 1930 (1976).
104. K. Oka, and S. Hara, *Yuki Gosci. Kagaku Kyokaishi* **37**, 25 (1979).
105. K. Oka, *Yakukaku Zasshi* **100**, 227 (1980).
106. C. W. Myers, J. W. Daly, and B. Malkin, *Bull. Am. Mus. Nat. Hist.* **161**, 307 (1978); *cf.* C. W. Myers, and J. W. Daly, *Bull. Am. Mus. Nat. Hist.* **157**, 177 (1976).
107. J. W. Daly, *in Proceedings of 4th Asian Symposium on Medical Plants and Spices*, Bangkok, Thailand, 15–19 Sept, 1980, a UNESCO Special Publication, 1981.
108. T. Tokuyama, and J. W. Daly, *Tetrahedron*, in press (1982).
109. A. Posada Arango, *Ann. Acad. Med.* Medellin **1**, 69 (1888).
110. C. Saffray, *Le Tour du Monde, Nouveau Journal des Voyages* **26**, 2 (1872).
111. S. H. Wassèn, *Etnografiska Museet Göteborg, Årstryck* 1955–1956, **73**.
112. B. Witkop, *Experientia* **27**, 1121 (1971).
113. T. Tokuyama, J. W. Daly, and B. Witkop, *J. Am. Chem. Soc.* **91**, 3931 (1969).
114. J. W. Daly, B. Witkop, D. Bommer, and K. Biemann, *J. Am. Chem. Soc.* **87**, 124 (1965).
115. T. Tokuyama, J. W. Daly, B. Witkop, I. L. Karle, and J. Karle *J. Am. Chem. Soc.* **90**, 1917 (1968).
116. I. L. Karle, and J. Karle, *Acta Crystallog., Sec. B* **25**, 428 (1969).
117. R. D. Gilardi, *Acta Crystallog., Sec. B* **26**, 440 (1970).
118. D. B. Gower, and N. Ahmad, *Biochem. J.* **104**, 550 (1967).
119. D. F. Johnson, and J. W. Daly, *Biochem. Pharmac.* **20**, 2555 (1971).
120. M. Y. Lorbora, and E. Nurimov, *Izv. Akad. Nauk. SSSR., Ser. Biol.* **54E** (1978).
121. G. A. Cordell, and J. E. Saxton, "The Alkaloids" (R. Manske, ed.), Vol. 20, p. 10–16, p. 280, 282. Academic Press, New York, 1981.
122. J. E. Saxton, "The Alkaloids" (R. Manske, ed.), Vol. 8, p. 581. Academic Press, New York, 1965; R. H. F. Manske, "The Alkaloids" (R. Manske, ed.), Vol. 7. Academic Press, New York, 1960; L. Marion, "The Alkaloids" (R. Manske, ed.), Vol. 2., p. 369. Academic Press, New York, 1952.
123. S. Ito, *in* "Natural Products Chemistry" (K. Nabanishi, T. Goto, S. Ito, S. Natori, S. Nozoe, eds.), Vol. II. Academic Press, New York, 1975.
124. M. Lounasmaa, and A. Nemes, *Tetrahedron* **38**, 223 (1982).
125. R. Imhof, E. Gössinger, W. Graf, H. Berner, L. Berner-Fenz, and H. Wehrli, *Helv. Chim. Acta* **55**, 1151 (1972).
126. R. Imhof, E. Gössinger, W. Graf, L. Berner-Fenz, H. Berner, R. Schaufelberger, and H. Wehrli, *Helv. Chim. Acta* **56**, 139 (1973).
127. H. Berner, L. Berner-Fenz, R. Binder, W. Graf, T. Grütter, C. Pascual, and H. Wehrli, *Helv. Chim. Acta* **53**, 2252 (1970).
128. L. Berner-Fenz, H. Berner, W. Graf, and H. Wehrli, *Helv. Chim. Acta* **53**, 2258 (1970).
129. W. Graf, H. Berner, L. Berner-Fenz, E. Gössinger, R. Imhof, and H. Wehrli, *Helv. Chim. Acta* **53**, 2267 (1970).
130. R. Imhof, E. Gössinger, W. Graf, W. Schnüriger, and H. Wehrli, *Helv. Chim. Acta* **54**, 2775 (1971).
131. E. Gössinger, W. Graf, R. Imhof, and H. Wehrli, *Helv. Chim. Acta* **54**, 2785 (1971).
132. W. Graf, E. Gössinger, R. Imhof, and H. Wehrli, *Helv. Chim. Acta* **54**, 2789 (19710.

133. W. Graf, E. Gössinger, R. Imhoff, and H. Wehrli, *Helv. Chim. Acta.* **55**, 1545 (1972).
134. U. Kerb, H. D. Berndt, U. Eder, R. Wiechert, P. Buchschacher, A. Furlenmeier, A. Fürst, and M. Müller, *Experientia* **27**, 759 (1971).
135. R. R. Schumaker, and J. F. W. Keana, *J. Chem. Soc., Chem. Comm.* 622 (19720.
136. R. R. Schumaker, and J. F. W. Keana, *J. Org. Chem.* **41**, 3840 (1976).
137. H. L. Sham, Ph.D. Thesis, University of Hawaii, 1980.
138. E. Pfenninger, D. E. Poel, C. Berse, H. Wehrli, K. Schaffner, and O. Jeger, *Helv. Chim. Acta* **51**, 772 (1968).
139. W. Klyne, and D. H. R. Barton, *J. Am. Chem. Soc.* **71**, 1500 (1949).
140. Ch. Meystre, K. Heusler, J. Kalvoda, P. Wieland, G. Anner, and A. Wettstein, *Helv. Chim. Acta* **45**, 1317 (1962); K. Heusler, J. Kalvoda, P. Wieland, and A. Wettstein, *Helv. Chim. Acta* **44**, 179 (1961).
141. E. Caspi, and Y. Shimizu, *J. Org. Chem.* **30**, 823 (1965).
142. H. Heymann, and L. F. Fieser, *J. Am. Chem. Soc.* **73**, 4054, 5252 (1951).
143. H. Heymann, and L. F. Fieser, *J. Am. Chem. Soc.* **74**, 5938 (1952).
144. L. F. Fieser, S. Rajagopalan, *J. Am. Chem. Soc.* **73**, 118 (1951).
145. L. F. Fieser, S. Rajagopalan, E. Wilson, and H. Tishler, *J. Am. Chem. Soc.* **73**, 4133 (1951).
146. I. L. Karle, *Proc. Natl. Acad. Sci. U.S.A.* **69**, 2932 (1972).
147. T. Tokuyama, K. Uenoyama, G. Brown, J. W. Daly, and B. Witkop, *Helv. Chim. Acta* **57**, 2597 (1974).
148. J. W. Daly, I. L. Karle, C. W. Myers, T. Tokuyama, J. A. Waters, and B. Witkop, *Proc. Natl. Acad. Sci. U.S.A.* **68**, 1970 (1971).
149. I. L. Karle, *J. Am. Chem. Soc.* **95**, 4036 (1973).
150. J. W. Daly, B. Witkop, T. Tokuyama, T. Nishikawa, and I. L. Karle, *Helv. Chim. Acta* **60**, 1128 (1977).
151. J. W. Daly, G. B. Brown, M. Mensah-Dwumah, C. W. Myers, *Toxicon* **16**, 163 (1978).
152. E. Winterfeldt, *Heterocycles* **12**, 1631 (1979).
153. Mothes, K. "The Problem of Chemical Convergence in Secondary Metabolism, Science and Scientists" (M. Kageyama, K. Nakamura, T. Oshima, and T. Uchida, eds.) Japan Scientific Soc. Press, Tokyo (1981).
154. T. Fukuyama, L. V. Dunkerton, M. Aratani, and Y. Kishi, *J. Org. Chem.* **40**, 2011 (1975).
155. E. J. Corey, J. F. Arnett, and G. N. Widiger, *J. Am. Chem. Soc.* **97**, 430 (1975).
156. B. Renger, H.-O. Kalinowsky, and D. Seebach, *Chem. Ber.* **110**, 1871 (1977).
157. E. J. Corey, M. Petrzilka, and Y. Ueda, *Tetrahedron Lett.* 4343 (1975).
158. E. J. Corey, M. Petrzilka, and Y. Ueda, *Helv. Chim. Acta* **60**, 2294 (1977).
159. E. J. Corey, L. S. Melvin, Jr., and M. F. Haslanger, *Tetrahedron Lett.* 3117 (1975).
160. M. Aratani, L. V. Dunkerton, T. Fukuyama, Y. Kishi, H. Kakoi, S. Sugiura, and S. Inoue, *J. Org. Chem.* **40**, 2009 (1975).
161. C. Bischoff, and E. Schröder, *J. Prakt. Chem.* **314**, 891 (1972).
162. F. T. Bond, J. E. Stemke, and D. W. Powell, *Synthetic Comm.* **5**, 427 (1975).
163. W. Kissing, and B. Witkop, *Chem. Ber.* **108**, 1623 (1975).
164. E. J. Corey, and R. D. Balanson, *Heterocycles* **5**, 445 (1976).
165. M. Harris, D.-S. Grierson, and H. P. Husson, *Tetrahedron Lett.* 1551 (1981).
166. K. Takahashi, A. E. Jacobson, C.-P. Mak, B. Witkop, A. Brossi, E. X. Albuquerque, J. E. Warnick, M. A. Maleque, A. Barosa, and J. V. Silverton, *J. Med. Chem.* **25**, 919–925 (1982).
167. T. A. Bryson, and C. A. Wilson, *Synthetic Comm.* 521 (1976).
168. S. A. Godleski, J. D. Meinhart, J. D. Miller, and S. V. Wallendael, *Tetrahedron Lett.* 2247 (1981).
169. L. E. Overman, *Tetrahedron Lett.* 1149 (1975).

170. R. J. Cvetovich, *Diss. Abstract Int. B* **39** (8), 3837 (1979); R. J. Cvetovich, Ph.D thesis, Columbia University, New York, 1978.
171. T. Ibuka, Y. Mitsui, K. Hayashi, H. Minakata, and Y. Inubushi, *Tetrahedron Lett.* 4425 (1981).
172. T. Ibuka, Y. Ito, Y. Mori, T. Aoyama, and Y. Inubushi, *Synthetic Comm.* **7,** 131 (1977).
173. T. Ibuka, H. Minakata, Y. Mitsui, K. Kinoshita, Y. Kawami, and N. Kimura, *Tetrahedron Lett.* 4073 (1980); T. Ibuka, H. Minakata, Y. Mitsui, K. Kinoshita, and Y. Kawami, *J. Chem. Soc., Chem. Comm.* 1198 (1980).
174. T. Ibuka, H. Minakata, Y. Mitsui, E. Tabushi, T. Taga, and Y. Inubushi, *Chem. Lett.* 1409 (1981).
175. T. Ibuka, H. Minakata, Y. Mitsui, T. Taga, and Y. Inubushi, *Symp. Chem. Nat. Prod. 23rd, 1980,* p. 351–358.
176. J. J. Venit, and P. Magnus, *Tetrahedron Lett.* 4815 (1980).
177. E. J. Corey, Y. Ueda, and R. A. Ruden, *Tetrahedron Lett.* 4347 (1975).
178. W. N. Speckamp, "Stereoselective Synthesis of Natural Products—Workshop Conference Hoechst" (Bartmann and Winterfeldt, eds.) Excerpta Medica, Amsterdam, 1979.
179. J. Dijkink, and W. N. Speckamp, *Tetrahedron Lett.* 4047 (1975).
180. H. E. Schoemaker, and W. N. Speckamp, *Tetrahedron* **34,** 163 (1978).
181. H. E. Schoemaker and W. N. Speckamp, *Tetrahedron Lett.* 1515 (1978).
182. H. E. Schoemaker, and W. N. Speckamp, *Tetrahedron Lett.* 4841 (1978).
183. H. E. Schoemaker, and W. N. Speckamp, *Tetrahedron* **36,** 951 (1980).
184. D. A. Evans, and E. W. Thomas, *Tetrahedron. Lett.* 411 (1979).
185. D. A. Evans, E. W. Thomas, and R. E. Cherpeck, *J. Am. Chem. Soc.,* **104,** 3695 (1982).
186. K. Takahashi, B. Witkop, A. Brossi, M. A. Maleque, and E. X. Albuquerque, *Helv. Chim. Acta* **65,** 252 (1982).
187. G. Ohloff, C. Vial, F. Näf, M. Pawlak, *Helv. Chim. Acta.* **60,** 1161 (1977).
188. J. J. Tufariello, and E. J. Trybulski, *J. Org. Chem.* **39,** 3378 (1974).
189. E. Gössinger, R. Imhof, and H. Wehrli, *Helv. Chim. Acta* **58,** 96 (1975).
190. E. J. Corey, and R. A. Ruden, *Tetrahedron Lett.* 1493 (1973).
191. M. Ahmed, G. C. Barley, M. T. W. Hearn, E. R. H. Jones, V. Thaller, and J. A. Yates, *J. Chem. Soc., Perkin Trans 1* 1493 (1974).
192. A. B. Holmes, R. A. Raphael, and N. K. Wellard, *Tetrahedron Lett.* 1539 (1976).
193. A. O. King, N. Okukado, and E. Negishi, *J. Chem. Soc., Chem. Comm.* 683 (1977).
194. E. J. Corey, and C. Rücker, *Tetrahedron Lett.* 719 (1982).
195. Y. Yamakado, M. Ishiguro, N. Ikeda, and H. Yamamoto, *J. Am. Chem. Soc.* **103,** 5568 (1981).
196. E. J. Corey, and H. A. Kirst, *Tetrahedron Lett.* 5041 (1968).
197. Y. Inubushi, and T. Ibuka, *Heterocycles 17,* 507 (1982).
198. G. E. Keck, R. R. Webb, and J. B. Yates, *Tetrahedron* **37** 4007 (1981).
199. J. W. Daly, and C. W. Myers, *Science* **156,** 970 (1967).
200. J. W. Daly, T. Tokuyama, G. Habermehl, I. L. Karle, and B. Witkop, *Justus Liebigs Ann. Chem.* **729,** 198 (1969).
201. W. K. Kissing, Doctoral Dissertation, Darmstadt, 1972.
202. Y. Inubushi, and T. Ibuka, *Heterocycles* **8,** 683 (1977).
203. T. Ibuka, Y. Inubushi, I. Saji, K. Tanaka, and N. Masaki, *Tetrahedron Lett.* 323 (1975).
204. T. Ibuka, N. Masaki, I. Saji, K. Tanaka, and Y. Inubushi, *Chem. Pharm. Bull.* (Jap.) **23,** 2779 (1975).
205. W. Oppolzer, C. Fehr, and J. Warneke, *Helv. Chim. Acta* **60,** 48 (1977).
206. K. Hattori, Y. Matsumura, T. Miyazaki, K. Maruoka, and H. Yamamoto, *J. Am. Chem. Soc.* **103,** 7368 (1981).

207. A. M. El-Abbady, M. El-Ashry, and S. H. Doss, *Can. J. Chem.* **47**, 1483 (1969).
208. G. Habermehl, H. Andres, and B. Witkop, *Naturwissenschaften* **62**, 345 (1975).
209. G. Habermehl, H. Andres, K. Miyahara, B. Witkop, and J. W. Daly, *Justus Liebigs Ann.* 1577 (1976).
210. R. F. Parcell, and F. P. Hauck, *J. Org. Chem.* **28**, 3468 (1963).
211. G. Habermehl, and H. Andres, *Liebigs Ann. Chem.* 800 (1977); see also the correction of ref. 209 in footnote 2.
212. G. B. Bennett, and H. Minor, *J. Heterocyclic Chem.* **16**, 633 (1979).
213. T. Ibuka, Y. Mori, and Y. Inubushi, *Tetrahedron Lett.* 3169 (1976).
214. T. Ibuka, Y. Mori, and Y. Inubushi, *Chem. Pharm. Bull.* (Jap.) **26**, 2442 (1978).
215. T. Ibuka, Y. Mori, T. Aoyama, and Y. Inubushi, *Chem. Pharm. Bull.* (Jap.) **26**, 256 (1978).
216. W. Oppolzer, and E. Flaskamp, *Helv. Chim. Acta* **60**, 204 (1977).
217. W. Oppolzer, W. Fröstl, and H. P. Weber, *Helv. Chim. Acta* **58**, 593 (1975).
218. W. Oppolzer, and W. Fröstl, *Helv. Chim. Acta* **58**, 593 (1975).
219. W. Oppolzer, and W. Fröstl, *Helv. Chim. Acta* **58**, 587 (1975).
220. W. Oppolzer, L. Biber, and E. Francotte, *Tetrahedron Lett.* 981 (1979).
221. L. E. Overman, and L. A. Clizbe, *J. Am. Chem. Soc.* **98**, 2352, 8395 (1976).
222. L. E. Overman, G. F. Taylor, and P. J. Jessup, *Tetrahedron Lett.* 3089 (1976).
223. L. E. Overman, G. F. Taylor, C. B. Petty, and P. J. Jessup, *J. Org. Chem.* **43**, 2164 (1978).
224. L. E. Overman, G. F. Taylor, K. N. Houk, L. N. Domelsmith, *J. Am. Chem. Soc.* **100**, 3182 (1978).
225. L. E. Overman, and P. J. Jessup, *Tetrahedron Lett.* 1223 (1977).
226. L. E. Overman, and P. J. Jessup, *J. Am. Chem. Soc.* **100**, 5179 (1978).
227. (a) L. E. Overman, R. L. Freerks, C. B. Petty, L. A. Clizbe, R. K. Ono, G. F. Taylor, and P. J. Jessup, *J. Am. Chem. Soc.* **103**, 2816 (1981); (b) L. E. Overman, L. A. Clizbe, R. L. Freerks, and C. Marlowe, *J. Am. Chem. Soc.* **103**, 2807 (1981).
228. D. T. Witiak, K. Tomita, R. J. Patch, and S. J. Enna, *J. Med. Chem.* **24**, 788 (1981).
229. C. W. Myers, and J. W. Daly, *Am. Mus. Novit.* **2674**, 1–24 (1979).
230. C. W. Myers, J. W. Daly, *Am. Mus. Novit.* **2692**, 1–23 (1980).
231. J. W. Daly, T. Tokuyama, R. J. Highet, and I. L. Karle, *J. Am. Chem. Soc.* **102**, 830 (1980).
232. T. Tokuyama, K. Shimada, M. Uemura, and J. W. Daly, *Tetrahedron Lett.* 2121 (1982).
233. L. E. Overman, and R. J. McCready, *Tetrahedron Lett.* 2355 (1982).
234. T. Tokuyama, J. W. Daly, and R. J. Highet, *Tetrahedron* (1982), submitted.
235. J. B. Hendrickson, E. Braun-Killer, and G. A. Toczko, *Tetrahedron* **37**, Suppl. 1, 359 (1981).
236. L. E. Overman, and K. L. Bell, *J. Am. Soc. Chem.* **103**, 1851 (1981).
237. T. Tokuyama, K. Uenoyama, G. B. Brown, J. W. Daly, and B. Witkop, *Helv. Chim. Acta* **57**, 2597 (1974).
238. T. L. MacDonald, *J. Org. Chem.* **45**, 193 (1980).
239. T. F. Spande, J. W. Daly, D. J. Hart, Y.-M. Tsai, and T. L. MacDonald, *Experientia* **37**, 1242 (1981).
240. F. J. Ritter, I. E. M. Rotgans, E. Talman, P. E. J. Verwiel, and F. Stein, *Experientia* **29**, 530 (1973).
241. G. G. Habermehl, and O. Thurau, *Naturwissenchaften* **67**, 193 (1980).
242. L. E. Overman, and C. Fukaya, *J. Am. Chem. Soc.* **102**, 1454 (1980).
243. L. E. Overman, and R. L. Freerks, *J. Org. Chem.* **46**, 2833, (1981).
244. T. A. Geissman, and R. M. Horowitz, *J. Am. Chem. Soc.* **72**, 1518 (1950).
245. L. E. Overman, and M. Kakimoto, *J. Am. Chem. Soc.* **101**, 1310 (1979).
246. L. E. Overman, and T. Yokomatsu, *J. Org. Chem.* **45**, 5229 (1980).
247. L. E. Overman, M. Kakimoto, and J. Okawara, *Tetrahedron Lett.* 4041 (1979).

248. L. E. Overman, M. Sworin, L. S. Bass, and E. J. Clardy, *Tetrahedron* **37**, 4041 (1981).
249. L. Velluz, G. Vales, and J. Mathieu, *Angew. Chem., Intl. Ed.* **6**, 778 (1967).
250. L. Crombie, *J. Chem. Soc.* 1007 (1955).
251. D. J. Hart, *J. Am. Chem. Soc.* **102**, 397 (1980).
252. D. J. Hart, *J. Org. Chem.* **46**, 367 (1981).
253. D. J. Hart, *J. Org. Chem.* **46**, 3576 (1981).
254. D. J. Hart, and K. Kanai, personal communication.
255. R. Fujimoto, Y. Kishi, and J. F. Blount, *J. Am. Chem. Soc.* **102**, 7156 (1980).
256. M. M. Gugelchuk, D. J. Hart, and Y.-M. Tsai, *J. Org. Chem.* **46**, 3671 (1980).
257. D. J. S. Tsai, and D. S. Matteson, *Tetrahedron Lett.* 2751 (1981).
258. R. Fujimoto, and Y. Kishi, *Tetrahedron Lett.* 4197 (1981).
259. R. B. Silberman, and M. A. Levy, *J. Org. Chem.* **45**, 81 (1980).
260. T. S. Spande, unpublished results.
261. R. V. Stevens, and A. W. M. Lee, *J. Chem. Soc., Chem. Comm.* 103 (1982).
262. J. C. Armande, and U. K. Pandit, *Tetrahedron Lett.* 897 (1977).
263. R. R. Fraser, and S. Passananti, *Synthesis,* 540 (1976).
264. B. Witkop, *Heterocycles* **17**, 431 (1982).
265. A. J. Hodgkin and A. F. Huxley, *J. Physiol.* **117**, 500–544 (1952).
266. W. A. Catterall and C. S. Morrow, *Proc. Natl. Acad. Sci. U.S.A.* **75**, 218–222 (1978).
267. A. Sastre and T. R. Podleski, *Proc. Natl. Acad. Sci. U.S.A.* **73**, 1355–1359 (1976).
268. R. Ray, C. S. Morrow, and W. A. Catterall, *J. Biol. Chem.* **253**, 7307–7313 (1978).
269. D. T. Campbell and B. Hille, *J. Gen. Physiol.* **67**, 309–323 (1976).
270. B. Hille, *Ann. Rev. Physiol.* **83**, 139–152 (1976).
271. G. S. Shotzberger, E. X. Albuquerque, and J. W. Daly, *J. Pharmacol. Exp. Ther.* **196**, 433–444 (1976).
272. E. X. Albuquerque, J. W. Daly, and B. Witkop, *Science* **172**, 995–1002 (1972).
273. E. X. Albuquerque, and J. W. Daly, *Recept. Recognition* **1**, 299–336 (1976).
274. E. X. Albuquerque, and J. E. Warnick, *J. Pharmacol. Exp. Ther.* **180**, 683–697 (1972).
275. E. X. Albuquerque, M. Sasa, B. P. Avner, and J. W. Daly, *Nature (London) New Biol.* **234**, 93–95 (1971).
276. W. A. Catterall, *Ann. Rev. Pharmacol. Toxicol.* **20**, 15–43 (1980).
277. B. I. Khodorov, *Progr. Biophys. Mol. Biol.* **37**, 49–89 (1981).
278. M. Lazdunski, M. Balerna, J. Barhanin, R. Chicheportiche, M. Fosset, C. Frelin, Y. Jacques, A. Lombet, J. Pouyssegur, J. F. Renaud, G. Romey, H. Schweitz, and J. P. Vincent, *in* "Neurotransmitters and Their Receptors" (U. Z. Littauer, Y. Dudai, I. Silman, V. I. Teichberg and Z. Vogel, eds.), p. 511–530. Wiley, New York, 1980.
279. E. Jover, F. Courand, and H. Rochat, *Biochem. Biophys. Res. Comm.* **95**, 1607–1614 (1980).
280. W. A. Catterall, and W. A. Risk, *Mol. Pharmacol.* **19**, 345–348 (1980).
281. W. A. Catterall, *J. Biol. Chem.* **252**, 8669–8676 (1977).
282. E. Bartels-Bernal, T. L. Rosenberry, and J. W. Daly, *Proc. Natl. Acad. Sci. U.S.A.* **74**, 951–955 (1977).
283. B. I. Khodorov, *Membr. Transp. Processes* **2**, 153–174 (1978).
284. B. I. Khodorov, E. Peganov, S. Revenko, and L. Shiskova, *Brain Research* **84**, 541–546 (1975).
285. G. B. Brown, S. C. Tieszan, J. W. Daly, J. E. Warnick, and E. X. Albuquerque, *Cell Mol. Neurobiol.* **1**, 19–40 (1981).
286. W. A. Catterall, C. S. Morrow, J. W. Daly, and G. B. Brown, *J. Biol. Chem.* **256**, 8922–8927 (1981).
287. W. A. Catterall, *J. Biol. Chem.* **250**, 4053–4059 (1975).

288. W. A. Catterall, *Proc. Natl. Acad. Sci. USA* **72**, 1782–1786 (1975).
289. N. G. Walter, Skeletal Muscle Pharmacology. Glossary of Muscle Constituents and Chemical Research Tools with Reference to Muscle Physiology and Neuromuscular Disorders. Excerpta Medica, Amsterdam, 1981.
290. B. I. Khodorov, B. Neumcke, W. Schwartz, and R. Stämpfli, *Biochim. Biophys. Acta* **648**, 93–99 (1981).
291. T. Narahasi, E. X. Albuquerque, and T. Deguchi, *J. Gen. Physiol.* **58**, 54–70 (1971).
291a. F. N. Quandt, and T. Narahashi, *Proc. Natl. Acad. Sci. U.S.A.* **79**, 6732–6736 (1982).
292. B. I. Khodorov, Modification of Voltage-sensitive Sodium Channels by Batrachotoxins, *in* "Structure and Function in Excitable Cells," Plenum Publishing Co., New York, 1983.
293. L.-Y. M. Huang, N. Moran, and G. Ehrenstein, *Proc. Natl. Acad. Sci. U.S.A.* **79**, 2082–2085 (1982).
294. B. Hille, *J. Gen. Physiol.* **69**, 497–515 (1977).
295. E. Gross, and B. Witkop, *Biochemistry* **4**, 2495 (1965). (Contains earlier references on structure and synthesis).
296. D. W. Urry, *Int. Rev. Neurobiol.* **21**, 311–334 (1979).
297. D. W. Urry, K. U. Prasad, and T. L. Trapane, *Proc. Natl. Acad. Sci. U.S.A.* **79**, 390 (1982).
298. B. Hille, *Biopys. J.* **22**, 283–294 (1978).
299. F. J. Sigworth, and B. C. Spalding, *Nature (London)* **283**, 293–295 (1980).
300. J. W. Daly, C. W. Myers, J. E. Warnick, and E. X. Albuquerque, *Science* **208**, 1383–1385 (1980).
301. C. Y. Kao, and F. A. Fuhrman, *Toxicon* **5**, 25–34 (1963).
302. Y. H. Kim, G. B. Brown, H. S. Mosher, and F. A. Fuhrman, *Science* **189**, 151–152 (1975).
303. C. Y. Kao, *Fed. Proc., Fed. Am. Soc. Exp. Biol.* **40**, 30–35 (1981).
304. S. Stuesse, and N. L. Katz, *Am. J. Physiol.* **224**, 55–61 (1973).
305. D. J. Cram, R. C. Helgeson, L. R. Sousa, J. M. Timko, M. Newcomb, P. Moreau, F. de Jong, G. W. Gokel, D. H. Hoffman, L. A. Domeier, S. C. Peacock, K. Madan, and L. Kaplan, *Pure Appl. Chem.* **43**, 327–349 (1975).
306. E. M. Kosower, (1981). Single Group Rotation II. "A Structural and Mechanistic Model for Ionic Channels in Biomembranes", Presented at Symposium on Structure and Dynamics of Nucleic Acids and Proteins, Univ. Calif., San Diego, La Jolla, Calif., 1981.
307. G. Romey, and M. Lazdunski, *Nature (London)* **297**, 79 (1982).
308. J. N. Langley, *J. Physiol.* **36**, 347 (1907).
309. Z. M. Bacq, "Chemical Transmission of Nerve Impulses." Pergamon Press, New York, 1975.
310. B. Holmstedt, Pages from the History of Research on Cholinergic Mechanisms" (P. G. Waser, ed.), 1–21. Raven Press, New York, (1975).
311. B. Katz, and R. Miledi, *J. Physiol.* **224**, 665–699 (1972).
312. A. Karlin, "Cell Surface and Neuronal Function" (C. W. Cotman, G. Poste and G. L. Nicolson, eds.), p. 191–260. Elsevier N. Holland, New York, 1980.
313. D. Colquhoun, "The Receptors," Vol. 1, p. 93–142. Plenum Press, New York, 1979.
314. P. R. Adams, *J. Membrane Biol.* **58**, 161–174 (1981).
315. E. X. Albuquerque, E. A. Barnard, T. H. Chiu, A. J. Lapa, J. O. Dolly, S. E. Jansson, J. Daly, and B. Witkop, *Proc. Natl. Acad. Sci. U.S.A.* **70**, 949–953 (1973).
316. J. O. Dolly, E. X. Albuquerque, J. M. Sarvey B. Mallick, and E. A. Barnard, *Mol. Pharmacol.* **13**, 1–14 (1977).
317. L. M. Masukawa, and E. X. Albuquerque, *J. Gen. Physiol.* **72**, 351–367 (1978).
318. E. Neher, and B. Sakmann, *Nature (London)* **260**, 799–801 (1976).
319. B. Katz, and S. Thesleff, *J. Physiol.* **138**, 63–80 (1957).
320. H. P. Rang, and J. M. Ritter, *Mol. Pharmacol.* **6**, 357–382 (1970).

321. A. Fessard, "Traité de Zoologie" (P. Grassé, ed.), Vol. 13A, 1143–1238. Masson, Paris, 1958.
322. J. P. Bourgeois, A. Ryter, A. Menez, P. Fromageot, P. Boquet, and J. P. Changeux, *FEBS Letters* **25**, 127–133 (1972).
323. H. C. Fertuck, and M. M. Salpeter, *J. Cell Biol.* **69**, 144–158 (1976).
324. E. X. Albuquerque, E. A. Barnard, C. W. Porter, and J. E. Warnick, *Proc. Natl. Acad. Sci. U.S.A.* **71**, 2818–2822 (1974).
325. C. L. Weill, M. G. McNamee, and A. Karlin, *Biochem. Biophys. Res. Commun.* **61**, 997–1003 (1974).
326. G. Weiland, D. Frisman, and P. Taylor, *Mol. Pharmacol.* **15**, 213–226 (1979).
327. J. B. Cohen, M. Weber, M. Huchet, and J.-P. Changeux, *FEBS Lett.* **26**, 43–47 (1972).
328. A. Sobel, M. Weber, and J.-P. Changeux, *Eur. J. Biochem.* **80**, 215–224 (1977).
329. A. Sobel, T. Heidmann, J. Hofler, and J.-P. Changeux, *Proc. Natl. Acad. Sci. U.S.A.* **75**, 510–514 (1978).
330. A. Y. Jeng, P. A. St. John, J. B. Cohen, *Biochim. Biophys. Acta* **646**, 411–421 (1981).
331. J. Cartaud, E. Benedetti, A. Sobel, and J.-P. Changeux, *J. Cell Sci.* **29**, 313–337 (1978).
332. M. W. Klymkowsky, and R. M. Stroud, *J. Mol. Biol.* **128**, 319–334 (1979).
333. J. Kistler, and R. M. Stroud, *Proc. Natl. Acad. Sci. U.S.A.* **78**, 3678–3682 (1981).
334. E. X. Albuquerque, and P. W. Gage, *Proc. Natl. Acad. Sci. U.S.A.* **75**, 1596–1599 (1978).
335. G. Kato, and J. P. Changeux, *Mol. Pharmacol.* **12**, 92-100 (1976).
336. R. S. Aronstam, A. T. Eldefrawi, I. N. Pessah, J. W. Daly, E. X. Albuquerque, and M. E. Eldfrawi, *J. Biol. Chem.* **256**, 2843–2850 (1981).
337. B. L. Ginsborg, and D. H. Jenkinson, *Hand. Exp. Pharmacol.* **42**, 229–420 (1976).
338. A. J. Eldefrawi, M. E. Eldefrawi, E. X. Albuquerque, A. C. Oliveira, N. Mansour, M. Adler, J. W. Daly, G. B. Brown, W. Burgermeister, and B. Witkop, *Proc. Natl. Acad. Sci. U.S.A.* **74**, 2172–2176 (1977).
339. M. E. Eldefrawi, A. T. Eldefrawi, N. A. Mansour, J. W. Daly, B. Witkop, and E. X. Albuquerque, *Biochemistry* **17**, 5474–5483 (1978).
340. J. Elliot, and M. A. Raftery, *Biochem. Biophys, Res. Commun.* **77**, 1347–1353 (1977).
341. E. K. Krodel, R. A. Beckman, and J. B. Cohen, *Mol. Pharmacol* **15**, 294–312 (1979).
342. M. E. Eldefrawi, A. T. Eldefrawi, R. S. Aronstam, M. A. Maleque, J. C. Warnick, and E. X. Albuquerque, *Proc. Natl. Acad. Sci. U.S.A.* **77**, 7458–7462 (1980).
343. E. X. Albuquerque, M.-C. Tsai, R. S. Aronstam, A. T. Eldefrawi, and M. E. Eldefrawi, *Mol. Pharmacol.* **18**, 167–178 (1980).
344. C. E. Spivak, M. A. Maleque, A. C. Oliveira, L. M. Masukawa, T. Tokuyama, J. W. Daly, and E. X. Albuquerque, *Mol. Pharmacol.* **21**, 351–361 (1982).
345. M. A. Maleque, C. Souccar, J. B. Cohen, and E. X. Albuquerque, *Mol. Pharmacol.* **22**, 636–647 (1982).
346. D. C. Medynski, and J. B. Cohen, *Soc. Neurosci. Abst.* **6** 779 (1980).
347. J. Elliott, and M. A. Raftery, *Biochemistry* **18**, 1868–1874 (1979).
348. R. R. Neubig, E. K. Krodel, N. D. Boyd, and J. B. Cohen, *Proc. Natl. Acad. Sci. U.S.A.* **76**, 690–694 (1979).
349. J. Eliott, S. M. Dunn, S. G. Blanchard, and M. A. Raftery, *Proc. Natl. Acad. Sci. U.S.A.* **76**, 2576–2579 (1979).
350. A. Sobel, T. Heidmann, J. Cartaud, and J.-P. Changeux, *Eur. J. Biochem.* **110**, 13–33 (1980).
351. P. A. St. John, S. C. Froehner, D. A. Goodenough, and J. B. Cohen, *J. Cell Biol.* **92**, 333–342 (1982).
352. T. Saitoh, R. Oswald, L. P. Wennogle, and J.-P. Changeux, *FEBS Lett.* **116**, 30–36 (1980).
353. D. S. Wise, J. Wall, and A. Karlin, *J. Biol. Chem.* **256**, 12624–12727 (1981).

354. D. S. Wise, B. P. Schoenborn, A. Karlin, *J. Biol. Chem.* **265**, 4124–4126 (1981).
355. P. R. Adams, *J. Physiol.* **268**, 291–318 (1977).
356. E. Neher, and J. H. Steinbach, *J. Physiol.* **277**, 153–176 (1978).
357. R. R. Neubig, and J. B. Cohen, *Biochemistry* **19**, 2770–2779 (1980).
358. N. D. Boyd, and J. B. Cohen, *Biochemistry* **19**, 5344–5353 (1980).
359. R. S. Aronstam, A. T. Eldefrawi, I. N. Pessah, J. W. Daly, E. X. Albuquerque, and M. E. Eldefrawi, *J. Biol. Chem.* **296**, 2848–2840 (1981).
360. C. Souccar, M. A. Maleque, J. W. Daly, and E. X. Albuquerque, *Fed. Proc.* **41**, 1299 (1982).
361. Y. Nimitkitpaisan, J. W. Daly, P. J. Jessup, L. Overman, J. E. Warnick, and E. X. Albuquerque, *Proc. Am. Soc. Neurochem.* **11**, 233 (1980); J. E. Warnick, P. J. Jessup, L. E. Overman, M. E. Eldefrawi, Y. Nimit, J. W. Daly, and E. X. Albuquerque, *Mol. Pharmacol.* **22**, 565–573 (1982).
362. E. X. Albuquerque, J. E. Warnick, M. A. Maleque, F. C. Kauffman, R. Tamburini, Y. Nimit, and J. W. Daly, *Mol. Pharmacol.* **19**, 411–424 (1981).
363. R. Tamburini, E. X. Albuquerque, J. W. Daly, and F. C. Kauffman, *J. Neurochem.* **37**, 775–780 (1981).
364. B. Witkop, *Chem. Unserer Zeit* **5**, 99–106 (1971).
365. B. Witkop, "Quimica y Farmacologia de la Batraciotoxina," Vol. 7, p. 51–100. Revista de la Facultad de Farmacia, Universidad de los Andes, Merida, Venezuela, 1970.
366. B. Witkop, *Kagaku* (Chemistry) **26**, 1095–1102 (1971).
367. B. Witkop, *La Recherche* **6**, 528–539 (1975).
368. G. B. Brown, and B. Witkop, *Israel J. Chem.* **12**, 697 (1971).
369. B. Witkop, *Kagaku* **8**, 605–619 (1977) (in Japanese).
370. B. Witkop, *Protein Nucleic Acid and Enzyme* **24**, 1521–1530 (1979) (in Japanese).
371. M. Vemura, K. Shimada, T. Tokuyama, and J. W. Daly, *Tetrahedron Lett.*, **23** (42) 4369–4370 (1982).
372. T. Tokuyama, K. Shimada, M. Vemura, and J. W. Daly, *Tehrahedron Lett.* **23** 2121 (1982).
373. T. Tokuyama, J. Yamamoto, J. W. Daly, and R. J. Highet, *Tetrahedron*, submitted.
374. P. Duhamel, and M. Kotera, *J. Org. Chem.* **47**, 1688 (1982).
375. A. J. Pearson, P. Ham, and D. C. Rees, *J. Chem. Soc., Perkin Trans. I* 489 (1982); *Tetrahedron Lett.* 4637 (1980).
376. G. E. Keck, and J. B. Yates, *J. Org. Chem.* **47**, 3590 (1982).
377. G. E. Keck, and J. B. Yates, *J. Am. Chem. Soc.* **104**, 5829 (1982).
378. M. Glanzmann, Ch. Karalai, B. Ostersehlt, U. Schön, Ch. Frese, and E. Winterfeldt, *Tetrahedron* **38**, 2805 (1982).
379. T. Ibuka, H. Minakata, Y. Mitsui, K. Hayashi, T. Taga, and Y. Inubushi, *Chem. Pharm. Bull.* (*Japan*) **30**, 2840 (1982).
380. J. W. Daly, *in* "Progress in the Chemistry of Organic Natural Products," (W. Herz, H. Grisebach, and G. W. Kirby, eds.), Vol. 41, 205–340. Springer Verlag, Wien, New York, 1980.
381. L. E. Overman and R. McCready, *Tetrahedron Lett.* 4887 (1982).
382. D. J. Hart and Y.-M. Tsai, *J. Am. Chem. Soc.* **104**, 1430 (1982).
383. D. J. Hart and Y.-M. Tsai, *J. Org. Chem.* **47**, 4403 (1982).
384. S. A. Godleski and D. Heacock, *J. Org. Chem.* **47**, 4820 (1982).

Addendum

After completion of this chapter the following publications dealing with the histrionicotoxins came to our attention. Two new histrionicotoxins have been isolated from the skin extracts of 1000 specimens of the Guayacana population of *D. histrionicus* (*373*): dihydrohistrionicotoxin

($R^1 = $ ⟍⟍⟍⟍ , $R^2 = $ ⟋⟍), which previously has been obtained by partial reduc-

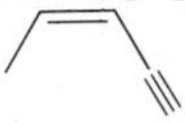

tion of histrionicotoxin (*150*), and Δ^{14}-*trans*-histrionicotoxin ($R^1 = $ ⟍⟍⟍ ,

$R^2 = $ ⟋⟍⟍). It is not entirely clear whether this latter trace alkaloid occurs naturally or is an artifact. Its structure determination is based on PMR, CMR and mass spectral data.

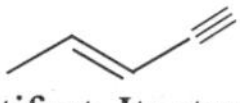

The isolation of larger amounts of HTX 259 ($R^1 = $ ⟋⟍⟍ , $R^2 = $ ⟋⟍) permitted

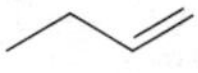

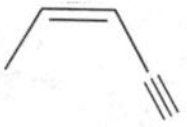

the confirmation of its structure by the use of PMR, CMR and mass spectral data (*373*). A description of the preparation of 8-deoxy-PHTX from HTX by hydrogenation, dehydration, and hydrogenation of the newly formed double bond is given (*373*).

Two preparations of functionalized 1-azaspiro[5.5]undecanes were reported. The first one (*374*, Scheme 78) starts with caprolactam, which, after N-methylation to **411**, is converted to the seven-membered cyclic enaminoester **412** in several steps. Bromination of this product followed by treatment with base yields the 2,2-disubstituted-*N*-methylpiperidine **413**, which after chain extension, Dieckmann condensation, and decarboxylation leads to *N*-methyl-1-azaspiro[5.5]undecan-7-one, **414**.

The second method uses a suitably substituted tricarbonyl (cyclohexadienylium) iron salt to produce 1-azaspiro[5.5]undecane derivatives (*375*, Scheme 79).

Starting with *p*-methoxyphenylbutanoic acid **415** the tricarbonyl [4(4'-methoxycyclohexa-2,4-dienylium)butyltosylate] iron hexafluorophosphate **416** is prepared. This salt is added to an excess of benzylamine yielding the spirocyclic tricarbonyl iron complex. Removal of the iron carbonyl with trimethylamine oxide and conversion of the dienol ether into the α,β-unsaturated ketone under acidic conditions leads to *N*-benzyl-1-azaspiro[5.5]undec-7-en-9-one, **417**.

A formal synthesis of PHTX has been published (*376*; Scheme 80) in which the key step is an intramolecular ene reaction of an acylnitroso compound (*198*). The requisite acylnitroso

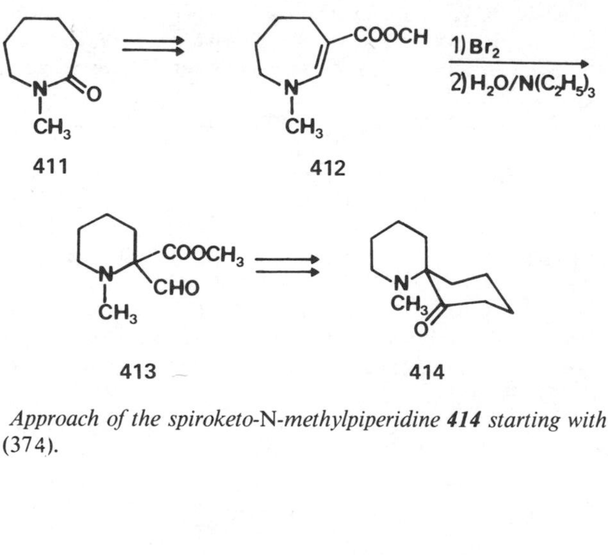

411   412

413   414

SCHEME 78. *Approach of the spiroketo-N-methylpiperidine **414** starting with N-methylcaprolactam **411** (374).*

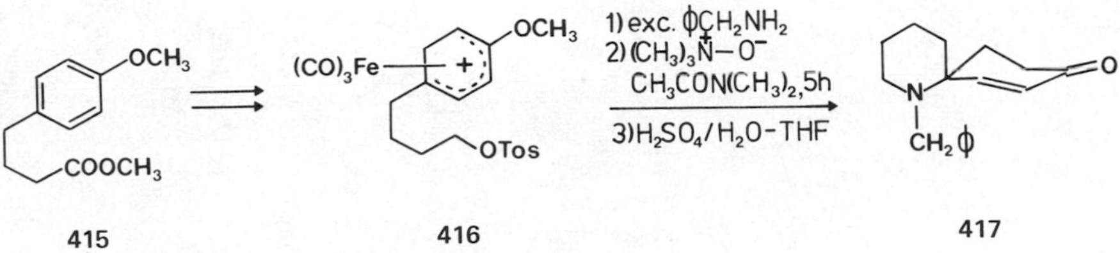

SCHEME 79. *The use of an iron tricarbonyl complex 416 as an intermediate in the synthesis of N-benzyl-1-azaspiro[5.5]undec-7-en-9-one, 417 (375).*

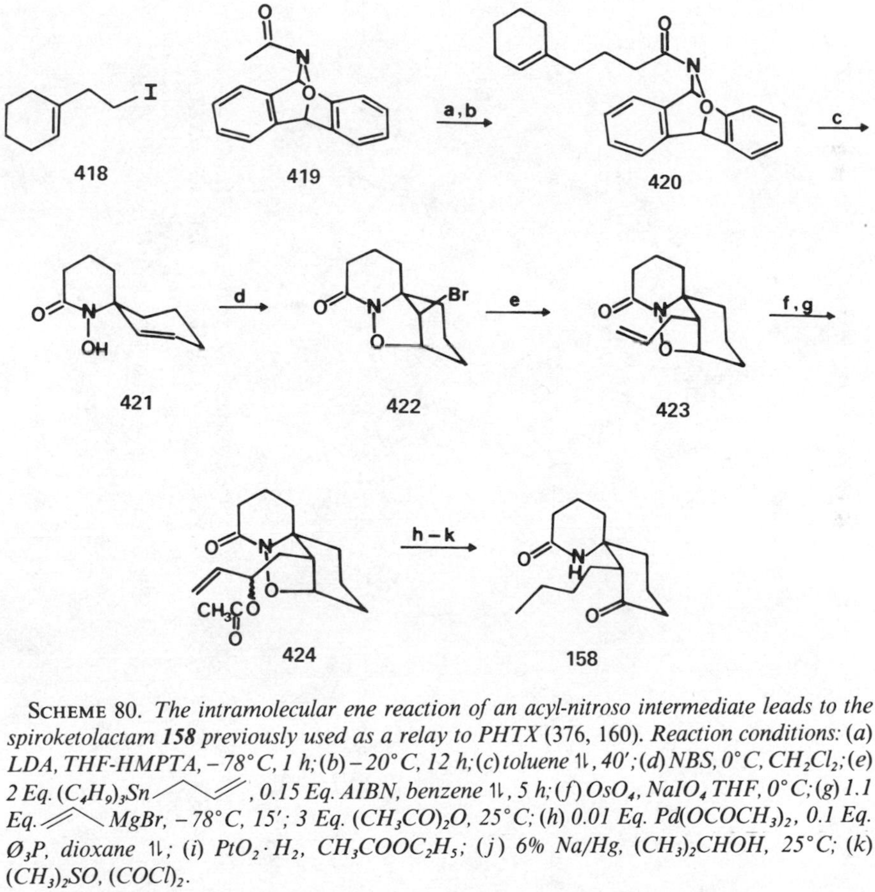

SCHEME 80. *The intramolecular ene reaction of an acyl-nitroso intermediate leads to the spiroketolactam 158 previously used as a relay to PHTX (376, 160). Reaction conditions: (a) LDA, THF-HMPTA, −78°C, 1 h; (b) −20°C, 12 h; (c) toluene ⅃, 40′; (d) NBS, 0°C, CH₂Cl₂; (e) 2 Eq. (C₄H₉)₃Sn⌒⌒, 0.15 Eq. AIBN, benzene ⅃, 5 h; (f) OsO₄, NaIO₄ THF, 0°C; (g) 1.1 Eq. ⌒MgBr, −78°C, 15′; 3 Eq. (CH₃CO)₂O, 25°C; (h) 0.01 Eq. Pd(OCOCH₃)₂, 0.1 Eq. Ø₃P, dioxane ⅃; (i) PtO₂·H₂, CH₃COOC₂H₅; (j) 6% Na/Hg, (CH₃)₂CHOH, 25°C; (k) (CH₃)₂SO, (COCl)₂.*

compound, an unstable intermediate, was prepared by alkylation of the anion of the Diels–Adler adduct of 9,10-dimethylanthracene with (nitrosocarbonyl)methane **419** with 2′-iodoethylcyclohex-1-ene **418**. Thermolysis of this product of alkylation, **420,** frees the desired nitrosocarbonyl compound, by retro-Diels–Adler reaction, which immediately undergoes an

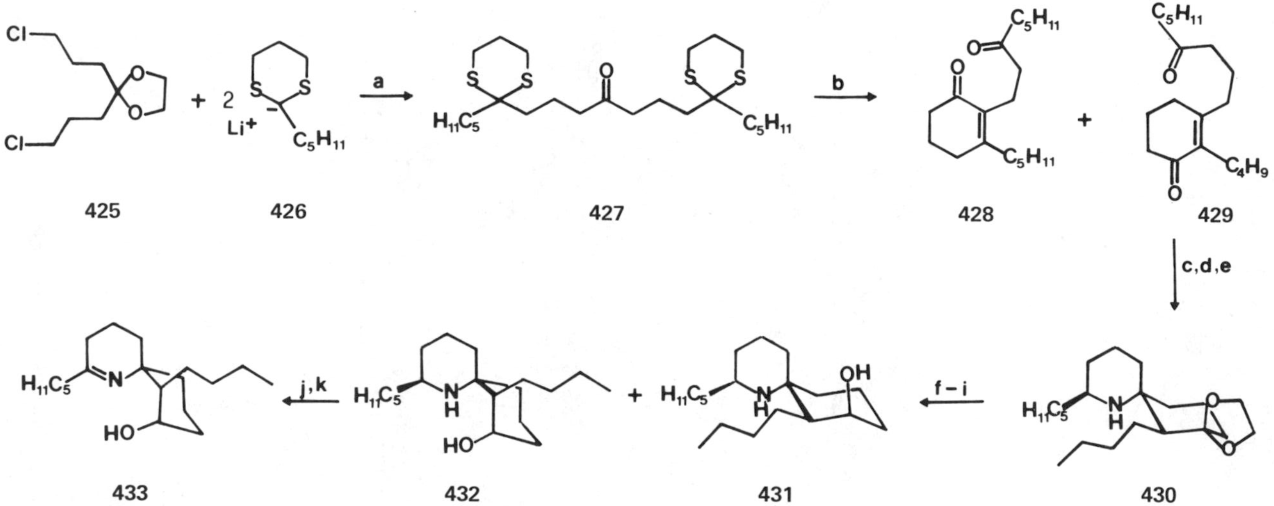

SCHEME 81. *In analogy to possible biogenetic schemes (Scheme 21, Section III.A.2) a triketone serves as an intermediate Δ¹-PHTX, 433, and PHTX, 130 (378). Reaction conditions: (a) THF, −25°C, 12; 20°C, 12 h; (b) CH₃CNHCl conc = 10:1, HOCH₂COOH; (c) NH₄⁺ CH₃COO⁻, CH₃OH; (d) NaBCNH₃, CH₃OH, 6 h; (e) benzene, (CH₂OH)₂, TosOH 1ʋ; (f) HCl conc, 65°C, 30'; (g) 5% (CH₃)₃COK, (CH₃)₃COH, 1ʋ, 3 h; (h) LAH (C₂H₅)₂O abs., 30'; (i) tl-chromatography; (j) (CH₃)₃COCl(C₂H₅)₂O, 30'; (k) (CH₃)₃COK, 30 h.*

intramolecular ene reaction (*198*) to the spirocyclic hydroxamic acid **421**. The hydroxamic acid moiety of **421** is used to introduce the oxygen function at the C-8 position with the correct stereochemistry by bromination with NBS to yield the tricyclic isoxazolidine **422**. Substitution of the bromine of **422** by the allyl group via treatment with tributyl allyl stannane (*377*) proceeds with inversion of the configuration at the C-7 position. The newly formed side chain in **423** is extended by cleavage of the double bond with osmium tetroxide–sodium periodate. The aldehyde so formed is converted to the allyl alcohol by addition of vinylmagnesium bromide. Acetylation finally leads to the allyl acetate **424** which, via a Pd(II) complex, is transformed to the diene. After reduction of this unsaturation the N–O bond is cleaved with sodium amalgan. Oxidation of the hydroxyl group yields the relay compound **158** which has previously been converted to PHTX (*160*).

2-Epi-PHTX, 2,7-epi-PHTX and Δ¹-decahydrohistrionicotoxin, an intermediate of earlier PHTX syntheses (*154, 160, 185, 186*) have been synthesized, following a possible biogenetic pathway (*378*) as outlined earlier (*152*) (See Scheme 21, Section III.A.2, Scheme 81). 6,10,14-Triketononadecane, prepared by alkylation of the 2-pentyl-1,3-dithianyl anion **426** with the ketal of 1,7-dichloroheptan-7-one **425** and consecutive hydrolysis. Prolonged treatment with acid yields two cyclohexenones, **428** and **429**. Addition of ammonium acetate to this mixture and reduction of the Schiff base so formed results in 30% of this aminohexenone, which by treatment with acid in the presence of ethylene glycol (to prevent retro-Michael or retro-Man-

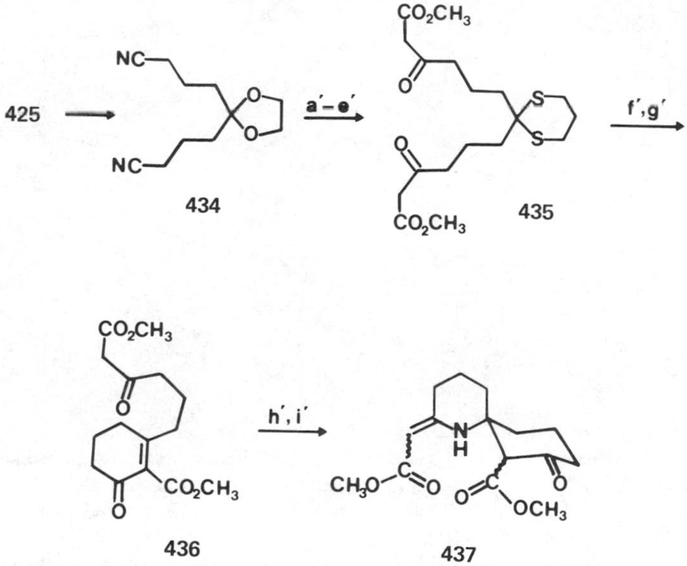

SCHEME 82. *In an improved modification of the preceding scheme the desired cyclohexenone* **436**, *an analog of* **429**, *is obtained by starting with the dicyano derivative* **434** (378). *Reaction conditions:* (*a′*) *KOH/H₂O₂;* (*b′*) *H⁺;* (*c′*) *HS(CH₂)₃SH, HOCOCH₃, BF₃ · 2(C₂H₅)₂O, 50°C, 1 h;* (*d′*) (⟨N⟩N—)₂ *CO, CH₂Cl₂;* (*e*) *CH₃COCHCO₂Mg, THF, 12 h;* (*f′*) *NCS, AgNO₃, H₂O, 15′;* (*g′*) [(CH₃)₂CH]₂NC₂H₅, *CH₃OH, HOOCCH₃, 4°C, 10d;* (*h′*) *NH₄NO₃, NH₃, CH₃OH, 5 h;* (*i*) *THF, 1.5 h, 0.1n HCl.*

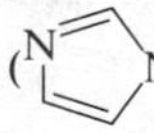

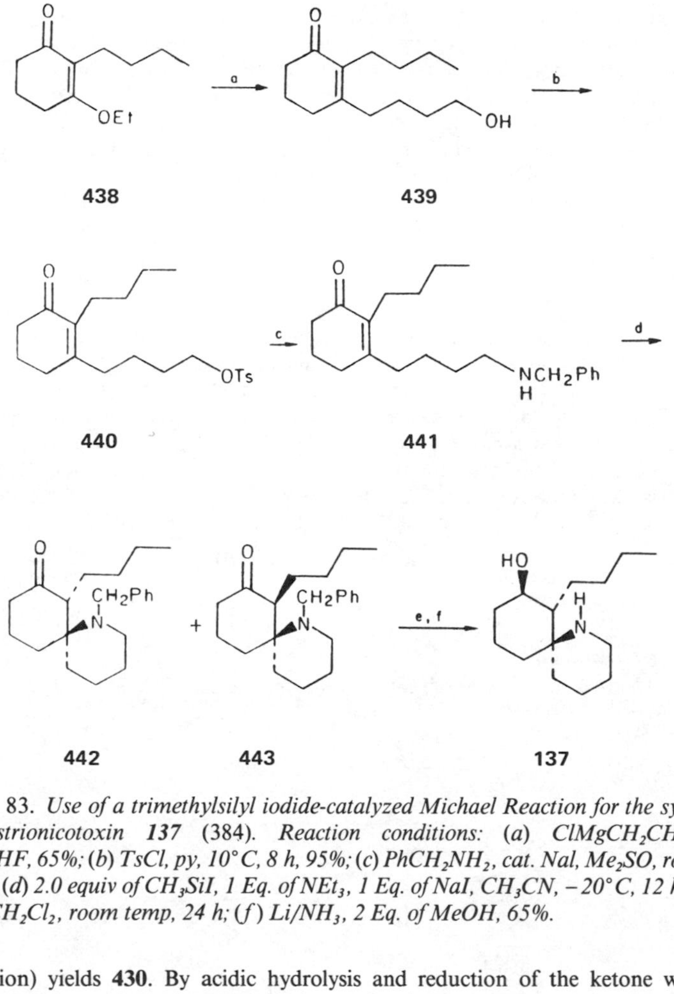

SCHEME 83. *Use of a trimethylsilyl iodide-catalyzed Michael Reaction for the synthesis of depentylhistrionicotoxin* **137** *(384). Reaction conditions: (a) ClMgCH₂CH₂CH₂CH₂ OMgCl, THF, 65%; (b) TsCl, py, 10°C, 8 h, 95%; (c) PhCH₂NH₂, cat. NaI, Me₂SO, room temp, 18 h, 72%; (d) 2.0 equiv of CH₃SiI, 1 Eq. of NEt₃, 1 Eq. of NaI, CH₃CN, −20°C, 12 h, 80%; (e) NaOMe, CH₂Cl₂, room temp, 24 h; (f) Li/NH₃, 2 Eq. of MeOH, 65%.*

nich reaction) yields **430**. By acidic hydrolysis and reduction of the ketone with LAH, 2-epi-PHTX **432** and 2,7-epi-PHTX **431** in a ratio of 1:2 are obtained. 2-epi-PHTX **432** is converted to Δ¹-PHTX **433** by treatment with *t*-butylhypochlorite followed by base. As mentioned before, **433** is an intermediate in earlier PHTX syntheses, and by reduction with aluminum hydride (*154*), yields PHTX **130**.

To change the ratio of the two possible cyclohexenones, which are obtained from δ-triketones, in favor of the precursor of the 1-azaspiro[5.5]undecane system, it is necessary to render the hydrogens in the methylene groups in the outward α-position of the first and the third carbonyl function more acidic (Scheme 82). Therefore, the di-β-keto acid ester **435** is prepared via the dicyano compound **434**. After removal of the thioketal function of **435**, ring closure under mild conditions leads to the desired cyclohexenone derivative **437**. This modified pathway has the additional advantage that is leads directly to an *sp²*-hybridized C atom in position 2.

The formal syntheses of PHTX and its 7,8-epimer mentioned in Section III,A,3 (*171, 174*) have been published in full detail (*379*).

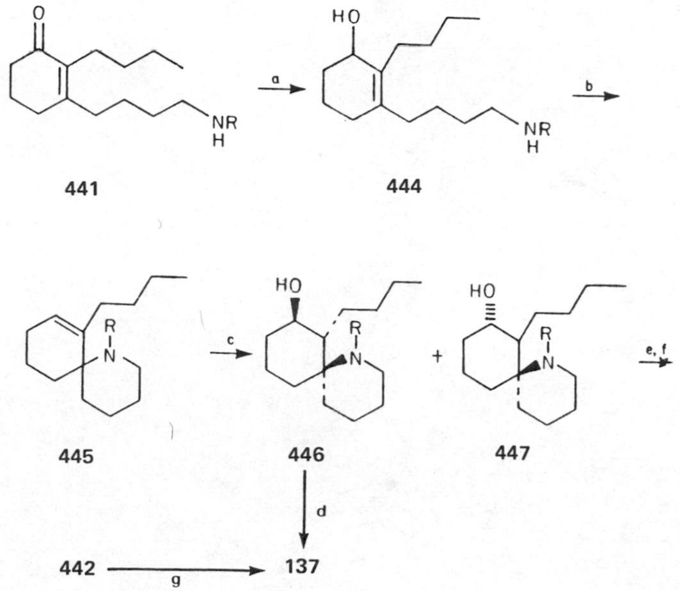

SCHEME 84. *Extension of the palladium-based intramolecular Michael addition* (168, *Scheme 30) to the preparation of depentyl histrionicotoxin* 137 (384). *Reaction conditions:* $R = CH_2Ph$; (a) *Dibal-H, PhCH$_3$,* $-50°C$, 5 h, 75%; (b) 1.5 Eq. of CH$_3$SiI, 1 Eq. of NEt$_3$, CH$_3$CN, $-20°C$, 40%; (c) 1.1 Eq. of BH$_3$ · Me$_2$S; NaOH, H$_2$O$_2$, diglyme, 80°C, 40%; (d)60 psi; H$_2$, Pd/C, ETOH, 85%; (e) Swern oxidation (COCl)$_2$, Me$_2$SO, NEt$_3$, $-50°C$, 90%; (f) NaOMe, CH$_2$Cl$_2$, room temp, 24 h; (g) Li/NH$_3$, 2 Eq. of MeOH, 65%.

The preparation of depentylhistrionicotoxin **137** utilizing the CH$_3$SiI-catalyzed Michael addition reaction is outlined in Scheme 83(*384*). The sequence is initiated by the reaction of the Normant Grignard reagent derived from 4-chlorobutanol with the known vinylogous ester **438** which on acidic workup yields the enone alcohol **439** in 65% yield. Tosylation of **439** provides **440** in 95% yield. Amination of the tosylate **440** with benzylamine yields **441**, which on treatment with 2 equiv. of CH$_3$SiI yields the crystalline ketones **442** and **443** (80%). After epimerization by base the equilibrium mixture contains **442**:**443** in a ratio of 13:1. The desirable ketone **442** is easily reducible to depentylhistrionicotoxin **137**.

The second CH$_3$SiI reaction serves as an adjunct (Scheme 84, *386*) to the palladium-based cyclization (Scheme 30, *168*) in promoting an S$_N$2' reaction of aminoallylic alcohols. The precursor for this reaction **444** was prepared by DiBAL-H reduction on the enone amine **441**. Treatment of **444** with 1.5 equiv. of CH$_3$SiI effected cyclization to **445** in 40% yield. Conversion of the spirocyclic olefin **445** to depentylhistrionicotoxin **137** was accomplished by hydroboration–oxidation to give the epimeric alcohols **446** and **447** in 40% yield. Alternatively the crude oxidation mixture is convertible to a 2:1 mixture of **442** and **443** by Swern oxidation. Epimerization and Li/NH$_3$ reduction again provides depentylhistrionicotoxin **137** (*384*).

Two excellent reviews on the histrionicotoxins have appeared. One of them (*379*) emphasizes the synthetic aspect, whereas the other, which contains a description of all the known alkaloids of neotropical frogs (*380*), encompasses synthesis, spectral data of natural and synthetic histrionicotoxins, as well as pharmacological data in detail.

——CHAPTER 6——

SIMPLE ISOQUINOLINE ALKALOIDS

JAN LUNDSTRÖM

Astra Pharmaceuticals, Södertälje, Sweden
and
Department of Pharmacognosy, Faculty of Pharmacy
Uppsala University, Uppsala, Sweden

I. Introduction

In his early review in Volume IV of this treatise, Reti (*1*) defined the simple isoquinoline alkaloids as isoquinolines having only one aromatic nucleus. Considerable progress in the field of isoquinoline alkaloids over the last decades has shown that this definition is no longer tenable. This review mainly considers isoquinoline alkaloids that do not contain additional cyclic structures except a methylenedioxy substituent. However, 1- and 4-phenyl-substituted and N-benzyl-substituted tetrahydroisoquinolines will be included because of their close structural relationship and the fact that these compounds do not easily fit into reviews of other classes of isoquinoline alkaloids.

Despite their simple chemical structure, much interest in the simple isoquinoline alkaloids has developed over the years and several new naturally occurring compounds have been found since the review by Reti. Biosynthetic considerations have certainly motivated many studies on this class of natural products. The search for variably substituted simple tetrahy-

THE ALKALOIDS, VOL. XXI
Copyright © 1983 by Academic Press, Inc.
ISBN 0-12-469521-3

droisoquinolines, as well as for their apparent progenitors, i.e., substituted phenethylamines, have been useful in the understanding of the biosynthesis of the tetrahydroisoquinoline structure. Further, the conditions for the formation of the isoquinoline structure in plants may be studied favorably with use of labeled substrates in species containing isoquinolines of the simple type, where further transformations of incorporated material into more complex alkaloidal structures do not occur. The synthesis of all the variably substituted simple isoquinoline alkaloids found in nature has also attracted much interest.

II. Simple Isoquinoline Alkaloids

A. Occurrence

All "true alkaloidal" simple isoquinolines found in plants are listed in Table I (structures **1–51**). Alkaloids already reviewed by Reti (*1*) are also included since new natural sources for most of them have been discovered. Alkaloidal conjugates with Krebs cycle acids are included in Tables IV and V and the isoquinolones in Table VII. Simple tetrahydroisoquinolines may also be formed in mammalian systems; these are, however, reviewed separately (see Chapter 7, M. A. Collins, this volume).

Simple isoquinoline alkaloids abundant in the Cactaceae outnumber those occasionally found in other plant families. Cactus alkaloids are therefore treated under separate headings and are listed in Table II.

1. Cactus Alkaloids

The earliest work in the field of cactus alkaloids suggested that cactus species contain mainly simple tetrahydroisoquinolines with a 6,7-dioxygenated or a 6,7,8-trioxygenated substitution pattern (*2*). However, studies in recent years have revealed the presence of oxygenated substitution patterns of great variety in the cactus tetrahydroisoquinolines (Table I). Ring-closing units (C-1 and its substituents) are usually one- or two-carbon units as represented by anhalamine (**23**) and anhalonidine (**30**), respectively. More unusual ring-closing units are, however, found as the five-carbon (isobutyl + C-1) unit in lophocerine (**20**) and the hydroxy-substituted two-carbon unit in the glucotetrahydroisoquinoline pterocereine (**41**).

The Cactaceae is a family that liberally elaborates β-phenethylamines (*3*). Tetrahydroisoquinolines co-occur with phenethylamine derivatives in the same species, and it is generally accepted that the tetrahydroisoquinolines evolve by cyclization of variously substituted phenethylamines with suitable ring-closing units.

a. Peyote Alkaloids. The cactus species appearing to be richest in alkaloids (Table II) is the Mexican peyote, in current chemical literature named *Lophophora williamsii* (Lem.) Coulter. Over the years this little cactus has enjoyed widespread interest, mainly because of its content of the hallucinogenic alkaloid mescaline (**52**). Mescaline was first isolated by Heffter in 1896 (*4*) and its chemical structure determined by Späth in 1919 to be 3,4,5-trimethoxyphenethylamine (*6*). Excellent reviews on the peyote constituents have been published relatively recently by Kapadia and Fayez (*7, 8*) and the historical, ethnobotanical, and pharmacological aspects have also been reviewed (*9–11*).

Mescaline	3-Demethylmescaline
52	**53**

The alkaloidal fraction of peyote contains, apart from simple tetrahydroisoquinolines, several phenethylamines mono-, di-, or tri-oxygenated in the aromatic nucleus, some of which are progenitors of the tetrahydroisoquinolines (*7, 12*). It is remarkable that all the tetrahydroisoquinolines of peyote are trioxygenated at the 6, 7, and 8 positions, and this fact led early to the assumption that they evolve by cyclization of 3,4,5-trioxygenated phenethylamines (*13*). The major tetrahydroisoquinolines are the phenolic anhalamine (**23**), anhalidine (**27**), anhalonidine (**30**), and pellotine (**36**), and the nonphenolic anhalinine (**25**), anhalonine (**29**), lophophorine (**32**) and *O*-methylanhalonidine (**33**). The elucidation of the structures and the first syntheses of these alkaloids were accomplished by the Austrian chemist Späth during the period 1919–1939 (*13–28*). Peyophorine (**39**), more recently isolated by Kapadia and Fales (*29*), contains the unusual *N*-ethyl substituent. The four isomeric alkaloids isoanhalamine (**24**), isoanhalidine (**28**), isoanhalonidine (**31**), and isopellotine (**37**) were identified in peyote by Lundström (*30*) using preparative gas chromatography (GLC) and gas chromatography–mass spectrometry. Biogenetically, these latter compounds are considered to evolve from the same phenethylamine precursor 3-demethylmescaline (**53**) as their counterparts **23, 27, 30**, and **36**, respectively, but with a cyclization involving the position para to the phenolic group instead of ortho (*30*).

TABLE I

SIMPLE ISOQUINOLINE ALKALOIDS

Number	Name	Structure	Molecular formula	Melting point (°C)	Species	Reference
1	Longimammatine	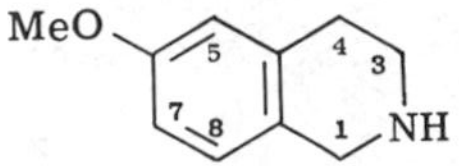 (structure)	$C_{10}H_{13}NO$	244–245.5 HCl: 238–239 (142)	Dolichothele longimamma	64
2	Weberidine	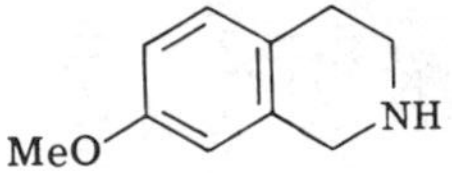 (structure)	$C_{10}H_{13}NO$	HCl: 233 HCl: 231–232 (142)	Pachycereus weberi	58
3	Longimammosine	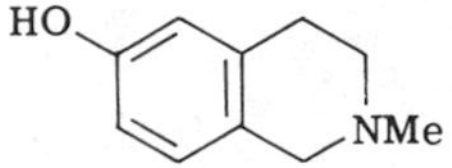 (structure)	$C_{10}H_{13}NO$	180–182 HCl: 234–235 HCl: 243–244 (142)	Dolichothele longimamma	64
4	Longimammidine	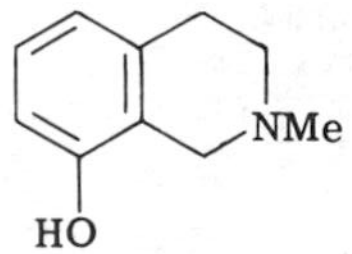 (structure)	$C_{10}H_{13}NO$	171–174 175.5–176 (142) HCl: 247–248.5 HCl: 243–244 (142)	Dolichothele longimamma	64
5	(−)-Longimammamine	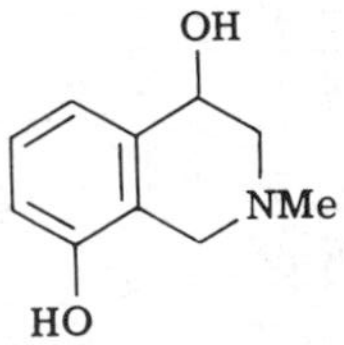 (structure)	$C_{10}H_{12}NO_2$	HCl: 224–228 HCl: 235–236.5 $[\alpha]_D^{25}$ −60° (65)	Dolichothele longimamma Dolichothele uberiformis	64 65
6	Uberine	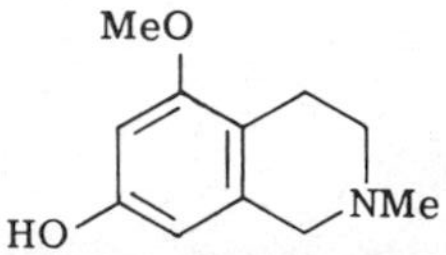 (structure)	$C_{11}H_{15}NO_2$	HCl: 263–267	Dolichothele uberiformis	65

No.	Name	Structure	Formula	mp (derivative)	Source	Ref.
7	Lemaireocereine	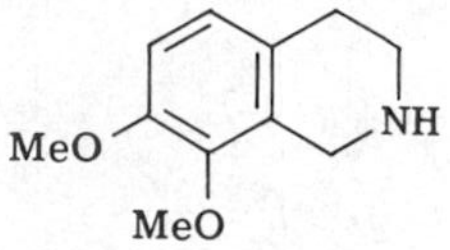 MeO, MeO, MeO, NH	$C_{11}H_{15}NO_2$	HCl: 180, 185 (58)	*Pachycereus weberi*	58
					Pachycereus pringlei	59
					Backebergia militaris	61
8	Hydrohydrastinine	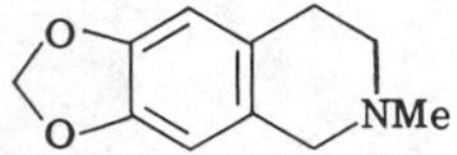 O, O, NMe	$C_{11}H_{13}NO_2$	66	*Corydalis tuberosa*	82
9	Corypalline	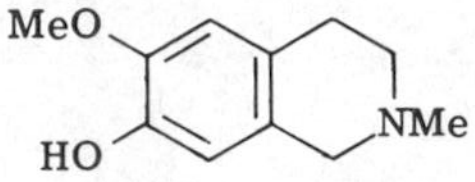 MeO, HO, NMe	$C_{11}H_{15}NO_2$	168 / Picrate: 178	*Corydalis aurea*	83
					Corydalis ophiocarpa	99
					Corydalis pallida	83
					Doryphora sassafras	84
					Thalictrum rugosum	87
10	Heliamine	MeO, MeO, NH	$C_{11}H_{15}NO_2$	HCl: 250 (57) / HCl: 248, 252 (54)	*Pachycereus pecten-aboriginum*	57
					Pachycereus weberi	58
					Pachycereus pringlei	59
					Backebergia militaris	60
11	*N*-Methylheliamine (*O*-Methylcorypalline)	MeO, MeO, NMe	$C_{12}H_{15}NO$	HCl: 210, 215 (58)	*Carnegiea gigantea*	256
					Nelumbo nucifera	102
					Pilosocereus guerreronis	69
					Pachycereus weberi	58
12	Salsolinol	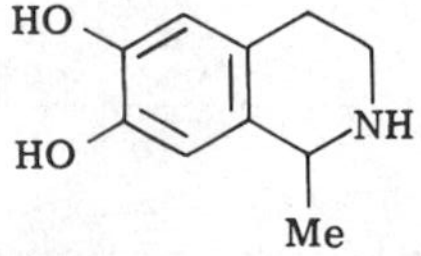 HO, HO, NH, Me	$C_{10}H_{13}NO_2$	Picrate: 92 (154)	Banana fruit (*Musa paradisiaca*)	92
					Cocoa powder (*Theobroma cacao*)	93

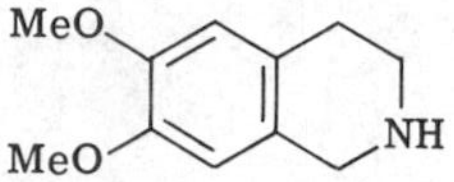

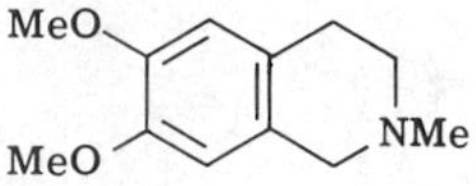

(Continued)

259

TABLE I (Continued)

Number	Name	Structure	Molecular formula	Melting point (°C)	Species	Reference
13	Salsoline		$C_{11}H_{15}NO_2$	218–221 HCl: 147–149 (155)	Alangium lamarckii	96
					Corispermum leptopyrum	97
					Desmodium tiliaefolium	78
					Echinocereus merkerii	68
					Genista purgens	101
					Pachycereus pecten-aboriginum	57
					Salsola arbuscula	73–75
					Salsola kalii	98
14	Isosalsoline	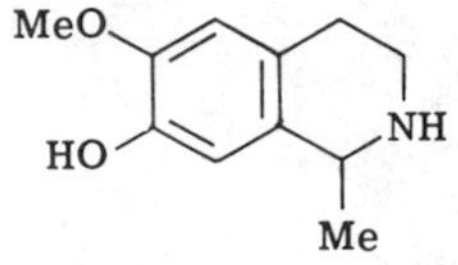	$C_{11}H_{15}NO_2$		Pachycereus pecten-aboriginium	57
15	Arizonine		$C_{11}H_{15}NO_2$	Salicylate: 208–210 (47)	Carnegiea gigantea	47
					Pachycereus pecten-aboriginium	57
16	Salsolidine	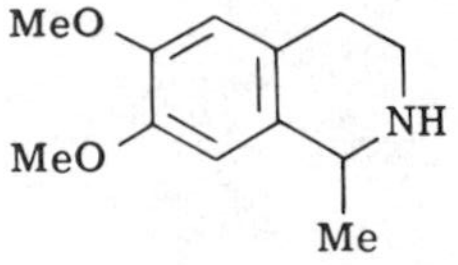	$C_{12}H_{17}NO_2$	69–70	Alhagi pseudalhagi	80, 80a
					Carnegia gigantea	47
					Corispermum leptopyrum	97
					Desmodium cephalotes	79
					Genista purgens	101

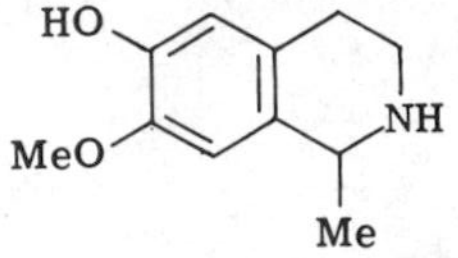

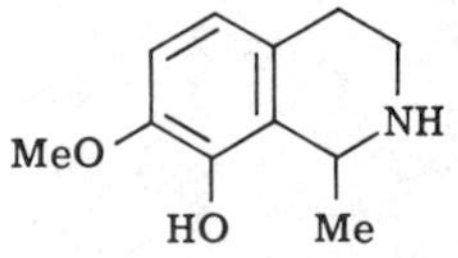

260

No.	Name	Structure	Formula	mp °C (deriv.)	Source	Ref.
16a	Dehydrosalsolidine	(MeO, MeO, =N, Me)	$C_{12}H_{15}NO_2$	195–197	Pachycereus pecten-aboriginum Salsola arbuscula Salsola kalii Carnegiea gigantea	56 73–75 98 256
17	N-Methylisosalsoline	(MeO, HO, NMe, Me)	$C_{12}H_{17}NO_2$	156–158 (88) HCl: 178–179 (76) $[\alpha]_D +33.5°$ $(c = 0.23, CHCl_3)$ (88)	Haloxylon articulatum Corydalis ambigua	76 88
18	Carnegine	(MeO, MeO, NMe, Me)	$C_{13}H_{19}NO_2$	HCl: 209–211 (46) Picrate: 213–215 (47)	Pachycereus pecten-aboriginum Carnegiea gigantea Haloxylon articulatum	41, 55 43, 44, 46, 47 76
19	Tepenine	(MeO, MeO, NMe, Me)	$C_{13}H_{19}NO_2$	—	Pachycereus tehauntepecanus	52
20	Lophocerine	(MeO, HO, NMe)	$C_{15}H_{23}NO_2$	Picrate: 192–193 (182) Styphnate: 171–172 (1)	Lophocereus schottii Pachycereus marginatus	37, 38, 39 42

(Continued)

TABLE I (*Continued*)

Number	Name	Structure	Molecular formula	Melting point (°C)	Species	Reference
21	R-(+)-Calycotomine	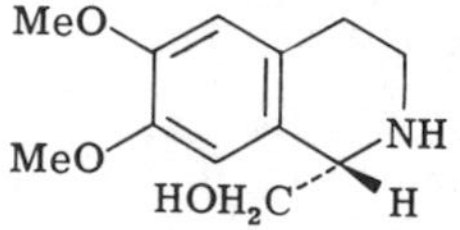	$C_{12}H_{17}NO_3$	139–141 HCl: 193 Picrate: 163–166 $[\alpha]_D +20°$ (H_2O)	*Calycotome spinosa*	80
22	Hydrocotarnine	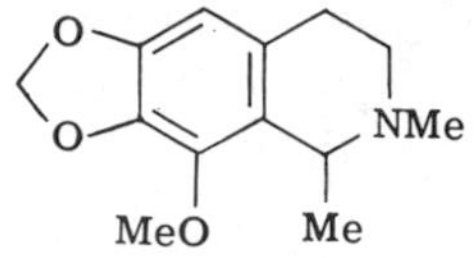	$C_{12}H_{15}NO_3$	55.5–56.5 HBr: 236–237	*Papaver somnifernum*	89
23	Anhalamine	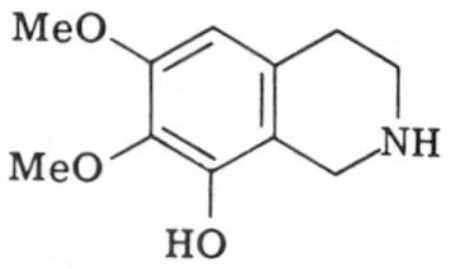	$C_{11}H_{15}NO_3$	189–191 (23) HCl: 277–278 (159)	*Lophophora williamsii* *Lophophora diffusa*	5, 23 34
24	Isoanhalamine	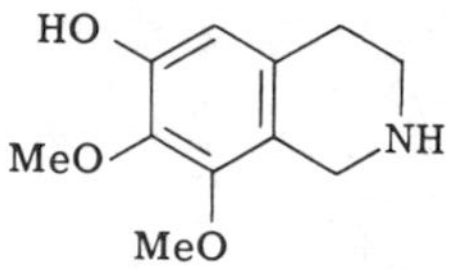	$C_{11}H_{15}NO_3$	HBr: 213–215	*Lophophora williamsii*	30
25	Anhalinine	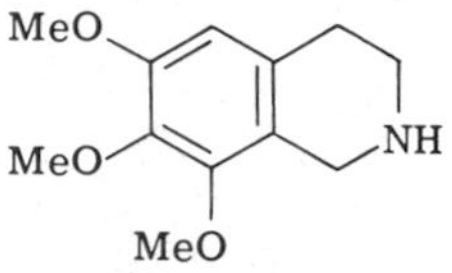	$C_{12}H_{17}NO_3$	61–63 HCl: 248–250 (1)	*Lophophora williamsii*	24
26	Nortehaunine	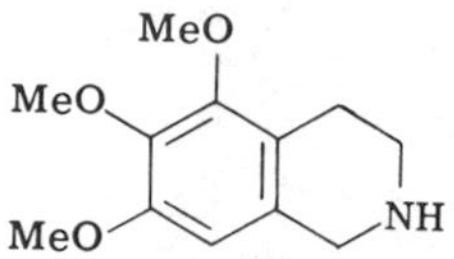	$C_{12}H_{17}NO_3$	HCl: 260, 268	*Pachycereus weberi*	58

No.	Name	Structure	Formula	Physical data	Source	Refs.
27	Anhalidine	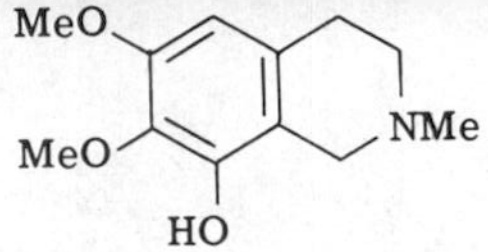	$C_{12}H_{17}NO_3$	131–133 HCl: 243 (159)	*Lophophora williamsii* *Pelecyphora aselliformis* *Stetsonia coryne*	25 66, 67 66
28	Isoanhalidine	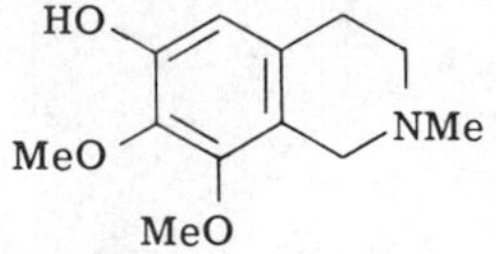	$C_{12}H_{17}NO_3$	HCl: 215–218	*Lophophora williamsii*	30
29	S-(−)-Anhalonine	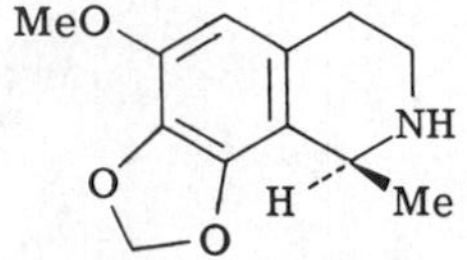	$C_{12}H_{15}NO_3$	85.5 HCl: 262–264 (179) $[\alpha]_D^{25}$ −56.3° (CHCl$_3$)	*Lophophora williamsii* *Lophophora diffusa*	4, 18, 23 34
30	Anhalonidine	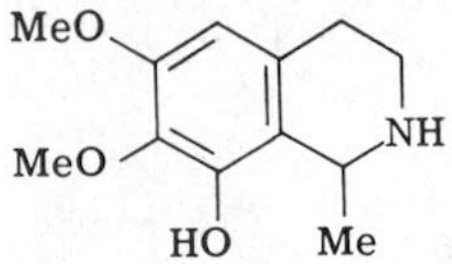	$C_{12}H_{17}NO_3$	160–161 (17) 161–161.5 (42) HCl: 248.5–250 (159)	*Lophophora williamsii* *Pachycereus weberi* *Trichocereus pachanoi* *Stetsonia coryne*	4, 17, 23 42 4 66
31	Isoanhalonidine	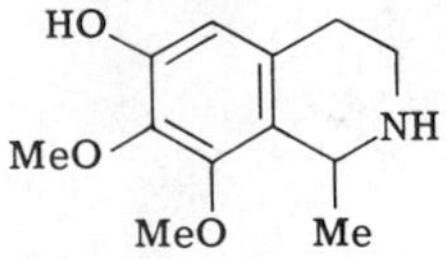	$C_{12}H_{17}NO_3$	HBr: 209–211	*Lophophora williamsii*	30
32	S-(−)-Lophophorine	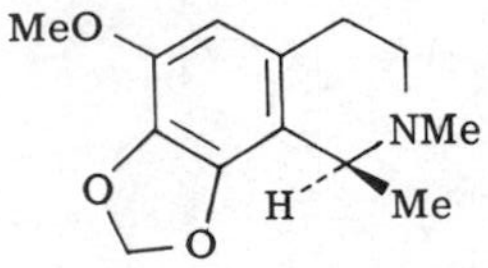	$C_{13}H_{17}NO_3$	bp: 140–145/0.05 mm Hg $[\alpha]_D^{25}$ −47.3° (CHCl$_3$) (159) HCl: 236–237 (179) Picrate: 162–163 (23)	*Lophophora williamsii* *Lophophora diffusa*	4, 23 34

(Continued)

TABLE I (*Continued*)

Number	Name	Structure	Molecular formula	Melting point (°C)	Species	Reference
33	S-(+)-O-Methyl-anhalonidine	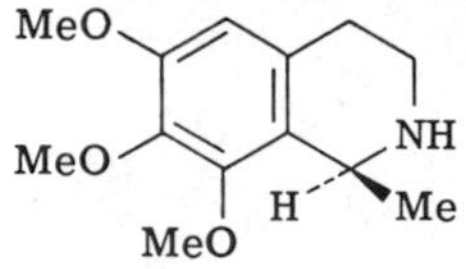	$C_{13}H_{19}NO_3$	bp: 140/0.05 mmHg HBr: 202–204 (*179*) $[\alpha]_D^{25} +20.7°$ (MeOH)	*Lophophora williamsii*	28
34	Tehaunine	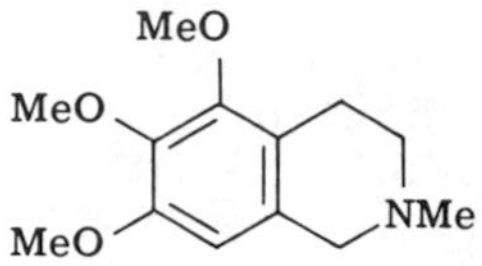	$C_{13}H_{19}NO_3$	HCl: 219–221 (*58*) HCl: 210 (*59*)	*Pachycereus tehauntepecanas* *Pachycereus weberi* *Pachycereus pringlei*	52 58 59
34a	Tehaunine *N*-oxide	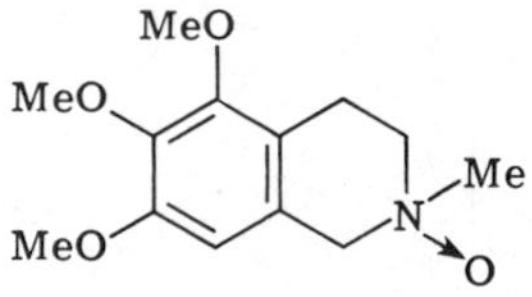	$C_{13}H_{19}NO_4$	185	*Pachycereus pringlei*	257
35	S-(+)-Gigantine	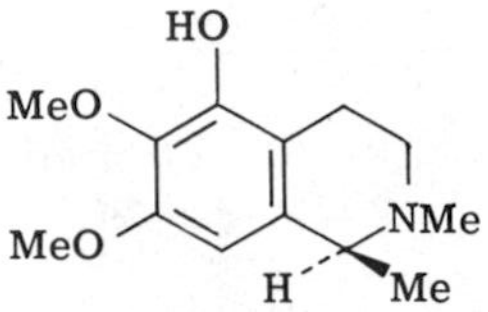	$C_{13}H_{19}NO_3$	151–152 (*46*) HCl: 221.5, 222.5 (*46*) $[\alpha]_D^{25} +27°$ (*c* = 1.99, CHCl$_3$)	*Carnegiea gigantea*	46, 47
36	Pellotine	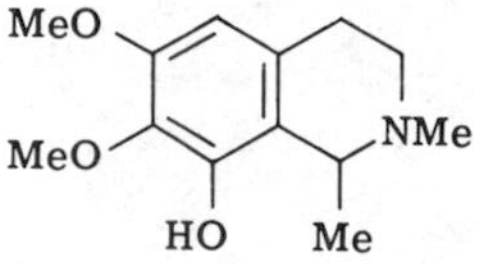	$C_{13}H_{19}NO_3$	111–112 (*23*) 116 (*1*) HCl: 240 (*58*)	*Lophophora williamsii* *Pelecyphora aselliformis* *Pachycereus weberi* *Lophophora diffusa*	4, 23 67 58 9

		Structure	Formula	Properties	Source	Ref
37	Isopellotine	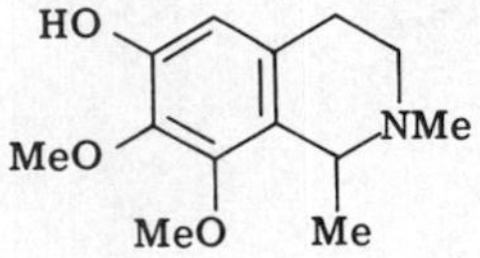 (HO, MeO, MeO, NMe, Me)	$C_{13}H_{19}NO_3$	HCl: 212–222	*Lophophora williamsii*	30
38	*O*-Methylpellotine	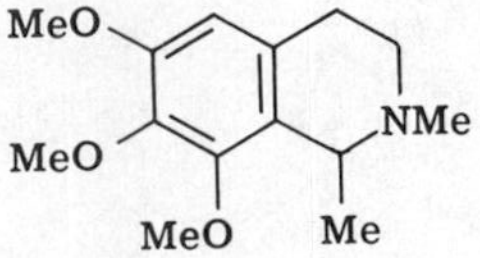 (MeO, MeO, MeO, NMe, Me)	$C_{14}H_{21}NO_3$		*Lophophora diffusa*	35
39	Peyophorine	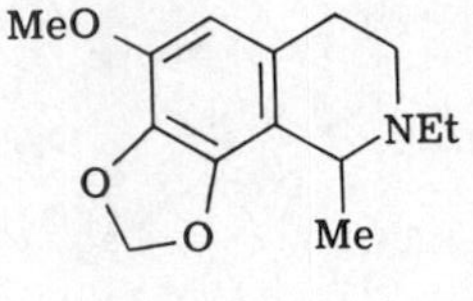 (MeO, O–CH₂–O, NEt, Me)	$C_{14}H_{21}NO_3$	Picrate: 155–156	*Lophophora williamsii*	29
40	Weberine	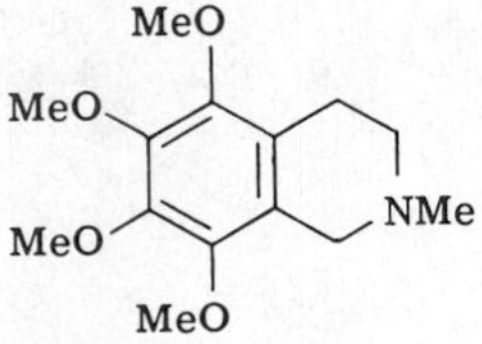 (MeO, MeO, MeO, MeO, NMe)	$C_{14}H_{21}NO_4$	HCl: 164–165 (58) HCl: 165–166 (151)	*Pachycereus weberi* *Pachycereus pringlei*	58 59
41	Pterocereine	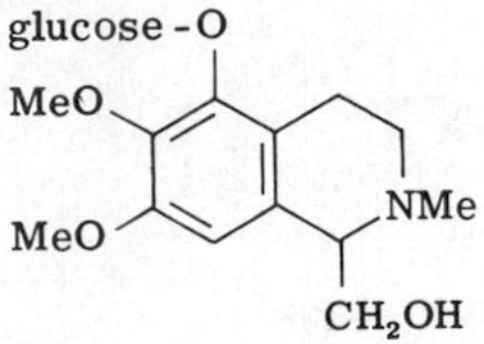 (glucose-O, MeO, MeO, NMe, CH₂OH)	$C_{19}H_{29}NO_9$	198–199 $[\alpha]_D^{26} -4.51°$ (1.35%, H_2O)	*Pterocereus gaumeri*	70

(Continued)

TABLE I (Continued)

Number	Name	Structure	Molecular formula	Melting point (°C)	Species	Reference
42	Deglucopterocereine	(structure)	$C_{13}H_{19}NO_4$	HCl: 247–248 $[\alpha]_D^{26} -1.04°$ (2.22%, H_2O)	*Pterocereus gaumeri*	70
42a	Deglucopterocereine *N*-oxide	(structure)	$C_{13}H_{19}NO_5$	210–213	*Pterocereus gaumeri*	257
43	Anhalotine (iodide)	(structure)	$C_{13}H_{20}NO_3I$	219–220	*Lophophora williamsii*	32
44	Lophotine (iodide)	(structure)	$C_{14}H_{20}NO_3I$	240–242	*Lophophora williamsii*	32
45	Peyotine (iodide)	(structure)	$C_{14}H_{22}NO_3I$	185–186	*Lophophora williamsii*	32

No.	Name	Structure	Formula	mp (°C)	Source	Ref.
46	3,4-Dihydro-6,7-dimethoxy-8-hydroxyisoquinoline	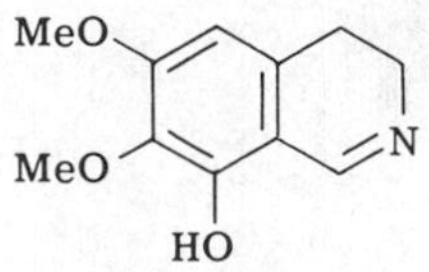	$C_{11}H_{13}NO_3$	159–165	*Lophophora williamsii*	*71*
47	3,4-Dihydro-6,7-dimethoxy-8-hydroxy-2-methyl-isoquinolinium inner salt	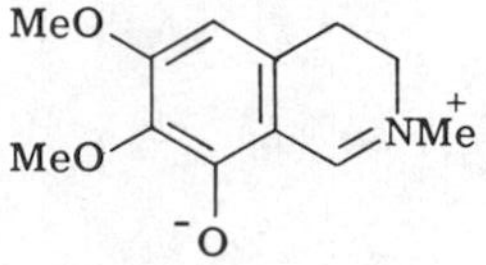	$C_{12}H_{15}NO_3$	95–104	*Lophophora williamsii*	*71*
48	3,4-Dihydro-6,7-dimethoxy-1-methyl-8-hydroxy-isoquinoline	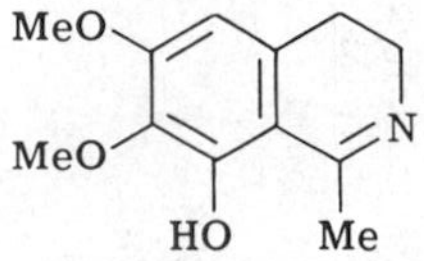	$C_{12}H_{15}NO_3$	—	*Lophophora williamsii*	*71*
49	3,4-Dihyro-6,7-dimethoxy-1,2-dimethyl-8-hydroxy-isoquinolinium inner salt	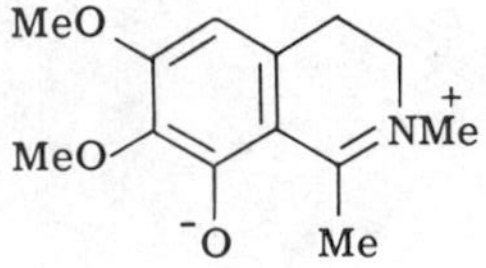	$C_{13}H_{17}NO_3$	—	*Lophophora williamsii*	*71*
50	Pycnarrhine		$C_{11}H_{15}NO_3$ Iodide: $C_{11}H_{14}NO_2I$	185–187 216–218	*Pycnarrhena longifolia*	*94* *176*
51	*N*-Methyl-6,7-dimethoxy-isoquinolinium chloride	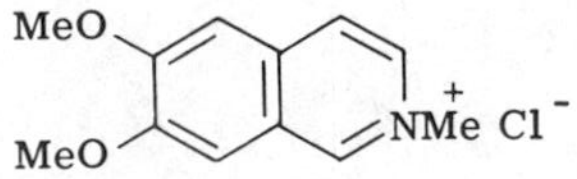	$C_{12}H_{14}NO_2Cl$	185.5–186.5	*Thalictrum revolutum*	*195*

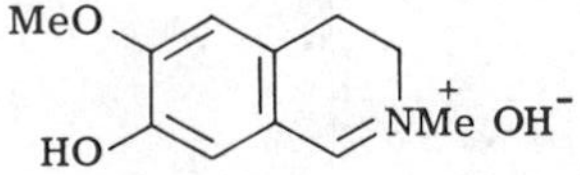

267

TABLE II
SIMPLE ISOQUINOLINE ALKALOIDS IN THE FAMILY OF CACTACEAE

Species	Alkaloid	Reference
Backeberga militaris (Andot)	Heliamine (**10**)	*60, 61*
Bravo ex Sanchez Mejorada	Lemaireocereine (**7**)	*61*
Carnegiea gigantea (Engelm.)	Arizonine (**15**)	*47*
Br & R	Carnegine (**18**)	*43, 44*
	Gigantine (**35**)	*46, 47*
	Salsolidine (**16**)	*45, 46, 47*
	Dehydrosalsolidine (**16a**)	*256*
	Heliamine (**10**)	*256*
Dolichothele longimamma (DC.)	Longimammamine (**5**)	*64*
Br & R	Longimammatine (**11**)	*64*
	Longimammidine (**4**)	*64*
	Longimammosine (**3**)	*64*
Dolichothele uberiformis (Zucc.)	Longimammamine (**5**)	*65*
Br & R	Longimammatine (**11**)	*65*
	Uberine (**6**)	*65*
Echinocereus merkerii Hildm.	Salsoline (**13**)	*68*
Lophocereus australis (K. Brand) Borg.	Pilocereine (**67**)	*42*
Lophocereus gatesii (M. E. Jones)	Pilocereine (**67**)	*42*
Lophocereus schottii (Engel.)	Lopocerine (**20**)	*37, 38, 39*
Br & R	Pilocereine (**67**)	*37, 39a*
Lophophora diffusa (Croizat)	Anhalamine (**23**)	*34*
H. Bravo	Anhalonidine (**30**)	*34*
	Lophophorine (**32**)	*34*
	Pellotine (**36**)	*9*
	O-methylpellotine (**38**)	*35*
Lophophora williamsii (Lem.)	N-Acetylanhalamine (**55**)	*32*
Coulter	N-Acetylanhalonine (**60**)	*32*
	Anhalamine (**23**)	*5, 23*
	Anhalidine (**27**)	*25*
	Anhalinine (**25**)	*24*
	Anhalonidine (**30**)	*4, 17, 23*
	Anhalonine (**29**)	*4, 18, 23*
	Anhalotine (**43**)	*32*
	N-Formylanhalamine (**54**)	*32*
	N-Formylanhalinine (**56**)	*32*
	N-Formylahalonidine (**57**)	*32*
	N-Formylanhalonine (**59**)	*32*
	N-Formyl-O-methyl-Anhalonidine (**58**)	*32*
	Isoanhalamine (**24**)	*30*
	Isoanhalidine (**28**)	*30*
	Isoanhalonidine (**31**)	*30*
	Lophophorine (**32**)	*4, 23*
	Lophotine (**44**)	*32*

TABLE II (*Continued*)

Species	Alkaloid	Reference
	O-Methylanhalonidine (**33**)	28
	O-Methylpeyoruvic acid (**64**)	33a
	O-Methylpeyoxylic acid (**66**)	33a
	Mescalotam (**62**)	32
	Pellotine (**36**)	4, 23
	Peyophorine (**39**)	29
	Peyoglutam (**61**)	32
	Peyoruvic acid (**65**)	32
	Peyotine (**45**)	32
	Peyoxylic acid (**63**)	33
	3,4-Dihydro-6,7-dimethoxy-8-hydroxy-2-methylisoquinolinium inner salt (**47**)	71
	3,4-Dihydro-6,7-dimethoxy-8-hydroxyisoquinoline (**46**)	71
	3,4-Dihydro-6,7-dimethoxy-8-hydroxy-1-methylisoquinoline (**48**)	71
	3,4-Dihydro-6,7-dimethoxy-1,2-dimethyl-8-hydroxyisoquinolinium inner salt (**49**)	71
Pachycereus marginatus (D.C.) Br & R	Lophocerine (**20**)	42
	Pilocereine (**67**)	42
Pachycereus pecten-aboriginum Br & R	Arizonine (**15**)	57
	Carnegine (**18**)	41, 55
	Heliamine (**10**)	57
	Isosalsoline (**14**)	57
	Salsoline (**13**)	57
	Salsolidine (**16**)	56
Pachycereus pringlei (S. Wats) Br & R	Heliamine (**10**)	59
	Lemaireocereine (**7**)	59
	Tehaunine (**34**)	59
	Weberine (**40**)	59
	Tehaunine N-oxide (**34a**)	257
Pachycereus tehauntepecanus (Mac Dough & H. Bravo)	Tepenine (**19**)	52
	Tehaunine (**34**)	52
Pachycereus weberi (Coult.) Backeb.	Anhalonidine (**30**)	42
	Heliamine (**10**)	58
	Lemaireocereine (**7**)	58
	N-Methylheliamine (**11**)	58
	Nortehaunine (**26**)	58
	Pellotine (**38**)	58
	Tehaunine (**34**)	58
	Weberidine (**2**)	58
	Weberine (**40**)	58

(Continued)

TABLE II (*Continued*)

Species	Alkaloid	Reference
Pelecyphora asselliformis Ehrenberg	Anhalidine (**27**)	*67*
	Pellotine (**36**)	*67*
Pilosocereus guerreronis Backeb. Byl et Rowl	*N*-Methylheliamine (**11**)	*69*
Pterocereus gaumeri Br & R, Mac Doug. & Mir.	Pterocereine (**41**)	*70*
	Deglucopterocereine (**42**)	*70*
	Deglucopterocereine N-oxide (**42a**)	*257*
Stetsonia coryne (SD.) Br & R	Anhalidine (**27**)	*66*
	Anhalonidine (**30**)	*66*
Trichocereus pachanoi Br & R	Anhalonidine (**30**)	*36*

The nonphenolic tetrahydroisoquinolines **29, 32,** and **33** have all been isolated in optically active form from peyote while the rapid racemization of the phenolic compounds anhalonidine (**30**) and pellotine (**36**) have made it difficult to investigate whether or not these occur optically active in the living plant (*27*). In a relatively recent report, however, the isolation of the optically active forms of **30** and **36** is described (*138*).

The total alkaloidal content of peyote has been estimated at 0.4% for the fresh plant (*10, 12*) and 3.4% for dried upper slices of peyote heads, the so-called mescal buttons (*10*). Reports are also available on the percentages

TABLE III
ALKALOID CONTENT IN PEYOTE (*Lophophora williamsii*)

Alkaloid	Content (%) in:	
	Alkaloid fraction of fresh plant (*12*)	Dried plant (*refs*)
Hordenine (**68**)	8	
N,N-Dimethyl-4-hydroxy-3-methoxyphenethylamine	0.5–2	
3-Demethylmescaline (**52**)	1–5	
Mescaline (**51**)	30	6 (*10*)
N-Methylmescaline	3	
Anhalamine (**23**)	8	0.1 (*5*)
Anhalinine (**25**)	0.5	0.01 (*24, 25*)
Anhalidine (**27**)	2	0.001 (*24, 25*)
Anhalonidine (**30**)	14	5 (*7*)
O-Methylanhalonidine (**33**)	0.5	
Pellotine (**36**)	17	0.74 (*5*)
Isopellotine (**37**)	0.5	
Anhalonine (**29**)	3	3 (*7*)
Lophophorine (**32**)	5	0.5 (*7*)
Peyophorine (**39**)	0.5	

of the most abundant alkaloids of peyote (Table III). The figures based on the dried plant material are obtained from mescal buttons collected from cacti growing in their natural habitat. The figures based on the alkaloidal fraction of peyote are obtained from fresh greenhouse-grown plants (*12*). In the latter study, seasonal variations in the alkaloidal composition were observed, whereby the contents of N-demethylated compounds, e.g., anhalamine (**23**) and anhalonidine (**30**), during the late autumn and winter were increased in favor of their N-methylated counterparts anhalidine (**27**) and pellotine (**36**). Peyote alkaloids not listed in Table III may be regarded as trace constituents (*12*).

Kapadia and co-workers have identified several minor constituents from nonbasic and polar extracts of peyote. These include the quaternary alkaloids (**43**), (**44**), and (**45**), (*31*) and the Krebs cycle acid conjugates of Tables IV (structures **54–60**) and V (structures **61–66**) (*32, 33, 33a*). The amino acid analogs (Table V) are from biogenetic considerations highly interesting. It has been demonstrated experimentally that the phenolic amino acids peyoxylic acid (**63**) and peyoruvic acid (**65**) play an important intermediary role in the biogenetic transformation of phenethylamines into tetrahydroisoquinolines (*33*).

The four dihydroisoquinoline derivatives **46–49** (Table I) isolated from peyote by Fujita *et al.* (*71*) may also be of biogenetic importance. Unfortunately two of these, **47** and **49**, were named anhalotine and peyotine respectively, names already assigned to the quaternary compounds **43** and **45**. In the past peyotine and pellotine have also been used interchangeably for **36** (*9, 10*).

TABLE IV

N-FORMYL- AND N-ACETYL DERIVATIVES OF TETRAHYDROISOQUINOLINES FOUND IN PEYOTE (*Lophophora williamsii*)[a]

Number	Compound	R_1	R_2	R_3	R_4	Molecular formula
54	N-Formylanhalamine	Me	H	H	COH	$C_{12}H_{15}NO_4$
55	N-Acetylanhalamine	Me	H	H	COMe	$C_{13}H_{17}NO_4$
56	N-Formylanhalinine	Me	Me	H	COH	$C_{13}H_{17}NO_4$
57	N-Formylanhalonidine	Me	H	Me	COH	$C_{13}H_{17}NO_4$
58	N-Formyl-O-methylanhalonidine	Me	Me	Me	COH	$C_{14}H_{19}NO_4$
59	N-Formylanhalonine	— CH_2 —		Me	COH	$C_{13}H_{15}NO_4$
60	N-Acetylanhalonine	— CH_2—		Me	COMe	$C_{14}H_{17}NO_4$

[a] From Kapadia and Fales (*32*).

TABLE V

ALKALOIDAL AMIDES AND AMINO ACIDS FOUND IN PEYOTE (*Lophophora williamsii*)

Number	Name	Structure	Molecular formula	Melting point (°C)	Reference
61	Peyoglutam	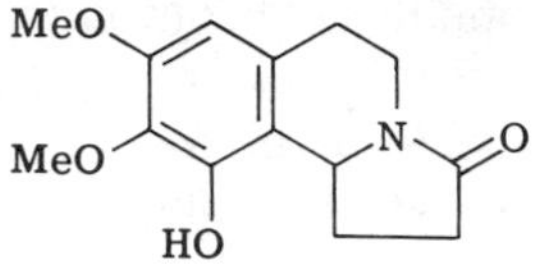	$C_{14}H_{17}NO_4$	217–219	*32*
62	Mescalotam	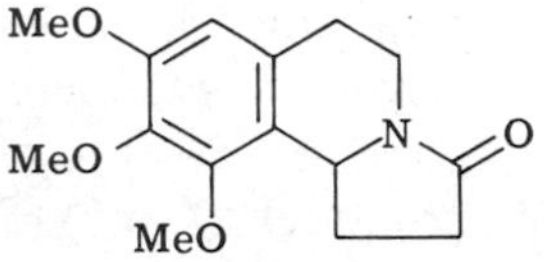	$C_{15}H_{19}NO_4$	—	*32*
63	Peyoxylic acid	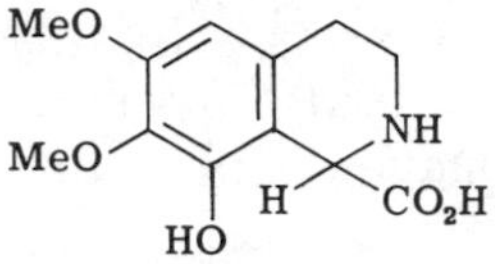	$C_{12}H_{15}NO_5$	237–238	*33*
64	*O*-Methylpeyoxylic acid	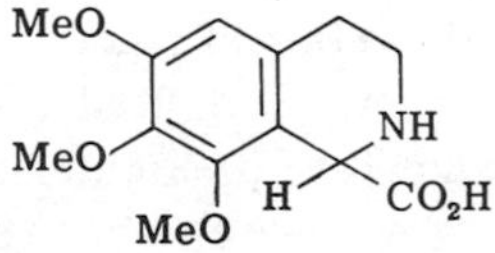	$C_{13}H_{17}NO_5$	238–240	*33a*
65	Peyoruvic acid	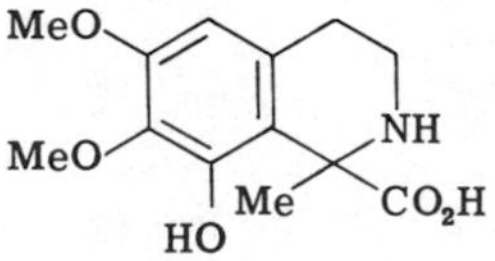	$C_{13}H_{17}NO_5$	233–234	*33*
66	*O*-Methylpeyoruvic acid	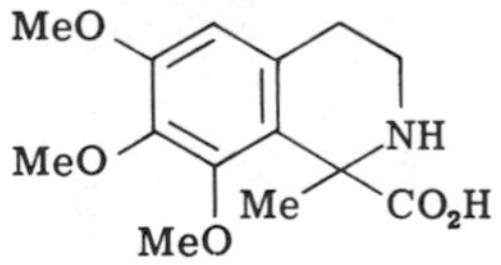	$C_{14}H_{19}NO_5$	245–246	*33a*

Lophophora diffusa, a species closely related to *L. williamsii,* contains the previously known peyote constituents anhalamine (**23**), anhalonidine (**30**), lophophorine (**32**) (*34*), and pellotine (**36**) (*9*) as well as the new compound *O*-methylpellotine (**38**) (*35*).

   b. Alkaloids in Giant Cacti. Several species of the giant cacti belonging to the subtribe Cereanae of the Cactaceae family (*36*) contain alkaloids. Iso-quinoline alkaloids so far are known to occur in species belonging to the genera *Lophocereus, Carnegiea,* and *Pachycereus.* Apart from the common 6,7-dioxygenated and 6,7,8-trioxygenated tetrahydroisoquinoline alkaloids, some of these species also contain the far less common 7,8-dioxygenated and 5,6,7-trioxygenated analogs.

The alkaloids of the "sinita" cactus *Lophocereus schottii (37–39a)* are unique among the isoquinoline alkaloids in having an isobutyl substituent at C-1, e.g., lophocerine (**20**) and the trimeric pilocereine (**67**). The latter may be formed *in vivo* from lophocerine by phenolic oxidative coupling (*40*). Pilocereine was first isolated in 1901 by Heyl who arrived at the empirical formula $C_{30}H_{44}O_4N_2$ (*41*). The isolation and structural determination of lophocerine and pilocereine was accomplished by Djerassi and co-workers (*37–39*). Pilocereine was also isolated from the related species *Lophocereus australis* and *Lophocereus gatesii,* and together with lophocerine, was isolated from one of the most common cacti of central Mexico *Pachycereus marginatus* known as "organo" (*42*).

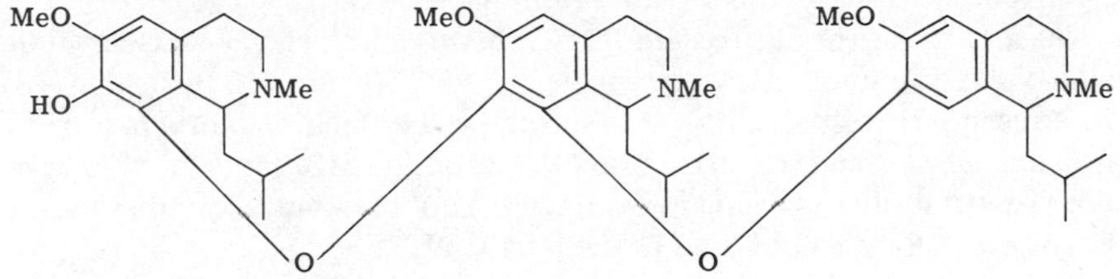

Pilocereine
67

The total alkaloidal content of *L. schottii* has been estimated at 3.7% of the dried plant (*37*). An investigation of the location of alkaloids in different tissues showed that the major portion resides in the skin (green epidemis), a minor portion in the cortex, and practically no alkaloids in the central core (*37*).

The alkaloids of the "saguaro" or giant cactus *Carnegiea gigantea* were first investigated in 1928 by Heyl (*43*) who isolated carnegine (**18**), the structure of which was elucidated by Späth (*44*). Salsolidine (**16**), the N-demethylated counterpart of carnegine, was isolated from this species some 40 years later (*45–47*).

Gigantine (**35**) was isolated from *C. gigantea* by Hodgkin *et al.* and at first assigned the structure 4-hydroxy-6,7-dimethoxytetrahydroisoquinoline (*48*). This structure was later questioned (*49*) and also rejected on the basis of its nonidentity with both the cis and trans isomers of the 4-hydroxy-6,7-di-methoxytetrahydroisoquinoline synthesized by Brossi *et al.* (*50*). The correct 5,6,7-trioxygenated structure (**35**) was later proved independently by both Kapadia *et al.* (*51*) and the original group (*46*).

Further work by Bruhn and Lundström (*47*) revealed the presence of the

7,8-dioxygenated arizonine (15) in the alkaloidal fraction of *C. gigantea.* These authors also reported the total alkaloidal content to be 0.6% in the fresh plant growing in its natural habitat and the major constituent (50%) of the alkaloid fraction to be salsolidine (16). Brown *et al.* (46) found carnegine (18) to be the major alkaloid (70% of the alkaloid fraction). Seasonal changes in the concentrations of these two compounds, like those described for peyote, may occur (12).

The first simple tetrahydroisoquinoline alkaloids encountered in nature with the 7,8-dioxygenated and the 5,6,7-trioxygenated substitution patterns were tepenine (19) and tehaunine (34), respectively; both compounds were isolated from *Pachycereus tehauntepecanus* (52). The 7,8-dioxygenated substitution pattern in isoquinoline alkaloids is unusual and, outside the family of Cactaceae, is only known in the benzylisoquinoline petalin (53), in cularine bases (54), and some of the protoberberines (54). The 5,6,7-trioxygenated pattern is also unusual although it is found in certain 1-benzyltetra-hydroisoquinolines in other plant families (54).

In early studies of *Pachycereus pecten-aboriginum,* Heyl (41) isolated the alkaloid "pectenine," shown by Späth (55) to be identical to carnegine (18). In a recent reinvestigation of this species, Bruhn and co-workers (56, 57) isolated salsolidine (16), arizonine (15), salsoline (13), and the two compounds isosalsoline (14) and heliamine (10) not previously found in nature. Carnegine (18) was not found in the later study.

The tallest mexican columnar cactus "candelabro" or *Pachycereus weberi* was recently reinvestigated by Mata and McLaughlin (58) and shown to contain a great variety of differently substituted simple tetrahydroisoquino-lines. Four new alkaloids were isolated, the 6-monooxygenated weberidine (2), the 7,8-dioxygenated lemaireocereine (7), the 5,6,7-trioxygenated norte-haunine (26), and the 5,6,7,8-tetraoxygenated weberine (40). Tetraoxygen-ated isoquinolines had not been found before. An additional four alkaloids, known from other species, were also isolated, namely heliamine (10), *N*-methylheliamine (11), pellotine (36), and tehaunine (34). The authors conclude that the presence of this variety of differently substituted tetrahy-droisoquinolines in the same species shows that this cactus has a highly evolved enzyme system for ring oxidation and O-methylation.

McLaughlin and co-workers have also studied the alkaloidal fraction of two other species closely related to *P. weberi.* In *Pachycereus pringlei* heliamine (10), tehaunine (34), lemaireocereine (7), and weberine (40) were identified (59). *Backebergia militaris* (60, 61) was shown to contain heli-amine (10) and lemaireocereine (7).

c. **Alkaloids in *Dolichothele* Species.** Cactus species belonging to the genus *Dolichothele* have ethnobotanic histories as narcotic "peyote" cacti (62) and early reports indicate that these species contain psychoactive or poisonous alkaloids (*2, 63*). Ranieri and McLaughlin investigated two *Dolichothele* species and isolated five new simple tetrahydroisoquinoline alkaloids (*64, 65*). The first species investigated, *Dolichothele longimamma* (*65*), contained the 6-monooxygenated longimammatine (**1**) and longimammosine (**3**), the 8-monooxygenated longimammidine (**4**), and the 4,8-dioxygenated longimammamine (**5**). *Dolichothele uberiformis* (*66*) was shown to contain longimammamine (**5**) and the 5,7-dioxygenated uberine (**6**).

A simple oxygenated substitutent at C-6 and C-8 as in the compounds (**1, 3–5**) are unusual for plant tetrahydroisoquinolines since tyrosine, the usual precursor, would give rise only to C-7 oxygenated derivatives. The simple tetrahydroisoquinolines **1–6** occur in these *Dolichothele* species together with phenethylamine and phenethanolamine derivatives such as hordenine (**68**), synephrine (**69**), and ubine (**70**) (*64, 65*), none of which appear to be their direct progenitors. The authors speculate that the 6- and 8-monooxygenated tetrahydroisoquinolines may be biosynthetically cyclized from *meta*-tyramines; the aromatic substituents may be added by oxidation after cyclization of the phenethylamine has occurred; or the aromatic ring could represent dopa which has been reduced at the para position.

68   **69**   **70**

d. Alkaloids in Other Cactus Species. In five species, containing mainly alkaloids of the phenethylamine type, some tetrahydroisoquinolines of the 6,7-dioxygenated and 6,7,8-trioxygenated types have also been identified. *Pelecyphora asseliformis* contains anhalidine (**27**) as one of the major alkaloids and pellotine (**36**) in trace amounts (*66, 67*). Anhalonidine (**30**) and anhalidine (**27**) were identified as trace constituents in the alkaloid fraction of *Stetsonia coryne* (*66*) and anhalonidine (**30**) also in *Trichocereus pachanoi* (*3*).

Salsoline (**13**) was identified for the first time in the family of Cactaceae in *Echinocereus merkerii* (*68*) and *N*-methylheliamine (i.e., *O*-methylcorypalline) (**11**) in *Pilosocereus guerreronis* (*69*).

The unusual glucoalkaloid pterocereine (**41**) was isolated as the major alkaloid in *Pterocereus gaumeri* (*70*). Acid employed in the extraction procedure decomposed the alkaloid (**41**) to the new phenolic alkaloid deglucopterocereine (**42**) and glucose. *β*-glucosidase also hydrolyzed the glucoalkaloid (*70*). The hydroxymethyl group at C-1 and the phenolic glucoside are both uncommon among tetrahydroisoquinoline alkaloids. The hydroxymethyl substituent was previously known only in calycotomine (**21**).

2. Alkaloids in Other Plants

a. Isoquinolines. Simple isoquinoline alkaloids are occasionally found in plant families other than the Cactaceae (Table VI). In the family of Chenopodiaceae, closely related to the Cactaceae, four species are known to contain simple tetrahydroisoquinolines. *Salsola arbuscula,* investigated by Orekhov and Proskurina in the 1930s (*72 – 75*), was the first known natural source of salsoline (**13**) and salsolidine (**16**). In more recent studies of this family, Carling and Sandberg (*76*) isolated *N*-methylisosalsoline (**17**) and carnegine (**18**) from *Haloxylon articulatum.* It may be pertinent to mention that another species of this genus *H. salicornum* (Moq.-Tand.) Boiss is rich in substituted β-phenethylamines (*77*).

Ghosal *et al.* have isolated salsoline (**13**) and salsolidine (**16**) from the three *Leguminosae* plants *Desmodium tiliaefolium* (*78*), *Desmodium cephalotes* (*79*), and *Alhagi pseudalhagi* (*80, 80a*), respectively. Similar to the situation in cactus species these Leguminosae species contain a variety of substituted β-phenethylamines. *Desmodium tiliaefolium* also contains simple indole bases and this species is unique with respect to the simultaneous occurrence of three alkaloidal types, namely, β-phenethylamine, simple tetrahydroisoquinoline and indole-3-alkylamine (*78*).

Calycotomine (**21**) was isolated by White from another Leguminosae species, *Calycotome spinosa* (*81*). The hydroxymethyl substitent of calycotomine at C-1 is uncommon among isoquinoline alkaloids and has only recently been found again in the cactus alkaloid pterocereine (**41**) (*70*).

Hydrohydrastinine (**8**) is known as a degradation product of hydrastine and cotarnine but as shown by Späth and Julian (*82*), it also occurs naturally in *Corydalis tuberosa.*

Corypalline (**9**) was first isolated by Manske (*83*) from *Corydalis pallida* and from seeds of *C. aurea.* More recently, corypalline (**9**) was isolated by Chen *et al.* (*84*) from the Atherospermataceae (Monimiaceae) species *Doryphora sassafras,* where it co-occurs with the isoquinolones doryanine (**79**) and doryfornine (**80**). The route by which corypalline (**7**) is formed in *D. sassafras* is not known, but a plausible one would be an *in vivo* reduction (*85*) of doryfornine (**80**), which in turn is formed by oxidation (see below) of more complex molecules (*86*). Another possibility suggested (*85*) would be that *D. sassafras* possesses a metabolic pathway leading directly to simple isoquinolines. Similar to the situation in *Doryphora,* Wu *et al.* (*87*) recently found corypalline (9) co-occurring with isoquinolones and more complex benzyl-isoquinolines in *Thalictrum rugosum*.

The optically active 1-methylcorypalline (17) was claimed to be first isolated as a natural product from *Corydalis ambigua* (88), but the racemic compound had earlier been found in *Haloxylon articulatum* (76) and given the name *N*-methylisosalsoline (17).

Hydrocotarnine (22) was isolated by Hesse (89) from opium (*Papaver somnifera*). The proposal that this alkaloid is a degradation product of a more complex alkaloid in *Papaver* (90) may, in light of recent knowledge in the biogenesis of simple isoquinolines, deserve consideration.

Salsolinol (12), known as a "mammalian alkaloid" (91), has been isolated from banana (92) and powdered cocoa (93). A major pathway of dopamine metabolism in the banana involves condensation with endogenous acetaldehyde to form salsolinol. This process accelerates during advanced stages of ripening due to increased production of acetaldehyde. In powdered cocoa (93), salsolinol is present at a concentration of 40 μg/g as determined by liquid chromatography.

Two quaternary alkaloids have been found. The dihydroisoquinoline pycnarrhine (50) was isolated from the Indonesian medicinal plant *Pycnarrhena longifolia* which also contains other alkaloids (94). The alkaloid fraction of *Thalictum revolutum* (95) mainly contains benzylisoquinoline-type alkaloids and as a minor constituent *N*-methyl-6,7-dimethoxyisoquinolinium chloride (51).

b. Isoquinolones. The isoquinolones can be considered to form a distinct group of alkaloids. The naturally occurring isoquinolones reported so far are presented in Table VII (structures 71–83). The substitution patterns found in these compounds are of the 6,7-dioxygenated or 5,6,7-trioxygenated types and compounds of both types may exist in the same species.

The first example of a naturally occurring isoquinolone was doryanine (79) isolated by Gharbo *et al.* (86) from the Australian sassafras tree *Doryphora sassafras*. Doryphornine (80) was later isolated from the same species (84). *Doryphora sassafras* mainly produces alkaloids derived from benzylisoquinoline units and the authors of the study (86) assume that the origin of the isoquinolones (79) and (80) most likely is a result of biochemical oxidation of these major products. Indirect support for this hypothesis was available from the presence of the isoquinolone (81) and the more complex hernandaline in the same species *Hernandia ovigera*, both possibly resulting from the oxidation of the dimeric benzylisoquinolineaporphine dimer thaliacarpine (104, 105).

Most of the compounds of Table VII appear to be minor constituents in plants mainly containing alkaloids of the more complex benzylisoquinoline

TABLE VI

SIMPLE ISOQUINOLINE ALKALOIDS IN PLANTS OF FAMILIES OTHER THAN THE CACTACEAE

Family/species	Alkaloid	Reference
Alangiaceae		
Alangium lamarchii Thw.	Salsoline (13)	96
Annonaceae		
Enantia polycarpa Engl. et Diels	Corydaldine (74)	110
Chenopodiaceae		
Corispermum leptopyrum L.	Salsolin (13)	97
	Salsolidine (16)	97
Haloxylon articalatum (Cav.) Bge	Carnegine (18)	76
	N-Methylisosalsoline (17)	76
Salsola arbuscula Pall.	Salsoline (13)	73–75
	Salsolidine (16)	73–75
Salsola kalii	Salsoline (13)	98
	Salsolidine (16)	98
Euphorbiaceae		
Euphorbia myrsinitis L.	1-Methyl-3-carboxy-6-hydroxytetrahydroisoquinoline (84)	118
Fumariaceae		
Corydalis ambigua Cham. et Schlecht	(+)-1-Methylcorypalline (*N*-Methylisosalsoline) (17)	88
Corydalis aurea	Corypalline (9)	83
Corydalis ophiocarpa Hook et Thoms	Corypalline (9)	99
Corydalis pallida	Corypalline (9)	83
Corydalis tuberosa	Hydrohydrastinine (8)	82
Fumaria schleicheri Soy.-Will	Oxyhydrastinine (72)	100
Hernandiaceae		
Hernandia ovigera L.	*N*-Methyl-6,7-dimethoxyisoquinolone (81)	104
Leguminosae		
Alhagi pseudalhagi (Bieb.) Desv.	salsolidine (16)	80, 80a
Calycotome spinosa Link	calycotomine (21)	81
Cassia siamena Lam.	siamine (83)	108, 109
Desmodium cephalotes Wall.	salsolidine (16)	79
Desmodium tiliaefolium G. Don	salsoline (13)	78
Genista purgens	salsoline (13)	101
	salsolidine (16)	101
Mucuna deeringiana (Bort.) Merr	(−)-1-Methyl-3-carboxy-6,7-dihydroxytetrahydroisoquinoline (86)	120
Mucuna mutisiana DC.	L-3-Carboxy-6-7-dihydroxytetrahydroisoquinoline (85)	119

TABLE VI (*Continued*)

Family/species	Alkaloid	Reference
Menispermaceae		
Pycnarrhena longifolia (Decne. ex Miq.) Beccari	Pycnarrhine (**50**)	*94*
Monimiaceae (Atherospermataceae)		
Doryphora sassafras Endlicher	Corypalline (**9**)	*84*
	Doryanine (**79**)	*84*
	Doryfornine (**80**)	*84*
Musaceae		
Musa paradisica	Salsolinol (**12**)	*92*
Nymphaceae		
Nelumbo nucifera	*N*-Methylheliamine (**11**)	*102*
Papaveraceae		
Papaver bracteatum Lindl. var Arya I	*N*-Methylcorydaldine (**75**)	*111*
Papaver somniferum L.	Hydrocotarnine (**22**)	*89*
Papaver urbanianum Fedde	*N*-Methylcorydaldine (**75**)	*112*
Ranunculaceae		
Thalictrum alpinum L.	Noroxyhydrastinine (**71**)	*113*
	N-Methyl-6,7-dimethoxyisoquinolone (**81**)	*113*
Thalictrum dioicum L.	*O*-Methylcorypalline (**11**)	*103*
Thalictrum flavum	Thalflavine (**76**)	*114*
Thalictrum fendleri Engelm. ex Gray	*N*-Methylcorydaldine (**75**)	*115*
	N-Methylthalidaldine (**77**)	*115*
Thalictrum foliolosum	Noroxyhydrastinine (**71**)	*116*
Thalictrum isopyroides C. A. Mey	*N*-Methyl-6,7-dimethoxy-isoquinolone (**81**)	*117*
Thalictrum minus L.	Thalactamine (**82**)	*106*
Thalictrum minus var. adiantifolium Hort.	Noroxyhydrastinine (**71**)	*107*
	Thalifoline (**73**)	*107*
Thalictum revolutum DC.	*N*-Methyl-6,7-dimethoxyisoquinolinium chloride (**51**)	*95*
Thalictrum rugosum Ait.	Corypalline (**9**)	*87*
	Noroxyhydrastinine (**71**)	*87*
	6,7-Methylenedioxyisoquinolone (**78**)	*87*
Rahmnaceae		
Ziziphus amphibia A. Cheval	Amphibin I (**87**)	
Sterculiaceae		
Theobroma cacao	Salsolinol (**12**)	*93*

TABLE VII
NATURALLY OCCURRING ISOQUINOLONES

Number	Name	Structure	Molecular formula	Melting point (°C)	Species	Reference
71	Noroxyhydrastinine		$C_{10}H_9NO_3$	182–183 (107) 185–186 183–184 (87)	Thalictrum alpinum Thalictrum foliolosum Thalictrum minus v. adiantifol. Thalictrum rugosum	113 116 107 87
72	Oxyhydrastinine		$C_{11}H_{11}NO_3$	—	Fumaria schleicheri	100
73	Thalifoline		$C_{11}H_{13}NO_3$	210–211	Thalictrum minus v. adiantifol.	107
74	Corydaldine		$C_{11}H_{13}NO_3$	—	Enantia polycarpa	110
75	N-Methylcorydaldine		$C_{13}H_{15}NO_3$	123–125 (112)	Papaver bracteatum Papaver urbanianum Thalictrum fendleri	111 112 115

No.	Name	Structure	Formula	mp (°C)	Source	Ref.
76	Thalflavine	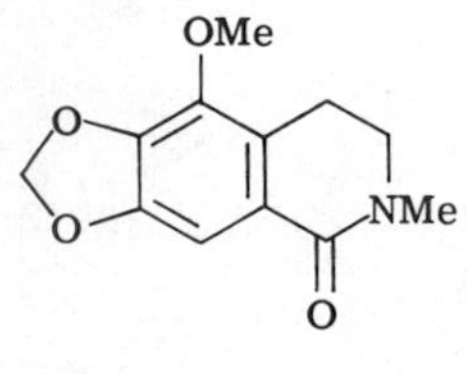	$C_{12}H_{13}NO_4$	—	*Thalictrum flavum*	114
77	*N*-Methylthalidaldine	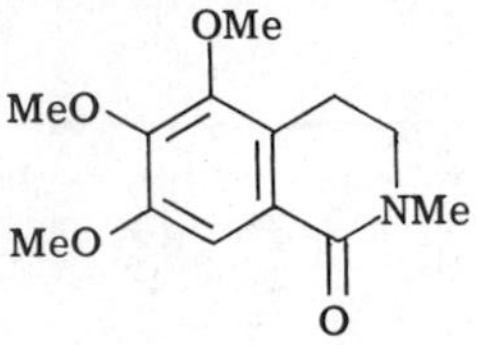	$C_{13}H_{17}NO_4$	(oil)	*Thalictrum fendleri*	115
78	6,7-Methylenedioxy-1-isoquinolone	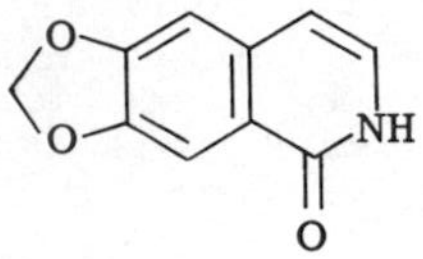	$C_{10}H_7NO_3$	268–270	*Thalictrum rugosum*	87
79	Doryanine	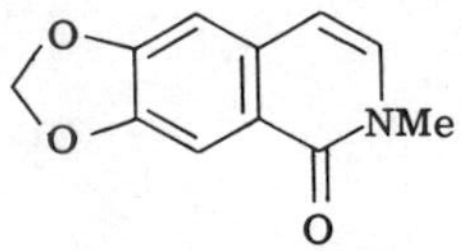	$C_{11}H_9NO_3$	160–162	*Doryphora sassafras*	86 84
80	Doryfornine	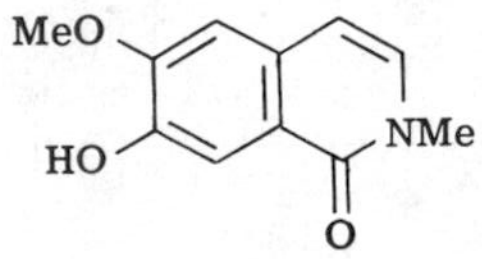	$C_{11}H_{11}NO_3$	215–217	*Doryphora sassafras*	84

(Continued)

TABLE VII (Continued)

Number	Name	Structure	Molecular formula	Melting point (°C)	Species	Reference
81	N-Methyl-6,7-dimethoxy-1-isoquinolone	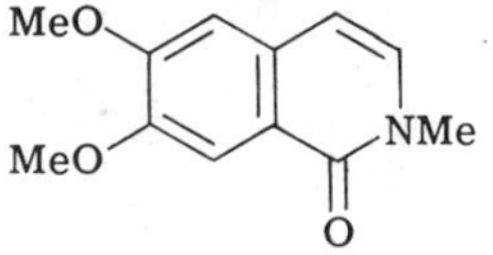	$C_{12}H_{13}NO_3$	112–113 (113)	Hernandia ovigera Thalictrum alpinum Thalictrum isopyroides	104, 105 113 117
82	Thalactamine	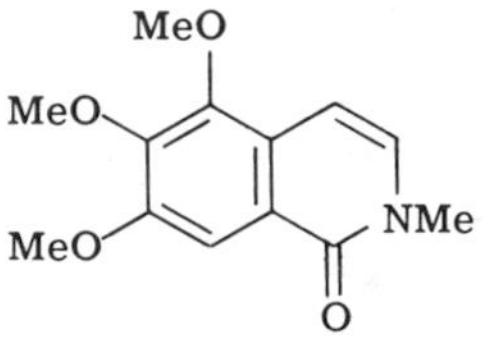	$C_{13}H_{15}NO_3$	112–114	Thalictrum minus	106
83	Siamine	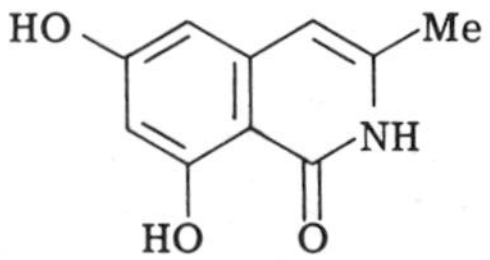	$C_{10}H_9NO_3$	268	Cassia siamena	108 109

type. However, in two varieties of *Thalictrum minus,* isoquinolones appear to be the major alkaloids. Mollow and Dutschewska isolated thalactamine (82) from a variety of *T. minus* growing near the Black Sea (*106*). Thalifoline (73) and noroxyhydrastinine (71) were isolated by Doskotch *et al.* (*107*) from *T. minus* var. *adiantifolium*. Based on the evidence previously cited, the latter authors anticipated the presence in *T. minus* of benzylisoquinoline alkaloids bearing noroxyhydrastinine and thalifoline moieties.

Siamine (83) isolated from *Cassia siamena* by Ahn and Zymalkowski (*108*) is an unusual isoquinolone containing two phenolic groups in a meta relationship and a methyl at C-3. This compound is probably not derived biogenetically from tyrosine but rather from structurally related polyketides that occur in the same plant (*109*).

c. **Alkaloidal Amino Acids.** Three 3-carboxyl-substituted simple tetrahydroisoquinoline alkaloids are known from natural sources (Table VIII; structures 84–86). Müller and Schütte (*118*) isolated the compound 84 from *Euphorbia myrsinitis* that is thought to be derived biogenetically from *m*-tyrosine by condensation with acetaldehyde or its equivalent.

L-Dopa (L-3,4-dihydroxyphenylalanine) occurs in various species of the legume genera and the seeds of some *Mucuna* species are especially rich in this amino acid (*119*). During extraction and isolation of L-dopa from *Mucuna* species, the two new alkaloids 85 and 86 were found as minor constituents. Bell and Nulu (*119*) isolated the amino acid 85 from the seeds of *M. mutisiana* and Daxenbichler *et al.* (*120*) found 86 in extracts of the velvet bean *M. deeringiana*. The compounds may be formed *in vivo* from L-dopa by condensation with formaldehyde and acetaldehyde or their equivalents to yield 85 and 86, respectively. The stereochemistry of the latter two compounds 85 and 86 of Table VIII is known (*119–121*).

d. **Amphibin I.** The interesting peptide alkaloid amphibin I (87) was isolated by Tschesche and associates (*122*) from *Ziziphus amphibia* A.

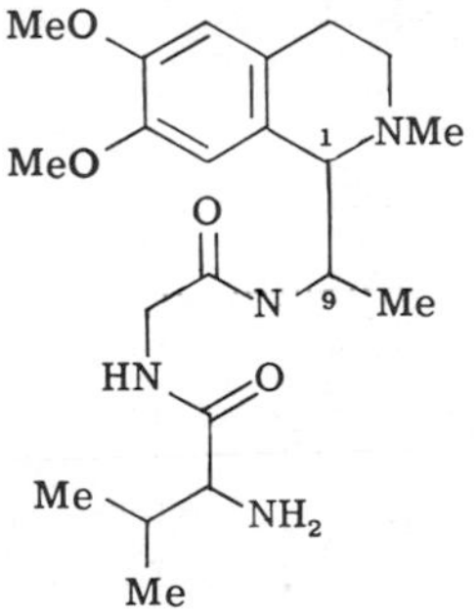

$C_{21}H_{34}N_4O_4$, $[\alpha]_D^{20}$: $-50°(c = 0,6,$ benzene)
2 HCl salt, $[\alpha]_D^{20}$: $-3.4°(c = 0,24,$ MeOH)
mp (dec) 175°

Amphibin I

87

TABLE VIII
ALKALOIDAL AMINO ACIDS

No.	Compound	Structure	Formula	Melting point (°C)	Species	Reference
84	1-Methyl-3-carboxy-6-hydroxytetrahydroisoquinoline		$C_{11}H_{13}NO_3$	269–271	*Euphorbia myrsinites*	*118*
85	S-(−)-3-carboxy-6,7-dihydroxytetrahydroisoquinoline		$C_{10}H_{11}NO_4$	286–288 $[\alpha]_D^{25}$ −114.9 (c = 1.65, 20% HCl)	*Mucuna mutisiana*	*119*
86	1S,3-(−)-3-carboxy-6,7-dihydroxy-1-methyltetrahydroisoquinoline		$C_{11}H_{13}NO_4$	280–281 (dec.) $[\alpha]_D^{25}$ −142.8 (c = 0.7, 6 N HCl)	*Mucuna deeringiana*	*119*

Cheval. This compound may be regarded as a simple tetrahydroisoquinoline alkaloid with a peptide chain at C-1. *Ziziphus* species contain both tetrahydroisoquinoline alkaloids and macrocyclic peptide alkaloids, and the presence of amphibin I in *Z. amphibia* might possibly indicate a biological connection between these two different classes of alkaloids (*122*).

Spectroscopic and synthetic work has shown that natural amphibin I consists of a mixture of two enantiomers with the configurations C-1 = S, C-9 = R and C-1 = R, C-9 = S (*123*). The terminal valine has in both cases the S configuration.

B. Isolation and Identification Procedures

1. Isolation Procedures and Analysis

The methods employed for isolation depend on the nature of the compounds, and specific conditions may therefore have to be devised for the selective isolation of a particular type. Usually, fresh or dried plant material is extracted with alcohol and the extract obtained is further fractionated by precipitation or extraction with variation of pH (*84*). Extraction columns may be useful in the purification of alkaloid fractions (*124*). The procedure of isolation may be oriented by subfractionation into phenolic and nonphenolic mixtures, preferably using ion-exchange materials (*124*). For further isolation of separate compounds, column chromatography (*47*), preparative thin-layer chromatography (*125*), and preparative gas chromatography (*30*) may be used.

A number of thin-layer (*124, 34, 64*), liquid (*255*), and gas (*126, 3, 127*) chromatographic methods have been reported for the separation and identification of simple isoquinoline alkaloids and related compounds. Spray reagents particularly useful for detection of these compounds on thin-layer chromatoplates are dansyl chloride (*128*) and fluorescamine (*129*) for nonphenolic products, and tetrazotized benzidine or di-*O*-anisidine for phenolic compounds (*128, 124*). Gibb's reagent is particularly useful for detection of phenolic compounds since it produces a characteristic blue color with phenols having free para positions (*46*).

Some of the products reviewed here have not been isolated in crystalline form from natural sources. In many studies on, e.g., cactus alkaloids, work has been guided by biogenetic considerations and the existence of certain compounds was anticipated or assumed, and it was a matter of proving or disproving their presence in given fractions. Gas chromatography – mass spectrometry is a useful analytical and diagnostic tool which has proved its practical value in the rapid identification of minute ($< 1\ \mu g$) amounts of a compound in a complex mixture. This technique has been used, e.g., by Agurell and Lundström (*130, 12, 30*) and Kapadia et al. (*29, 32, 33*) in the study of trace constituents of peyote and as a rapid screening method for alkaloids in small-sized plant materials.

Mass fragmentography is another advanced method for the identification of trace substituents in complex mixtures where the mass spectrometer is used as a gas chromatographic detector, continuously monitoring one to eight selected mass numbers of compounds eluted from the gas chromatograph (131). This method has been used by Lindgren et al. for identification of biosynthetic intermediates (132) and trace alkaloids (69).

Mass-analyzed ion kinetic energy spectrometry (MIKES) is a new approach to mixture analysis that has been applied to the direct detection of alkaloids in plant materials. MIKES is a particular approach to mass spectrometry/mass spectrometry in which both separation of individual constituents and their subsequent identification are accomplished by mass spectrometry (133–135). A great advantage with this method is that the usual work-up procedures may be minimized or eliminated e.g., direct analyses of dry plant material are possible (136). Unger et al. (137) has used this method to screen nine related Mexican columnar cactus species for the presence of a related series of tetrahydroisoquinoline alkaloids.

UV, CD, and ORD spectra of simple isoquinoline alkaloids have been reviewed (138). Mass spectra and nuclear magnetic reasonance (NMR) spectra are briefly outlined below.

2. Mass Spectroscopy

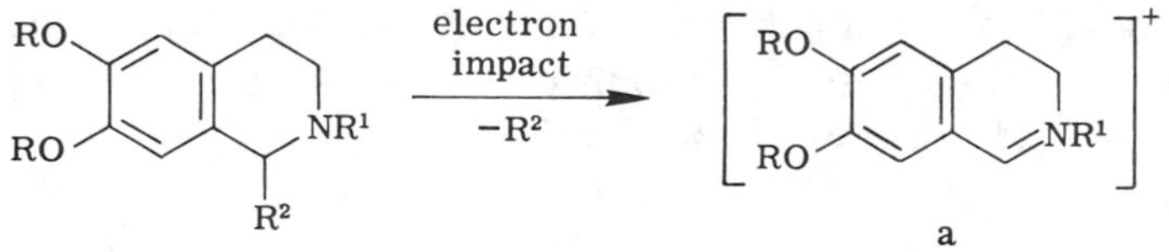

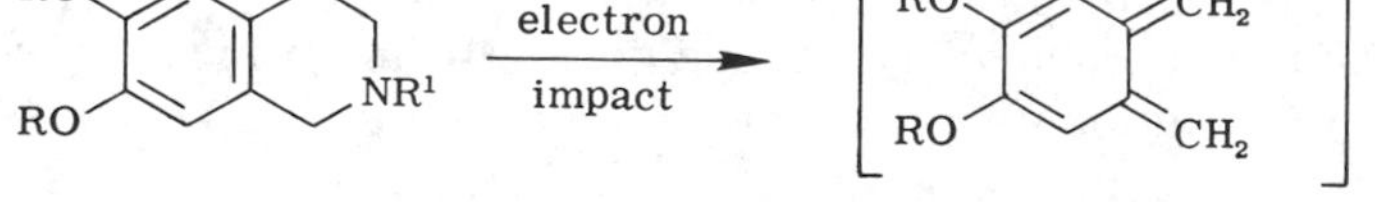

Electron impact mass spectra of simple tetrahydroisoquinolines bearing C-1 substituents exhibit molecular ions of very low intensities and molecular peaks are often less than 1% of the base peaks. The primary reaction involves expulsion of the C-1 substituent (139) giving the highly stabilized dihydroisoquinolinium ion species a. Abundant peaks in spectra of C-1-unsubstituted tetrahydroisoquinolines are due to ions (species b) formed by collapse of the heterocyclic system through the retro-Diels-Alder reaction (7). Molecular ions are of medium-high intensities with the latter type of compounds (30).

To locate the molecular ion in C-1-substituted tetrahydroisoquinolines it is desirable to run either a chemical ionization mass spectrum (140) or a field desorption mass spectrum (141). In the former technique, quasimolecular ions at m/z ($M^+ + 1$) or ($M^+ - 1$) having relatively long halftimes are generated. As an example (140), the chemical ionization mass spectrum of O-methylpellotine (38) gives the ions m/z 252 (59%, M + 1), 250 (98%, M − 1), and 236 (100%, M − 15). Field desorption mass spectrometry of simple tetrahydroisoquinolines results in base peaks of M or M + 1 (141). The spectrum (141) of salsolidine (16) shows the ions m/z 207 (100%, M^+) and 208 (17%, M + 1).

The primary reactions exhibited by 3,4-dihydroisoquinolones may be exemplified by the spectrum (electron impact) of thalifoline (71) (107).

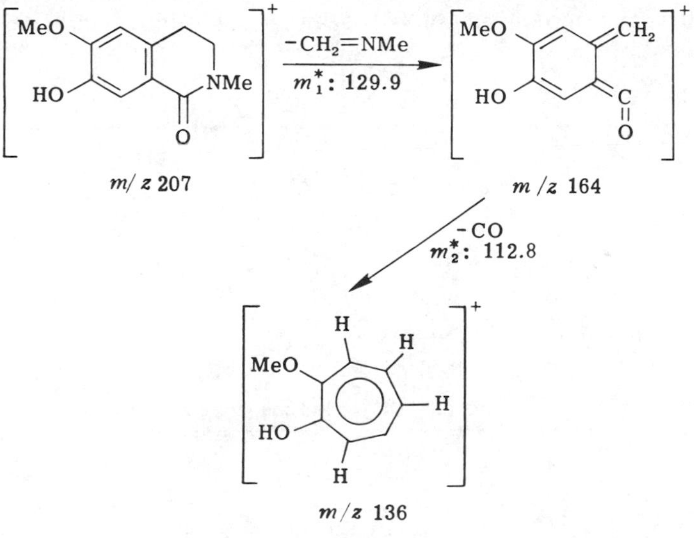

m/z 207 m/z 164

m/z 136

Further fragmentations are dependent upon ring substituents (107, 115). The aromatic isoquinolones exhibit primary fragmentations similar to those of the 3,4-dihydroisoquinolones (84, 87). Distinct metastable peaks (m^*) have been observed in the spectra of the isoquinolones (107).

3. NMR Spectroscopy

Proton magnetic resonance (^{1}H-NMR) spectra are particularly useful in determining the number and position of aromatic substituents in simple tetrahydroisoquinoline alkaloids. This is illustrated in a study by Schenker *et al.* (*142*) on monosubstituted tetrahydroisoquinolines from which the data on the fully methylated compounds are taken (Table IX). The 5- and 8-substituted derivatives can be distinguished from their 6- and 7-substituted isomers by the multiplicity of the signals of the aromatic protons; the 5- and 8-substituted compounds show a triplet (t) and two doublets (d), whereas the 6- or 7-substituted derivatives give rise to a doublet of doublets (dd) and two doublets. The aromatic substitution patterns may be further distinguished by the magnitude of the difference between the chemical shifts of the methylene groups at positions 1 and 4.

The ^{1}H-NMR spectra of salsolinol (**12**) and isosalsolinol (**12a**) illustrate the possibility of distinguishing 6,7- and 7,8-disubstituted compounds (*143*). The spectrum of salsolinol (**12**) contains two aromatic singlet signals at τ 3.20 and τ 3.17 (tertiary butanol methyl τ 8.70) with no apparent coupling between the two protons. The spectrum of isosalsolinol (**12a**) contains a two-proton AB quartet due to two aromatic proton signals at τ 3.21 and τ 3.05 with a coupling constant of 8 Hz (*143*). The natural abundance carbon

TABLE IX
^{1}H-NMR Spectral Data of Monosubstituted N-Methyltetrahydroisoquinolines[a]

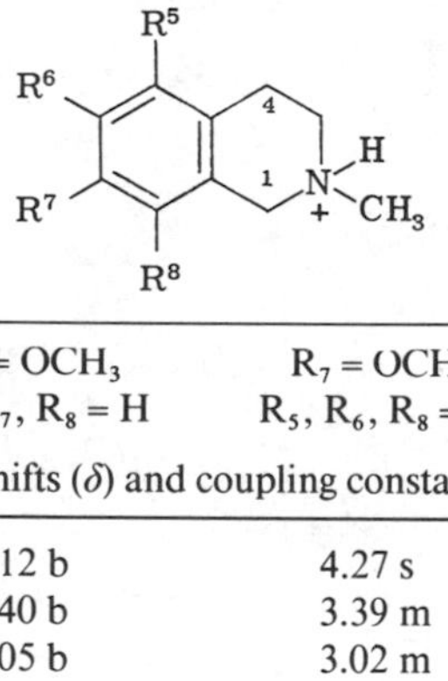

Group	$R_5 = OCH_3$ $R_6, R_7, R_8 = H$	$R_6 = OCH_3$ $R_5, R_7, R_8 = H$	$R_7 = OCH_3$ $R_5, R_6, R_8 = H$	$R_8 = OCH_3$ $R_5, R_6, R_7 = H$
	Chemical shifts (δ) and coupling constants (Hz)[b]			
CH-1	4.26 s	4.12 b	4.27 s	4.13 s
CH-3	3.38 m	3.40 b	3.39 m	3.36 m
CH-4	2.90 m	3.05 b	3.02 m	3.09 m
CH-5		6.77 d	7.12 d	6.86 d
			$J_o = 9$	$J_o = 8$
CH-6	6.71 d		6.83 dd	7.23 t
	$J_o = 8$		$J_m = 2.5$	
CH-7	7.20 t	6.88 dd		6.80 d
		$J_m = 2.5$		$J_o = 8$
CH-8	6.85	7.07 d	6.73 d	
	$J_o = 8$	$J_o = 9$		
OCH₃	3.76 s	3.72 s	3.71 s	3.79 s
NCH₃	2.79 s	2.83 s	2.82 s	2.85 s

[a] From Schenker *et al.* (*142*).
[b] Solvent: DMSO-d_6; Internal standard: Tetramethylsilane.

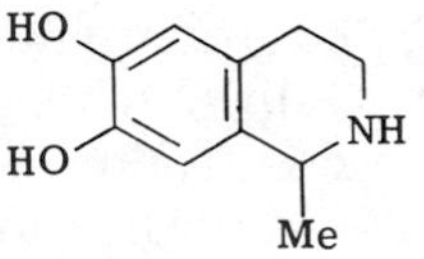

12 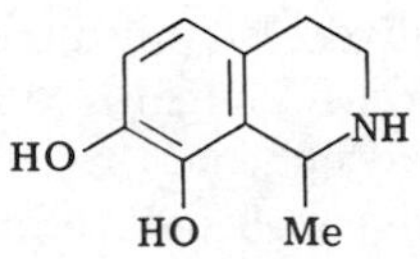   **12a**

magnetic resonance (^{13}C NMR) spectra have been used lately in structure elucidations of simple tetrahydroisoquinolines [e.g., pterocereine (**41**) and deglucopterocereine (**42**)] (*70*). The spectral data (*144*) for salsolidine (**16**)

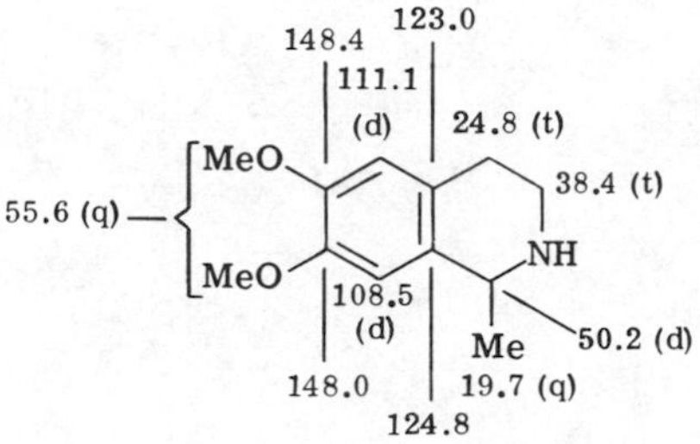

may serve as an example. ^{13}C-NMR spectra of simple isoquinoline alkaloids have been reviewed recently in this treatise (*145*).

C. SYNTHESIS

A wide variety of methods are available for the synthesis of isoquinolines (*54*). Some of the methods most frequently used for the preparation of the naturally occurring simple tetrahydroisoquinoline alkaloids are outlined below.

1. The Pictet–Spengler Cyclization

In this method, tetrahydroisoquinolines may be formed directly by condensation of an appropriately substituted phenethylamine with an aldehyde under acidic conditions (*146, 147*). Thus, the first synthesis of anhalamine (**23**) was realized by Späth and Röder (*16*) through condensation of 3,4-dimethoxy-5-benzyloxyphenethylamine (**88**) with formaldehyde.

In a later study, Brossi *et al.* (*148*) showed that the cyclization of (**88**) proceeds in both possible directions yielding equal amounts of anhalamine (**23**) and isoanhalamine (**24**):

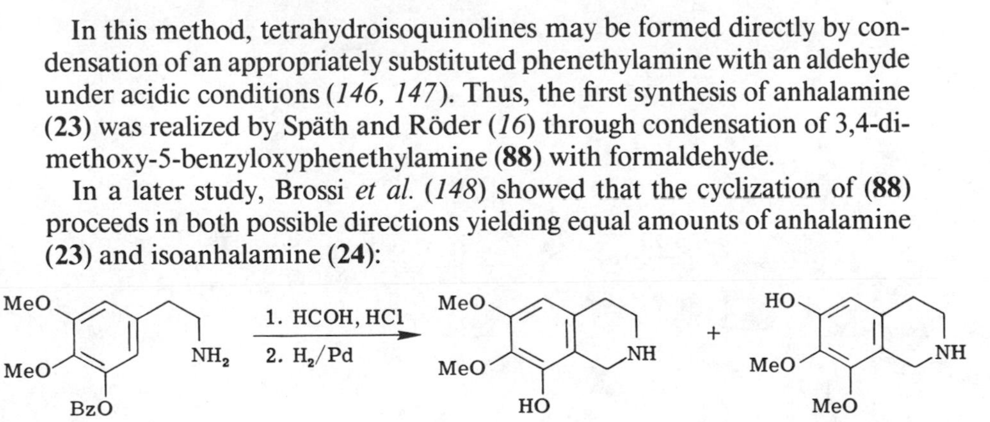

88   **23**   **24**

4-Hydroxytetrahydroisoquinolines may also be obtained by this procedure, and Ranieri and McLaughlin (65) recently synthesized longimammamine (5) from L-phenylephrine (89). Catalytic hydrogenation (149) of the two cyclization products 90 and 5 afforded the two other Dolichotele alkaloids 3

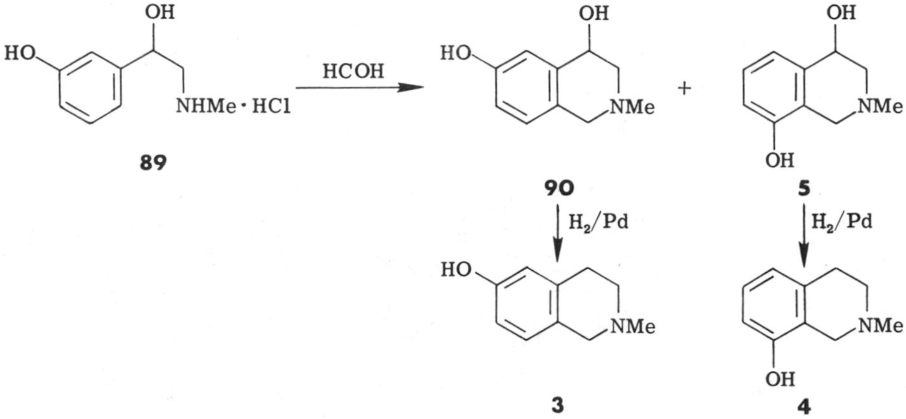

and 4 (65). Activating hydroxy or methoxy groups para and ortho to the ring-closing site facilitates the reaction. Thus, Castrillon (150), using the Eschweiler–Clarke reaction to effect N,N-dimethylation of mescaline (52), obtained anhalinine (25) resulting from Pictet–Spengler cyclization. Hordenine (68) and 2,5-dimethoxyphenethylamine were not cyclized under these conditions (150). Takahashi and Brossi (151) recently synthesized the newly discovered cactus constituent weberine (40) from 2,3,4,5-tetramethoxyphenethylamine (91) using the Pictet–Spengler cyclization.

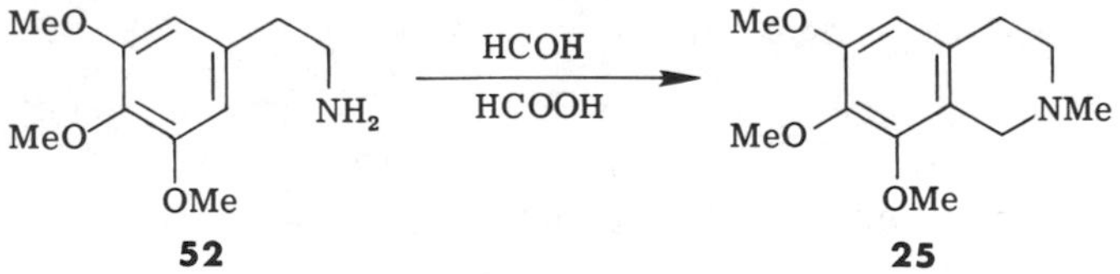

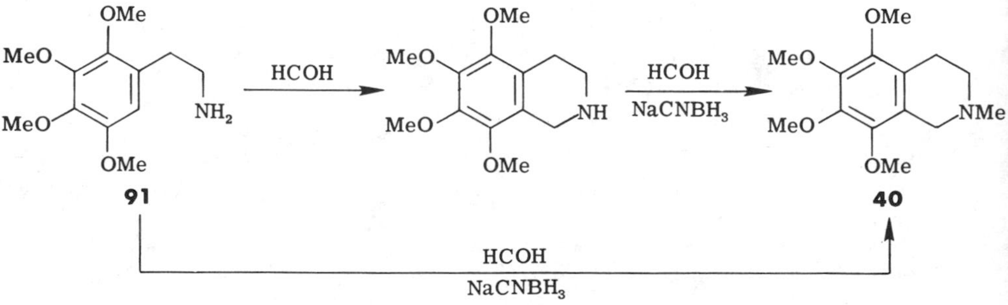

The Pictet–Spengler condensation has been regarded as a "biogenetic" type reaction and it was proposed early on (*146, 152, 153*) that analogous condensations of phenethylamine derivatives lead to tetrahydrisoquinoline alkaloids in plants. In their classical experiment lending support to this proposal, Schöpf and Bayerle (*154*) achieved the condensation of acetaldehyde with dopamine (**92**) under physiological conditions of pH, temperature, and concentration to produce salsolinol (**12**) in 83% yield. Kovacs and Fodor (*155*) further showed that 3-hydroxy-4-methoxyphenethylamine (**93**) under the same conditions yielded salsoline (**13**):

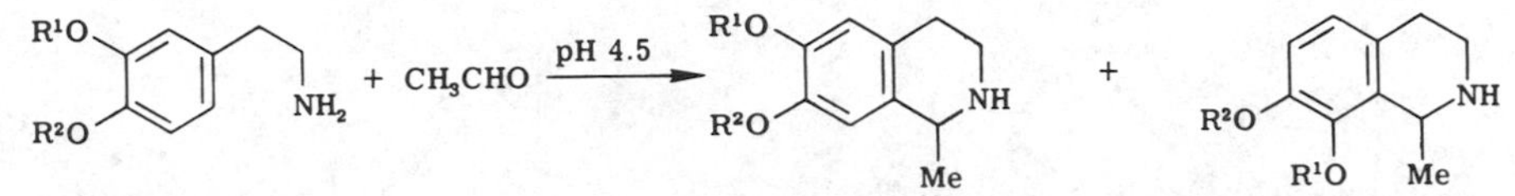

92 $R^1 = R^2 = H$ Dopamine
93 $R^1 = H, R^2 = Me$

12 $R^1 = R^2 = H$ Salsolinol
13 $R^1 = H, R^2 = Me$ Salsoline

12a $R^1 = R^2 = H$ Isosalsolinol
15 $R^1 = H, R^2 = Me$ Arizonine

Later reinvestigations of these reactions have shown that cyclization not only takes place para to the activating phenolic group, but also ortho to a lesser extent. Condensation of dopamine with acetaldehyde at pH 4.5 results in formation of salsolinol (**12**) and isosalsolinol (**12a**) in the ratio ~ 10:1 (*143*), and the same cyclization of **92** yields salsoline (**13**) and arizonine (**15**) in the approximate ratio 20:1 (*47*).

Cyclization of the trioxygenated compound **53** under the same conditions takes place primarily ortho to the free phenolic group yielding anhalonidine (**30**) and isoanhalonidine (**31**) in the ratio 30:1, and in fact, these products occur naturally in the peoyte cactus in the same ratio (*30*). Similarly,

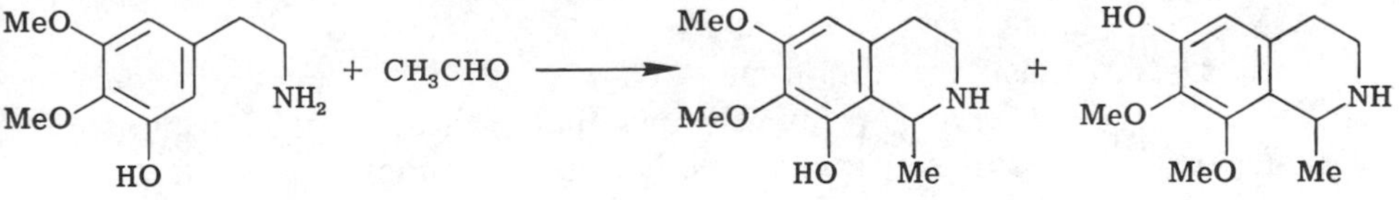

53

30

31

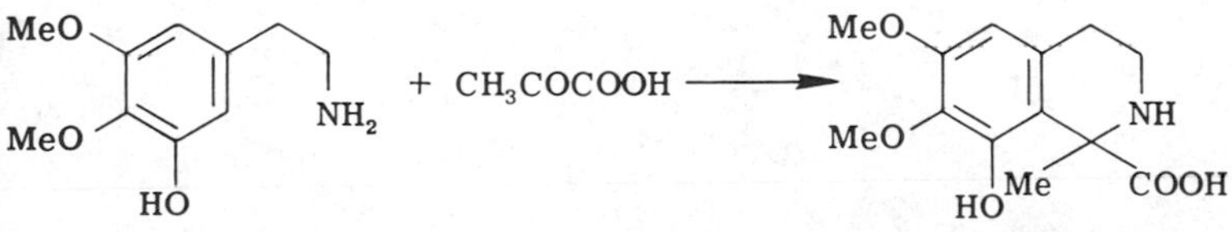

53

65

condensation under physiological conditions of **53** with pyruvic acid gives the amino acid **65** in quantitative yields (*33*). The formation of small amounts of an iso compound analogous to **31** has not yet been investigated.

To achieve 7,8-dioxygenated tetrahydroisoquinolines in the Pictet–Spengler reaction, Kametani *et al.* (*156*) made use of phenethylamines where one ortho position is blocked by bromine (e.g., **94**). Arizonine (**15**) has thus been synthesized (*47*) from 2-bromo-5-hydroxy-4-methoxyphenethylamine (**94**).

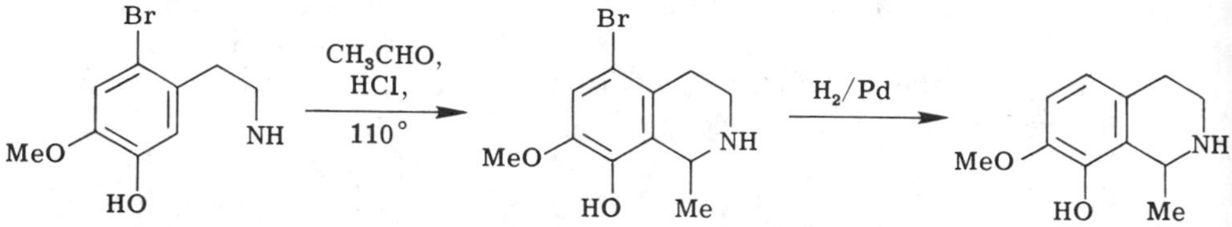

2. The Bischler–Napieralski Cyclization

The Bischler-Napieralski reaction is one of the most frequently used methods for the preparation of simple isoquinoline alkaloids (*157*). This reaction consists of the cyclodehydration, using phosphorus oxychloride and phosphorus pentoxide, etc., of an amide derived from a substituted β-phenethylamine to yield the corresponding 3,4-dihydroisoquinoline. Amides with electron-releasing groups on the aromatic ring are readily cyclized, while electron-withdrawing groups hinder the reaction. The synthesis of gigantine (**35**) may serve as an example of this reaction (*49*). Späth used the

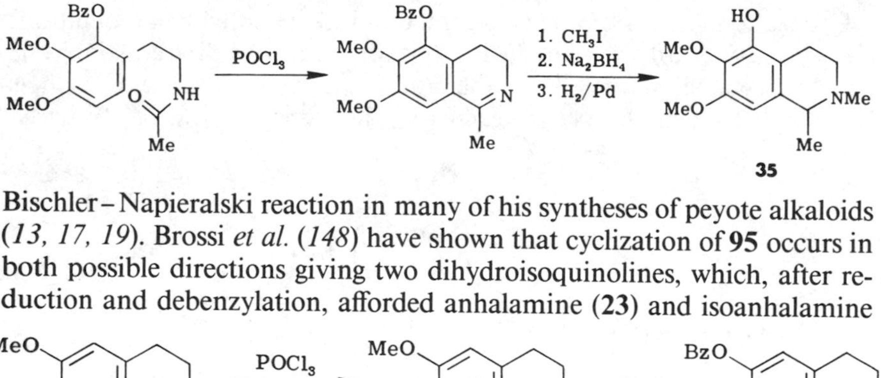

Bischler–Napieralski reaction in many of his syntheses of peyote alkaloids (*13, 17, 19*). Brossi *et al.* (*148*) have shown that cyclization of **95** occurs in both possible directions giving two dihydroisoquinolines, which, after reduction and debenzylation, afforded anhalamine (**23**) and isoanhalamine

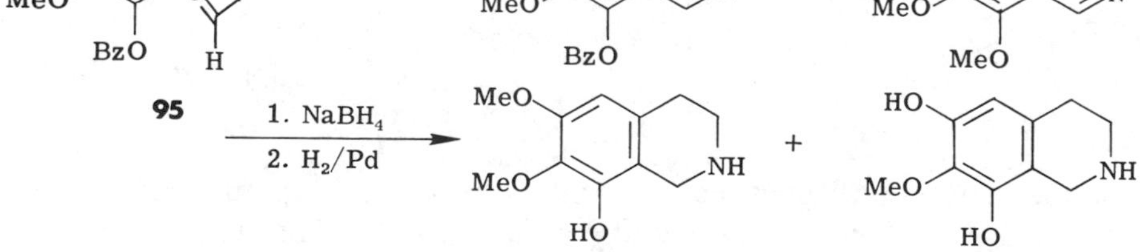

(24). Isoquinolones (e.g., **73** and **80**) have also been synthesized by use of the Bischler–Napieralski reaction (*84*).

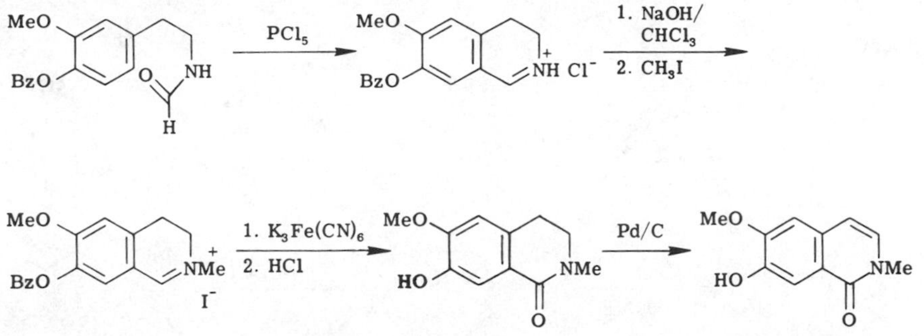

73   **80**

Recent studies on the mechanism of the Bischler–Napieralski reaction suggest that the formation of the dihydroisoquinolines occur via imidoyl chlorides (**96**) and thus dehydration or loss of carbonyl oxygen takes place

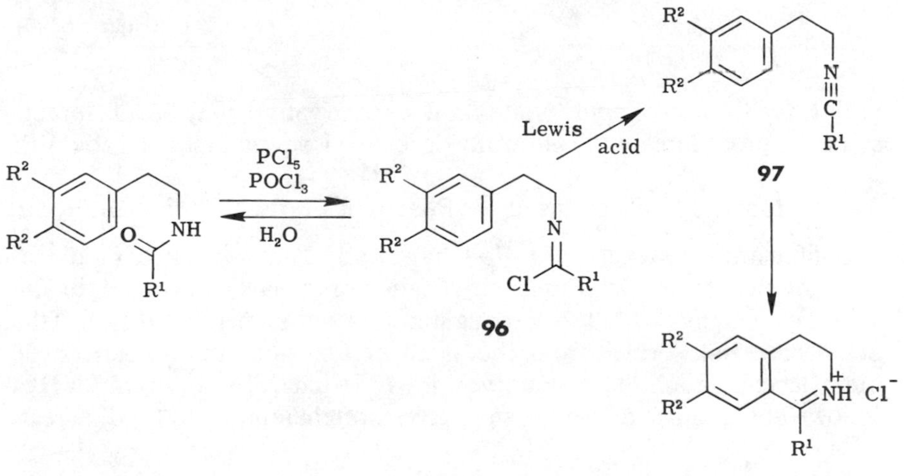

96

97

prior to ring closure (*158*). The rate of cyclization is enhanced by the addition of a Lewis acid that can be explained by the formation of a nitrilium salt (**97**). The two-step reaction required milder conditions (20–50°C) compared to the drastic classical conditions of refluxing at 100–200°C.

Brossi *et al.* (*159*) have used a modification of the Bischler–Napieralski cyclization in the synthesis of 6,7-dimethoxy-8-hydroxy-substituted tetrahydroisoquinolines of peyote. In this reaction the urethane **98** was cyclized with polyphosphoric acid (PPA) to the lactam or isoquinolone **99**, which, in good yield, underwent selective demethylation in the 8 position. After oxidation of the tetrahydroisoquinoline **101** to the imine **102**, this may be functionalized at C-1. Cyclization of appropriate urethane derivatives may be a favorable synthetic procedure for the naturally occurring isoquinolones of Table VII.

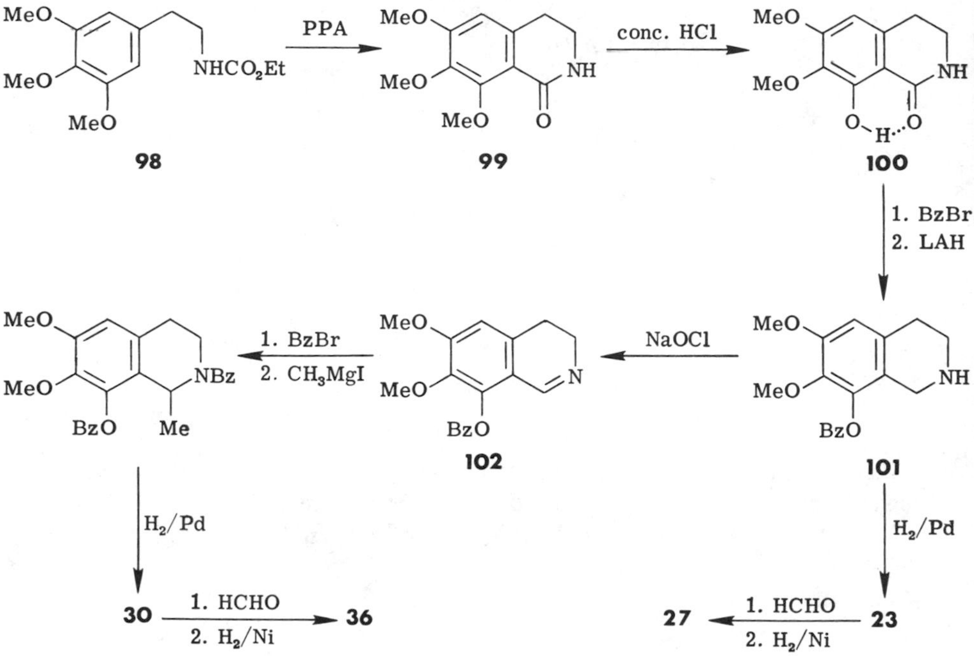

3. The Bobbitt Modifications of the Pomeranz–Fritsch Cyclization

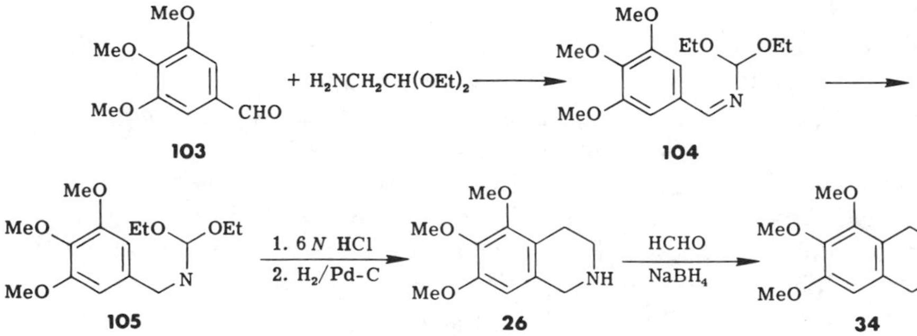

Bobbitt and co-workers (*160 – 164*) have worked out a variety of modifications of the Pomeranz–Fritsch cyclization that have been used in the synthesis of many naturally occurring simple isoquinolines. Reduction of the Schiff base **104**, formed from the aldehyde **103** and aminoacetaldehyde diethylacetal, affords the secondary amine **105** which, on standing in 6 N HCl followed by another reduction step, gives nortehaunine (**26**) (*160*). Treatment of the Schiff base **106** with an alkyl Grignard reagent affords an 1-alkyltetrahydroisoquinoline as in one synthesis (*47*) of arizonine (**15**).

Starting with an acetophenone (**107**) also yields 1-methyltetrahydroiso-quinolines as in the synthesis (*165*) of anhalonidine (**30**).

Mannich reaction of an appropriate phenol with formaldehyde and a suitably substituted aminoacetal may be a more direct route to the interme-diate benzylaminoacetal as in a synthesis (*164*) of anhalidine (**27**). The

cyclization proceeds through 4-hydroxytetrahydroisoquinolines, and such compounds (eg., **108**) can be isolated if the last reduction step is omitted (*162*).

4. Other Synthetic Procedures

Many additional methods have been reported for the formation of the tetrahydroisoquinoline alkaloids (*54, 166*); of these, only two syntheses of biogenetic interest and two additional methods that were reported very recently will be mentioned below.

Tertiary amine oxides have been suggested as intermediates in the formation of heterocyclic rings during alkaloid biogenesis (*167*). The *in vitro* formation of tetrahydroisoquinolines from *N*-oxides have been demonstrated (*168, 169*). In formic acid solvent, sulfur dioxide effects the dehydrative cyclization of the *N*-oxide **109** to yield the 2-methyltetrahydroisoquinoline **11**. In a more nucleophilic solvent such as water, demethylation to the secondary amine **111** rather than cyclization takes place. The 3-methyl-sub-

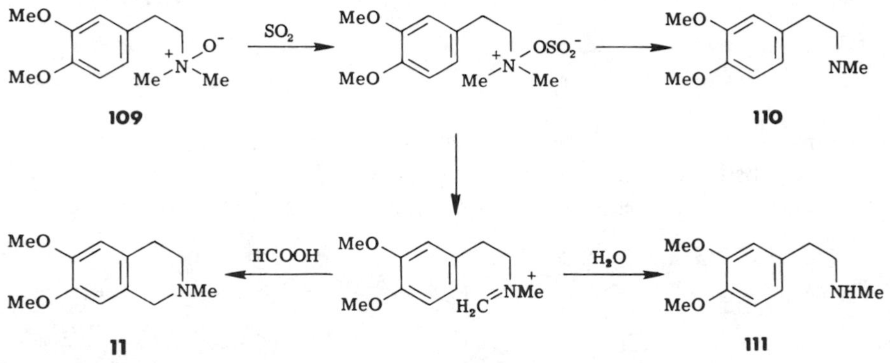

stituted dihydroisoquinolone siamine (**83**) was obtained in high yields from ammonia treatment of either the dihydroxyisocoumarin **112** or the isocoumarin carbonic acid **113** (*109*). These syntheses may have biogenetic implications since polyketides similar to **112** and **113** co-occur with siamine in the same plant (*109*).

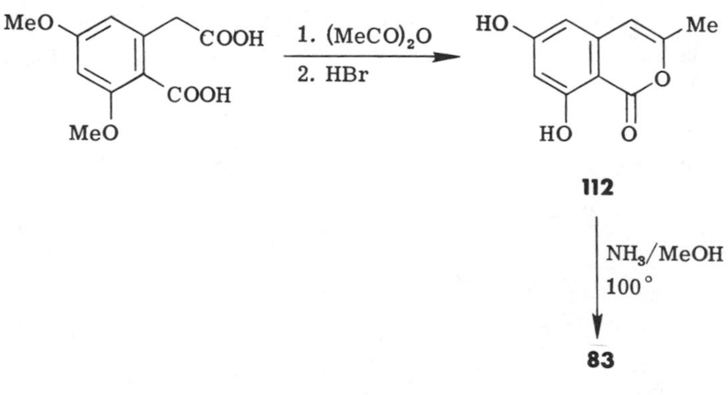

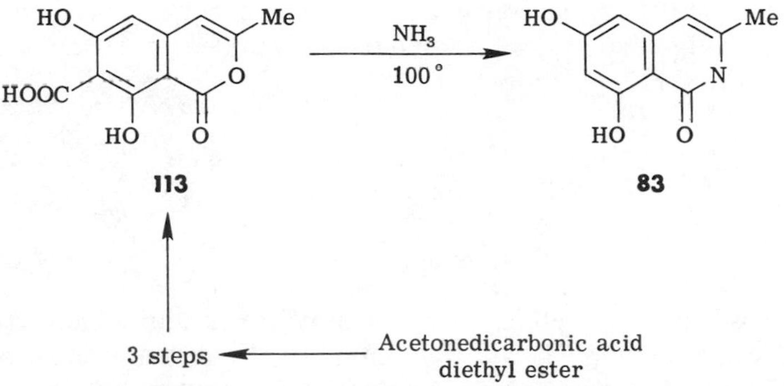

113 → **83**

NH₃, 100°

3 steps ← Acetonedicarbonic acid diethyl ester

Iida *et al.* (*170*) recently reported a facile synthesis of doryanine (**79**) from the homophthalic acid **114**.

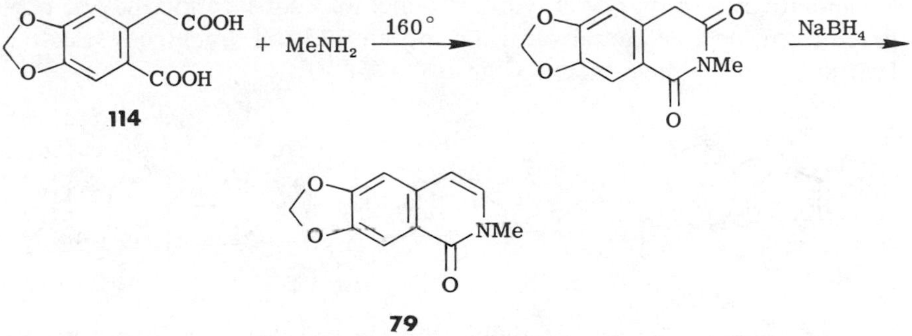

114 + MeNH₂ →(160°)→ →(NaBH₄)→

79

Isoquinolones may be obtained by cyclization of β-phenethylisocyanates (**115**) with magic methyl (methyl fluorosulfonate). This is the first case in which magic methyl has been used as an initiator for a reaction beyond the methylation reaction (*171*).

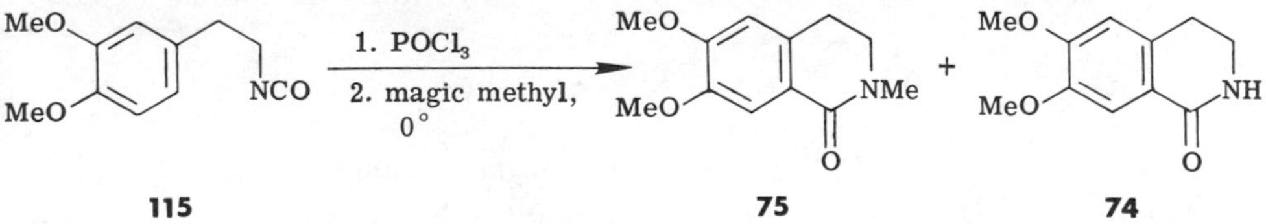

115 →(1. POCl₃; 2. magic methyl, 0°)→ **75** + **74**

5. Selective O-Methylations

Specifically substituted phenolic isoquinolines sometimes may be readily prepared by selective hydrolysis from di- or trimethoxylated analogs. Studies on such partial O-demethylations were performed almost exclusively by Brossi and his associates (*172–180*).

Corypalline (**9**) may be conveniently prepared from 6,7-dimethoxy-3,4-dihydroisoquinoline (**116**) which preferentially undergoes HBr hydrolysis at the 7-methoxyl (*176*).

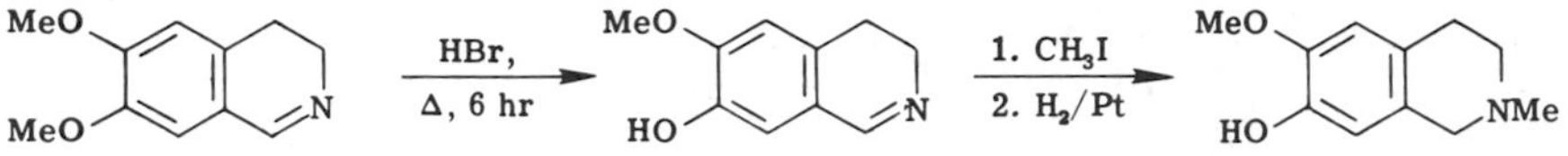

A series of other dimethoxylated 3,4-dihydroxyisoquinolines have been studied similarly (*178*). It was thus established that preferential cleavage occurs at the 5-methoxyl with 5,6- and 5,8-dimethoxy isomers, at the 6-methoxyl with the 6,8 isomer, at the 7-methoxyl with the 5,7 isomer, and at the 8-methoxyl with the 7,8 isomer.

Derivatives carrying the 6-hydroxy-7-methoxy substitution may be prepared from the 4-ketotetrahydroisoquinoline **117** that undergoes selective HBr hydrolysis at the 6-methoxyl group (*175*).

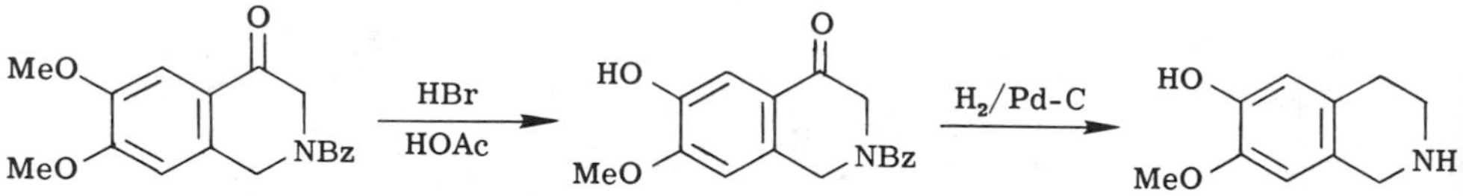

The isoquinolone **75** is preferentially ether-cleaved at the 7-methoxyl group with methionine in methanesulfonic acid (*171*) to yield thalifoline (**73**). The

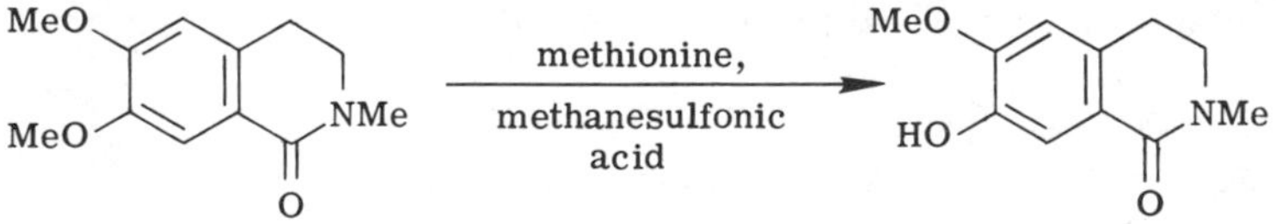

selective hydrolysis (*159*) of the trimethoxylated isoquinolone **118** with concentrated HCl occurs via an oxonium ion (**119**).

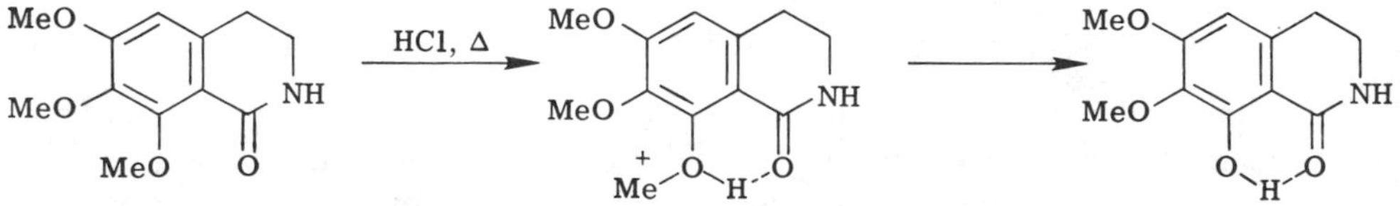

When O-methylpellotine (**38**) was refluxed in 20% HCl for 24 hr, the triphenol **120** was obtained (*174*). A shorter reaction time preferentially cleaves the 7-methoxyl group.

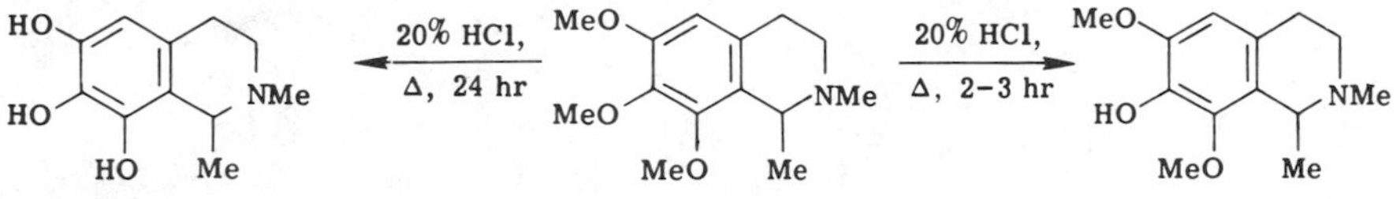

With an intermediate reflux time, the two methoxyl groups at C-7 and C-8 of O-methylanhalonidine (**33**) can be hydrolyzed, thus allowing a convenient synthesis (*179*) of anhalonine (**29**) and lophophorine (**32**). The methoxy

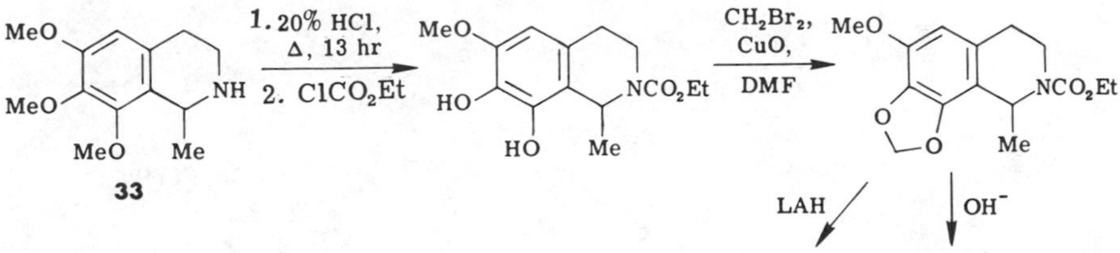

group in hydrocotarnine (**22**) can be cleaved selectively (*180*) with trimethyl-silyl iodide in the presence of 1,4-diazabicyclo[2.2.2]octane (DABCO):

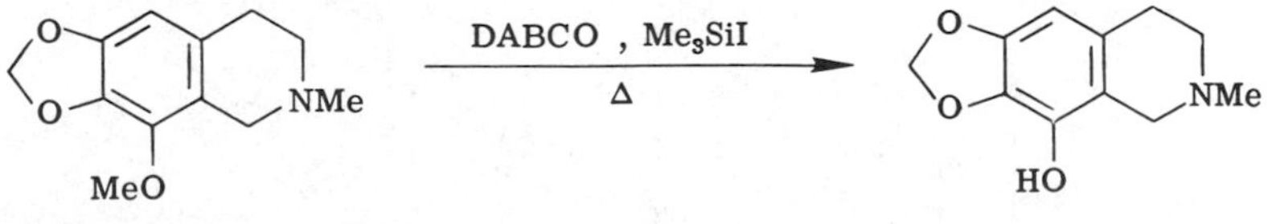

6. Degradations

Chemical degradation of simple isoquinoline alkaloids, used previously as a tool in structure determination work, now is used exclusively for the location of the label in biosynthetic studies.

Oxidation of the methiodide of anhalamine **121** with potassium permanganate to the phthalic anhydride **122** involves loss of C-3 and a label in this position (*181*). A label at C-1 may be determined by Kuhn–Roth oxidation of the derivative **38**. The degradations used by O'Donovan and Horan (*182*) in their biosynthetic studies on lophocerine (**21**) further illustrate this technique.

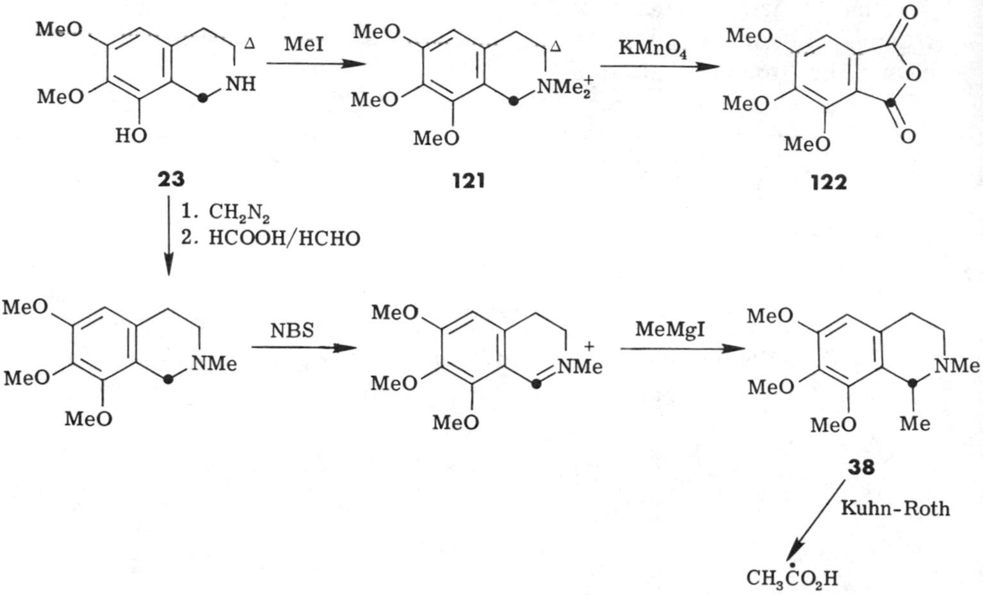

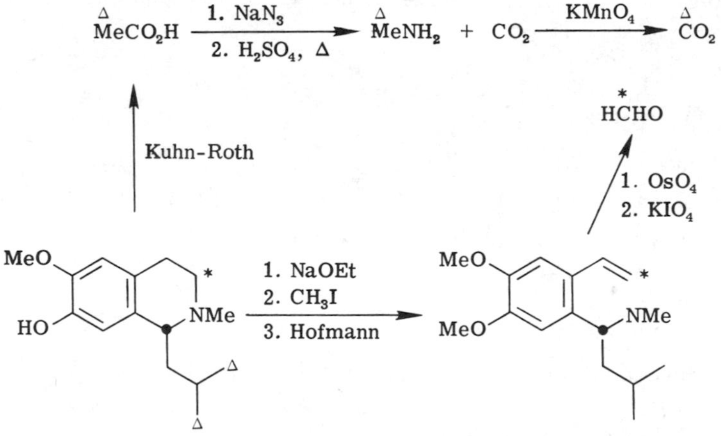

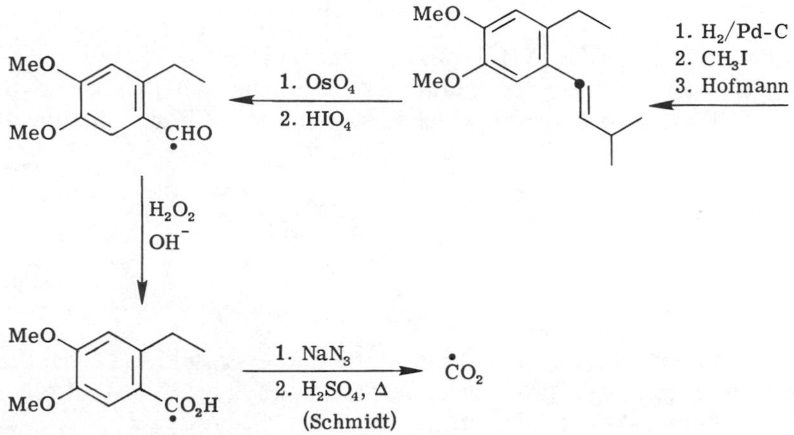

7. Stereoselective Synthesis

The optically active forms of various simple isoquinoline alkaloids have been obtained by resolution of the racemic mixtures using (+)- or (−)-tartaric acid or similar procedures (*27*). Some examples of stereoselective synthesis have been reported. Kametani and Okawara (*183*) achieved optically active salsolidine (**16**) by reduction of the optically active dihydroisoquinoline **123** with sodium borohydride followed by hydrogenolysis. The optical purity of the obtained *S*-(−)-salsolidine was 36–44%. *R*-(+)-Salsolidine was obtained after reduction of the antipode of **123** (R₁ = *R*-(+)-PhCHMe). The alkaloid

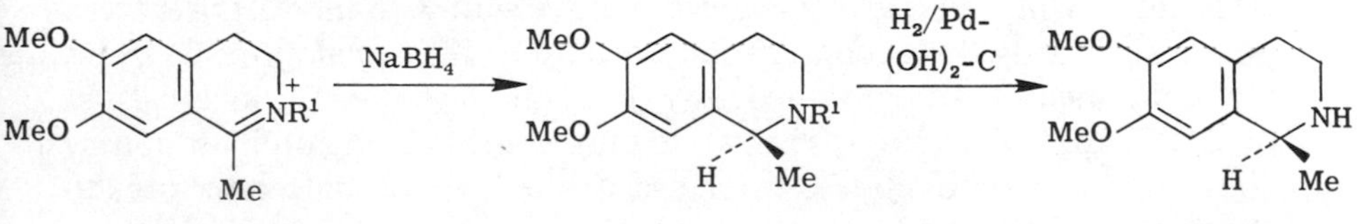

R¹ = (*S*)-(−)-PhCHMe

123

S-(−)-Salsolidine

16

86 was synthesized stereoselectively by Pictet–Spengler condensation of L-dopa with acetaldehyde (*120, 121*).

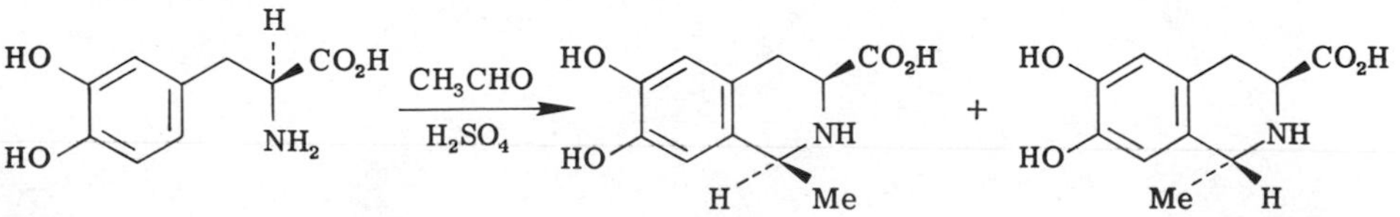

86 (86%)

(5.7%)

D. Absolute Configuration

Battersby and Edwards (*184*) showed that the naturally occurring (−)-salsolidine (**16**) possesses the *S* configuration by oxidizing its *N*-formyl derivative **124** to the amino diacid **125**, which was also obtained from L-alanine. It

124 **125**

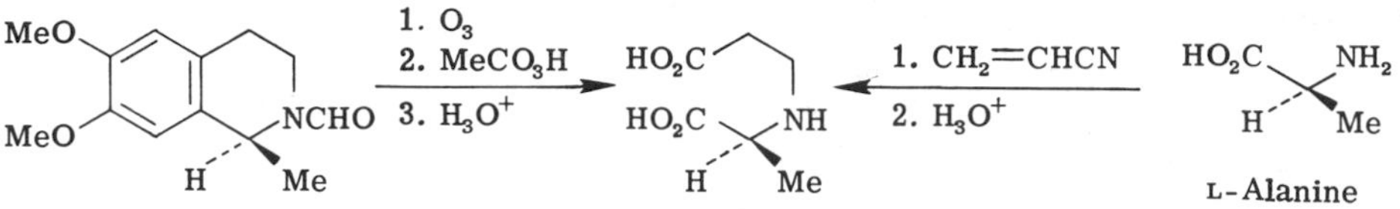

follows from this result that the naturally occurring (+)-salsoline (**13**) has the opposite absolute configuration since one must methylate (*75*) (−)-salsoline in order to obtain (−)-salsolidine (**16**).

Naturally occurring (+)-calycotomine (**21**) was shown to possess the *R* configuration because the *p*-toluenesulfonamide **126** obtained from (+)-calycotomine was levorotatory and the corresponding *p*-toluensulfonamide from (−)-salsolidine was dextrorotatory.

21 **126**

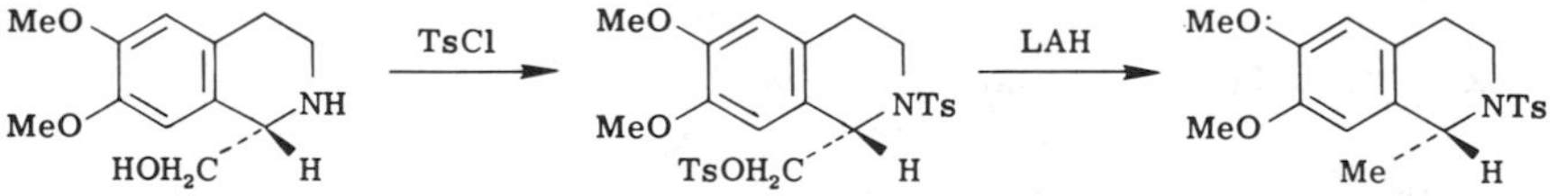

It has been found that the molecular rotations of bases of type **127**, where R^1 is not a highly polar group, show a positive shift with increasing solvent polarity (*184*). By this generalization, carnegine (**18**), anhalonine (**29**), and lophophorine (**32**) were assigned (*184*) the same *S* configuration as salsolidine (**16**). Gigantine (**35**) was shown to possess the same *S* configuration by converting it via reductive cleavage of its diethylphosphate ester (**46**) to (*S*)-carnegine (**18**). An X-ray analysis by Brossi *et al.* (*179*) confirmed the *S*

127

16 : R = H
18 : R = Me

13

35

29

32

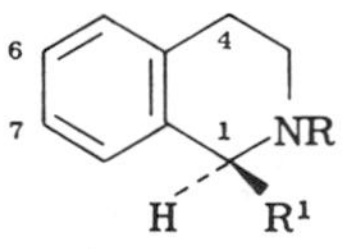

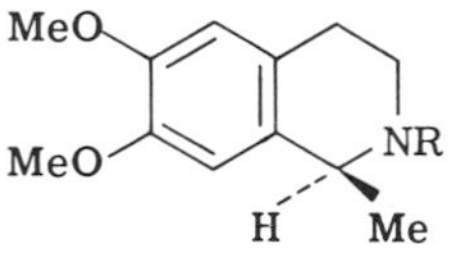

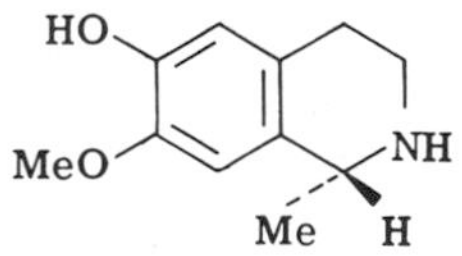

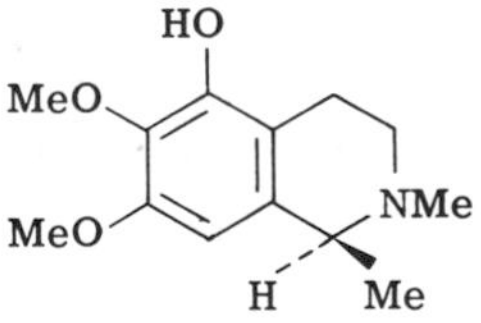

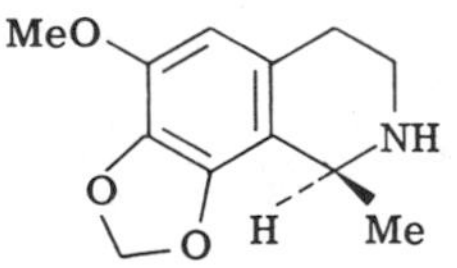

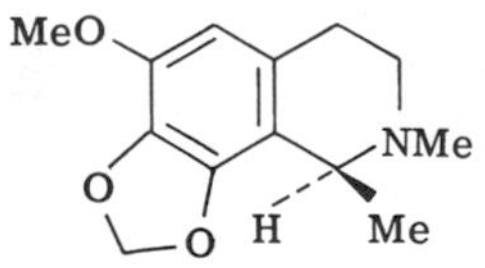

configuration of (–)-anhalonine (**29**) and also showed that the absolute
configuration of (+)-*O*-methylanhalonidine (**33**) exhibits the same *S* configuration. The method of optical rotatory shifts is inconclusive for (+)-*O*-
methylanhalonidine (**33**), since it shows the same molecular rotation in both
chloroform and 1 *M* HCl (*179*). The X-ray study further showed that the
piperidine ring in (–)-anhalonine (**29**) has the conformation **128** referred to
as M helicity and in (+)-*O*-methylanhalonidine (**33**), the conformation **129**
referred to as P helicity (*138*).

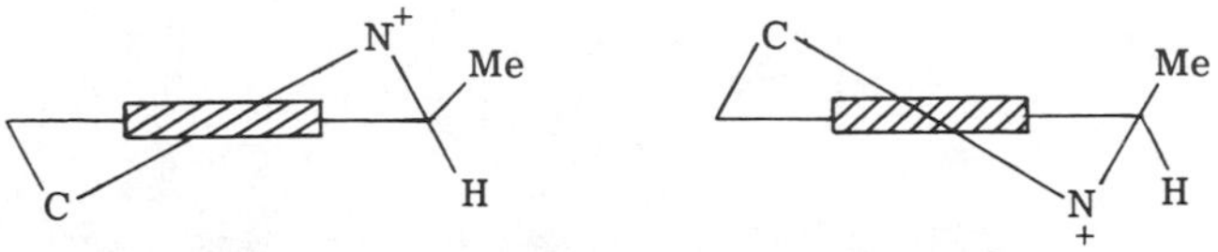

Späth reported (*19*) that the alkaloids anhalonidine (**30**) and pellotine (**36**)
obtained from peyote were optically inactive. The optically active forms of
pellotine (**36**) were later prepared after tartaric acid resolution (*27*). However, these compounds rapidly racemized on standing which suggested that
the alkaloids may be optically active in the plant but racemized during the
isolation procedure.

By use of ion-exchange chromatography, optically active (–)-anhaloni-
dine (**30**) and (–)-pellotine (**36**) have now been obtained from peyote (*138*).
In a study on the chiroptical properties of 1-methyl-substituted tetrahydroisoquinolines, Cymerman Craig *et al.* (*138*) suggest that (–)-anhalonidine
(**30**) and (–)-pellotine (**36**) both possess the *R* configuration. The co-occur-
rence of closely related bases having opposite configurations in *Salsola*

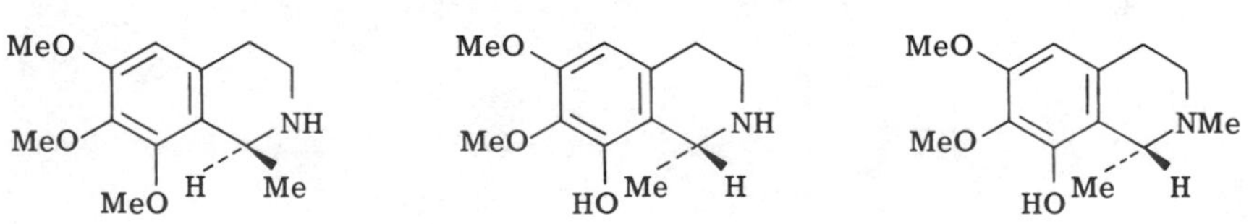

arbuscula [(–)-salsoline and (+)-salsolidine] and in peyote [e.g., (–)-anha-
lonidine and (+)-*O*-methylanhalonidine] may have biogenetic implications.

III. 1-Phenyl-, 4-Phenyl-, and *N*-Benzyl-Substituted Simple Isoquinoline Alkaloids

A. 1-Phenylisoquinolines: The Cryptostylines

Five 1-phenyl-substituted isoquinolines have been found by Leander and
co-workers (*185, 187*) in the family of Orchidaceae (Table X). It is remark-

TABLE X

1-PHENYL-SUBSTITUTED TETRAHYDROISOQUINOLINE ALKALOIDS IN *Cryptostylis* SPECIES; *S*-(+) FORMS IN *Cryptostylis fulva* SCHLR[a]; *R*-(−) FORMS IN *Cryptostylis erythroglossa* HAYATA[b]

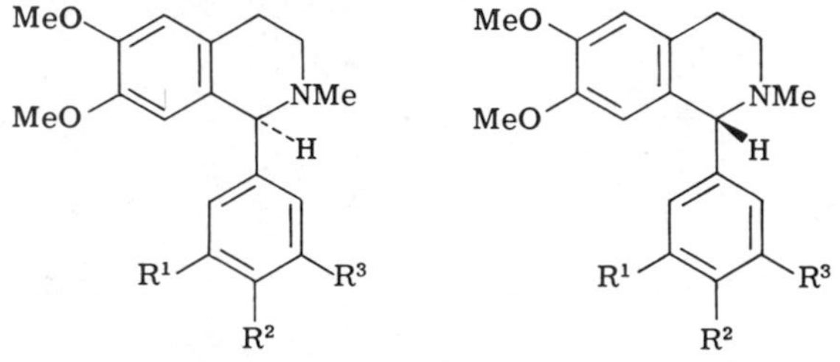

S-(+)-   *R*-(−)-

Number	Compound	R^1	R^2	R^3	Formula	Melting point (°C)	ORD data
130	*S*-(+)-Cryptostyline I	H	O—CH$_2$—O	O—CH$_2$—O	C$_{19}$H$_{21}$NO$_4$	101–102 $[\alpha]_D^{20}$	+56 (c = 2.7, CHCl$_3$)
130	*R*-(−)-Cryptostyline I	H	O—CH$_2$—O	O—CH$_2$—O		101–102 $[\alpha]_D^{22}$	−56 (c = 0.4, CHCl$_3$)
131	*S*-(+)-Cryptostyline II	H	OMe	OMe	C$_{20}$H$_{25}$NO$_4$	117–118 $[\alpha]_D^{25}$	+58 (c = 0.28, CHCl$_3$)
131	*R*-(−)-Cryptostyline II	H	OMe	OMe		116–117 $[\alpha]_D^{22}$	−58 (c = 0.4, CHCl$_3$)
132	*S*-(+)-Cryptostyline III	OMe	OMe	OMe	C$_{21}$H$_{27}$NO$_5$	126–129 $[\alpha]_D^{25}$	+51 (c = 0.15, CHCl$_3$)
132	*R*-(−)-Crytostyline III	OMe	OMe	OMe		128–130 $[\alpha]_D^{22}$	−52 (c = 0.3, CHCl$_3$)

133 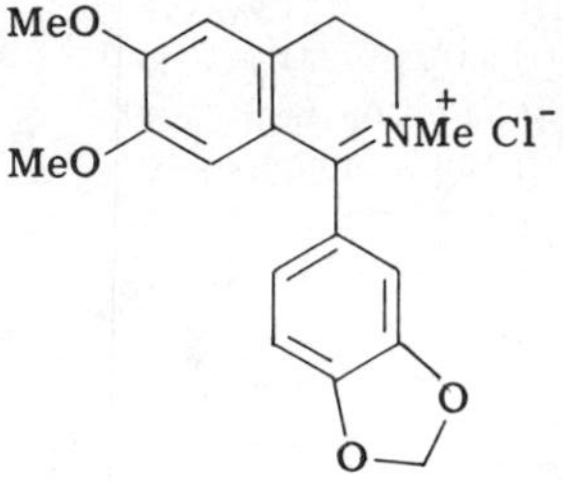 1-(3,4-Methylenedioxyphenyl)-6,7-dimethoxy-2-methyl-3,4-dihydroisoquinolinium chloride

$C_{19}H_{20}NO_4$, mp of iodate 208–209°

Source: *Cryptostylis erythroglossa*

134 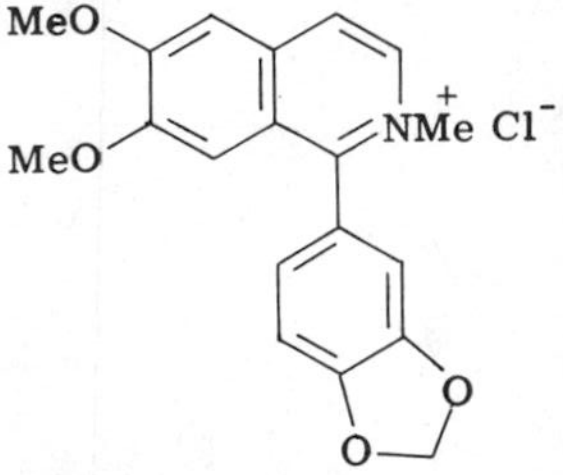 1-(3,4-Methylenedioxyphenyl)-6,7-dimethoxy-2-methylisoquinolinium chloride

$C_{19}H_{18}NO_4$, mp of picrate 218–223°

Source: *Cryptostylis erythroglossa*

[a] From Leander *et al.* (*186*).

[b] From Agurell *et al.* (*187*).

305

able that both stereoisomers of the tertiary amine alkaloids cryptostyline I, II, and III occur naturally. The dextrorotatory forms were found in *Cryptostylis fulva* Schlr. (*185, 186*) and the levorotatory forms in *Cryptostylis erythroglossa* Hayata (*187*). The two quaternary alkaloids **133** and **134** were found in *C. erythroglossa* (*187*).

The structures of cryptostylines I–III were elucidated mainly by NMR and mass spectroscopic methods. The mass spectra showed the facile loss of C-1 substituents of simple tetrahydroisoquinolines, and in all cases base peaks of *m/z* 206 were obtained.

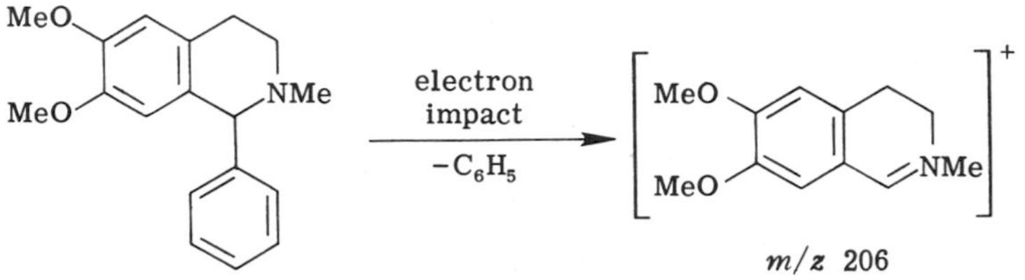
MeO
MeO
NMe
electron
impact
$-C_6H_5$
MeO
MeO
NMe
+
m/z 206

In the NMR spectrum of cryptostyline I, the C-1 singlet at δ 4.2 was shifted to δ 5.9 in the methiodide salt indicating that the phenyl group was attached at C-1 (*186*). The structures were proved by synthesis in which the appropriate 3,4-dimethoxybenzamides (**135**) were cyclized by Bischler–Napier-

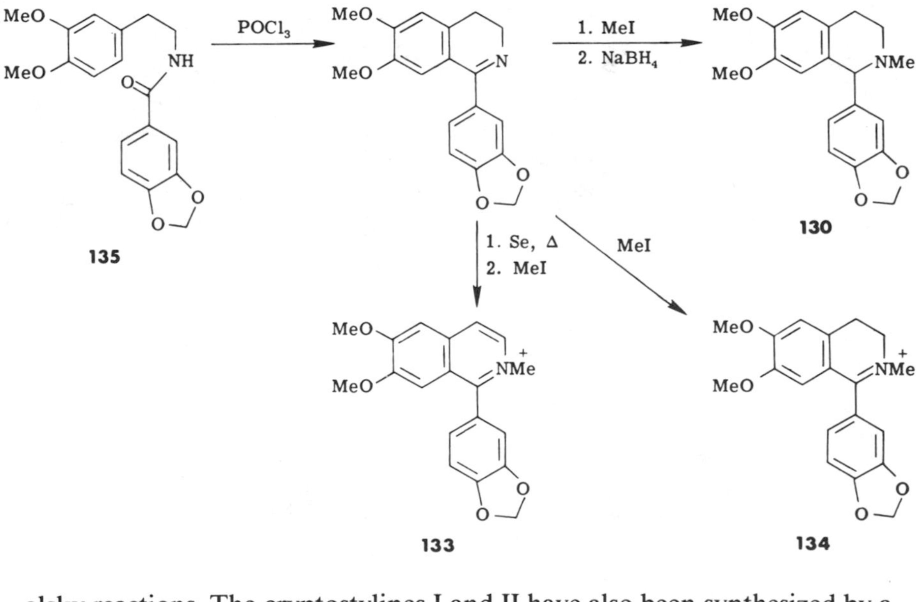
MeO
MeO
NH
O
O
O
POCl$_3$
N
1. MeI
2. NaBH$_4$
NMe
1. Se, Δ
2. MeI
MeI
+
NMe

135 **130** **133** **134**

alsky reactions. The cryptostylines I and II have also been synthesized by a method of inner α-amidoalkylation (*188*).

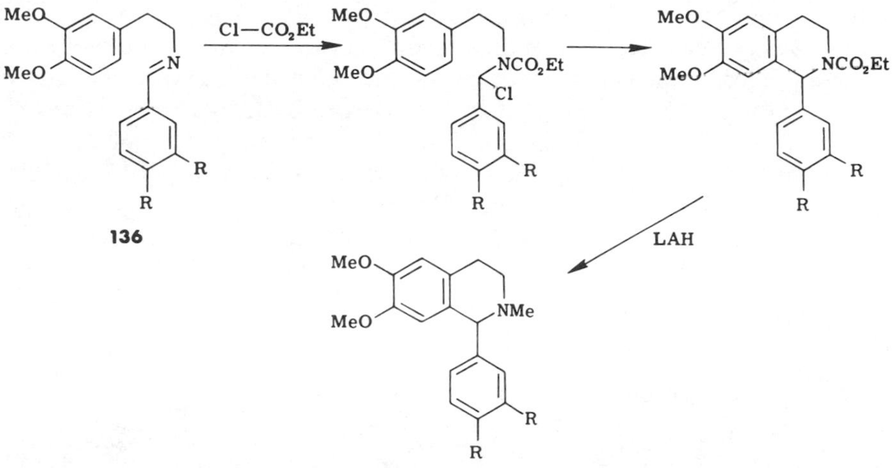

136

Two independent X-ray crystallographic analyses of the cryptostylines have been published. Brossi and Teitel (*189, 190*) showed in a study on (−)-cryptostyline II that the levorotatory forms possess the *R*-configuration. This result was confirmed by Westin (*191*) and Leander *et al.* (*192*) in a study on the methiodide of (+)-cryptostyline I.

B. A 4-ARYLISOQUINOLINE: CHERYLLINE

A phenolic 4-phenyl-substituted tetrahydroisoquinoline alkaloid (−)-cherylline has been isolated by Brossi *et al.* (*193*) from *Crinum powelli* var. *alba* and other *Crinum* species (Amaryllidaceae).

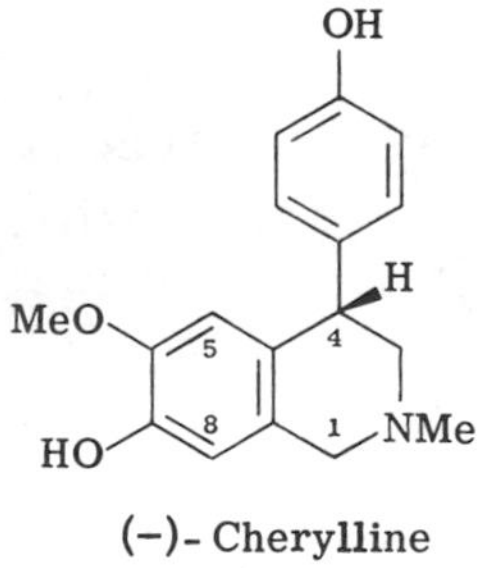

$C_{17}H_{19}NO_3$ mp 217 -218°, HCl 238 -239°
MW 285 $[\alpha]_D^{26}$ −69°(c = 0.20, MeOH)

(−)- Cherylline

137

Spectral data were most informative in the structure elucidation of cherylline. The UV spectrum had maxima at 285 and 280 nm which underwent a bathochromic shift to 299 nm on addition of base. The NMR spectrum in DMSO-d_6 showed one methoxyl at δ 3.51 (s), one *N*-methyl at δ 2.24 (s), an A_2B_2 pattern at δ 6.91 and 6.64 characteristic of a 1,4-disubstituted aromatic ring, and two one-proton singlets at δ 6.49 and 6.23 indicative of two para-orientated protons on a second aromatic ring. Because of the shielding effect of the phenyl group at C-4, the signal at δ 6.23 was attributed to the proton at C-5 and the one at δ 6.49 to the proton at C-8. On addition of a drop of NaOD in D_2O, both of the latter singlets were shifted upfield to δ 6.06 representing a shift of 10 and 26 Hz, respectively. Since the proton at C-8 underwent the largest shift, the phenolic function must be at C-7.

Synthetic work (193, 194) confirmed the structure 137. The total synthesis (194) involved an interesting selective demethylation step of the intermediate 139.

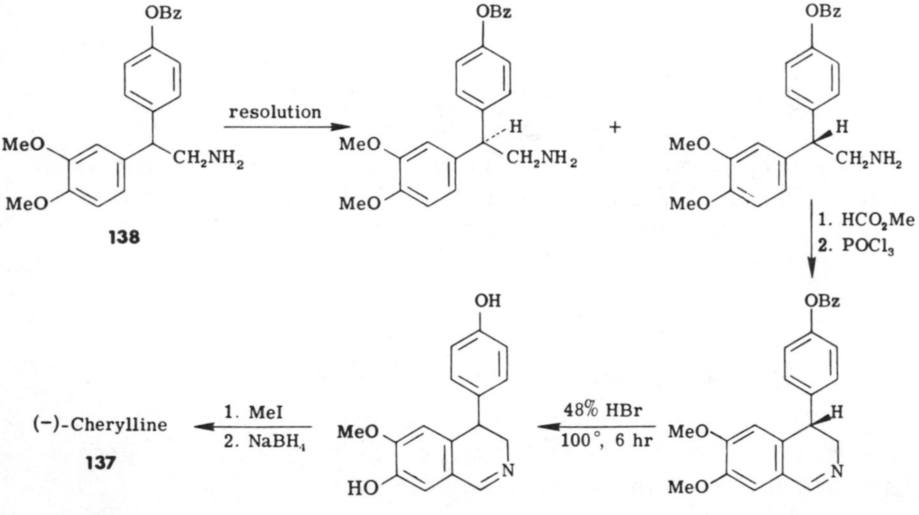

The ORD and CD curves of cherylline have been recorded and the absolute configuration (S) was assigned by analogy with the ORD curves of 4-aryltetralines (193). The absolute configuration was confirmed by an X-ray crystallographic study (193) of the O,O-dimethyl-N-p-bromobenzamide derivative 140.

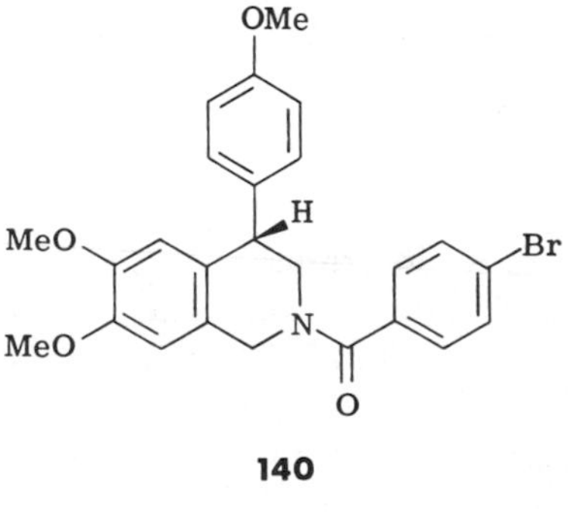

140

Schwartz and Scott (*195*) have worked out a biogenetically patterned synthesis of (±)-cherylline involving a quinone methide intermediate (**141**).

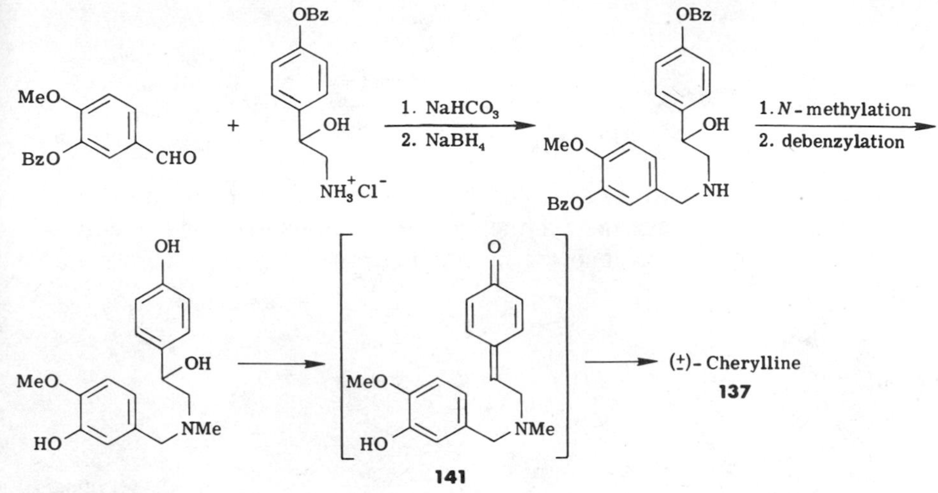

A similar synthesis involving a quinone methide intermediate has been reported Kametani *et al.* (*196*).

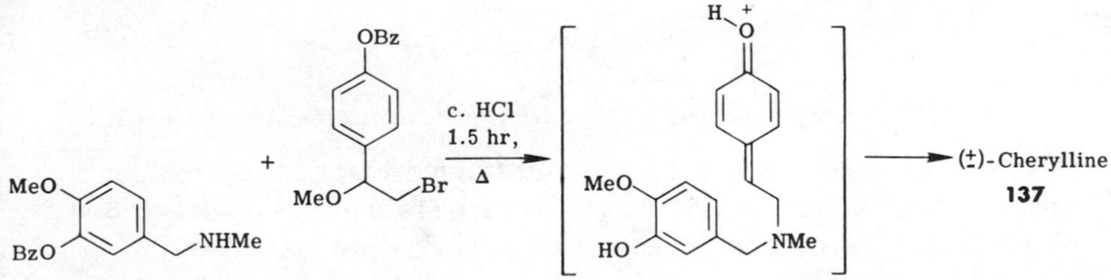

p-Quinone methide ketals may be prepared from *p*-quinone monoketals and α-trimethylsilylamides or phosphoranes (e.g., **142**). A total synthesis of (±)-cherylline based on this method has been presented by Hart *et al.* (*197*).

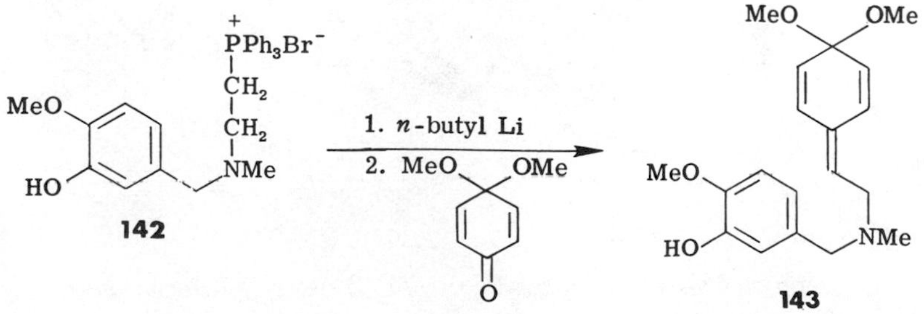

$$143 \xrightarrow{BF_3Et_2O} \text{[structure]} \xrightarrow[DMF]{EtS\ Na} (\pm)\text{-Cherylline}\quad 137$$

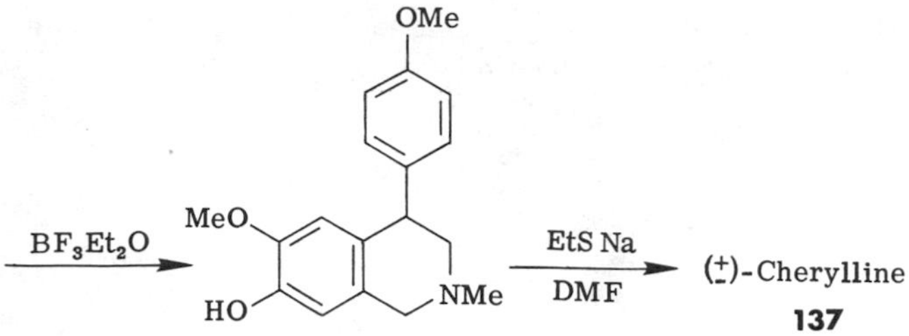

Irie *et al.* (*171*) recently have published a synthesis in which the isocyanate **144** is cyclized with magic methyl (methyl fluorosulfonate) to the *N*-methyl-isoquinoline lactam **145** followed by regioselective cleavage of aromatic methoxy groups.

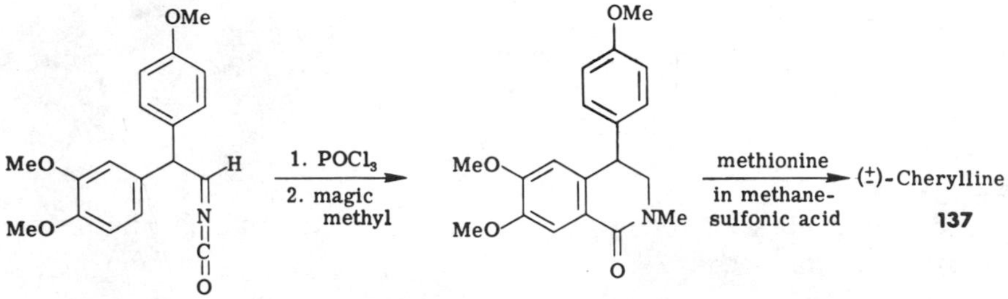

$$144 \xrightarrow[\substack{2.\ \text{magic} \\ \text{methyl}}]{1.\ POCl_3} 145 \xrightarrow[\substack{\text{in methane-} \\ \text{sulfonic acid}}]{\text{methionine}} (\pm)\text{-Cherylline}\quad 137$$

144   **145**

C. *N*-BENZYLTETRAHYDROISOQUINOLINES: SENDAVERINE AND CORGOINE

Two naturally occurring *N*-benzyl-substituted tetrahydroisoquinolines are known. Manske (*198*) isolated sendaverine (**146**) from *Corydalis aurea* Willd. (Fumariaceae) and Ibragumov *et al.* (*199*) found corgoine (**147**) in *Corydalis gortschakovii.*

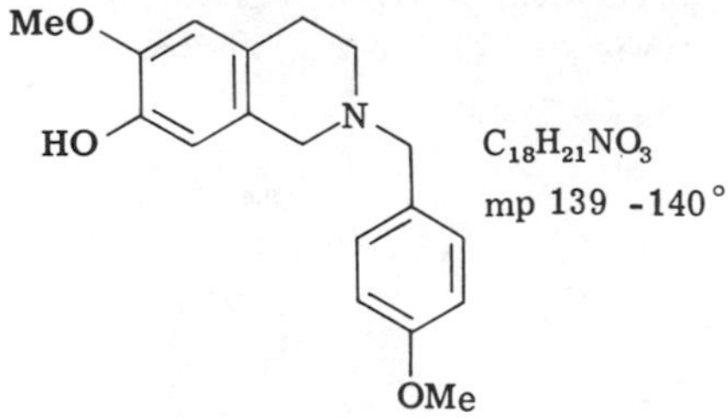

$C_{18}H_{21}NO_3$
mp 139 –140°

$C_{17}H_{19}NO_3$
mp 190 –191°,
HCl: 239°-240°

Sendaverine
146

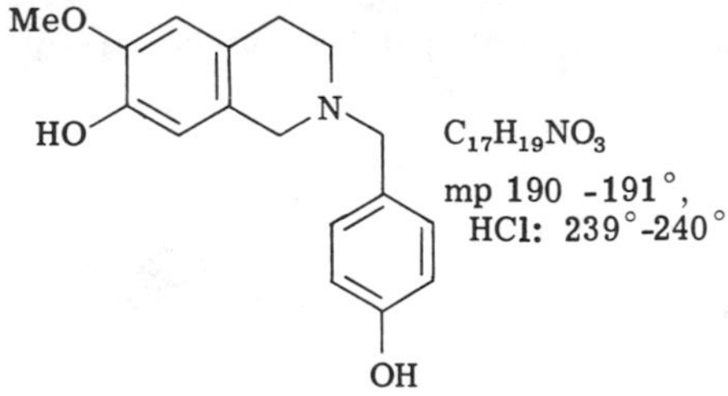

Corgoine
147

Proof of the structure of sendaverine was provided by its synthesis (*200*). N-Alkylation of the imine **148** with 4-methoxybenzylchloride **149**, followed by reduction and debenzylation furnished a product identical with natural sendaverine (**146**).

148 **149**

Sendaverine **146**

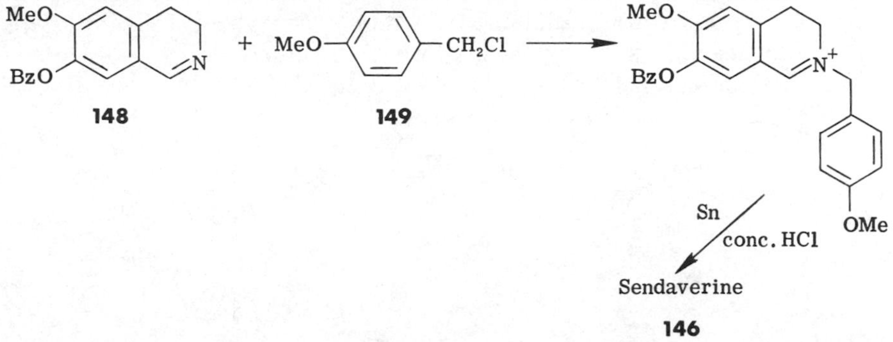

Two recently published syntheses of sendaverine utilize the lactones **150** (*201*) and **151** (*202*) which are condensed with 4-methoxybenzylamine (**152**).

150 **152**

Sendaverine **146**

151 **152**

Sendaverine **146**

153 Corgoine **147** **154**

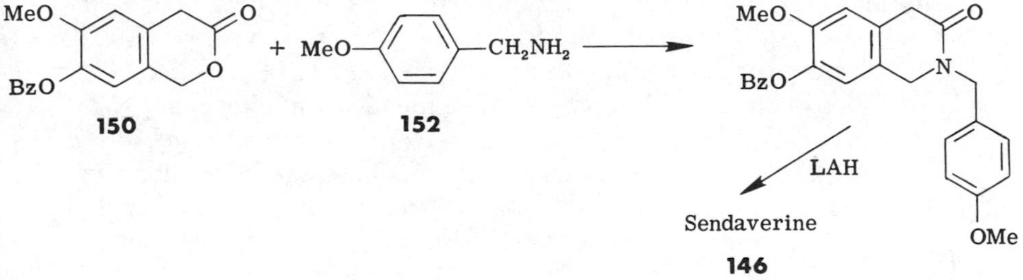

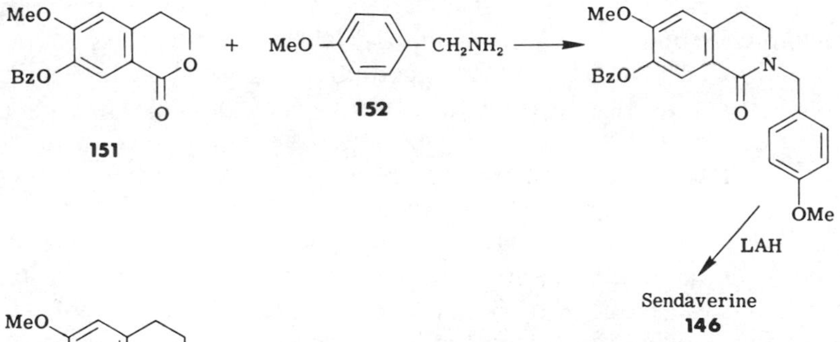

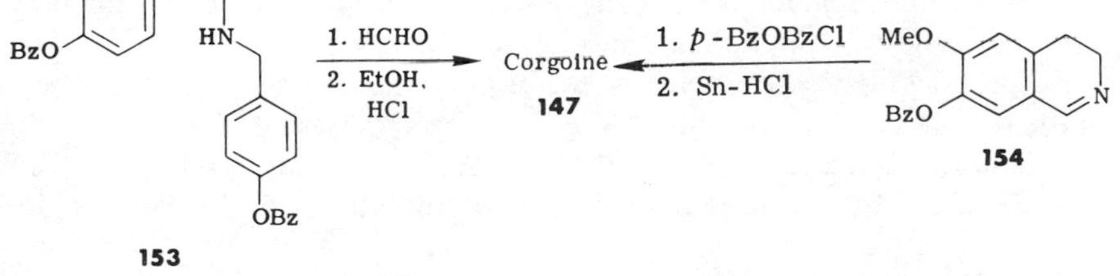

The structure of corgoine has been confirmed by three syntheses: Pictet–Spengler cyclization of the substituted phenethylamine **153** (*203*), reduction of the *N*-benzyl quaternary salt derived from the isoquinoline **154** (*204*) or by heating the tetrahydroisoquinoline **155** with *p*-hydroxybenzyl alcohol (*196*). The latter reaction probably occurs through an intermediate *p*-benzoquinone methide which adds to the basic nitrogen of **155**.

IV. Biogenesis

The biogenesis of the simple tetrahydroisoquinoline alkaloids has been studied rather extensively. Most studies concerned the biosynthesis of mescaline and the related tetrahydroisoquinolines of the peyote cactus *Lophophora williamsii* (for reviews see Refs *7, 12, 205*). Other cactus alkaloids studied were gigantine (**35**) in *Carnegiea gigantea* (*45*), salsoline (**13**) in *Echinocereus merkerii* (*206*), and lophocerine (**20**) and pilocereine (**67**) in *Lophocereus shottii* (*182, 40, 207*) and *Marginatocereus* (*Pachycereus*) *marginatus* (*208*). The biosynthesis of the orchid alkaloid cryptostyline I (**130**) has been studied in *Cryptostylis erythroglossa* (*187, 209*) and the alkaloidal amino acid 6-hydroxytetrahydroisoquinoline-3-carboxylic acid (**84**) in *Euphorbia myrsinites* (*210*).

The methods used have mainly involved feeding the various plants with postulated precursors to the alkaloids, suitably labeled with ^{3}H or ^{14}C atoms at specific sites. The identification of trace compounds as intermediates of the biosynthetic pathways has also been most informative (*30, 32, 33, 130, 132*). More recently, the results of some elegant *in vitro* studies on an *O*-methyltransferase isolated from the peyote cactus have supported earlier findings (*205, 211, 211a*).

Earlier it was a matter of general agreement that the simple tetrahydroisoquinoline alkaloids arose in nature from tyrosine and the first experimental evidence of this was reported by Battersby and Garratt (*212*) and by Leete (*213*) who found significant incorporations of tyrosine into the phenolic alkaloids pellotine (**36**) and anhalonidine (**30**).

Further detailed experiments on the origin of the phenethylamine moiety of the peyote tetrahydroisoquinolines have been carried out by Lundström and Agurell (*214, 215, 216, 181, 12*) and by Paul and his co-workers (*217, 205, 211*), and there is now substantial experimental evidence for common biosynthetic routes from tyrosine (**156**) via dopamine (**92**) to the phenethyl-amine mescaline (**52**) and then to the main phenolic tetrahydroisoquinoline alkaloids of the peyote cactus. Dopamine is O-methylated to 4-hydroxy-3-methoxyphenethylamine (**157**) which then undergoes hydroxylation to the key intermediate 4,5-dihydroxy-3-methoxyphenethylamine (**158**). The position of further O-methylation of this intermediate is all important in

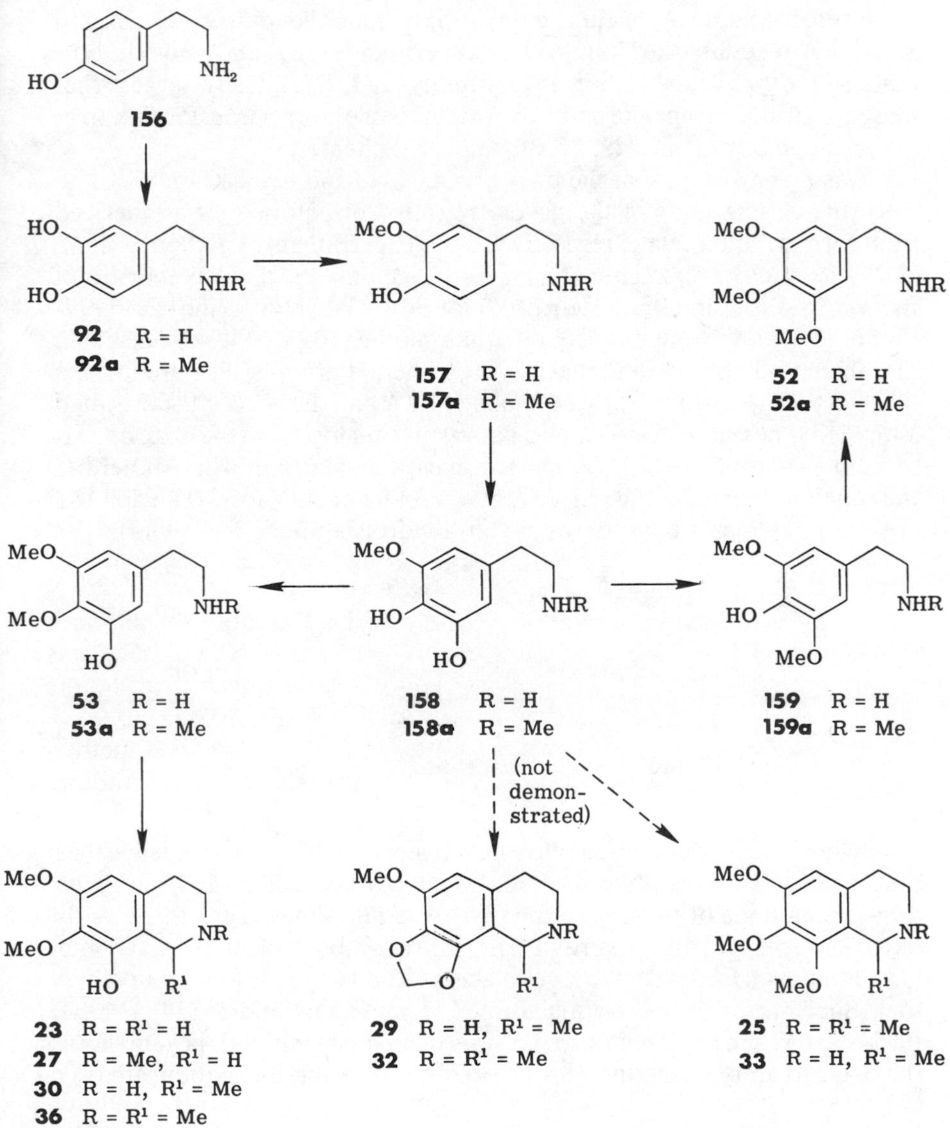

156

92   R = H
92a R = Me

157 R = H
157a R = Me

52 R = H
52a R = Me

53 R = H
53a R = Me

158 R = H
158a R = Me

159 R = H
159a R = Me

(not demonstrated)

23 R = R¹ = H
27 R = Me, R¹ = H
30 R = H, R¹ = Me
36 R = R¹ = Me

29 R = H, R¹ = Me
32 R = R¹ = Me

25 R = R¹ = Me
33 R = H, R¹ = Me

determining the biogenetic fate of the molecule. meta-O-Methylation will yield (**159**) and subsequently mescaline (**52**). On the other hand, if O-methylation proceeds at the *p*-hydroxy group, 3-demethylmescaline (**53**) is formed which, with suitable ring-closing units, gives rise to such tetrahydroisoquinolines as anhalamine (**23**) and anhalonidine (**30**). The N-methylated phenolic alkaloids pellotine (**36**) and anhalidine (**27**) are apparently formed in parallel routes involving *N*-methyl-substituted phenethylamine derivatives (*181*).

The precise pathways leading to the nonphenolic alkaloids **25, 29, 32,** and **33** are not yet established, but feeding experiments (*181*) and, more importantly, *in vitro* O-methylation experiments (*211, 211a*) may suggest that these are formed in separate pathways rather than by direct transformation of the phenolic compounds **23, 27, 30,** and **36** (Scheme 1).

It was first assumed that the one-carbon (C-1) and two-carbon (C-1 and C-9) ring-closing units of the peyote tetrahydroisoquinolines are derived from formate and acetate, respectively, but experiments by Battersby *et al.* (*218, 219*) showed that neither formate nor acetate were direct precursors of these units. Leete and Braunstein (*220*) found a relatively specific incorporation of [3-^{14}C]pyruvate into C-9 of anhalonidine (**30**), and it was suggested that the tetrahydroisoquinolines having a methyl group at C-1 are formed through the intermediacy of acetyl coenzyme A and the *N*-acetyl-substituted 3-demethylmescaline (**160**). Such a pathway is analogous to the biosynthesis of simple *β*-carbolines (*221, 222*). However, it was later shown that neither the *N*-acetyl derivative **160** (*33, 181, 206, 222a*) nor its *N*-formyl analog **161** (*181*) are direct precursors of simple tetrahydroisoquinoline alkaloids.

160   **161**

Kapadia and co-workers (*33*) have now collected convincing evidence that *α*-keto acids such as glyoxylic (**162**) and pyruvic acids (**163**) serve as ring-closing units of the peyote tetrahydroisoquinolines. The amino acids peyoxylic acid (**63**) and peyoruvic acid (**65**) have been identified in peyote (*33*). Moreover, feeding of the labeled acids **63** and **65** to the plant resulted in their efficient conversion to anhalamine (**23**) and anhalonidine (**30**), respectively (*33*). When peyoruvic acid (**65**) was incubated with fresh peyote slices, the 3,4-dihydroisoquinoline **48** was isolated from the incubation mixture (*33*).

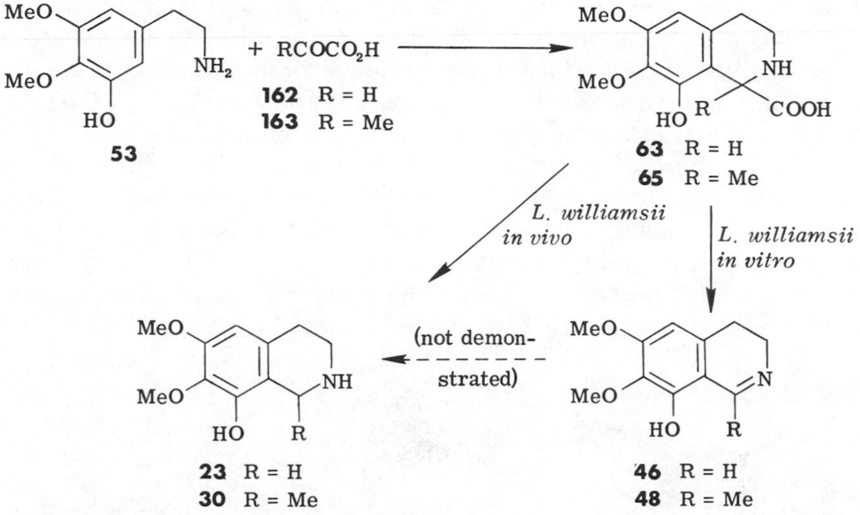

162 R = H
163 R = Me

63 R = H
65 R = Me

23 R = H
30 R = Me

46 R = H
48 R = Me

53

These results strongly suggest that simple tetrahydroisoquinolines such as anhalamine (23) and anhalonidine (30) are biogenetically derived from 3-demethylmescaline (53) which condenses with glyoxylic or pyruvic acid to give peyoxylic (63) or peyoruvic acid (65). These amino acids undergo oxidative decarboxylation to supply 3,4-dihydroisoquinolines (46 and 48), which are then stereospecifically reduced to the tetrahydroisoquinolines. The presence of the dihydroisoquinolines 46–49 (Table I) in peyote (71) might support such a pathway.

Studies on the cacti *Lophocereus shottii* (*182, 207*) and *Marginatocereus marginatus* (*208*) have shown that the five-carbon unit in lophocerine (20)

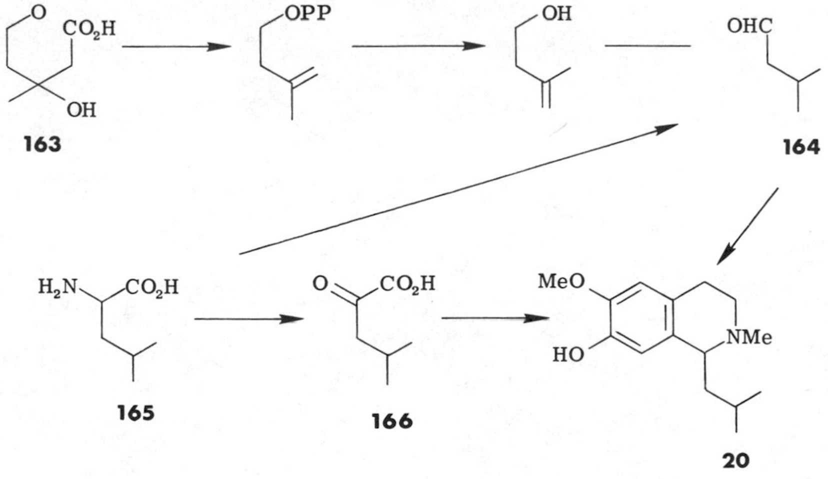

163

164

165

166

20

probably arises independently from both mevalonic acid (163) and leucine
(165). A common intermediate in the formation of the isoquinoline would be
3-methyl butanal (164), which has been found to be specifically incorporated
(207) into lophocerine (20). It has also been suggested that leucine (165) may
be incorporated in lophocerine (20) via the keto acid 166 in a pathway
equivalent to those of the peyote tetrahydroisoquinolines (223). Pilocereine
(67) arises *in vivo* (40) by oxidative coupling of lophocerine (20).

Some experiments on the biosynthesis of the 1-aryl-substituted tetrahydroisoquinoline cryptostyline I (130) have been carried out with *Cryptostylis
erythroglossa* (187, 209). It was shown that dopamine (92), ^{14}C-labeled at the
β-position, serves as the origin of both the aromatic rings of 130. Of the two

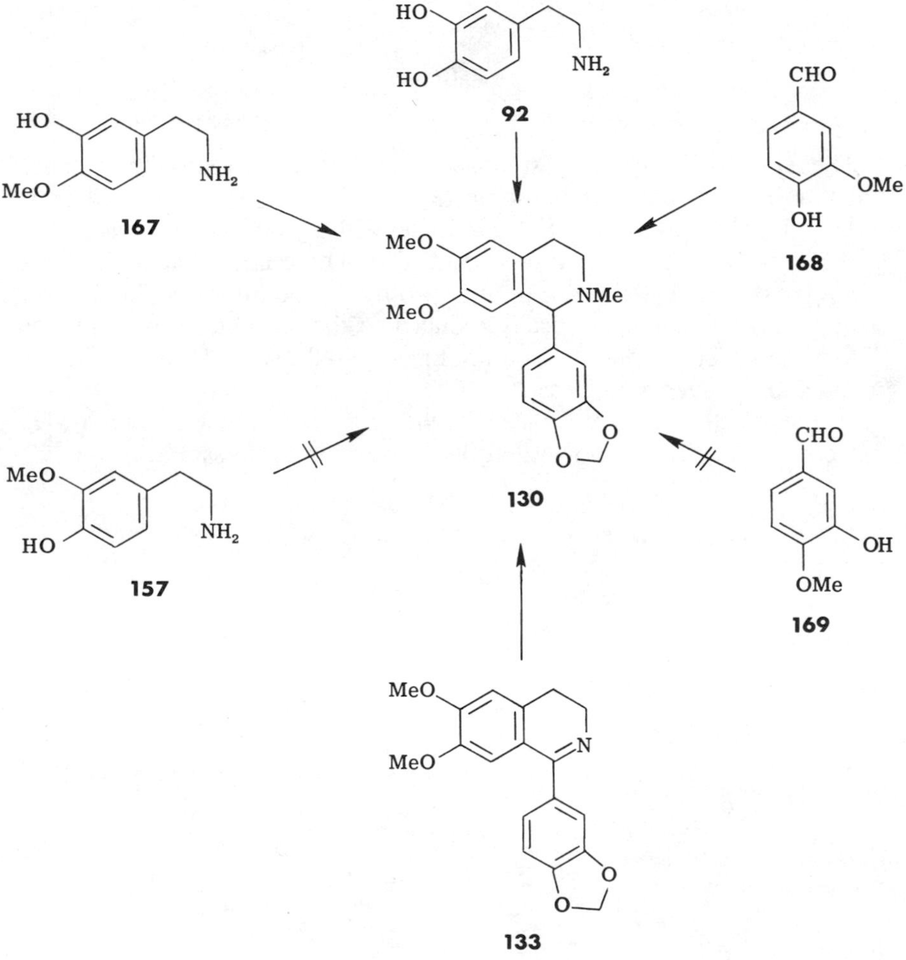

mono-O-methylated derivatives of dopamine 157 and 167, only 3-hydroxy-4-methoxyphenethylamine (167) was incorporated in 130, suggesting that the ring closure to the tetrahydroisoquinoline is facilitated by the *p*-hydroxy group. Vanilline (168) and isovanilline (169) were tested as precursors of the 1-phenyl group, and the former showed low but apparently significant incorporation into 130; however, degradations to ensure a specific incorporation in cryptostyline I was not carried out in this case. The labeled dihydroisoquinoline 133 was efficiently incorporated in (−)-cryptostyline I when fed to the plant, indicating useful information for future studies. The results obtained thus far in the biosynthesis of the cryptostyline alkaloids are not conclusive regarding the formation of the tetrahydroisoquinoline unit.

Looking to the future, some interesting questions remain to be answered regarding the biogenesis of the simple tetrahydroisoquinoline alkaloids. The sequence of hydroxylations of the aromatic ring and O-methylations have only been studied in detail for the peyote alkaloids, and similar studies in plants producing alkaloids with different substitution patterns, such as the alkaloids of the giant cacti, may be rewarding. Of special interest also is the biogenesis of the monooxygenated *Dolichothele* alkaloids, e.g., longimammatine 1 and longimammidine 4, since the alkaloidal amino acid 84 of *Euphorbia myrsinites* (*210*) has been shown to origniate from *m*-tyrosine (170).

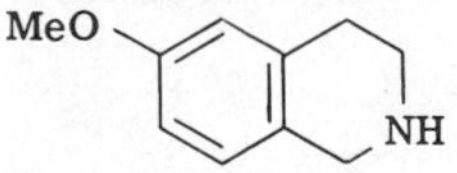

1

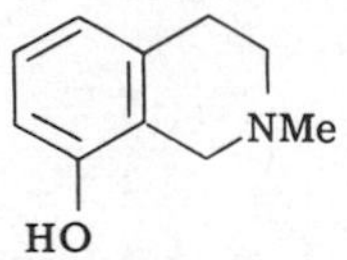

4

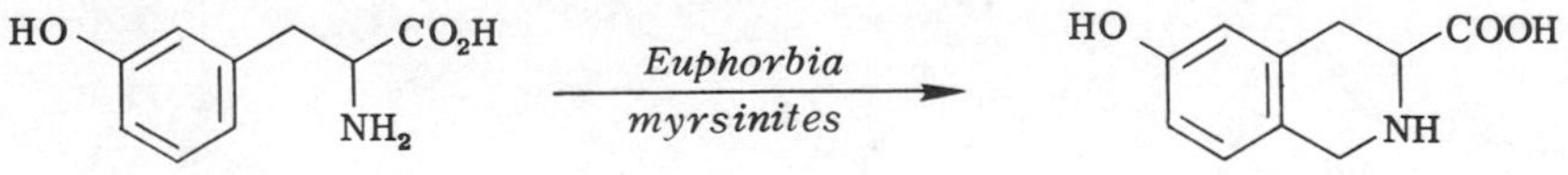

170   *Euphorbia myrsinites*   84

The origin of the ring-closing units also requires further study. In some instances it would appear as if the second building block is incorporated as an aldehyde (route a) as in the biosynthesis of, e.g., lophocerine (20), whereas in other alkaloids, e.g., the peyote tetrahydroisoquinolines, the second unit is incorporated in the form of a keto acid (route b) (*223*). With regard to the second alternative (route b), the oxidative decarboxylation to a dihydroisoquinoline followed by a stereospecific reduction to yield an optically active tetrahydroisoquinoline will have to be verified further. Comparisons may be drawn from experiences with the benzylisoquinoline alkaloids in the opium

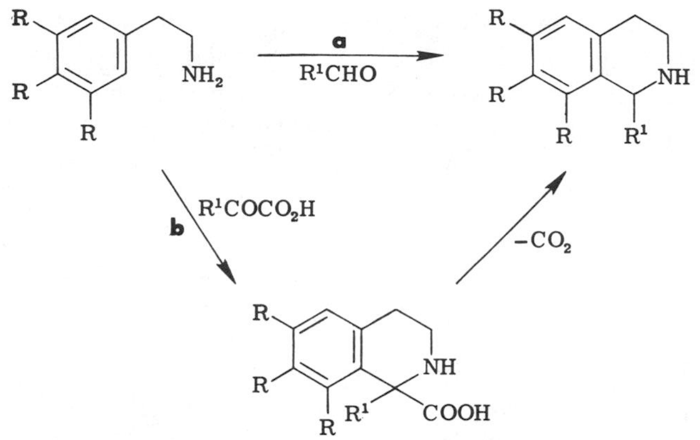

R = H, HO or MeO

poppy. *In vivo* studies (*224, 225*) have suggested that these are formed via route b and thus involve keto acids as ring-closing units. However, recent *in vitro* studies (*226*) have shown that dopamine combines with 3,4-dihydroxy-phenylacetaldehyde (**171**) to yield directly (*S*)-norlaudanosoline (**172**). The enzyme catalyzing this reaction, (*S*)-norlaudanosoline synthase, has been partly purified and characterized and has been found to be present in several species producing benzylisoquinoline alkaloids (*226*). Similar enzymes may be present in peyote and other plants producing simple tetrahydroisoquino-line alkaloids.

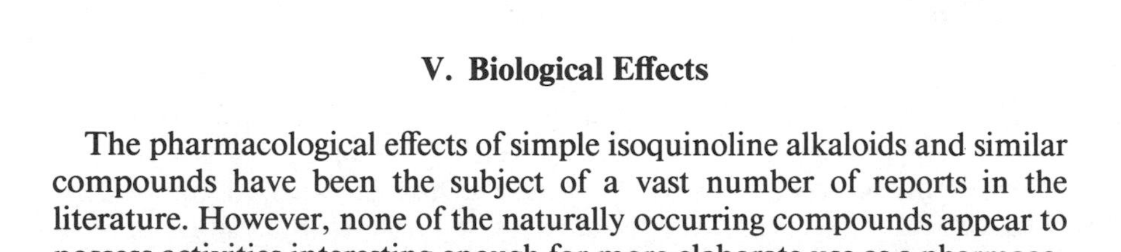

V. Biological Effects

The pharmacological effects of simple isoquinoline alkaloids and similar compounds have been the subject of a vast number of reports in the literature. However, none of the naturally occurring compounds appear to possess activities interesting enough for more elaborate use as a pharmacological tool or as a drug.

With regard to hallucinogenic properties similar to those of the peyote phenethylamine alkaloid mescaline (**52**), studies in man have shown that the main peyote tetrahydroisoquinolines are devoid of such action (*11*). Based on tests on squirrel monkeys and cats, the saguaro alkaloid gigantine (35) has, been suggested to possess hallucinogenic properties (48), but this has not been verified in studies in man.

Simple tetrahydroisoquinolines have been shown to have both central and peripheral pharmacological effects (227). They may have central nervous system depressant or stimulant and convulsant properties (227). They can also cause vasopressor or depressor actions, various effects on the smooth muscle (227, 228), and lipid-mobilizing activity (229). Simple dihydro- and tetrahydroisoquinolines have been shown to be potent inhibitors of cyclic nucleotide phosphodiesterases (230–232). Catecholamine-derived tetrahydroisoquinolines such as salsolinol (12) can act as false neurotransmittors, i.e., they can be taken up and stored in adrenergic neurons and on release either act as agonists (233) or antagonists (234) on adrenergic receptors. Dopamine-related tetrahydroisoquinolines have been studied as inhibitors of dopamine uptake into rat brain slices, and 6,7-dihydroxytetrahydroisoquinoline and S-(−)-salsolinol were found to be potent inhibitors while R-(+)-salsolinol was less effective (235). 6,7-Dihydroxytetrahydroisoquinoline is a directly acting sympathomimetic agent, but can also act indirectly on adrenergic receptors by releasing norepinephrine from adrenergic neurons (236). It was recently demonstrated that salsolinol can act as a dopamine antagonist in radioreceptor assays, in cell culture, and in vivo (237).

Salsolinol and similar dopamine-related compounds are substrates of catechol-O-methyltransferase (COMT) (238) and these substances may therefore also inhibit the COMT-mediated O-methylation of catecholamines. Methyl substituents at the nitrogen or at C-4 of 6,7-dihydroxytetrahydroisoquinolines do not significantly change their inhibitory action on COMT, but eliminate their releasing effect on norepinephrine from adrenergic neurons (238). The formation of salsolinol and similar catecholamine-derived tetrahydroisoquinolines in mammals will be reviewed elsewhere in this treatise (Chapter 7).

Heliamine (10) and similar compounds have been reported to inhibit the growth of sarcoma 45 in rats by 60–79%, while there was little or no activity against Walker carcinosarcoma or Erhlich ascites carcinoma (239). Several other isoquinoline alkaloids have shown antitumor activity, and a few (emetine, thalicarpine) have even undergone clinical trials (240). The isoquinolone corydaldine has analgetic and antirheumatic activities but results in human clinical trials were disappointing (241).

A simple tetrahydroisoquinoline derivative, debrisoquine (173) is used in the treatment of hypertension (242). The main metabolite in man of debrisoquine is the 4-hydroxy derivative 174 (243) and it has been shown that the ratio between excreted debrisoquine and its metabolite 174 is polymorphically distributed (244). Poor metabolizers of debrisoquine also exhibit a decreased ability to oxidize some other drugs, e.g., guanoxin and phenacetin (245) and nortriptyline (246). A debrisoquine hydroxylation test may therefore be useful in predicting an individual's ability to metabolize certain types of drugs (246). The antidepressant drug nomifensine (175) (247) has a close chemical relationship to the 4-phenyl-substituted tetrahy-

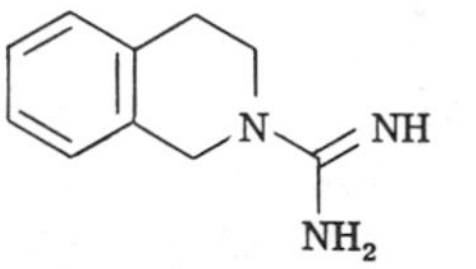

Debrisoquine
173

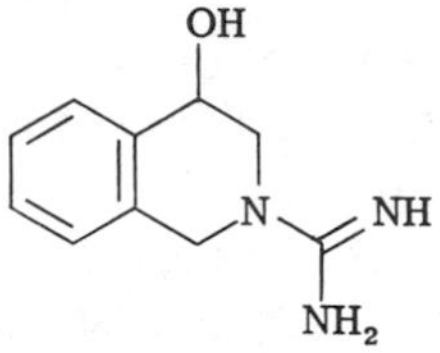

174

droisoquinoline alkaloid cherylline (137). Nomifensine shares with the classic tricyclic antidepressants the common property of inhibiting the uptake of norepinephrine into brain nerve endings, but differs in that it is also a potent inhibitor of dopamine uptake by nerve endings obtained from brain regions that are rich in dopamine (248).

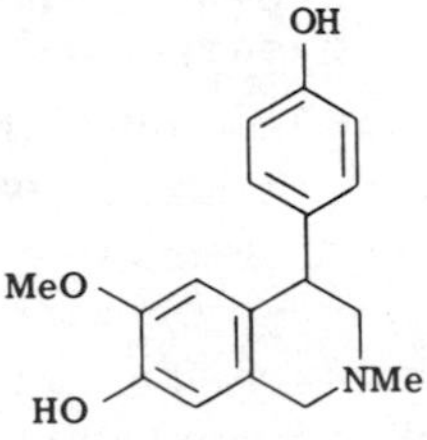

Cherylline
137

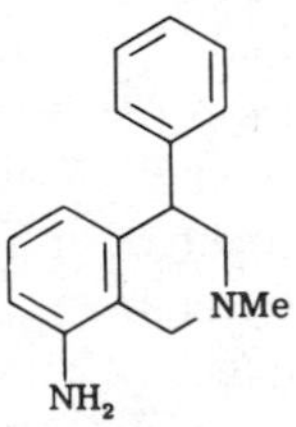

Nomifensine
175

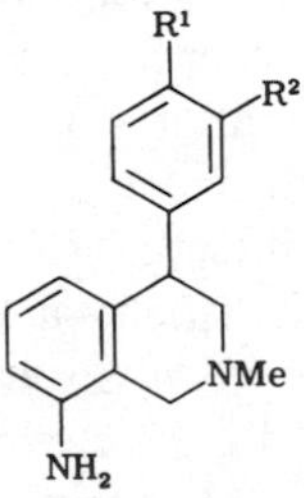

176 $R^1 = R^2 = HO$
177 $R^1 = HO, R^2 = H$
178 $R^1 = MeO, R^2 = HO$
179 $R^1 = HO, R^2 = MeO$

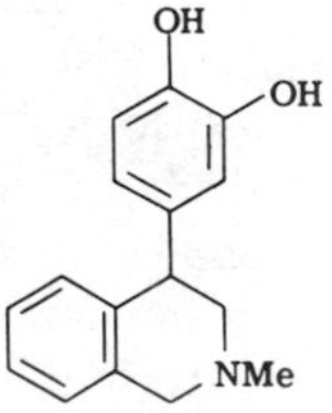

180

The dihydroxy derivative of nomifensine (176) is a potent dopamine agonist directly acting on postsynaptic dopamine receptors (249). Neither nomifensine itself nor its three known metabolites 177, 178, and 179 possess dopamine-like receptor stimulant properties (250). Although $3^1,4^1$-dihydroxynomifensine (176) has not yet been reported as a metabolite, it is possible that this compound is an active metabolite responsible for some of the actions of the drug. The 8-amino group is apparently not essential for dopamine-like activity as the derivative 180 is equipotent to 176 in a renal artery model (251).

The function of secondary metabolites such as simple isoquinoline alkaloids in the plants that produce them remains obscure (*252*). However, a recent report suggests that several simple isoquinolines have growth-regulating activity (*253*). Salsoline (**12**), *O*-methylanhalonidine (**33**), and pellotine methyl iodide (**45**) showed strong growth inhibition in a bean second node bioassay. Several other simple isoquinolines showed slight-to-moderate effects and a few were phytotoxic (*253*).

Some alkaloids are known to be insecticidal (*252*). The alkaloids of the sinita cactus *Lophocereus schottii* are toxic to most *Drosophila* species except *D. pachea,* which is the only species that breeds in old stems of the cactus (*254*).

REFERENCES

1. L. Reti, *in* "The Alkaloids", Vol. 4, (R. H. F. Manske, ed.) p. 7. Academic Press, New York, 1954.
2. L. Reti, *Fortschr. Chem. Org. Naturst.* **6**, 242 (1950).
3. S. Agurell, *Lloydia* **32**, 206 (1969).
4. A. Heffter, *Chem. Ber.* **29**, 216 (1896).
5. A. Heffter, *Chem. Ber.* **27**, 2975 (1894) and **34**, 3004 (1901).
6. E. Späth, *Monatsh. Chem.* **40**, 129 (1919).
7. G. Kapadia and M. B. E. Fayez, *J. Pharm. Sci.* **59**, 1699 (1970).
8. G. Kapadia and M. B. E. Fayez, *Lloydia* **36**, 9 (1973).
9. J. B. Bruhn and B. Holmstedt, *Econ. Bot.* **28**, 353 (1974).
10. Anonymous, *Bull. Narc.,* U.N. Dept. of Social Affairs, **11**, 16 (1959).
11. A. T. Shulgin, *Lloydia* **36**, 46 (1973).
12. J. Lundström, *Acta Pharm. Suec.* **8**, 275 (1971).
13. E. Späth, *Monatsh. Chem.* **42**, 97 (1921).
14. E. Späth, *Monatsh. Chem.* **40**, 129 (1919).
15. E. Späth, *Monatsh. Chem.* **42**, 263 (1921).
16. E. Späth and H. Röder, *Monatsh. Chem.* **43**, 93 (1922).
17. E. Späth, *Monatsh. Chem.* **43**, 477 (1922).
18. E. Späth and J. Gangl, *Monatsh. Chem.* **44**, 103 (1923).
19. E. Späth and J. Passl., *Chem. Ber.* **65**, 1778 (1932).
20. E. Späth and F. Boschan, *Monatsh. Chem.* **63**, 141 (1933).
21. E. Späth and F. Becke, *Chem. Ber.* **67**, 266 (1934).
22. E. Späth and F. Becke, *Chem. Ber.* **67**, 2100 (1934).
23. E. Späth and F. Becke, *Monatsh. Chem.* **66**, 327 (1935).
24. E. Späth and F. Becke, *Chem. Ber.* **68**, 501 (1935).
25. E. Späth and F. Becke, *Chem. Ber.* **68**, 944 (1935).
26. E. Späth and F. Keszter, *Chem. Ber.* **68**, 1663 (1935).
27. E. Späth and F. Keszter, *Chem. Ber.* **69**, 755 (1936).
28. E. Späth and J. Bruck, *Chem. Ber.* **72**, 334 (1939).
29. G. J. Kapadia and H. M. Fales, *J. Pharm. Sci.* **57**, 2017 (1968).
30. J. Lundström, *Acta Chem. Scand.* **26**, 1295 (1972).
31. G. J. Kapadia, N. J. Shah, and T. B. Zalucky, *J. Pharm. Sci.* **57**, 254 (1968).
32. G. J. Kapadia and H. M. Fales, *Chem. Commun.* 1688, (1968).

33. G. J. Kapadia, G. S. Rao, E. Leete, M. B. E. Fayez, Y. N. Viashnav, and H. M. Fales, *J. Amer. Chem. Soc.* **92** 6943 (1970).
33a. G. J. Kapadia, G. S. Rao, W. H. Hussain, and B. K. Chowdhury, *J. Heterocycl. Chem.* **10**, 135 (1973).
34. J. S. Todd, *Lloydia* **32**, 395 (1969).
35. J. G. Bruhn and S. Agurell, *Phytochemistry* **14**, 1441 (1975).
36. N. L. Britton and J. N. Rose, *"The Cactaeae,"* Vols I–IV. Carnegie Inst. Washington D.C., 1919–1923.
37. C. Djerassi, N. Frick, and L. E. Geller, *J. Amer. Chem. Soc.* **75**, 3632 (1953).
38. C. Djerassi, T. Nakano, and J. M. Bobbitt, *Tetrahedron* **2**, 58 (1958).
39. C. Djerassi, H. W. Brewer, C. Clarke, L. J. Durham, *J. Amer. Chem. Soc.* **84**, 3210 (1962).
39a. L. G. West, J. L. McLaughlin, and W. H. Earle, *Phytochemistry* **14**, 291 (1975).
40. D. G. O'Donnovan and H. Horan, *J. Chem. Soc. C* 1737, (1969).
41. G. Heyl, *Arch. Pharm.* **239**, 451 (1901).
42. C. Djerassi, C. R. Smith, S. P. Marfey, R. N. McDonald, A. J. Lemin, S. K. Figdor, and H. Estrada, *J. Amer. Chem. Soc.* **76**, 3215 (1954).
43. G. Heyl, *Arch. Pharm.* **266**, 668 (1928).
44. E. Späth, *Ber.* **62**, 1021 (1929).
45. J. G. Bruhn, U. Svensson, and S. Agurell, *Acta Chem. Scand.* **24**, 3775 (1970).
46. J. G. Bruhn, U. Svensson, and S. Agurell, *Acta Chem. Scand.* **24**, 3775 (1970).
46. S. D. Brown, J. E. Hodgkins, J. L. Massingill, Jr., and M. G. Reinecke, *J. Org. Chem.* **37**, 1825 (1972).
47. J. G. Bruhn and J. Lundström, *Lloydia* **39**, 197 (1976).
48. J. E. Hodgkins, S. D. Brown and J. L. Massingill, *Tetrahedron Lett.* 1321, 1967.
49. S. D. Brown, J. L. Massingill, and J. E. Hodgkins, *Phytochemistry* **7**, 2031 (1968).
50. G. Grethe, M. Uskoković, T. Williams, and A. Brossi, *Helv. Chim. Acta*, **50**, 2397 (1976).
51. G. J. Kapadia, M. B. E. Fayez, M. L. Sethi, and G. S. Rao, *J. Chem. Soc., Chem. Commun.* 856, 1970.
52. J. Weisenborn. Details of isolation and structure elucidation have never been published. Kapadia *et al.* (ref. 51) have reported the synthesis.
53. G. Grethe, M. Uskokovic, and A. Brossi, *J. Org. Chem.* **33**, 2500 (1968).
54. M. Shamma. "The Isoquinoline Alkaloids." Academic Press, New York, 1972.
55. E. Späth and F. Kuffner, *Ber.* **62**, 2242 (1929).
56. J. G. Bruhn and J-E. Lindgren, *Lloydia* **39**, 175 (1976).
57. J. Strömbom and J. G. Bruhn, *Acta Pharm. Suec.* **15**, 127 (1978).
58. R. Mata and J. L. McLaughlin, *Phytochemistry* **19**, 673 (1980).
59. R. Mata and J. L. McLaughlin, *Planta Med.* **38**, 180 (1980).
60. R. Mata and J. L. McLaughlin, *J. Pharm. Sci.* **69**, 94 (1980).
61. S. Pummangura and J. L. McLaughlin, *J. Nat. Prod.* **44**, 498 (1981).
62. R. E. Shultes, *Science* **163**, 245 (1969).
63. L. Lewin, *Ber. Dtsch. Bot. Ges.* **12**, 283 (1894).
64. R. L. Ranieri and J. L. McLaughlin, *J. Org. Chem.* **41**, 319 (1976).
65. R. L. Ranieri and J. L. McLaughlin, *Lloydia* **40**, 173 (1977).
66. S. Agurell, J. G. Bruhn, J. Lundström, and U. Svensson, *Lloydia* **34**, 183 (1971).
67. J. G. Bruhn and C. Bruhn, *Econ. Bot.* **27**, 241 (1973).
68. S. Agurell, J. Lundström, and A. Masoud, *J. Pharm. Sci.* **58**, 1413 (1969).
69. J-E. Lindgren and J. G. Bruhn, *Lloydia* **39**, 464 (1976).
70. Y. A. H. Mohamed, C.-J. Chang, and J. L. McLaughlin, *J. Nat. Prod.* **42**, 197 (1979).
71. M. Fujita, H. Itokawa, J. Inoue, Y. Nozo, N. Goto, and K. Hasegawa, *J. Pharm. Soc. Jpn* **92**, 482 (1972).
72. A. Orekhov and N. Proskurina, *Ber.* **66**, 841 (1933).

73. A. Orekhov and N. Proskurina, *Ber.* **67**, 878 (1934).
74. N. Proskurina and A. Orekhov, *Bull. Soc. Chim. Fr.* **4**, 1265 (1937).
75. N. Proskurina and A. Orekhov, *Bull. Soc. Chim. Fr.* **6**, 144 (1939).
76. C. Carling and F. Sandberg, *Acta Pharm. Suec.* **7**, 285 (1970).
77. K.-H. Michel and F. Sandberg, *Acta Pharm. Suec.* **5**, 67 (1968).
78. S. Ghosal and R. S. Srivastava, *Phytochemistry* **12**, 195 (1973).
79. S. Ghosal and R. Mehta, *Phytochemistry* **13**, 1628 (1974).
80. S. Ghosal and R. S. Srivastava, *J. Pharm. Sci.* **62**, 1555 (1973).
80a. S. Ghosal, R. S. Srivastava, S. K. Bhattacharya, and P. K. Debnath, *Planta Med.* **26**, 318 (1974).
81. E. P. White, N.Z. *J. Sci. Technol.*, Sec. B. **25**, 137 (1944).
82. E. Späth and P. L. Julian, *Ber.* **64**, 1131 (1931).
83. R. H. F. Manshe, *Can. J. Res., Sect. B.* **15**, 159 (1937).
84. C. R. Chen, J. L. Beal, R. W. Doskotch, L. A. Mitscher, and G. H. Svoboda, *Lloydia* **37**, 493 (1974).
85. A. Urzua and B. K. Cassels, *Lloydia* **41**, 98 (1978).
86. S. A. Gharbo, J. L. Beal, R. H. Schlessinger, M. P. Cava and G. H. Svoboda, *Lloydia* **28**, 237 (1965).
87. W-N. Wu, J. L. Beal, and R. W. Doskotch, *J. Nat. Prod.* **43**, 143 (1980).
88. S. Naruto and H. Kaneko, *Phytochemistry* **12**, 3008 (1973).
89. O. Hesse, *Ann. Chem. Suppl.* **8**, 261 (1872).
90. H.-G. Boit, "Ergebnisse dér Alkaloid-Chemie bis 1960." Akademi-Verlag, Berlin 1961.
91. S. Teitel and A. Brossi, *Lloydia* **37**, 196 (1974).
92. R. M. Riggin, M. J. McCarthy, and P. T. Kissinger, *J. Agric. Food Chem.* **24**, 189 (1976).
93. R. M. Riggin and P. T. Kissinger, *J. Agric. Food Chem.* **24**, 900 (1976).
94. J. Siwon, R. Verpoorte, T. Van Beek, H. Meerburg, and A. Baerhiem Svendsen, *Phytochemistry* **20**, 323 (1981).
95. J. Wu, J. L. Beal, W.-N. Wu, R. W. Doskotch, *J. Nat. Prod.* **43**, 270 (1980).
96. B. Achari, E. Ali, P. G. Dastidar, R. R. Sinha, and S. C. Pakrashi, *Planta Med. Suppl.* **5-7** (1980).
97. K. Drost-Karbowska, *Acta Pol. Pharm.* **34**, 421 (1977).
98. K. Drost, *Diss. Pharm.* (Poland) **13**, 167 (1961); *BA* **37**, 7361 (1962).
99. C. Tani, N. Nagakura and C. Kuriyama, *Yakugaku Zasshi,* **98**, 1658 (1978).
100. S. s. Markosyan, T. A. Tsulikyan and V. A. Mnatsakanyan, *Arm. Khim. Zh.* **29**, 1053 (1976).
101. R. Barca, J. Dominguez, and I. Ribas, *An. R. Soc. Esp. Fis. Quim. Ser. B* (Madrid) **55**, 717 (1959); *CA* **54**, 14289 (1960).
102. T.-H. Yang and C.-M. Chen, *J. Chin. Chem. Soc.* (*Taipei*) **17**, 235 (1970).
103. M. Shamma and A. S. Rothenberg, *Lloydia* **41**, 169 (1978).
104. M. P. Cava and K. Bessho, B. Douglas, S. Markey, and J. A. Weisback, *Tetrahedron Lett.* 4279 (1966).
105. M. P. Cava and K. T. Buck, *Tetrahedron* **25**, 2795 (1969).
106. N. M. Mollow and H. B. Dutschewska, *Tetrahedron Lett.* 1951 (1969).
107. R. W. Doskotch, P. L. Schiff, Jr, and J. L. Beal, *Tetrahedron* **25**, 469 (1969).
108. B. Z. Ahn and F. Zymalkowski, *Tetrahedron Lett.* 821 (1976).
109. B. Z. Ahn, U. Degen, C. Lienjayetz, P. Pachaly, and F. Zymalkowski, *Arch. Pharm.* **311**, 569 (1978).
110. A. Jössang, M. Leboeuf, and A. Cavé, *Planta Med.* **32**, 249 (1977).
111. H. G. Theuns, J. E. G. van Dam, J. M. Luteijn, and C. A. Salemink, *Phytochemistry* **16**, 753 (1977).
112. V. Preininger and V. Tosnarová, *Planta Med.* **23**, 233 (1973).

113. W.-N. Wu, J. L. Beal, and R. W. Doskotch, *J. Nat. Prod.* **43**, 372 (1980).
114. S. Kh. Umarov, F. Z. Ismailov, and S. Yu. Yunusov, *Khim. Prir. Soedin.* **6**, 444 (1970); *CA* **74**, 1042e (1971).
115. M. Shamma, and Sr. M. A. Podczasy, *Tetrahedron* **27**, 727 (1971).
116. S. K. Chattopadhyay, A. B. Ray, D. J. Slatkin, J. E. Knapp, and P. L. Schiff, *J. Nat. Prod.* **44**, 45 (1981).
117. S. Abdizhabbarova, S. Kh. Maekh, S. Yu. Yunusov, M. Yagudaev, and D. Kurbahov, *Khim. Prir. Soedin.* 472 (1978); *CA* **89**, 176370.
118. P. Müller and H. R. Schütte, *Z. Naturforsch. B.* **23**, 491 (1968).
119. E. A. Bell, J. R. Nulu, and C. Cone, *Phytochemistry* **10**, 2191 (1971).
120. M. E. Daxenbichler, R. Kleiman, D. Weisleder, C. H. Van Etten, and K. D. Carlson, *Tetrahedron Lett.* 1801 (1972).
121. A. Brossi, A. Focella, and S. Teitel, *Helv. Chim. Acta* **55**, 15 (1972).
122. R. Tschesche, C. Spilles, and G. Eckhardt, *Chem. Ber.* **107**, 1329 (1974).
123. R. Tschesche, J. Moch, and C. Spilles, *Chem. Ber.* **108**, 2247 (1975).
124. J. Lundström and S. Agurell, *J. Chromatog.* **30**, 271 (1967).
125. J. Lundström and S. Agurell, *Acta Pharm. Suec.* **8**, 261 (1971).
126. J. Lundström and S. Agurell, *J. Chromatog.* **36**, 105 (1968).
127. G. J. Kapadia and G. S. Rao, *J. Pharm. Sci.* **54**, 1817 (1965).
128. J. L. McLaughlin and A. G. Paul, *Lloydia* **29**, 315 (1966).
129. R. L. Ranieri and J. L. McLaughlin, *J. Chromatog.* **111**, 234 (1975).
130. S. Agurell and J. Lundström, *Chem. Commun.* 1688 (1968).
131. C. G. Hammar, B. Holmstedt and R. Ryhage, *Anal. Biochem.* **25**, 532 (1968).
132. J.-E. Lindgren, S. Agurell, J. Lundström, and U. Svensson, *FEBS Lett.* **13**, 21 (1971).
133. T. L. Kruger, J. F. Litton, R. W. Kondrat, and R. G. Cooks, *Anal. Chem.* **48**, 2113 (1976).
134. R. W. Kondrat, R. G. Cooks, *Anal. Chem.* **50**, 81A (1978).
135. T. L. Kruger, R. G. Cooks, J. L. McLaughlin, and R. L. Ranieri, *J. Org. Chem.* **42**, 4161 (1977).
136. R. W. Kondrat, R. G. Cooks, J. L. McLaughlin, *Science* **199**, 978 (1978).
137. S. E. Unger, R. G. Cooks, R. Mata, and J. L. McLaughlin, *J. Nat. Prod.* **43**, 288 (1980).
138. J. Cymerman Craig, S.-Y. C. Lee, R. P. K. Chan and I. Y.-F. Wang, *J. Amer. Chem. Soc.* **99**, 7996 (1977).
139. H. Budzikiewicz, C. Djerassi, and D. H. Williams, "Structure Euricidation of Natural Products by Mass Spectrometry." Holden-Day, San Francisco, 1964.
140. H. M. Fales, G. W. A. Milne, and M. L. Westal, *J. Amer. Chem. Soc.* **91**, 3682 (1969).
141. G. W. Wood, N. Mak, and A. M. Hogg, *Anal. Chem.* **48**, 981 (1976).
142. F. Schenker, R. A. Schmidt, T. Williams, and A. Brossi, *J. Heterocycl. Chem.* **8**, 665 (1971).
143. G. S. King, B. L. Goodwin, and M. Sandler, *J. Pharm. Pharmacol.* **26**, 476 (1974).
144. S. P. Singh, S. S. Parmar, V. I. Stenberg, and S. A. Farnum, *J. Heterocycl. Chem.* **15**, 541 (1978).
145. D. W. Hughes and D. B. MacLean, *in* "The Alkaloids," Vol. 18, p. 217. Academic Press, New York, 1981.
146. A. Pictet and I. Spengler, *Ber.* **44**, 2030 (1911).
147. W. M. Whaley and T. R. Govindachari, *Org. React.* **6**, 151 (1951).
148. A. Brossi. F. Schenker, R. Schmidt, R. Banziger and W. Leimgruber, *Helv. Chim. Acta* **49**, 403 (1966).
149. J. M. Bobbitt and J. C. Sih, *J. Org. Chem.* **33**, 856 (1968).
150. J. A. Castillon, *J. Amer. Chem. Soc.* **74**, 558 (1952).
151. K. Takahashi and A. Brossi, *Heterocycles* **19**, 691 (1982).
152. E. Winterstein and G. Trier, "Die Alkaloide," p. 307. Bornträger, Berlin, 1910.

153. R. Robinson, *J. Chem. Soc.* 894 (1917).
154. C. Schöpf and H. Bayerle, *Ann.* **513**, 190 (1934).
155. Ö. Kovacs, and G. Fodor, *Ber.* **84**, 795 (1931).
156. T. Kametani, S. Shibuya, and M. Satoh, *Chem. Pharm. Bull.* (Japan) **16**, 953 1968.
157. W. M. Whaley and T. R. Govindachari, *Org. React.* **6**, 74 (1960).
158. S. Nagubandi and G. Fodor, *J. Heterocycl. Chem.* **17**, 1457 (1980).
159. A. Brossi, F. Schenker, and W. Leimgruber, *Helv. Chim. Acta* **47**, 2089 (1964).
160. J. M. Bobbitt, J. M. Kiely, K. L. Khanna, and R. Ebermann, *J. Org. Chem.* **30**, 2247 (1965).
161. J. M. Bobbitt, D. N. Roy, A. Marchand, and C. W. Allen, *J. Org. Chem.* **32**, 2225 (1967).
162. J. M. Bobbitt and J. C. Sih, *J. Org. Chem.* **33**, 856 (1968).
163. J. M. Bobbitt, A. S. Steinfield, K. H. Weisgraber, and S. Dutta, *J. Org. Chem.* **34**, 2478 (1969).
164. J. M. Bobbitt and C. P. Dutta, *J. Org. Chem.* **34**, 2001 (1969).
165. M. Takido, K. L. Khanna, and A. G. Paul, *J. Pharm. Sci.* **59**, 271 (1970).
166. M. Shamma and J. L. Moniot, "Isoquinoline Alkaloids Research, 1972–1977." Plenum Press, New York, 1978.
167. A. R. Battersby, *Proc. Chem. Soc.* **189** (1963).
168. P. A. Bather, J. R. Lindsay Smith, R. O. C. Norman, and J. S. Sadd, *Chem. Commun.* 1116 (1969).
169. J. R. Lindsay Smith, R. O. C. Norman, and A. G. Rowley, *Chem. Commun.* 1238 (1970).
170. H. Iida, N. Katoh, M. Narimiya, and T. Kikuchi, *Heterocycles* **6**, 2017, (1977).
171. H. Irie, A. Shiina, T. Fushimi, J. Katakawa, N. Fuijii, and H. Yajima, *Chem. Lett.,* 875 (1980).
172. A. Brossi, and R. Borer, *Monatsh. Chem.* **96**, 1409 (1965).
173. H. Bruderer and A. Brossi, *Helv. Chim. Acta* **48**, 1945 (1965).
174. A. Brossi, M. Baumann and R. Borer, *Monatsh. Chem.* **96**, 25 (1965).
175. G. Grethe, V. Toomé, H. L. Lee, M. Uskokovic, and A. Brossi, *J. Org. Chem.* **33**, 504 (1968).
176. A. Brossi, J. O'Brien, S. Teitel, *Org. Prep. Proced.* **2**, 281 (1970).
177. G. Grethe, H. L. Lee, M. R. Uskokovic, and A. Brossi, *Helv. Chim. Acta* **53** 874 (1970).
178. A. Brossi, and S. Teitel, *Helv. Chim. Acta* **53**, 1779 (1970).
179. A. Brossi, J. F. Blount, J. O'Brien, S. Teitel, *J. Amer. Chem. Soc.* **93**, 6248 (1971).
180. J.-I. Minamikawa and A. Brossi, *Can. J. Chem.* **57**, 1720 (1979).
181. J. Lundström, *Acta Pharm. Suec.* **8**, 485 (1971).
182. D. G. O'Donovan and H. Horan, *J. Chem. Soc. C.* 2791 (1968).
183. T. Kametani and T. Okawara, *J. Chem. Soc. Perkin Trans.* 579 (1977).
184. A. R. Battersby and T. P. Edwards, *J. Chem. Soc.* 1214 (1960).
185. K. Leander and B. Lüning, *Tetrahedron Lett.* 1393 (1968).
186. K. Leander, B. Lüning and E. Ruusa, *Acta Chem. Scand.* **23**, 244 (1969).
187. S. Agurell, I. Granelli, K. Leander, B. Lüning and J. Rosenblom, *Acta Chem. Scand. Ser. B.* **28**, 239 (1974).
188. A. P. Venkov and N. M. Mollow, *Dokl. Bolg. Akad. Nauk.* **32**, 895 (1979).
189. A. Brossi and S. Teitel, *Helv. Chim. Acta* **54**, 1564 (1971).
190. J. F. Blount, V. Toome, S. Teitel and A. Brossi, *Tetrahedron* **29**, 31 (1973).
191. L. Vestin, *Acta Chem. Scand.* **26**, 2305 (1972).
192. K. Leander, B. Lüning and L. Westin, *Acta Chem. Scand.* **27**, 710 (1973).
193. A. Brossi, G. Grethe, S. Teitel, W. C. Wildman and D. T. Bailey, *J. Org. Chem.* **35**, 1100 (1970).
194. A. Brossi and S. Teitel, *J. Org. Chem.* **35**, 3559 (1970).
195. M. A. Schwartz and S. W. Scott, *J. Org. Chem.* **36**, 1827 (1971).

196. T. Kametani, K. Takahashi, and C. V. Loc, *Tetrahedron* **31**, 235 (1975).
197. D. J. Hart, P. A. Cain and D. A. Evans, *J. Amer. Chem. Soc.* **100**, 1548 (1978).
198. R. H. F. Manske, *Can. J. Res. Sect. B.* **16**, 81 (1938).
199. M. V. Ibragumov, M. S. Yunusov, and S. Yu. Yunusov, *Khim. Prir. Soedin.* **6**, 638 (1970); *CA* **74**, 54046 r (1971).
200. T. Kametani and K. Ohkubo, *Tetrahedron Lett.* 4317 (1965).
201. T. Kametani, Y. Enomoto, K. Takahashi, and K. Fukumoto, *J. Chem. Soc. Perkin Trans. 1,* 2836 (1979).
202. M. Masood, P. K. Minocha, K. P. Tiwari, *Curr. Sci.* **49**, 510 (1980).
203. T. Kametani, K. Takahashi, C. V. Loc, and M. Hirata, *Heterocycles* **1**, 247 (1973).
204. H. Suguna, and B. R. Pai, *Indian J. Chem.* **12**, 1141 (1974).
205. A. G. Paul, *Lloydia* **36**, 36 (1973).
206. I. J. McFarlane and M. Slaytor, *Phytochemistry* **11**, 235 (1972).
207. D. G. O'Donovan and E. Barry, *J. Chem. Soc. Perkin Trans. 1,* 2528 (1974).
208. H. R. Schütte and G. Seelig, *Liebigs Ann. Chem.* **730**, 186 (1969).
209. S. Agurell, I. Granelli, K. Leander and J. Rosenblom, *Acta Chem. Scand. Ser. B.* **28**, 1175 (1974).
210. P. Müller, H. R. Schütte, *Biochem. Physiol. Pflanz.* **162**, 234 (1971).
211. G. P. Basmadjian and A. G. Paul, *Lloydia* **34**, 91 (1971).
211a. G. P. Basmadjian, S. F. Hussain, and A. G. Paul, *Lloydia* **41**, 375 (1978).
212. A. R. Battersby and S. Garratt, *Quart. Rev.* **15**, 272 (1961).
213. E. Leete, *J. Amer. Chem. Soc.* **88**, 4218 (1966).
214. J. Lundström and S. Agurell, *Tetrahedron Lett.* 4437 (1968).
215. J. Lundström, *Acta Chem. Scand.* **25**, 3489 (1971).
216. J. Lundström and S. Agurell, *Acta Pharm. Suec.* **8**, 261 (1971).
217. K. L. Khanna, M. Takido, H. Rosenberg, and A. G. Paul, *Phytochemistry* **9**, 1811 (1970).
218. A. R. Battersby, R. Binks, and R. Huxtable, *Tetrahedron Lett.* 563 (1967).
219. A. R. Battersby, R. Binks, and R. Huxtable, *Tetrahedron Lett.* 6111 (1968).
220. E. Leete and J. D. Braunstein, *Tetrahedron Lett.* 451 (1969).
221. K. Stolle and D. Gröger, *Arch. Pharm.* **301** 561 (1968).
222. M. Slaytor and I. J. McFarlane, *Phytochemistry* **7**, 605 (1968).
222a. I. J. McFarlane, and M. Slaytor, *Phytochemistry* **11**, 229, (1972).
223. J. Staunton, *Planta Med.* **36**, 1 (1979).
224. M. L. Wilson and C. J. Coscia, *J. Amer. Chem. Soc.* **97**, 431 (1975).
225. A. R. Battersby, R. C. F. Jones and R. Kazlauskas, *Tetrahedron Lett.* 1873 (1975).
226. M. Rueffer, H. El-Shagi, N. Nagakura, and M. H. Zenk, *FEBS Lett.* **129**, 5 (1981).
227. A. M. Hjort, E. J. de Beer, and D. W. Fassett, *J. Pharmacol. Exp. Ther.* **62**, 195 (1938); D. W. Fassett and A. M. Hjort, *ibid.* **63**, 253 (1938); A. M. Hjort, E. J. de Beer, J. S. Buck, and L. O. Randall, *ibid.* **76**, 263 (1942).
228. L. Simon, J. Povszasz, P. Gibiszer, P. Katalin, and S. Talpas, *Pharmacia* **34**, 439 (1979).
229. E. Toth, G. Fassina, and E. S. Soncin, *Arch. Int. Pharmacodyn. Ther.* **169**, 375 (1967).
230. K. Masayasu, I. Waki, and I. Kimura, *J. Pharmacobio-Dyn.* **1**, 145 (1978).
231. R. Van Inwegen, Ph. Salaman, V. St. Georgiew, and I. Weinryb, *Biochem. Pharmacol.* **28**, 1307 (1979).
232. V. St. Gerogiew, R. P. Carlson, R. Van Inwegen, and A. Khandwala, *J. Med. Chem.* **22**, 348 (1979).
233. C. Mytilineou, G. Cohen, and R. Barrett, *Eur. J. Clin. Pharmacol.* **25**, 390 (1974).
234. O. S. Lee, J. E. Mears, J. J. Bardin, D. D. Miller, and D. R. Feller, *Fed. Proc., Fed. Am. Soc. Exp. Biol.* **32**, 723 Abstr. (1973).
235. G. Cohen, R. E. Heikkila, D. Dembiec, D. Sang, S. Teitel, and A. Brossi, *Eur. J. Clin. Pharmacol.* **29**, 292 (1974).

236. I. I. Simpson, *J. Pharmacol. Exp. Ther.* **192**, 365 (1975).
237. D. R. Britton, C. Rivier, T. Shier, F. Bloom, and W. Vale, *Biochem. Pharmacol.* **31**, 1205 (1982).
238. E. E. Smissman, J. R. Reid, D. A. Walsh, and R. T. Borchardt, *J. Med. Chem.* **19**, 127 (1976).
239. A. A. Chachoyan, B. T. Garibdzhanyan, and E. Z. Markaryan., *Biol. Zh. Arm.* **25**, 102 (1972). *CA* **78**, 52538m (1973).
240. J. L. Hartwell. *Cancer Treat. Rep.* **60**, 1031 (1976).
241. A. Brossi, H. Besendorf, L. A. Pirk and A. Rheiner, Jr., *Med. Chem.* **5**, 281 (1965).
242. M. H. Luria and E. D. Ries, *Curr. Ther. Res.* **7**, 289 (1965); A. H. Kitchin and R. W. D. Turner, *Br. Med. J.* **2**, 728 (1966).
243. M. Angelo, L. G. Dring, R. Lancaster, A. Mahgoub, R. I. Smith, *Biochem. Soc. Trans.* **4**, 704 (1976).
244. A. Mahgoub, J. R. Idle, L. G. Dring, R. Lancaster, and R. L. Smith, *Lancet* **2**, 584 (1977).
245. T. P. Sloan, A. Mahgoub, R. Lancaster, J. R. Idle, and R. L. Smith, *Br. Med. J.* **2**, 655 (1978).
246. B. Mellström, L. Bertilsson, J. Säwe, H-U. Schulz, and F. Sjöqvist, *Clin. Pharmacol. Ther.* **30**, 189 (1981).
247. I. Hoffmann, G. Ehrhart and K. Schmitt, *Arzneim.-Forsch.* **21**, 1045 (1971).
248. P. Hunt, M. L. Kannengiesser, I. P. Raynand, *J. Pharm. Pharmacol.* **26**, 370 (1974); V. Schacht and W. Heptner, *Biochem. Pharmacol.* **23**, 3413 (1974); J. Tuomisto, *Eur. J. Pharmacol.* **42**, 101 (1977).
249. J. A. Doat, G. N. Woodruff and K. J. Watling, *J. Pharm. Pharmacol.* **30**, 495 (1978); B. Costall and R. J. Naylor, *J. Pharm. Pharmacol.* **30**, 514 (1978).
250. C. Braestrup and J. Scheel-Krüger, *Eur. J. Pharmacol.* **38**, 305 (1976); B. Costall, R. J. Naylor, J. G. Cannon and T. Lee, *Eur. J. Pharmacol.* **41**, 307 (1977).
251. J. N. Jacob, N. E. Nichols, J. D. Kohli, and D. Glock, *J. Med. Chem.* **24**, 1013 (1981).
252. T. Robinson, *Science* **184**, 430 (1974).
253. N. B. Mandava, J. F. Worley, and G. J. Kapadia, *J. Nat. Prod.* **44**, 94 (1981).
254. H. W. Kircher, W. B. Heed, J. S. Russel, and J. Grove, *J. Insect Physiol.* **13**, 1869 (1967).
255. J. Strömbom and J. G. Brahn, *J. Chromatog.* **147**, 513 (1978).
256. S. Pummangura, J. L. McLaughlin, D. V. Davies, and R. G. Cooks, *J. Nat. Prod.* In press (1983).
257. S. Pummangura, Y. A. H. Mohamed, C.-J. Chang, and J. L. McLaughlin, *Phytochemistry.* In press (1983).

——Chapter 7——

MAMMALIAN ALKALOIDS

Michael A. Collins

Department of Biochemistry and Biophysics,
Loyola University of Chicago,
Stritch School of Medicine,
Maywood, Illinois

I. Introduction

During the past decade, evidence has progressively accumulated indicating that isoquinolines and β-carbolines, normally considered to be plant substances, are trace chemical constituents in mammalian cells and fluids. In the face of a myriad of exquisitely regulated, enzymatically mediated biosynthetic processes, the most basic explanation for the origin of mammalian alkaloids is still valid: that is, direct nonenzymatic condensation of metabolic or exogenous carbonyl compounds with endogenous "biogenic" amines. However, in this relatively new area, many probing questions need resolution and clarification with regard to the nature of the substrates (particularly the carbonyl components), the reaction mechanisms including whether enzyme catalysis is involved, the stereochemistry of the alkaloid products, and even the extent, if any, of artifactual alkaloid formation during

THE ALKALOIDS, VOL. XXI

complex isolation procedures. Although "mammalian alkaloids," a phrase apparently coined in 1974 by Teitel and Brossi (1), have been discussed briefly in an earlier chapter in this series, this is the first chapter devoted entirely to the topic.

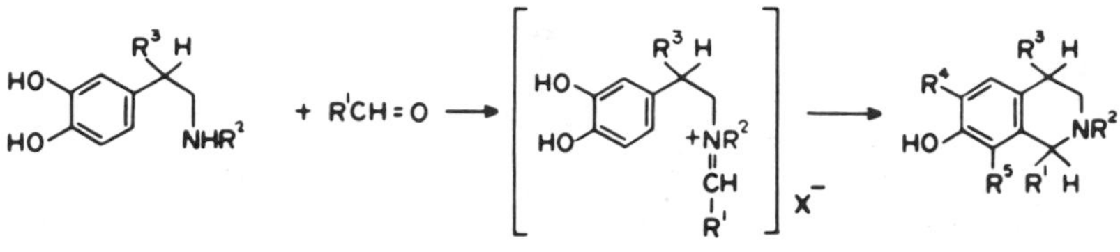

CA

Schiff base
or
Iminium Salt (R^2=Me)

TIQ

1a Dopamine (R^2=R^3=H)
  b Norepinephrine (R^2=H; R^3=OH)
  c Epinephrine (R^2=Me; R^3=OH)

2a 1-Demethyl salsolinol (R^1=R^2=R^3=R^5=H; R^4=OH)
  b Salsolinol (R^1=Me; R^2=R^3=R^5=H; R^4=OH)
  c Isosalsolinol (R^1=Me; R^2=R^3=R^4=H; R^5=OH)
  d (R^2=Me; R^1=R^5=H; R^3=R^4=OH)
  e (R^2=Me; R^1=R^4=H; R^3=R^5=OH)
  f (R^1=R^2=Me; R^3=R^4=OH; R^5=H)
  g (R^1=R^2=Me; R^3=R^5=OH; R^4=H)
  h (R^1=R^2=R^5=H; R^3=R^4=OH)

SCHEME 1

Alkaloids detected to date in experiments with mammalian systems have been limited to simple 1,2,3,4-tetrahydroisoquinolines (TIQs), simple 1,2,3,4-tetrahydro and possibly aromatic β-carbolines (TBCs), 1-benzyl-substituted TIQs, and aromatic oxygenated berbines. Their apparent amino compound substrates are, specifically, the catecholamines: dopamine (**1a**), (−)-norepinephrine (**1b**) and (−)-epinephrine (**1c**), or the indoleamines:

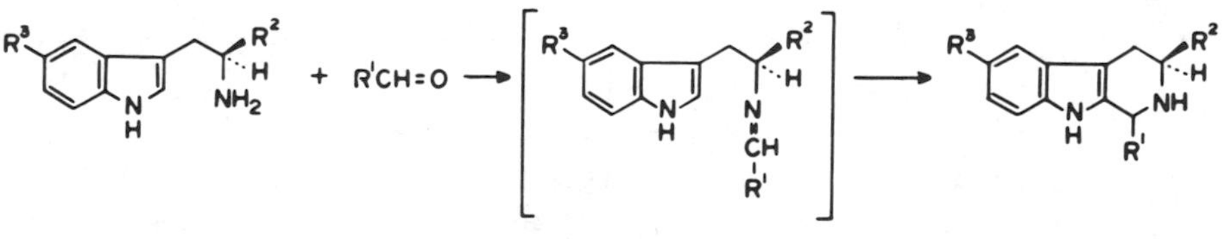

Indoleamine

Schiff base

TBC

3a Tryptamine (R^2=R^3=H)
  b Serotonin (R^2=H; R^3=OH)
  c Tryptophan (R^2=COOH; R^3=H)
  d 5-Hydroxytryptophan (R^2=COOH; R^3=OH)
  e 5-Methoxytryptamine (R^2=H; R^3=OMe)
  f 5-Methoxytryptophan (R^2=COOH; R^3=OMe)

4a (R^1=Me; R^2=COOH; R^3=H)
  b (R^1=Me; R^2=COOH; R^3=OH)
  c (R^1=R^2=R^3=H)
  d (R^1=R^2=H; R^3=OMe)
  e (R^1=Me; R^2=R^3=H)
  f (R^1=R^2=H; R^3=OH)

SCHEME 2

tryptamine (**3a**) and serotonin (**3b**) (Schemes 1 and 2). These amines (biogenic amines) are concentrated in nerve cells, where they may function as transmitters or modulators, as well as in endocrine and mast cells. The amino acid precursors of these biogenic amines—L-dopa (**5**), tryptophan (**3c**), or 5-hydroxytryptophan (**3d**)—also could be physiological Pictet–Spengler substrates. The carbonyl substrates are aldehydes, α-keto acids, and possibly activated ketones. Additionally, there is evidence that certain mammalian alkaloids could form intramolecularly following enzymatic oxidation of an *N*-methylamino moiety within an appropriate phenethylamine or indolethylamine.

Biochemical Aspects of the Physiological Pictet–Spengler Cyclization

Historically, mammalian alkaloid studies evolved from the "physiological" applications of the classic Pictet–Spengler (*2*) or Winterstein–Trier (*3*) reactions investigated extensively by chemists interested in plant alkaloids in the 1930s (*4*). With the realization that serotonin and dopamine were discrete transmitters in mammalian nervous system, it was speculated that β-carboline or isoquinoline formation might occur in mammalian systems, particularly in connection with a disease state where accumulation of biogenic amines or carbonyl substrates was likely (*5–7*). Parenthetically, however, one of the earliest speculations on mammalian alkaloids focused on the possible formation of 1-methyl-6-methoxy-1,2,3,4-TBC not through an amine/aldehyde condensation but via a Bischler–Napieralski-type cyclization of *N*-acetyl-5-methoxytryptamine (melatonin) in the pineal gland (*8*).

The chemistry of the Pictet–Spengler reaction, reviewed by Whaley and Govindachari (*4*), showed that the condensation at neutral pH involved formation of a Schiff base as an intermediate, and that the isoquinoline ring only formed appreciably from this intermediate when the phenyl ring was activated para to the position of cyclization. Schiff bases originating from tryptamine were sufficiently activated by the indole ring to form TBCs directly.

Condensations of catecholamines and formaldehyde or acetaldehyde often afforded TIQ products originating from an ortho and para cyclization of the intermediate Schiff base (*4,9,10*). Specifically, epinephrine's reaction with formaldehyde provided equivalent amounts of the 4,6,7- and 4,7,8-triols (**2d** and **2e**) at neutral pH values (*10*). Only the 4,6,7-isomer was formed at pH < 1, suggesting that a different cyclization mechanism, possibly involving a phenoxide anion, was operating at physiological pH ranges. The condensation of acetaldehyde with epinephrine (*10*) or with norepinephrine

(*11*) afforded four discernible products based on high-performance liquid chromatography (HPLC) analysis. Only the cis and trans epimers of the 4,6,7-triol **2f** were evident at pH < 1, but two epimers of **2g** constituted 60% of the total TIQs at pH 5.5 and 26% at pH 7 (*10*). Condensation of acetaldehyde with dopamine gave 10% of the ortho ring-closed isosalsolinol (**2c**, 7,8-dihydroxy-1-methyl-TIQ) at pH 4.5 and trace amounts at higher pH values (*12*). A similar pH-dependent study of the formation of TBC isomers from indolylethylamine precursors has not been reported.

The second-order rate of alkaloid formation with acetaldehyde was shown to decrease in the following rank: dopamine > dopa > epinephrine > norepinephrine > serotonin (*13, 14*), and formaldehyde reacted some 14–70 times faster with catecholamines than acetaldehyde (*9, 13, 14*). The half-life of the epinephrine's condensation at pH 6.5 was < 0.5 min with formaldehyde, and 35 min with acetaldehyde (*10*), and lower pH dramatically decreased the reaction rate.

II. Chemical Considerations

Many of the mammalian alkaloids have been synthesized following methods discussed in detail by Whaley and Govindachari (*14*) and Shamma and Moniet (*15*), employing either the Pictet–Spengler (neutral or acid-catalyzed) condensation, the Bischler–Napieralski cyclization, or variations of the Pomerantz–Fritsch reaction (*16*).

Considering first the Pictet–Spengler approach, the cis (1*S*, 3*S*) and trans (1*R*, 3*S*) epimers of 3-carboxysalsolinol (**6** and **7**) were isolated from condensations of L-dopa (**5**) and acetaldehyde, carried out under nitrogen in 0.5 *N* sulfuric acid (*17*). The structure of the major (95%) product was shown to be the cis isomer by single crystal X-ray analysis and by chemical oxidation of **6** with mercuric acetate followed by reduction with sodium borohydride, affording only traces of **7** (Scheme 3). The corresponding 6-methoxy alkaloids were obtained in lower yields from the 3-*O*-methyl ether of L-dopa, and drastic conditions were required for their formation (*16*). Likewise, the epimeric TBC derivatives **4a** and **4b** were obtained from tryptophan or

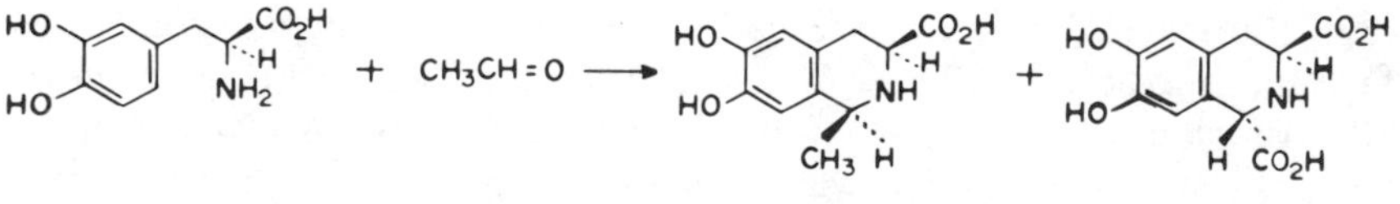

5   6   7

SCHEME 3

5-hydroxytryptophan with acetaldehyde, and were characterized (*18*). The cis isomers were the nearly exclusive products.

As mentioned previously, the four tetrahydroisoquinoline-triol products of epinephrine's reaction with acetaldehyde, or two isomers with formaldehyde, have been separated and identified (*10*). 4-Hydroxylated TIQs related to biogenic amines have been synthesized by Kametani's group (*19*) and by Quessy and Williams (*20*), by condensation of appropriate 1-phenyl-2-aminoethanols with various carbonyl compounds in neutral nonaqueous media such as ethanol or acetone.

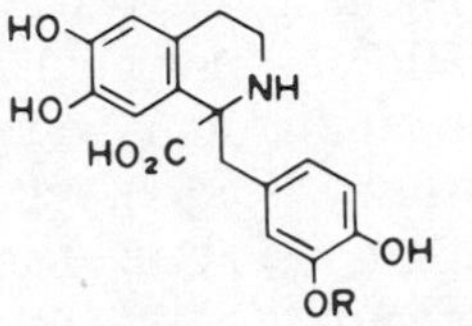

8a NLCA (R=H)
b MNLCA (R=Me)

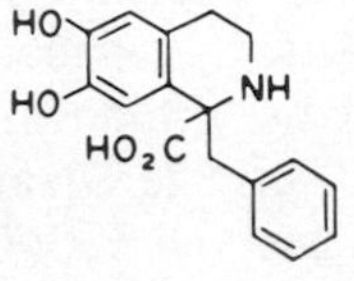

9 DNLCA

1-Carboxy-1-benzyl-TIQs of types **8** and **9** were prepared in good yield by neutral condensation of phenylpyruvic acid derivatives with dopamine (*21*) or with noradrenaline (*22*). 1-Carboxy-TBCs were similarly synthesized in this one-step procedure from phenylpyruvic acids and tryptamine, serotonin, or 5-methoxytryptamine. Silica gel in catalytic quantities was found advantageous in promoting the condensation of norepinephrine and phenylpyruvic acids (*22*). Analogous condensations have been carried out with glyoxylic acid as the carbonyl substrate. In this case, the cis and trans isomers of the respective 1-carboxylated phenolic TIQ-4-ols were obtained from norepinephrine or epinephrine and glyoxylic acid (*23*), with tetrahydroxybenzazepin-2-ones as by-products. A preferred method for the preparation of TBC was found in the condensation of glyoxylic acid with tryptamine followed by reductive decarboxylation (*24*).

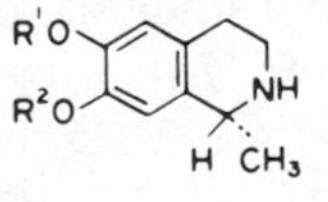

10 R-(+)-Salsoline (R^1=H; R^2=Me)
11 R-(+)-Isosalsoline (R^1=Me; R^2=H)

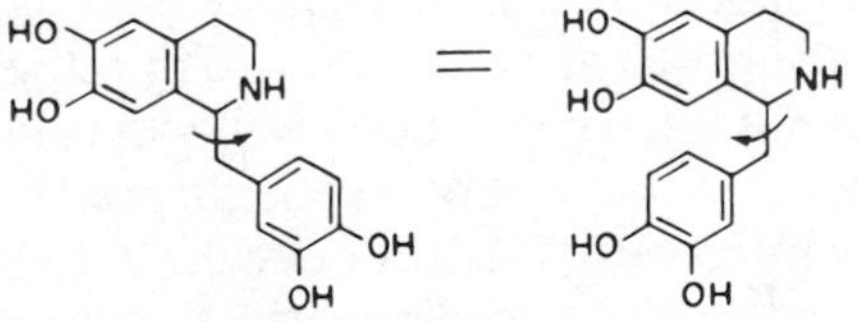

12 (±)-Tetrahydropapaveroline (THP)

Through Bischler–Napieralski synthetic routes, the optical isomers of salsolinol (**2b**), salsoline (**10**), isosalsoline (**11**) and tetrahydropapaveroline

(THP, **12**) have been made available for neurobiological and neurochemical investigations (*25, 26*). [The structure of the THP **12** (Fig. 2) is shown in two different presentations, indicating that the aromatic moiety of the 1-benzyl substituent can rotate freely around the C-1—CH_2 axis.] The oxygen-protected dihydroisoquinolines were synthesized and reduced with appropriate reducing agents, and the racemic TIQs were resolved with optically active acids. After removal of the oxygen-protecting groups, good yields of the *R* and *S* enantiomers were obtained. The optically active TIQs were further characterized as *N*-methyl derivatives and, in the cases of *S*-(−)-salsolinol and *S*-(−)-THP, by conversion to (*S*)-carnegine and (*S*)-laudanosoline, respectively. The 3′-*O*-methyl ether of THP was prepared in good yield by O-benzylation of the amide from dopamine and 3-methoxy-4-hydroxyphenylacetic acid, Bischler–Napieralski cyclization, reduction, and O-debenzylation with aqueous HCl in methanol (*27*); the usual catalytic O-debenzylation procedure with palladium/carbon gave low yields.

Cyclization of phenolic aminoacetals in aqueous HCl at room temperature, a variation of the Pomerantz–Fritsch synthesis, was found to be a facile synthetic route to simple phenolic TIQ-4-ols or, following catalytic reduction, to simple TIQ plant alkaloids (*28, 29*). TIQ-triols related to norepinephrine were prepared by this route (*30, 31*), and 1-benzyl-substituted TIQ-4-ols could be obtained by cyclization of the appropriate aminoacetals in formic acid/acetone or sulfuric acid/acetone mixtures at room temperature (*32*).

Selected isopavinan and pavinan alkaloids potentially derivable from THP in mammalian cells (*33*) were synthesized in racemic and optically active forms (*34*). The route shown in Scheme 4 involved the preparation of *N*-acyl-4-methoxy-1-benzyl-TIQs such as **13**. Conversion to the respective isopavinans **14** in good yield was achieved with catalytic amounts of methanesulfonic acid in acetonitrile. Thermolytic loss of methanol from **11** gave a 1,2-dihydroisoquinoline which afforded the pavinan **15** on cyclization. O-Demethylation of **14** and **15** with refluxing HBr afforded the corresponding tetrols (*26*) representing tetracyclic analogs of THP. This preparation of isopavinan and pavinan alkaloids is superior to earlier methods which gave multiple products and proceeded in rather low yield (*35*).

It is not known whether the alkaloids in mammalian systems (Sections V and VI) that have chiral centers are present as optically active entities. Brossi has pointed out (*36*) that optically active mammalian alkaloids would be expected if formation involved stereospecific (enzymatic) catalysis of the type recently recognized to occur *in vitro* with cell-free extracts from alkaloid-producing plants (*37*). Evidence that enzymes may be involved in mammalian alkaloidal transformations is discussed in Section VII. The endogenous salsolinol, salsoline, or THP that some laboratories have reported in mammalian tissues may in fact be the result of condensation of

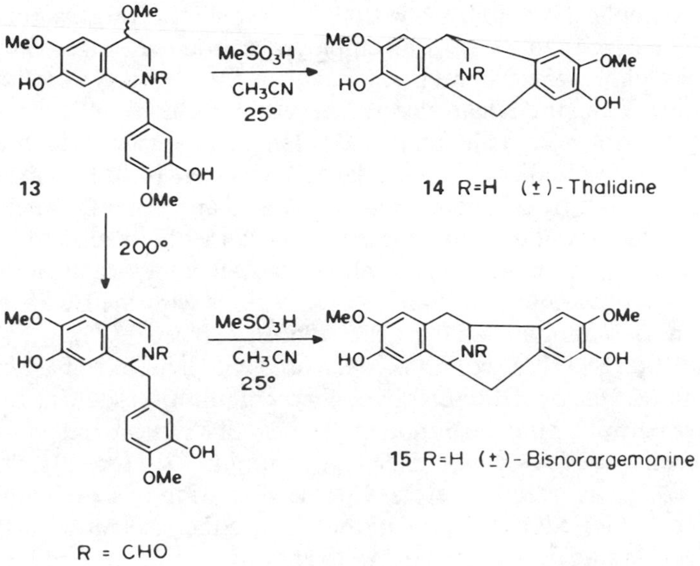

SCHEME 4

dopamine with α-ketocarboxylic acids which are recognized cellular intermediates, rather than with metabolically derived aldehydes, whose levels are usually very low in cells. Oxidative decarboxylation to 3,4-dihydroisoquinolines, followed by enzymatic reduction to the appropriate optically active TIQ (*38*), would nicely mimic the biosynthesis of plant TIQ alkaloids (*39*). Although experimental support for the oxidative decarboxylation of 1-carboxy-substituted TIQs by plant enzymes has been provided (*40*), definite evidence for the formation and particularly the reduction of the proposed 3,4-dihydroisoquinoline intermediates in mammalian systems is still lacking (*11*).

III. Analytical Separation and Detection of Mammalian Alkaloids

The early research methodologies in this area depended almost entirely on conventional thin-layer chromatography (TLC) or paper chromatography with oxidation or fluorescence as the basis of detection. Radiolabeled substrates permitted greater sensitivity in some situations. In most cases, these approaches have been supplanted by HPLC with its greater resolving power, and, when coupled with electrochemical (EC) detection, greater sensitivity (if radiolabels are not being used) for hydroxylated alkaloids.

The first application of HPLC with the mammalian TIQ alkaloids was in 1977 when Riggin and Kissinger employed cation-exchange columns to analyze catechol extracts of rat brain and urine for THP and salsolinol (*41*), and Nijm *et al.* reported in an abstract on cation-exchange HPLC assays for salsolinol in urines from alcoholics (*42*). The metabolism of TIQs by liver *O*-methylases was studied with a similar HPLC system (*43*), as were the rates of TIQ formation from aldehydes and biogenic amines (*44*). The electrochemical detector was used in these studies because of its excellent response (low picogram) to the catecholic alkaloids. An analytical study of the performance of various cation-exchange columns with the EC detector in separations of mammalian TIQs has also been published (*45*).

Reversed-phase HPLC columns, when used with a mobile phase employing organic acid/alcohol mixtures or buffers containing ion-pairing sulfonate salts, have permitted the separation of a variety of TIQs with resolution that has not been possible on cation-exchange columns. Most recently, St. Claire *et al.* (*45*) demonstrated the reversed-phase separation of six dopamine-derived (catecholic) TIQs and dopamine using either isocratic or gradient elution with aqueous acetic acid/2-propanol mobile phase combinations. Alternatively, a mixture of seven catecholic and phenolic TIQ derivatives (salsolinol and six 1-benzyl TIQs) were chromatographed isocratically on C-18 reversed-phase columns with a mobile phase of phosphate buffer (pH 4.85) containing 20% methanol and octyl sulfonate (OSA) counter ion, but as shown in the left chromatogram (Figure 1A, 3-*O*-methyl-NLCA (norlaudanosoline carboxylic acid) and 4′-hydroxy-DNLCA (3′,4′-deoxy-NLCA) were not separable (*46*). The chromatogram on the right (Fig. 1B) shows a complete reversed-phase separation of the three endogenous CAs (**1a,b,c**) and their nine acetaldehyde condensation products discussed in Section I,A using phosphate buffer mobile phase (pH 5.2) with 6% methanol and OSA (*11*).

Further applications of reversed-phase HPLC in biological assays of TIQs are with studies of potential condensation products in the brains of rats treated with ^{14}C-labeled dopamine, in which radioactive detection was employed (*47*). Standards of salsolinol, salsoline, and THP were separated from a large variety of CA and CA metabolites using ammonium phosphate/acetonitile gradient elutions. Hirst *et al.* (*48*) separated and analyzed salsolinol in the urines of alcohol-treated rats or humans with reversed-phase HPLC/EC.

A limited number of reports on the isolation of TBCs from mammalian tissues have been based on reversed-phase HPLC. Rommelspacher *et al.* (*49*) used C-8 reversed-phase columns and gradient elution to isolate 6-hydroxy-TBC from human and rat tissues. An apparent oxidized derivative and 1-methyl-TBC were isolated from human urine with similar HPLC columns, using an isocratic mobile phase (*50*). An interesting application was the use of

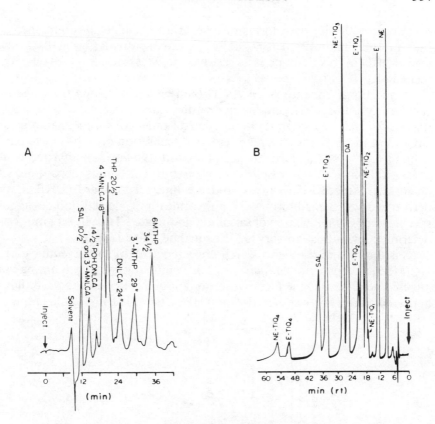

FIG. 1. Reversed-phase HPLC separation of derived TIQs of dopamine or CAs using isocratic elution with aqueous phosphate/methanol/OSA-mobile phase combinations. (A) Mixture of seven catecholic and phenolic TIQ derivatives of dopamine; (B) reversed-phase separation of three endogenous CAs and nine TIQ condensation products with acetaldehyde.

C-18 reversed-phase resin ("Sep-Pac" cartridges) to isolate TBCs from rat brain prior to GC-mass spectral analysis (51). In contrast to the TIQs, the EC detector has not been widely used in HPLC measurements of (hydroxylated) TBCs.

The second important general method of analysis of mammalian alkaloids is based on gas chromatography with either electron capture (GC/ECD) or mass spectrometric detection (GC/MS). It was logical that these techniques be extended to the biogenic amine-derived alkaloids following their initial utilization with the biogenic amines. GC/ECD of simple mammalian TIQs was reported in brief in 1971 (52) and subsequently in greater detail (53, 54) when the approach was used to measure salsolinol in the brains of rodents treated with ethanol, with and without enzyme inhibitors (55, 56). Sandler et *al.* utilized GC with both forms of detection to establish the presence of salsolinol and THP (**12**) in urines of Parkinsonian patients (*57*). A disadvantage of ECD in this situation is its relatively low response to 1-benzyl-TIQs related to THP (*53*).

An additional advance in GC/ECD (and GC/MS, below) is the use of capillary columns. Origitano has shown that the short glass capillary columns effectively separate the O-methylated isomers of simple mammalian TIQs (*58*). Figure 2 shows a representative separation on a 10-m capillary column (145°) of standards of salsolinol and its two O-methylated congeners, salsoline and isosalsoline, in a mixture containing dopamine, its analogous *O*-methyl derivatives, and the internal standard (DHBA), all derivatized with heptafluorobutyryl anhydride for EC detection. The stereospecificity of O-methylation of salsolinol and related TIQs in rat brain was determined with this capillary GC/EC method (*59, 131*).

As is the case with other research areas requiring precise identification, GC/MS has become the *sine qua non* of the methodology in the mammalian alkaloid field, despite the fact that important quantitative differences have appeared recently between GC/MS and HPLC data. The first applications of

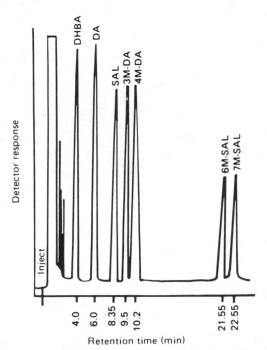

FIG. 2. GC/EC chromatogram showing separation of salsolinol and its two O-methylated congeners, salsoline and isosalsoline, on a short glass capillary column (HFB-derivatives).

GC/MS in mammalian TIQ studies was the Sandler *et al.* report mentioned above (*57*), and a study by Turner *et al.* (*60*) on THP in the brain of dopa-treated rats. The evidence for 1-carboxy-1-benzylated TIQs in animal and human systems was first obtained with GC/MS assays (*61, 62*). The metabolism of THP and related alkaloids by liver enzymes was investigated in some detail by GC/MS (*63*). The technique is the basis for the determination of salsolinol and salsoline in human fluids (*85, 86*).

GC/MS with both conventional and, more recently, capillary columns has been the most applicable method of analysis for mammalian β-carbolines. In the mass fragmentographic mode with conventional columns, Barker *et al.* has identified several TBCs as endogenous constituents in rat tissues (*65, 66*). Similarly, Airaksinen's group has determined the TBC levels in human fluids (*24, 67*). Capillary GC/MS methodology for endogenous TBCs in rat or human samples has been pioneered recently by Faull *et al.* (*51*) and by Beck *et al.* (*68, 111*). As with the TIQs, the capillary columns provide increased resolution, reduced absorption of derivatives (and thus greater sensitivity), and reduced possibility of "false" identification.

Two additional analytical approaches for mammalian alkaloids are radioenzyme assay using catechol *O*-methyltransferase (COMT), and UV laser spectrometry. The first technique, widely used by neurochemists for high-sensitivity analysis of CAs, was developed for the estimation of salsolinol in mammalian tissues by two different research groups (*69, 70*). Based on the enzymatic transfer of a ^{14}C-labeled methyl group from *S*-adenosylmethionine to catechols by the enzyme COMT and subsequent TLC radioanalysis, the method is efficient and very sensitive, but is limited to certain catecholic TIQs.

The UV laser spectrofluorimetric approach was developed Shoemaker *et al.* for the detection of β-carbolines in rat brain (*71*). Certain β-carbolines exhibit yellow-green fluorescence when irradiated *in situ* at 325 nm with the He–Cd laser. Although novel, the method would appear to lack specificity and has not been exploited further.

IV. Production of 1,2,3,4-Tetrahydroisoquinolines *in Vitro*

TIQs of the 1-benzyl series were the subjects of the first investigations. Critical to an appreciation of these and later approaches is an understanding of the roles of the enzymes monoamine oxidase (MAO) and aldehyde dehydrogenase (AldDH) in biogenic amine metabolism. MAO, found almost exclusively on or in mitochondrial membrane, is the principal enzyme deaminating biogenic amines to (biogenic) aldehydes (Scheme 5). As transient intermediates, biogenic aldehydes are rapidly oxidized to acids by

AldDH, the major form of which is also extensively associated with mito-chondria. [In some organs, the α-hydroxy "biogenic" aldehydes originating from metabolism of epinephrine and norepinephrine and their O-methyl congeners are largely reduced to alcohols by aldehyde reductase (AldR).] When the disposition of the aldehyde is limited or prevented or when the biogenic amine is in excess, condensation reactions forming alkaloidal derivatives can occur.

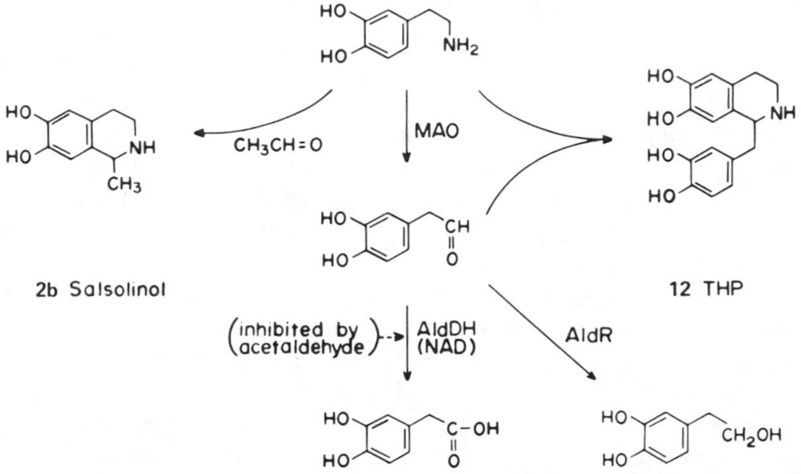

2b Salsolinol   12 THP

SCHEME 5

Holtz and co-workers examined this possibility by incubating excess (millimolar) amounts of dopamine with liver mitochondrial preparations. They concluded, based on pharmacological and limited chemical evidence, that the alkaloid THP (Scheme 5) was a significant product (72). Haluska and Hoffman found that if the mitochondrial or whole-liver incubations con-tained physiological (micromolar) concentrations of ^{14}C-labeled dopamine, THP as determined by thin layer chromatography (TLC) was a minor (< 3%) dopamine metabolite (73). When NAD, the pyridine nucleotide cofactor for AldDH, was included in liver or brain homogenates with millimolar quanti-ties of dopamine in order to favor oxidation of the biogenic aldehyde, again very little THP was formed (74). Agents which could interact with, trap, or reduce the Schiff base intermediate also were shown to reduce the formation of THP in such homogenates (75). Most effective was cysteine, followed by glutathione and ascorbate.

The amounts of THP formed during incubation of excess dopamine with rat liver and brain (either with or without added NAD) could be increased 6–12% by adding acetaldehyde or ethanol (76, 77). Presumably, the acetaldehyde competitively inhibited AldDH, elevating the biogenic aldehyde and promoting its condensation with dopamine (Scheme 5). Based on this experimental finding and the knowledge of THP intermediacy in the plant biosynthesis of morphine and other isoquinoline alkaloids, Davis and co-workers hypothesized that THP and particularly its possible morphine-related alkaloid metabolites were addictive neuromodulators in human alcoholism (76, 77). This hypothesis generated considerable controversy when it was proposed (78), and it still remains unproved. Other work from Davis's laboratory (79) has indicated that the formation of a similar THP-related alkaloid, apparently produced in homogenates containing millimolar amounts of norepinephrine, was enhanced by the addition of barbiturates, which are known inhibitors of AldR enzyme in Scheme 5. More recently, in studies on the metabolism of (millimolar) ^{14}C-labeled dopamine in rat liver slices, THP accounted for 12% of the deaminated dopamine products after a 2-hr incubation (80), but added ethanol or acetaldehyde did not significantly increase this percentage. A two-dimensional TLC/electrophoresis technique was used in these radiochemical experiments.

The initial mammalian alkaloid experiments also viewed simple aldehydes such as acetaldehyde or formaldehyde as potential co-substrates for TIQ formation. Two or more 4-hydroxylated-1-methyl-TIQ isomers derived from endogenous norepinephrine and epinephrine were separated by TLC from the catechol extracts of intact (cow) adrenal glands that had been perfused briefly with acetaldehyde in pH 7 phosphate buffer (9), formaldehyde perfusion resulted in formation of the analogous demethyl congeners. Subsequently, the analytical sensitivity was improved by using "blood levels" of ^{14}C-labeled acetaldehyde (81). Due to a structural resemblance to certain psychoactive and/or toxic plant alkaloids, it was postulated that the simple TIQs could play an important role in the complex pathophysiology of alcohol dependence if they accumulated during alcohol abuse (9). Similarly, salsolinol (2b), the major condensation product of dopamine and acetaldehyde, was formed (in competition with THP) in incubations of millimolar amounts of dopamine in mammalian tissue homogenates or tissue slices with acetaldehyde or ethanol (82).

V. Tetrahydroisoquinolines in Human and Animal Models

Salsolinol (2b) and THP (12) were measured in the catechol extracts of urines from patients undergoing treatment with L-dopa for Parkinsonism (57). Acute ethanol ingestion appeared to increase the levels of urinary salsolinol but not of THP. With similar GC/MS methodology, several substituted berbine alkaloids (**16a, 16b**), derivatives of THP, were reported to be detectable in the urines of patients given L-dopa (*83*).

16a R^1=H; R^2=OH
  b R^1=OH; R^2=H

    Alcoholic patients during detoxification as well as nonalcoholic volunteers excreted salsolinol and O-methylated salsolinols **10** and **11,** according to two different laboratories employing HPLC and confirmatory GC/ECD in one situation (*84*) or GC/MS in the other (*85, 86*). In both studies the catechol, salsolinol, was significantly elevated in the alcoholic urines on the first day of admission when blood acetaldehyde levels were substantial. However, urinary salsoline/isosalsoline levels in alcoholics were increased above the average levels of volunteers in only one of these studies (*84*).

    Similarly, the cerebrospinal fluids (CSF) from detoxified alcoholics and control individuals were stated to contain about equal quantities of O-methylated alkaloids (*86*). The levels of the same alkaloids, estimated in autopsy specimens, appeared to be no different in brain tissues of intoxicated alcoholics and nonalcoholics, but were significantly lower in similar tissues of abstaining alcoholics (*87*). Since brain dopamine may be lowered by chronic alcoholism (*88*), this result would suggest an increased rate of TIQ formation during alcohol consumption in alcoholics (*11*).

    In small animal studies, brain regions of untreated rats were reported to contain substantial levels (30–50 ng/g tissue) of salsolinol, according to three different GC/MS reports (*89–91*), and chronic alcohol ingestion (180 days) resulted in a 2.5- to 5-fold increase in this TIQ (*91*). Salsoline and isosalsoline were also quantitated in two of these studies (*89, 91*). Using different but sufficiently sensitive techniques, however, other investigators have been unable to demonstrate endogenous rodent brain levels of these TIQ alkaloids (*41, 56, 92–95*). In one of these studies, chronic alcohol treatment did result in the appearance of the phenolic alkaloid, salsoline, in a dopamine-rich brain region of the mouse (*56*). In three other studies, salsolinol was not detectable in brain regions of rats given ethanol acutely or chronically (*92, 93, 95*). However, if inhibitors of acetaldehyde metabolism and O-methylation or amine deamination were preadministered, the catecholic alkaloid was readily detectable (*93, 95*).

Small quantities of salsolinol have been detected in neonatal rats (*69*) and in rat fetuses (*96*) by means of a specific radioenzymatic technique. Traces (10 ng/g) of 1-demethylsalsolinol (**2a**) have been estimated in rat brain (*97*); about 15% may have formed artifactually by a formaldehyde condensation reaction during the isolation procedure. Radiochemical and fluorescence histochemical evidence was obtained indicating endogenous levels of TIQ-4,6,7-triols (condensation products of norepinephrine or epinephrine and formaldehyde) in the adrenals of rats treated repeatedly with methanol (*98, 99*). Employing GC/MS analysis, THP was detected in low levels in the brains of rats fed L-dopa but ethanol administration did not cause elevations (*60*). THP was also apparent by HPLC analysis along with salsolinol and its 1-demethyl analog **2a** in urinary extracts from rats treated with L-dopa but not with ethanol (*41*). Radioactive THP was reported to be recovered from the perfusates of rat brain following the administration of [^{14}C]dopamine into a dopamine-rich rat brain region (*100*), and an ethanol dose did not change its proportions. However, another study failed to provide evidence for endogenous radioactive THP or salsolinol after injection of [^{3}H]dopamine into the brains of rats, either with or without ethanol pretreatment (*47*).

1-Carboxylated TIQs (condensation products of dopamine with substituted phenylpyruvic acids) structurally related to THP have been detected and assayed in animal tissues and urines. Norlaudanosoline carboxylic acid (NLCA **8a**) and its 3′-*O*-methyl ether metabolite (MNLCA **8b**) were present in the urines of L-dopa-treated Parkinsonian patients (*61*). Untreated volunteers also excreted small amounts of MNLCA, and the same 1-carboxy-1-benzyl-TIQ was stated to be present in varying but low amounts in rat and human brain tissues (*101*). A related alkaloid resulting from condensation of dopamine with phenylpyruvic acid, 3′,4′-deoxy-NLCA (DNLCA **9**), was a trace constituent in urine from infants with the disease of phenylketonuria (*62*). Low rat brain levels of DNLCA were also raised by treatment with the amino acid L-phenylalanine. Confirmation of these GC/MS results using equally sensitive HPLC analysis was not obtained, however (*103*). More recently, HPLC evidence has been obtained for simple 1-carboxy condensation products of pyruvic acid with dopamine (and serotonin) in monkey CSF (*64*).

VI. Mammalian β-Carboline Formation

In vitro formation of β-carboline has been shown in incubations of tryptamines with potential formaldehyde donors (methyl folates) as de scribed in the following section. There have been few studies on β-carboline production from other aldehydes in isolated mammalian systems.

However, various β-carbolines have been detected and assayed in untreated intact animals. 1,2,3,4-TBC (**4c**) was measured in whole rat brain (*65, 66*), rat forebrain (*49*), and rat adrenal glands (*66*). Levels of 6-methoxy-TBC (**4d**) have been obtained from whole rat brain (*66*), rat adrenal gland (*66*) and chicken pineal gland (*103*). Early evidence for radioactive urinary 1-methyl-TBC (**4e**) was found in rats treated with [U-¹⁴C]-5-methoxytryptamine (**3e**), ethanol, and inhibitors of amine deamination and acetaldehyde metabolism (*104*). The apparent condensation product of serotonin and formaldehyde was estimated in rat brain and platelets (*49, 105*). Spectroscopic and chromatographic evidence was reported for a fluorescent aromatic β-carboline, possibly **17a** in several different rat brain areas (*71, 106*), suggesting that oxidation of a TBC occurred *in vivo* (or during isolation). Similarly, investigators isolated an alkaloid possibly like **17b** from lens tissue of old animals that was not apparent in young lens (*107*). It was suggested that the β-carboline arose from a 1,3-dicarboxy-TBC derived from tryptophan in lens protein.

17a R¹=R²=H
  b R¹=R²=COOH
  c R¹=Me; R²=H
  d R¹=H; R²=COOH

18

Focusing on human studies, there was an early suggestion (*108*) that 3-ethyl-7-hydroxy-TBC (**18**) was present in the urine of humans treated with 6-hydroxy-α-ethyltryptamine. TBC **4c** and/or two different aromatic fluorescent derivatives have been detected in human platelets (*67*), plasma (*109*), and urine (*110*). Administration of 5-hydroxytryptophan to human volunteers resulted in 6-hydroxy-TBC (**4f**) excretion (*105*). The 1-methyl-TBC **4e** and its fully aromatized derivative **17c** (harman) were isolated from the urines of volunteers given moderate quantities of ethanol (*50*). Finally, analogous to the TIQ studies (Section V), TBC **4e** and its 6-hydroxy analog were detected and measured in the body fluids of alcoholics in early ethanol detoxification (*68, 111*) and levels declined during the week of hospitalization. The same β-carbolines were present in lower quantities (based on creatinine excretion rates) in the urines of healthy volunteers, and **4e** was not increased after acute ethanol consumption.

VII. Alternative Modes of Mammalian Alkaloid Formation

Formaldehyde is a normal intermediate in the metabolism of methanol and in the N-demethylation of various drugs, and may arise normally from the "one-carbon pool," via 5-methyltetrahydrofolate (5-MTHF) (*112*). The radioactive products from incubations of soluble rat brain extracts with either dopamine or certain indoleamines and ^{14}C-labeled 5-MTHF were shown to be the respective "formaldehyde-derived" TIQ (*113, 114*) or TBC (*115, 116*). Originally the products were misidentified as *N*-methyl biogenic amines (*117, 118*). Rate studies indicated that the generation but not the condensation of formaldehyde was enzyme-catalyzed. This pathway thus represents a potential route for several of the endogenous alkaloids discussed in Sections V and VI.

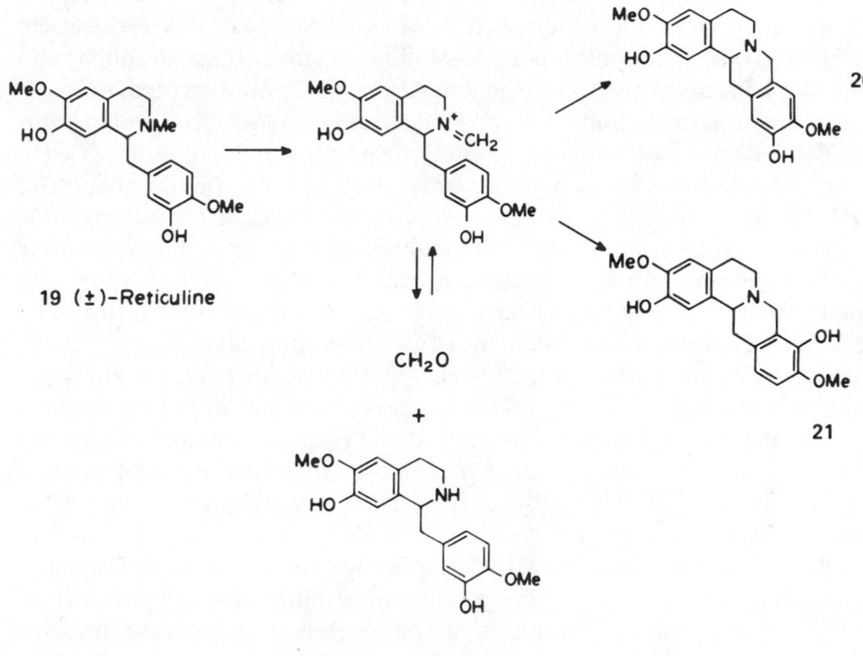

19 (±)-Reticuline

CH₂O

+

19a (±)-*N*-norreticuline

SCHEME 6

Enzyme catalysis in the intramolecular condensation of certain mammalian alkaloids is suggested in the formation of berbines from *N*-methyl-substituted TIQ as shown in Scheme 6. The berbine (±)-coreximine (**20**) was found in the urine of rats treated with (±)-reticuline (**19**) (*119*). After treatment with ^{14}C-*N*-methyl-labeled (±)-reticuline, traces of labeled (±)-coreximine were found in the rat urine, indicating that the berbine methylene bridge originated from the radioactive *N*-methyl group of reticuline (*120*). Additionally, a rat liver microsomal membrane fraction (the primary location for N-demethylation) incubated with (±)-reticuline (**19**) catalyzed the formation of (±)-coreximine (**20**) and its 9,10-hydroxymethoxy isomer (±)-scoulerine (**21**), together with the expected major N-demethylation product, (±)-*N*-norreticuline (**19a**).

The formation of berbines was stimulated by NADPH, $MgCl_2$, oxygen, and nicotinamide, suggestive of a mixed function oxidase as the drug-metabolizing system. If the substrate lacked an *N*-methyl group (i.e., THP or *N*-norreticuline), berbines were not produced. The berbine alkaloid formation might be occurring during interaction with N-demethylating enzyme systems via a Schiff base, without the intermediacy of formaldehyde. To clarify this, mixed microsomal metabolism experiments were carried out with hexadeuteroreticuline followed by MS analysis (*121*). The results were ambiguous but did not rule out the possibility that the berbines resulted from direct cyclization of an enzyme-generated iminium ion intermediate.

Berbine formation from a 1-benzyl-substituted TIQ was reported to occur in soluble enzyme preparations (nonmicrosomal) which contained *S*-aden osylmethionine (SAM), a biological methylating cofactor. Incubation of THP with rat brain or (microsome-free) liver supernatants containing SAM resulted in the appearance (GC/MS analysis) of the isomeric tetrahydroxy berbines **16a** and **16b**, and traces of mono-O-methylated derivatives (*83*). The liver "benzyl TIQ methyltransferase" activity, which was destroyed by boiling, was purified somewhat by chromatography and various kinetic characteristics were determined (*122*). The two major berbines also were found in the urines of THP-treated rats and of humans on L-dopa therapy. To date, the results are appealing but not at all conclusive for methyltransferase activity, since free formaldehyde [arising from the soluble enzymes of the one-carbon pool or from SAM (*123*)] could be condensing with THP in these experiments (*132*).

TBC formation from acetaldehyde and serotonin was reported to be faster in fresh than in boiled brain homogenates, suggesting some form of catalysis (*124*). A comparison of the ratios of endogenous β-carboline/tryptamine versus 1-demethylsalsolinol/dopamine in rat brain led Barker *et al.* (*97*) to conclude that the formation of TBC was a catalyzed process.

VIII. Metabolic Studies of Mammalian Alkaloids

A. O-METHYLATION AND CONJUGATION

Considerable effort has been expended to determine the metabolism of mammalian TIQs. With the catecholic species, studies showed that 1-demethylsalsolinol, 4,6,7-TIQ-triol **2d**, salsolinol (**2b**), and THP (**12**) were substrates for O-methylation by catechol *O*-methyltransferase (COMT) in a semipurified state or in brain or liver homogenates (*125, 126, 127*). The importance of O-methylation *in vivo* was supported by the fact that the rates of disappearance of salsolinol and THP following injection into the brains of rats were decreased 2- and 4-fold, respectively, when COMT action was blocked by the inhibitor pyrogallol (*128*).

Studies on the orientation of the O-methylation reaction revealed that, contrary to the rather specific meta-O-methylation pattern with the precursor catecholamines, the catecholic TIQs were enzymatically O-methylated to a large or even exclusive extent on the para or 7-hydroxy group. 1-De-methylsalsoline was a predominant product when 1-demethylsalsolinol (**2a**) was incubated with rat liver COMT (*125*). Similar incubations of THP or its optical isomers resulted in considerable 7-O-methylation, particularly for the *R*(−)-THP stereoisomer, and regioselective O-methylation was also noted for tetrahydroxyberbines (*129*). *In vivo* results supported these *in vitro* findings; the nearly exclusive (95%) O-methylated product of salsolinol given into the rat brain was salsoline (**10**) (*59, 130*). Tetrahydroisoquinoline-triol **2h** gave 50% 7-O-methylation, while 1-carboxysalsolinol was 100% O-methylated on the 7-hydroxy group to produce 1-carboxysalsoline **22** (*131*). The 3,4-dihydroisoquinoline (**23a**) gave no apparent O-methylation products in brain, possibly because it exists as the quinoidal tautomer (**23b**) in the physiological pH range. The O-methylation of THP in the intact rat brain also involved the 7- and 3′-hydroxy groups to a significant degree (*132*).

Catechols are enzymatically conjugated to varying extents in mammalian systems with sulfate or glucuronic acid moieties. There have been no definitive studies on the mammalian TIQs, with the exception of the indications that, based on hydrolysis experiments with sulfatase, salsolinol injected in rats was ∼99% (sulfate) conjugated in the liver and ∼25% conjugated in the brain (*89*). Salsoline/isosalsoline, largely formed from administered salsolinol, was totally sulfated in the liver and 7% conjugated in brain. The knowledge that 61% of the excreted products of a dopamine-related isoquinoline, trimetaquinol, injected in the guinea pig is O-conjugated

22   23a   23b

with glucuronic acid (*133*) indicates that this pathway may be equally important for mammalian alkaloids in certain species.

B. OXIDATION

The oxidation of TIQs in mammalian systems has not been explored to any degree. Potential sites for oxidation include the amine function, the aromatic ring, and, in the case of 1-benzyl substituted TIQ, the methylene group of the benzyl substituent. TIQ are inhibitors of amine oxidases, specifically monoamine oxidase (MAO), and it is possible that they also serve as substrates. Oxidative decarboxylation (Scheme 7) of 1-carboxy-1-benzyl-

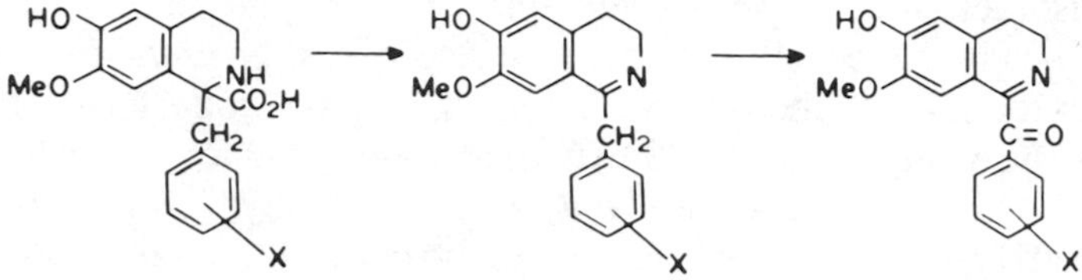

SCHEME 7. *Oxidation of 1-benzyl-substituted TIQ.*

TIQs similar in structure to those found in animals and humans formed 3,4-dihydroisoquinolines and 1-benzoyl derivatives. This was achieved with fungal laccase or peroxidase, or with aqueous bicarbonate in air (*50, 134*). THP was reported to undergo facile air oxidation readily at pH 7.5 to the dihydrodibenz[*b,g*]indolizine **24** and its dimer (*135*). *In vivo*, 8-methoxy-*N*-methyl-1,2,3,4-TIQ was converted in dogs to several fluorescent urinary species which could be oxidized isoquinoline derivatives (*136*).

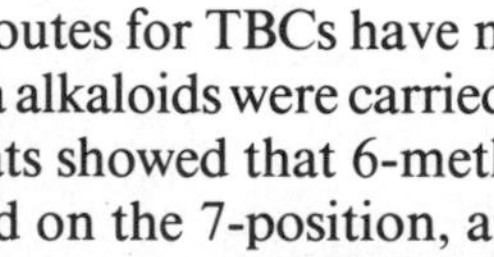

24

Mammalian metabolic routes for TBCs have not been clarified, although studies with related harmala alkaloids were carried out at the beginning of the century (*137*). Studies in rats showed that 6-methoxy-TBC was extensively demethylated, hydroxylated on the 7-position, and excreted in the form of conjugates (*138*). The 6-hydroxy-1-methyl TBC in human urine was conjugated 25% in controls and ~55% in alcoholics (*68*). Oxidation of the tetrahydropyridine ring was not reported but in view of evidence for the coexistence of TBCs with aromatic β-carbolines in mammals, this possibility should be reinvestigated. The formation of a 3,4-dihydro alkaloid from a TBC may be the limiting step, since the mammalian oxidation of a 3,4-dihydro-β-carboline such as harmaline or harmalol to a fully aromatic species is well established (137, 139).

IX. Effects of Mammalian Alkaloids on Biochemical Processes

A. ENZYMES

Numerous studies have been done on the effects of mammalian alkaloids of five or six relevant neuronal enzymes. These are described in detail elsewhere (7) and will be summarized here. A substantial number of TIQs and TBCs have been shown to inhibit monoamine oxidase (MAO) in vitro (126, 140–146). In intact animals only the mammalian β-carbolines, like the related harmala alkaloids, have shown significant MAO-inhibiting properties (145–147). The sole in vivo demonstration of MAO inhibition by TIQs is a study in which salsolinol and 1-demethylsalsolinol behaved like effective inhibitors of this enzyme in promoting ^{3}H-labeled norepinephrine accumulation in the atria of reserpinized mice (148).

Catechol O-methyltransferase (COMT) was inhibited in vitro by most catecholic TIQs (126, 142, 149) but the significance or extent of this effect in vivo is probably negligible. Tyrosine hydroxylase, the rate-controlling enzyme in catecholamine biosynthesis, was inhibited in brain or adrenal extracts by simple TIQs (150, 151) or by NLCA (8a) and DNLCA (9) (62, 142). Evidence for in vivo inhibition of the brain enzyme by DNLCA also was obtained. Other enzymes which have been reported to be inhibited or altered in vitro by 1-benzyl-TIQs and the berbines include dopamine β-hydroxylase, (Na$^+$,K$^+$)-ATPase, and adenylate cyclase (142, 152–156). Several dihydroisoquinolines were inhibitors of solubilized cyclic nucleotide phosphodiesterases (157, 158).

B. STORAGE, UPTAKE, AND RELEASE OF BIOGENIC AMINES

These important dynamic processes have been studied with various neuronal preparations in the presence of alkaloids. The simple TIQs derived from the catecholamines have been shown in a number of studies to inhibit catecholamine storage and re-uptake, and to induce release of the catecholamine neurotransmitters from storage granules; this is reviewed elsewhere in detail (7, 159). 1-Benzyl-TIQs such as THP may be less effective as inhibitors of catecholamine uptake (*160*), but THP (as well as TBC) was able to induce the release of radioactive dopamine in the intact rat brain (*161*). The berbine isomers (**16a, 16b**) were also effective in causing dopamine release *in vitro* (*162*). Considerable interest has been shown in the TBCs as inhibitors of serotonin uptake in brain studies (*163–167*), and the current view is that the serotonin increases observed following treatment of animals with β-carbolines depends upon this inhibitory action for some isomers (*146*).

C. Receptors

With the β-carbolines, much interest has been generated by the observation that a component isolated from the urine and identified as the ethyl ester of β-carboline-3-carboxylic acid (**17d**), bound with very high affinity (4–7 nM) to the receptor in the mammalian nervous system for benzodiazepine drugs (*110*). For several years neuroscientists have been searching for endogenous binding agents or inhibitors for this receptor and β-carbolines are now considered likely possibilities as inhibitors (*168*). Other mammalian β-carbolines that have been reported to be potent inhibitors of benzodiazepine binding are β-carboline itself (**17a**) and 1-methyl-β-carboline (*169–171*). The affinities of the TBCs for the receptor were several orders of magnitude less than those of aromatic alkaloids (*169–171*). Other neuronal or hormonal receptors have not been studied to any extent with the β-carboline alkaloid class.

Mammalian metabolic routes for TBCs also have not been clarified, although studies with related harmala alkaloids were carried out at the beginning of the century (*137*). Studies in rats showed that 6-methoxy-TBC was extensively demethylated, hydroxylated on the 7-position, and excreted in the form of conjugates (*138*). The 6-hydroxy-1-methyl TBC in human urine was conjugated 25% in controls and ~55% in alcoholics (*68*). Oxidation of the tetrahydropyridine ring was not reported but in view of evidence for the coexistence of TBCs with aromatic β-carbolines in mammals, this possibility should be reinvestigated. The formation of a 3,4-dihydro alkaloid from a TBC may be the limiting step, since the mammalian oxidation of a 3,4-dihydro-β-carboline such as harmaline or harmalol to a fully aromatic species is well established (*137, 139*).

X. Physiological and Behavioral Effects of Mammalian Alkaloids

The mammalian alkaloids have been shown to have a wide range of physiological and behavioral actions. These include stimulation of alcohol preference in rodents, the production of analgesia and narcosis, endocrine and opiate-like effects, alterations in motor activity (rodents), modulation of stress, effects of body temperature on learning or memory, and last, mutagenic and cytotoxic effects. Some of the most significant actions will be emphasized.

Several mammalian TIQs and one TBC were reported to induce large increases in preference for alcohol when they were injected in trace quantities acutely or chronically into the brains of "nonpreferring" rats (179–181). THP appeared to be the most potent, but salsolinol, 3-carboxysalsolinol, and TBC were relatively effective. The TIQ-triol 2h was ineffective. The preference effects have been replicated to some extent in one laboratory (182) but not in another and they remain controversial (183).

Treatment of animals with varying amounts of THP, salsolinol, or 3-carboxysalsolinol resulted in analgesia which, in some cases, was prevented by pretreatment with the opiate receptor antagonist naloxone (177, 184). These results have lent support to the idea that the isoquinolines interact with the endogenous opiate receptor in vivo (185). Injections of TBC or the 6-hydroxy derivative caused dose-related analgesia in rodents (49, 186). This β-carboline effect was thought to be due to interaction with the serotonergic nervous system that has been implicated in some forms of analgesia (146).

Several mammalian TIQs have been shown to induce hyperactivity and aberrant motor behavior when injected in microgram amounts into the brains of rats (187, 188). In mouse studies, salsolinol appeared to cause hyperexcitability or enhance the hyperexcitable state seen during withdrawal from chronic exposure to ethanol (189). Repeated injection of THP into rat brain produced behavioral tremors and increased the propensity for seizures (180, 182). In contrast, mammalian TBCs usually have inhibited motor activity and caused catalepsy when administered to rats or mice (49, 186, 190).

Enhancement of the mutagenicity of benzo[a]pyrene by β-carboline has been reported (191, 192), but the alkaloid alone was not a mutagen. Intercalation with DNA was implicated in this co-mutagenic effect of β-carboline (193, 194). There is evidence that some of the mammalian isoquinolines have cytotoxic actions. Repeated injections of a mixture of 1,2-dimethyl-TIQ-triols, products of the condensation of epinephrine and acetaldehyde, caused extensive damage to the adrenergic nervous system in adult or neonatal rats (195, 196). Liver damage has also been observed in rats with these alkaloids (197). DNCLA (9), chronically injected into neonatal rats prior to myelination, caused reduced levels of myelin and of brain serotonin (198). The protracted increase in alcohol preference (discussed above) some 9 or 10 months after termination of THP infusion suggested some long-term neurotoxic effects in rats (182).

XI. Summary

A growing number of analytical chemical reports have now established that TIQ and TBC alkaloid derivatives of dopamine, tryptamine, and serotonin are endogenous constituents in mammals. The recent publication of a 2-day symposium at the Salk Institute in La Jolla California, solely on the subject of mammalian alkaloids, attests to the upsurge of interest in this neuroscience area (cf., refs. *11, 38, 47, 48, 51, 87,* and *132*). Still largely unappreciated are the condensation possibilities of a "third" biogenic amine class, namely, histamine and its precursor amino acid, histidine. For example, it has been known for some time that pyridoimidazole products result from physiological condensations of formaldehyde and histamine (*199*), and that the heterocyclic derivatives can inhibit histaminase, the principal deaminating enzyme of histamine (*200*). Furthermore, histamine/aldehyde condensation products appear to be alkaloidal constituents in animals (*201*). However, recent interest in this potentially interesting class of biogenic amine derivatives is clearly lacking.

As emphasized, there is very little knowledge about the stereochemical nature of those mammalian TIQs and TBCs which are apparent derivatives of carbonyl compounds other than formaldehyde. The endogenous alkaloids need to be isolated for optical rotation determination and certain standards, particularly the 1-carboxy derivatives, are required in optically pure form. The overall problem poses a formidable challenge to analytical, organic and alkaloid chemists, but the answer would clarify whether sterospecific enzyme catalysis is involved in mammalian alkaloid biosynthesis (*38*).

The possible physiological roles of mammalian alkaloids are still open to question, debate, and research solutions. While several avenues of investigation were reviewed, one possibility that remains completely unexplored is that the mammalian TIQs or TBCs have their *own* neuronal receptors. The lack of research on alkaloid receptors is in part due to the absence of ^{3}H-labeled alkaloid ligands of high specific activity (>10 Ci/mmole). Another physiological possibility that was referred to was neuropathology due to mammalian alkaloids, but no mechanism was discussed. Speculatively, dihydroisoquinolines, in addition to being possible intermediates in stereochemical reductive pathways, could bind covalently and irreversibly by virtue of their tautomeric quinoidal forms (which are isoelectronic with established neurotoxins) to membrane nucleophiles (*202*). With many unanswered questions of a chemical, metabolic, and physiological nature, the field of mammalian alkaloids represents fertile ground for the insights and contributions of alkaloid chemists.

ACKNOWLEDGMENTS

The assistance of Laura Colagrossi in the preparation of the manuscript and the support of the ADAMHA AA00266 are acknowledged.

REFERENCES

1. S. Teitel and A. Brossi, *Heterocycles* **1**, 73 (1973).
2. A. Pictet and T. Spengler, *Chem. Ber.* **44**, 2030 (1911).
3. E. Wintersteiner and G. Trier, "Die Alkaloide," p. 307. Bornträger, Berlin, 1910.
4. W. Whaley and T. Govindachari, *Org. React. (N.Y.)* **6**, 151 (1951).
5. G. Cohen, *Biochem. Pharmacol.* **25**, 1123 (1976).
6. R. Deitrich and V. E. Erwin, *Annu. Rev. Pharmacol. Toxicol.* **20**, 55 (1980).
7. C. Melchior and M. A. Collins, *CRC Crit. Rev. Toxicol.* **9**, 313 (1982).
8. W. McIsaac, *Postgrad. Med.* **30**, 111 (1961).
9. G. Cohen and M. A. Collins, *Science (Washington, D.C.)* **167**, 1749 (1970).
10. H. A. Bates, *J. Org. Chem.* **46**, 4931 (1981).
11. M. A. Collins, J. Hannigan, T. Origitano, D. Moura, and W. Osswald, *in* "Beta-Carbolines and Tetrahydroisoquinolines" (F. Bloom, J. Barchas, M. Sandler and E. Usdin, eds.), p. 155. Liss, New York, 1982.
12. G. S. King, B. Goodwin, and M. Sandler, *J. Pharm. Pharmacol.* **26**, 476 (1974).
13. J. Robbins, *Clin. Res.* **16**, 554 (1968).
14. T. M. Kenyhercz and P. T. Kissinger, *J. Pharm. Sci.* **67**, 112 (1978).
15. M. Shamma and J. Moniet, "Isoquinoline Alkaloids Research." Plenum Press, New York, 1978.
16. W. Gensler, *Org. Reac. (N.Y.)* **6**, 191 (1951).
17. A. Brossi, A. Focella, and S. Teitel, *Helv. Chim. Acta* **55**, 15 (1972).
18. A. Brossi, A. Focella, and S. Teitel, *J. Med. Chem.* **16**, 418 (1973).
19. T. Kametani, S. Shibuya, and M. Satoh, *Chem. Pharm. Bull.* (Tokyo) **16**, 953 (1968).
20. S. Quessy and L. Williams, *Aust. J. Chem.* **32**, 1317 (1979).
21. C. Wilson and C. J. Coscia, *J. Am. Chem. Soc.* **97**, 431 (1975).
22. T. Hudlicky, T. M. Kutchan, G. Shen, V. E. Sutliff, and C. J. Coscia, *J. Org. Chem.* **46**, 1738 (1981).
23. J. Forneau, C. Gaignault, R. Jaquier, O. Stoven, and M. Davy, *Chim. Ther.* **4**, 67 (1969).
24. I. Kari, P. Peura, and M. Airaksinen, *Biomed. Mass Spectrom.* **7**, 549 (1980).
25. S. Teitel, J. O'Brien, W. Pool, and A. Brossi, *J. Med. Chem.* **17**, 134 (1974).
26. S. Teitel, J. O'Brien, and A. Brossi, *J. Med. Chem.* **15**, 845 (1972).
27. A. Brossi, K. Rice, C. Mak, J. Reden, A. Jacobson, Y. Nimitkitpaisan, P. Skolnick, and J. Daly, *J. Med. Chem.* **23**, 648 (1980).
28. J. M. Bobbitt, *Adv. Heterocycl. Chem.* **15**, 99 (1973).
29. J. M. Bobbitt, D. N. Roy, A. Marchand, and C. Allen, *J. Org. Chem.* **32**, 2225 (1967).
30. M. Collins and F. Kernozek, *J. Heterocycl. Chem.* **9**, 1437 (1972).
31. R. Sarges, *J. Heterocycl. Chem.* **11**, 599 (1974).
32. R. Elliott, F. Hewgill, E. McDonald, and P. McKenna, *Tetrahedron Lett.* **21**, 4633 (1980).
33. J. Reden, W. Ripka, K. Rice, and A. Brossi, *in* "Biological Effects of Alcohol" (H. Begleiter, ed.), p. 69. Plenum Press, New York, 1980.
34. K. Rice, W. Ripka, J. Reden, and A. Brossi, *J. Org. Chem.* **45**, 601 (1980).

35. S. Dyke, R. Kinsman, P. Warren, and A. White, *Tetrahedron* **34,** 241 (1978).
36. A. Brossi, *Heterocycles* **3,** 343 (1975).
37. M. Rueffer, H. El-Shagi, N. Nagakura, and M. H. Zenk, *FEBS Lett.* **129,** 5 (1981).
38. A. Brossi, *in* "Beta-Carbolines and Tetrahydroisoquinolines" (F. Bloom, J. Barchas, M. Sandler and E. Usdin, eds.), p. 125. Liss, New York, 1982.
39. G. Kapadia, G. Subba Rao, E. Leete, M. Fayez, Y. Vaishnav, and H. Fales, *J. Am. Chem. Soc.* **92,** 6943 (1970).
40. I. Coutts, M. R. Hamblin, E. J. Tinley, and J. M. Bobbitt, *J. Chem. Soc. I.* 2744 (1979).
41. R. Riggin and P. T. Kissinger, *Anal. Chem.* **49,** 530 (1977).
42. W. Nijm, R. Riggin, G. Teas, P. Kissinger, G. Borge, and M. A. Collins, *Fed. Proc., Fed. Am. Soc. Exp. Biol.* **36,** 334 (1977).
43. L. R. Meyerson, J. Cashaw, K. McMurtrey, and V. Davis, *Biochem. Pharmacol.* **28,** 1745 (1979).
44. K. McMurtrey, J. L. Cashaw, and V. Davis, *J. Liq. Chromatogr.* **3,** 663 (1980).
45. R. T. St. Claire III, G. Ansari, and C. W. Abell, *Anal. Chem.* **54,** 186 (1982).
46. M. A. Collins, unpublished results.
47. W. T. Shier, L. Koda, and F. Bloom, *in* "Beta-Carbolines and Tetrahydroisoquinolines" (F. Bloom, J. Barches, M. Sandler, and E. Usdin, eds.), p. 191. Liss, New York, 1982.
48. M. Hirst, M. A. Adams, S. Okamoto, C. W. Gowdey, D. R. Evans, and J. M. LeBarr, *in* "Beta-Carbolines and Tetrahydroisoquinolines" (F. Bloom, J. Barches, M. Sandler, and E. Usdin, eds.), p. 81. Liss, New York, 1982.
49. Rommelspacher, H. Honecker, M. Barbey, and B. Meinke, *Arch. Pharmacol.* **310,** 35 (1979).
50. H. Rommelspacher, St. Strauss, and J. Lindemann, *FEBS Lett.* **109,** 209 (1980).
51. K. F. Faull, R. B. Holman, G. R. Elliott, and J. D. Barchas, *in* "Beta-Carbolines and Tetrahydroisoquinolines" (F. Bloom, J. Barchas, M. Sandler and E. Usdin, eds.), p. 135, Liss, New York, 1982.
52. M. A. Collins, M. G. Bigdeli, and F. Kernozek, *Pharmacologist* **13,** 654 (1971).
53. M. Bigdeli and M. A. Collins, *Biochem. Med.* **12,** 55 (1975).
54. P. J. O'Neill and R. G. Rahwan, *J. Pharm. Sci.* **66** 893 (1977).
55. M. A. Collins and M. G. Bigdeli, *Life Sci.* **16,** 585 (1975).
56. M. Hamilton, M. Hirst and K. Blum, *in* "Biological Effects of Alcohol" (H. Begleiter, ed), p. 73. Plenum Press, New York, 1980.
57. M. Sandler, S. Carter, K. Hunter, and G. Stern, *Nature (London)* **241,** 439 (1973).
58. T. C. Origitano, Ph.D. Dissertation, "Neurochemical Studies on the Metabolism and Effects of Catecholic Isoquinolines," Loyola University of Chicago, 1981.
59. T. Origitano and M. A. Collins, *Life Sci.* **26,** 2061 (1980).
60. A. J. Turner, K. Baker, S. Algeri, A. Frigerio, and S. Garattini, *Life Sci.* **14,** 2247 (1974).
61. C. J. Coscia, W. Burke, G. Jamroz, J. Lasala, J. McFarlane, J. Mitchell, M. O'Toole, and M. Wilson, *Nature (London)* **269,** 617 (1977).
62. J. Lasala and C. J. Coscia, *Science (Washington, D.C.)* **203,** 283 (1979).
63. J. L. Cashaw, K. McMurtrey, K. Brown, and V. Davis, *J. Chromatogr.* **99,** 567 (1974).
64. M. A. Collins, K. Dahl, W. Nijm, and L. Major, *Soc. Neurosci. Symp.* **8,** 277 (1982).
65. S. A. Barker, R. Harrison, G. Brown, and S. T. Christian, *Biochem. Biophys. Res. Commun.* **87,** 146 (1979).
66. S. A. Barker, R. Harrison, J. Monte, G. Brown, and S. T. Christian, *Biochem. Pharmacol.* **30,** 9 (1981).
67. I. Kari, R. Peura, and M. Airaksinen, *Med. Biol.* **57,** 412 (1979).
68. O. Beck, T. Bosin, A. Lundman, and S. Borg, *Biochem. Pharmacol.* **31,** 2517 (1982).

69. C. Nesterick and R. G. Rahwan, *J. Chromatogr.* **164**, 205 (1979).
70. R. Dean, D. P. Henry, R. Bowsher, and R. B. Forney, *Life Sci.* **27**, 403 (1980).
71. D. W. Shoemaker, J. T. Cummins, and T. Bidder, *Neuroscience* **3**, 233 (1978).
72. P. Holtz, K. Stock, and E. Westermann, *Nature* (*London*) **203**, 656 (1964).
73. P. Halushka and P. Hoffman, *Biochem. Pharmacol.* **17**, 1873 (1968).
74. M. Walsh, V. E. Davis and Y. Yamanaka, *J. Pharmacol. Exp. Ther.* **174**, 388 (1970).
75. S. Alivisatos, F. Ungar, O. Callaghan, L. Levitt, and B. Tabakoff, *Can. J. Biochem.* **51**, 28 (1973).
76. V. E. Davis, M. Walsh, and Y. Yamanaka, *J. Pharmacol. Exp. Ther.* **174**, 401 (1970).
77. V. E. Davis and M. Walsh, *Science* (*Washington, D.C.*) **167**, 1005 (1970).
78. M. Seevers, *Science* (*Washington, D.C.*) **170**, 1113 (1970).
79. V. E. Davis, J. Cashew, B. McLaughlin, and T. Hamlin, *Biochem. Pharmacol.* **23**, 1877 (1974).
80. A. W. Tank, H. Weiner, and J. A. Thurman, *Ann. N.Y. Acad. Sci.* **273**, 219 (1976).
81. G. Cohen, *Biochem. Pharmacol.* **20**, 1757 (1971).
82. Y. Yamanaka, M. Walsh and V. E. Davis, *Nature* (*London*) **227**, 1143 (1970).
83. J. L. Cashaw, K. McMurtrey, H. Brown, and V. E. Davis, *J. Chromatogr.* **99**, 567 (1974).
84. M. A. Collins, W. Nijm, G. Borge, G. Teas, and C. Goldfarb, *Science* (*Washington, D.C.*) **206**, 1184 (1979).
85. B. Sjöquist, S. Borg, and H. Kvande, *Subst. Alcohol/Actions Mis.* **2**, 73 (1981).
86. B. Sjöquist, S. Borg and H. Kvande, *Subst. Alcohol/Actions Mis.* **2**, 163 (1981).
87. B. Sjöquist, A. Eriksson, and B. Winblad, *in* "Beta-Carbolines and Isoquinolines" (F. Bloom, J. Barches, M. Sandler and E. Usdin, eds.), p. 57. Liss, New York, 1982.
88. A. Carlson, R. Adolfson, S. Aquilonius, C. Gottfries, L. Oreland, L. Svennerholm, and B. Winblad, *in* "Ergot Compounds and Brain Function" (M. Goldstein, ed.), p. 295. Raven Press, New York, 1980.
89. B. Sjöquist and Magnusson, *J. Chromatogr.* **183**, 17 (1980).
90. G. Smythe, M. Duncan, and J. Bradshaw, *IRCS Med. Sci.* (Biochem.) **9**, 472 (1981).
91. B. Sjöquist, S. Liljequist, and J. Engel, *J. Neurochem.* **39**, 259 (1982).
92. P. O'Neill and R. Rahwan, *J. Pharmacol. Exp. Ther.* **200**, 306 (1975).
93. M. A. Collins and M. G. Bigdeli, *Life Sci.* **16**, 585 (1975).
94. T. Origitano, J. Hannigan, and M. A. Collins, *Brain Res.* **224**, 446 (1981).
95. M. A. Collins and M. G. Bigdeli, *in* "Alcohol Intoxication and Withdrawal: Experimental Studies II" (M. M. Gross, ed.), p. 79. Plenum Press, New York, 1975.
96. C. Nesterick and R. Rahwan, *Pharmacologist* **22**, 211 (1980).
97. S. A. Barker, J. Monte, L. Tolbert, G. Brown, and S. Christian, *Biochem. Pharmacol.* **30**, 2461 (1981).
98. G. Cohen and R. Barrett, *Fed. Proc., Fed. Am. Soc. Exp. Biol.* **28**, 288 (1969).
99. M. A. Collins and G. Cohen, *Fed. Proc., Fed. Am. Soc. Exp. Biol.* **29**, 608 (1970).
100. H. Weiner, *Alcohol. Clin. Exp. Res.* **2**, 127 (1978).
101. M. P. Galloway, W. J. Burke, A. Kosloff, D. Lieberman, J. Mitchell, and C. J. Coscia, *Soc. Neurosci. Symp.* **4**, 315 (1978).
102. C. J. Coscia, private communication.
103. I. Kari, *FEBS Lett.* **127**, 277 (1980).
104. W. McIsaac, *Biochim. Biophys. Acta* **52**, 607 (1961).
105. H. Honecker and H. Rommelspächer, *Naunyn-Schmiedeberg's Arch. Pharmacol.* **305**, 135 (1978).
106. D. W. Shoemaker, J. T. Cummins, T. Bidder, H. Boettger, and M. Evans, *Naunyn-Schmiedeberg's Arch. Pharmacol.* **310**, 227 (1980).
107. J. Dillon, A. Spector, and K. Nakanishi, *Nature* (*London*) **259**, 422 (1976).

108. R. V. Heinzelman and J. Szmuszkovics, *Prog. Drug Res.* **6**, 75 (1963).
109. T. Bidder, D. W. Showmaker, H. Boettger, M. Evans, and J. T. Cummins, *Life Sci.* **25**, 157 (1979).
110. C. Braestrup, M. Nielsen, and C. Olsen, *Proc. Natl. Acad. Sci. U.S.A.* **77**, 2288 (1980).
111. J. Allen, O. Beck, S. Borg, and R. Skroeder, *Eur. J. Mass Spectrom. Biochem. Med. Environ. Res.* **1**, 171 (1980).
112. A. J. Turner, *Biochem. Pharmacol.* **26**, 1009 (1977).
113. E. Meller, H. Rosengarten, A. J. Friedhoff, R. Stebbins, and R. Silber, *Science* (*Washington, D.C.*) **187**, 171 (1975).
114. W. Vanderheuvel, V. Gruber, L. Mandel, and R. Walker, *J. Chromatogr.* **114**, 476 (1975).
115. R. J. Wyatt, E. Erdelyi, J. DoAmaral, G. Elliott, J. Renson, and J. Barchas, *Science* (*Washington, D.C.*) **187**, 853 (1975).
116. L. Hsu and A. J. Mandell, *J. Neurochem.* **24**, 631 (1975).
117. P. Laduron, *Nature* (*London*), *New Biol.* **238**, 212 (1972).
118. P. Banerjie and S. Snyder, *Science* (*Washington, D.C.*) **182**, 74 (1973).
119. T. Kametani, M. Ihara and K. Takahashi, *Chem. Pharm. Bull* (Tokyo) **20**, 1587 (1972).
120. T. Kametani, M. Takemura, M. Ihara, K. Takahashi, and K. Fukumoto, *J. Am. Chem. Soc.* **98**, 1956 (1976).
121. T. Kametani, Y. Ohta, M. Takemura, M. Ihara, and M. Fukumoto, *Bioorg. Chem.* **6**, 249 (1977).
122. L. Myerson and V. E. Davis, *Fed. Proc., Fed. Am. Soc. Exp. Biol.* **34**, 508 (1975).
123. E. Meller, H. Rosengarten, and A. J. Friedhoff, *Life Sci.* **14**, 2167 (1974).
124. S. E. Saheb and R. Dajani, *Comp. Gen. Pharmacol.* **4**, 225 (1973).
125. C. R. Creveling, N. Morris, H. Shimizu, H. Ong, and J. Daly, *Mol. Pharmacol.* **8**, 398 (1972).
126. A. C. Collins, J. Cashaw, and V. E. Davis, *Biochem. Pharmacol.* **22**, 2337 (1973).
127. J. Rubenstein and M. A. Collins, *Biochem. Pharmacol.* **22**, 2928 (1973).
128. C. Melchior, A. Mueller and R. Deitrich, *Biochem. Pharmacol.* **29**, 657 (1980).
129. L. Myerson, J. Cashaw, K. McMurtrey and V. E. Davis, *Biochem. Pharmacol.* **28**, 1745 (1979).
130. M. Bail, J. Miller and G. Cohen, *Life Sci.* **26**, 2051 (1980).
131. T. Origitano and M. A. Collins, *Trans. Am. Soc. Neurochem.* **11**, 217 (1980).
132. V. E. Davis, J. Cashaw, K. McMurtrey, S. Ruchirawat, and Y. Nimit, in "Beta-Carbolines and Tetrahydroisoquinolines" (F. Bloom, J. Barchas, M. Sandler and E. Usdin, eds.), p. 99. Liss, New York, 1982.
133. T. Meshi, M. Otsuka, and Y. Sato, *Biochem. Pharmacol.* **19**, 2937 (1970).
134. J. M. Bobbitt, C. Kilkarin, and P. Wiriyachitra, *Heterocycles* **4**, 1645 (1976).
135. C. Mak and A. Brossi, *Heterocycles* **12**, 1413 (1979).
136. J. deSilva, N. Strojny and N. Munno, *J. Pharm. Sci.* **62**, 1066 (1973).
137. F. Flury, *Arch. Exp. Pathol. Pharmakol.* **64**, 105 (1911).
138. B. T. Ho, D. Taylor, K. E. Walker, and W. McIsaac, *Zenobiotica* **2**, 349 (1972).
139. A. Villeneuve and T. Sourkes, *Rev. Can. Biol.* **25**, 231 (1966).
140. T. Yamanaka, *Jpn. J. Pharmacol.* **21**, 833 (1971).
141. L. Myerson, K. McMurtrey and V. E. Davis, *Biochem. Pharmacol.* **25**, 1013 (1976).
142. C. J. Coscia, W. Burke, M. Galloway, A. Kosloff, J. LaSala, J. McFarlane, J. Mitchell, M. O'Toole, and B. Roth, *J. Pharmacol. Exp. Ther.* **212**, 91 (1980).
143. S. Katz and G. Cohen, *Res. Commun. Chem. Pathol. Pharmacol.* **13**, 217 (1976).
144. B. T. Ho, W. McIsaac and K. Walker, *J. Pharm. Sci.* **57**, 1365 (1968).
145. E. Meller, E. Friedman, J. Schweitzer and A. J. Friedhoff, *J. Neurochem.* **28**, 955 (1977).
146. N. Buckholtz, *Life Sci.* **27**, 893 (1980).

147. N. Buckholtz and W. O. Boggan, *Biochem. Pharmacol.* **26**, 1991 (1977).
148. G. Cohen and S. Katz, *J. Neurochem.* **25**, 719 (1975).
149. E. Smissman, J. Reid, D. A. Walsh, and R. Borchardt, *J. Med. Chem.* **19**, 127 (1976).
150. R. Patrick and J. Barchas, *J. Neurochem.* **23**, 7 (1974).
151. C. Weiner and M. A. Collins, *Biochem. Pharmacol.* **27**, 2699 (1978).
152. L. Myerson, K. McMurtrey and V. E. Davis, *Neurochem. Res.* **3**, 239 (1978).
153. H. Sheppard and C. Burghardt, *Biochem. Pharmacol.* **27**, 113 (1978).
154. H. Sheppard, C. Burghardt and S. Teitel, *Mol. Pharmacol.* **12**, 854 (1976).
155. R. Miller, A. Horn, L. Iverson and R. Pinder, *Nature (London)* **250**, 238 (1974).
156. H. Sheppard and C. Burghardt, *Res. Commun. Chem. Pathol. Pharmacol.* **8**, 527 (1974).
157. M. T. Piascik, P. Osei-Gyimah, D. Miller, and D. Feller, *Eur. J. Pharmacol.* **48**, 393 (1978).
158. R. Van Inwegen, P. Salaman, V. St. Georgier, and I. Weinrylo, *Biochem. Pharmacol.* **28**, 1307 (1979).
159. G. Cohen, *in* "Progress in Clinical and Biological Research" (C. Sharp and L. Abood, eds.), Vol. 27, p. 73. Liss, New York, 1979.
160. W. T. Shier, J. Rossier, L. Koda, and R. Bloom, *in* "Alcohol and Aldehyde Metabolizing Systems." (R. Thurman, ed.), p. 807. Academic Press, Inc., New York, 807 (1980).
161. C. Melchior, C. Simpson, and R. Myers, *Brain Res. Bull.* **3**, 63 (1978).
162. H. Alpers, B. McLaughlin, W. Nix, and V. E. Davis, *Biochem. Pharmacol.* **24**, 1391 (1975).
163. H. Lomulainin, J. Tuomisto, M. M. Airaksinen, I. Kari, P. Peura, and L. Pollari, *Acta Pharmacol. Toxicol.* **46**, 299 (1980).
164. K. J. Keller, G. Elliott, R. Holman, J. Vernikos-Danellis, and J. Barchas, *J. Pharmacol. Exp. Ther.* **198**, 619 (1978).
165. H. Rommelspacher, P. Bade, H. Cope, and G. Kossmehl, *Naunyn-Schmiedeberg's Arch. Pharmacol.* **292**, 93 (1976).
166. H. Rommelspacher, S. Strauss, and K. Rehse, *J. Neurochem.* **30**, 1573 (1978).
167. L. Tuomisto and J. Tuomisto, *Arch. Pharmacol.* **279**, 371 (1973).
168. I. L. Martin, *Trends Neurosci.* **3**, 299 (1980).
169. K. Fehske, H. Borbe, W. Miller, H. Rommelspacher, and U. Wollert, *Naunyn-Schmiedeberg's Arch. Pharmacol.* **313**, R34 (1980).
170. M. Airaksinen and E. Mikkonen, *Naunyn-Schmiedeberg's Arch. Pharmacol.* **313**, R34 (1980).
171. H. Rommelspacher, C. Nanz, H. Borbe, K. Fehske, W. Miller and U. Wollert, *Naunyn-Schmiedeberg's Arch. Pharmacol.* **314**, 97 (1980).
172. D. R. Feller, R. Venkatraman and D. Miller, *Biochem. Pharmacol.* **24**, 1357 (1975).
173. M. Hamilton and M. Hirst, *Eur. J. Pharmacol.* **39**, 237 (1976).
174. Y. Nimitkipaisan and P. Skolnick, *Life Sci.* **23**, 375 (1978).
175. P. Laidlaw, *J. Physiol.* **40**, 480 (1910).
176. L. Tampier, H. Alpers, and V. E. Davis, *Res. Commun. Chem. Pathol. Pharmacol.* **17**, 731 (1977).
177. R. Fertel, J. Greenwald, R. Schwartz, L. Wong, and J. Bianchine, *Res. Commun. Chem. Pathol. Pharmacol.* **27**, 3 (1980).
178. J. M. Lasala, T. J. Cicero and C. J. Coscia, *Biochem. Pharmacol.* **29**, 57 (1980).
179. C. M. Melchior and R. D. Myers, *Pharmacol. Biochem. Behav.* **7**, 19 (1977).
180. R. D. Myers and C. M. Melchior, *Science (Washington, D.C.)* **196**, 554 (1977).
181. R. D. Myers and M. Oblinger, *Drug Alcohol Depend.* **2**, 469 (1977).
182. C. Duncan and R. Deitrich, *Pharmacol. Biochem. Behav.* **13**, 265 (1980).
183. J. D. Sinclair and R. D. Myers, *Subst. Alcohol/Actions Mis.* **3**, 5 (1982).
184. A. Marshall, M. Hirst, and K. Blum, *Experientia* **33**, 754 (1977).

185. M. G. Hamilton and M. Hirst, *Subst. Alcohol/Actions Mis.* **1**, 121 (1980).
186. H. Rommelspacher, H. Kauffmann, C. Cohnitz, and H. Cope, *Naunyn-Schmiedeberg's Arch. Pharmacol.* **298**, 83 (1977).
187. B. Costall, R. J. Naylor, and R. M. Pinder, *Eur. J. Pharmacol.* **31**, 94 (1975).
188. B. Costall, R. J. Naylor and R. M. Pinder, *Eur. J. Pharmacol.* **39**, 153 (1976).
189. K. Blum, J. Eubanks, J. E. Wallace, H. Schwertner, and W. Morgan, *Ann. N.Y. Acad. Sci.* **273**, 234 (1976).
190. R. Green, J. D. Barchas, G. Elliott, J. Carman, and R. J. Wyatt, *Pharmacol. Biochem. Behav.* **5**, 383 (1976).
191. T. Fujino, H. Fugiki, M. Nagao, T. Yahagi, Y. Seino, and T. Sugimura, *Mutat. Res.* **58**, 151 (1978).
192. J. Pezzuto, P. Lau, Y. Luh, P. Moore, G. Wogan, and S. Hecht, *Proc. Natl. Acad. Sci. U.S.A.* **77**, 1427 (1980).
193. R. Leavitt, C. Legraverend, D. Nevert, and O. Pelkonen, *Biochem. Biophys. Res. Commun.* **79**, 1167 (1977).
194. K. Hayashi, M. Nagao, and T. Sugimura, *Nucleic Acids Res.* **4**, 3679 (1977).
195. I. Azevedo and W. Osswald, *Naunyn-Schmiedeberg's Arch. Pharmacol.* **300**, 139 (1977).
196. W. Osswald, J. Polonia, and M. Polonia, *Naunyn-Schmiedeberg's Arch. Pharmacol.* **289**, 275 (1975).
197. D. Moura, I. Azevedo, and W. Osswald, *J. Pharm. Pharmacol.* **29**, 255 (1977).
198. M. Druse-Manteuffel, M. A. Collins, D. Tonetti, C. Waddell, and P. Patel, *Soc. Neurosci. Symp.* **7**, 511 (1981).
199. K. Eliassen and O. V. Sjaastad, *Acta Pharmacol. Toxicol.* **26**, 482 (1968).
200. B. Mondovi, A. Scioscia-Santora, G. Rotilio, and M. Costa, *Enzymologia* **28**, 228 (1965).
201. V. Erspamer, T. Vitali, M. Roseghini, and J. M. Cei, *Experientia* **19**, 346, (1963).
202. M. A. Collins, *Trends Pharmacol. Sci.* **3**, 373 (1982).

INDEX